Praise for Peter Watson

'In drawing together evidence from complex strands of archaeology, climatology, genetics and religious symbolism, Watson is compulsively speculative' *Independent*

'Synthesizers like Watson play a valuable role in disseminating and linking up specialist research findings' *TLS*

'The past, Peter Watson argues in this magnificent history of sixteen and a half millennia, is a whole series of foreign countries – and explaining the differences between them helps account for just about everything we take for granted in the here and now ... Impossible, of course, to summarise this massive book in a small review. Sufficient, perhaps, to say that the year's first necessary read is here'

Word Magazine

'Nobody who reads it could ever look at an Indian antiquity in a Western museum – or auction house – in the same way again ... A triumph of investigative reporting' *Sunday Times*

'An enthralling account ... no thriller with manufactured irony could provide a better ending' *Time*

'Gripping ... both shocking and compelling' *Observer*

'Watson has an excellent eye for vivid detail ... One is left gasping in admiration' *The Times*

Peter Watson has been a senior editor at the *Sunday Times*, New York correspondent of *The Times* and a columnist for the *Observer.* He has published three exposés in the world of art and antiquities, and is the author of several books of cultural and intellectual history. From 1997 to 2007 he was a Research Associate at the McDonald Institute for Archaeological Research at the University of Cambridge.

The Great Divide

History and Human Nature in the Old World and the New

Peter Watson

Phoenix

To Kathrine

A PHOENIX PAPERBACK

First published in Great Britain in 2012
by Weidenfeld & Nicolson
This paperback edition published in 2013
by Phoenix,
an imprint of Orion Books Ltd,
Orion House, 5 Upper St Martin's Lane,
London WC2H 9EA

An Hachette UK company

1 3 5 7 9 10 8 6 4 2

A CIP catalogue record for this book is available from the British Library.

ISBN 978-0-7538-2845-8

Typeset by Input Data Services Ltd, Bridgwater, Somerset

Printed and bound by CPI Group (UK) Ltd,
Croydon, CRO 4YY

Contents

Author's Note

The Aztecs 'as evil as Nazis'

In 2009 the British Museum in London staged an exhibition entitled 'Moctezuma: Aztec Ruler' which did not go down at all well in some quarters. Critics felt aggrieved that the ruler's name had been changed from Montezuma to Moctezuma, the former spelling having been 'in satisfactory use' for 500 years. But beyond that, these critics found the quality of Aztec craftsmanship to be poor, no better than *bric-à-brac* to be found in London's Portobello Road, a popular antiques bazaar. The art critic of the London *Evening Standard* found that, compared with the achievements of Donatello and Ghiberti (i.e., broadly contemporary European artists), the Aztec material 'was pretty feeble stuff', that there was 'no art' in the 'barbarism of the Aztec world', that many of the masks were of the 'utmost hideousness', gruesome and grotesque fetishes of a cruel culture. The London *Mail on Sunday* was equally forthright. In an article headlined, 'British Museum Artefacts "As Evil As Nazi Lampshades Made From Human Skin",' Philip Hensher, a writer listed as one of the 100 most influential people in Britain, wrote: 'If there is a more revoltingly inhumane and despicable society known to history than the Aztecs, I really don't care to know about it.' On top of the moral and aesthetic ugliness of the Aztecs, this critic concluded that 'It is difficult to imagine a museum display that gives off such an overwhelming sense of human evil as this one.'

Strong words, but there are other ways of looking at the civilisations of the New World. In two recent books, for example, the authors stress the ways in which the ancient Americans outstripped their Old World counterparts. Gordon Brotherston, in his *The Book of the Fourth World*, describes the Mesoamerican calendar as 'demanding greater chronometric sophistication than the West was at first capable of'. Charles Mann, in his excellent book, *1491: New Revelations of the Americas Before Columbus*, points out not only that the Mesoamerican 365-day calendar was more accurate than its contemporaries in Europe, but that the population of Tiwanaku (in ancient Bolivia) reached 115,000 in AD 1000, *five centuries* ahead of Paris, that Wampanoag Indian families were more loving than the families of the English invaders, that Indians were cleaner than the British or French they came into contact with, that Indian moccasins were 'so much more comfortable and waterproof' than mouldering English boots, that the Aztec empire was bigger by far than any European state, and that Tenochtitlán had botanical gardens when none existed in Europe.

Such ad hoc individual comparisons, though interesting enough on the face of it, may or may not mean anything in the long run. After all, there is no getting away from the fact that it was the Europeans who sailed westward and 'discovered' the Americas and not the other way round. There is also no getting away from the fact that, over the last thirty years, a body of knowledge has built up which does indeed confirm that, in some significant respects, the ancient New World was very different from the Old World.

The most telling of these differences falls in the realm of organised violence. In research for this book, I counted twenty-nine titles published in the last thirty years – one a year – devoted to human sacrifice, cannibalism and other forms of ritual violence. Here, for example, are the titles published since the millennium: *The Taphonomy of Cannibalism*, 2000; *Ritual Sacrifice in Ancient Peru*, 2001; *Victims of Human Sacrifice in Multiple Tombs of the Ancient Maya*, 2003; *Cenotes, espacios sagrados y la práctica del sacrificio humano en Yucatán*, 2004; *Human Sacrifice, Militarism and Rulership*, 2005; *Human Sacrifice for Cosmic Order and Regeneration*, 2005; *Meanings of Human Companion Sacrifice in Classic Maya Society*, 2006; *Sacrificio, tratamiento ritual del cuerpo humano en la Antigua sociedad maya*, 2006; *Procedures in Human Heart Extraction and Ritual Meaning*, 2006; *New Perspectives on Human Sacrifice and Ritual Body Treatment in Ancient Maya Society*, 2007; *The Taking and Displaying of Human Body Parts as Trophies by Amerindians*, 2007; *Bonds of Blood: Gender, Lifecycle and Sacrifice in Aztec Culture*, 2008; *Los Origines de Sacrificio Humano en Mesoamerica Formativo*, 2008; *Walled Settlements, Buffer Zones and Human Decapitation in the Acari Valley, Peru*, 2009; *Blood and Beauty: Organised Violence in the Art and Archaeology of Mesoamerica and Central America*, 2009. Jane E. Buikstra, an expert on Mayan mortuary techniques, has calculated that the number of scholarly papers on Mayan ritual violence has grown from about two a year before 1960 to fourteen a year in the 1990s, a rate of publication that continued at least until 2011. On top of this, research into ritual violence in pre-Columbian North America has also grown. According to John W. Verano, professor of anthropology at Tulane University in New Orleans, each year brings a significant new discovery. Again, it is not so much the level of violence that fascinates researchers, so much as its organised nature and the specific forms of brutality that existed, and different New World attitudes and practices in regard to the associated pain.

It was an awareness of these seemingly strange yet significant differences between the hemispheres, and a desire to make sense of the context, that sparked the idea for this book. Initially, my ideas were worked out in discussion with Rebecca Wilson, my editor at Weidenfeld & Nicolson in London, but they have since benefited greatly from the energies of Alan Samson, publisher at W&N. I would also like to thank the indexer Helen Smith, and the following specialist scholars – archaeologists, anthropologists, geographers – for their input, some of whom have read all or parts of the typescript, and have corrected errors and made suggestions for improvements: Ash Amin, Anne Baring, Ian Barnes, Peter Bellwood, Brian Fagan, Susan Keech McIntosh, Chris Scarre, Kathy Tubb, Tony Wilkinson and Sijia Wang. Needless to say, such errors and omissions as remain are the sole responsibility of the author.

I would also like to thank the staffs of several research libraries: The Haddon Library of Archaeology and Anthropology, in the University of Cambridge; the Institute of Archaeology Library, in the University of London; the London Library, St James's Square, London; the Library of the School of Oriental and African Studies, also in the University of London.

From time to time, instead of repeating the phrases 'Old World'/ 'New World', I have varied the wording and employed 'western'/ 'eastern' hemisphere or 'the Americas'/'Eurasia'. This is simply for the sake of variety (and, occasionally, strict accuracy), and nothing ideological is implied by this usage.

I have sometimes used BC to date sites, or events, and sometimes BP (before the present). This respects the wishes of the researchers whose work is being discussed.

This is a book that concentrates on the *differences* between Old World and New World peoples. This is not to deny that there are also many *similarities* between the civilisations that existed in both hemispheres before the Europeans 'discovered' America. In fact, investigation of these similarities has thus far been the chief interest of archaeologists. For readers who wish to explore these similarities, they are referred to an appendix available online at *www.orionbooks.co.uk/thegreatdivide*.

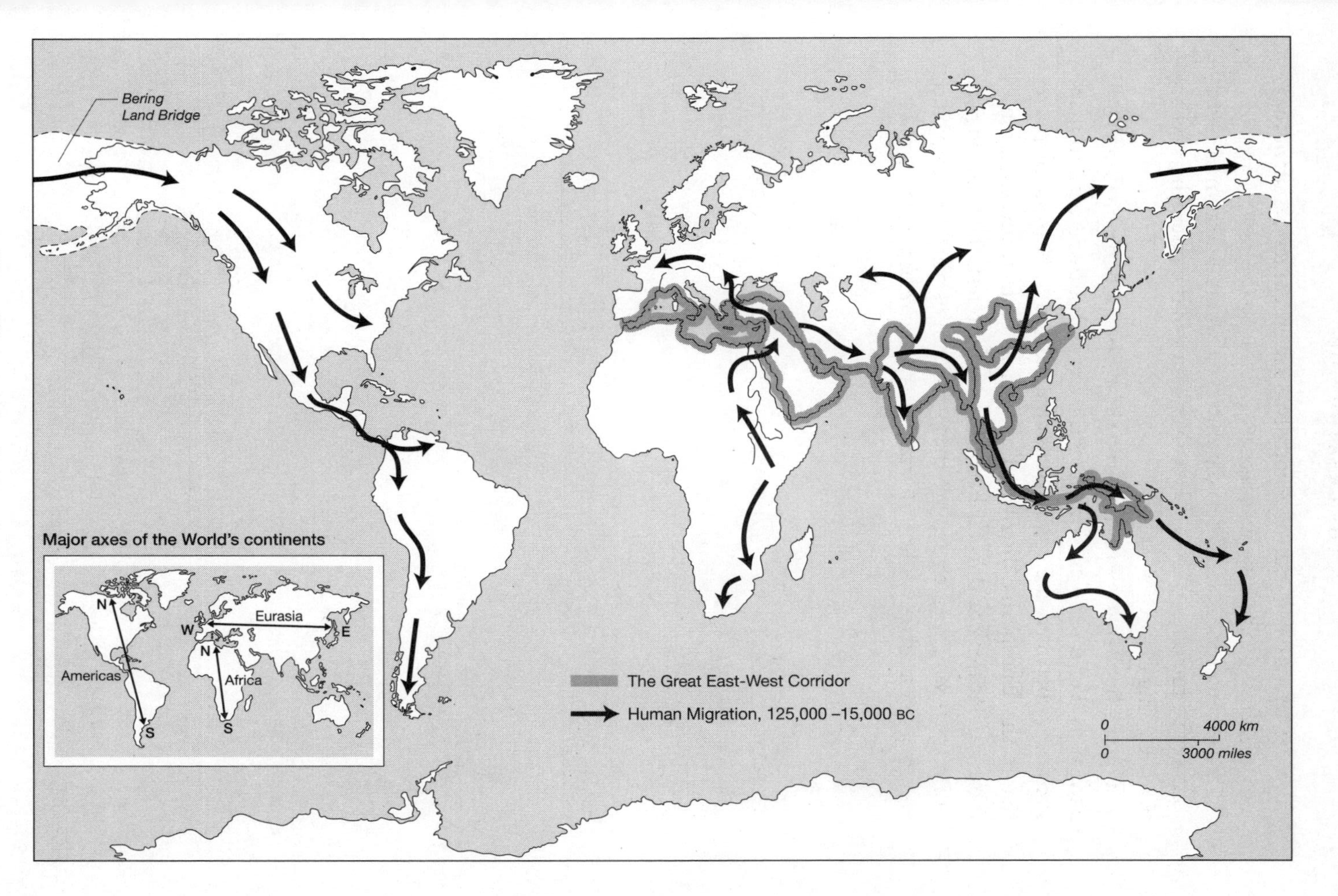

Map 1: Human migration, 125,000–15,000 BC

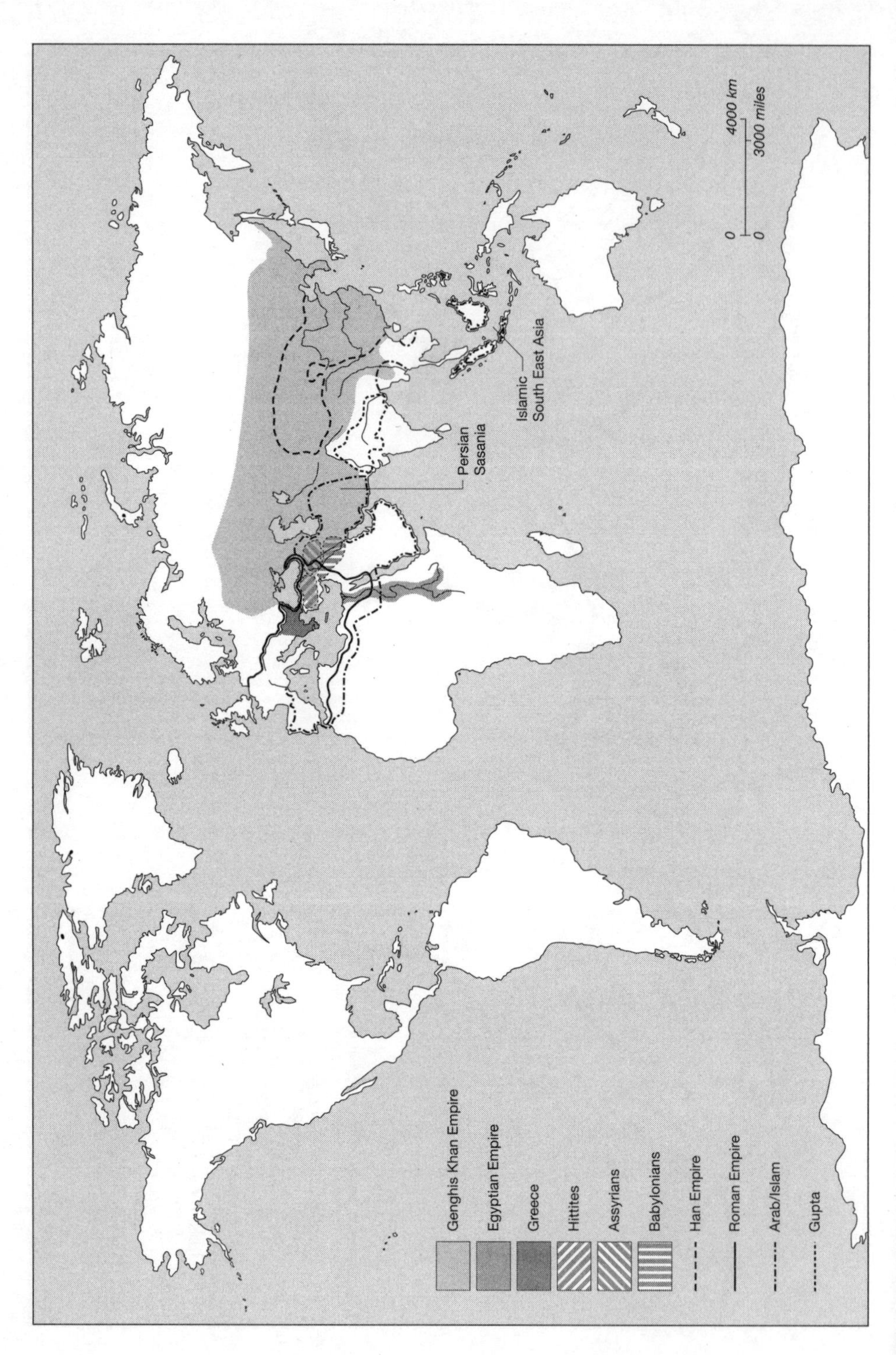

Map 2: The extent of the major old world ancient civilisations

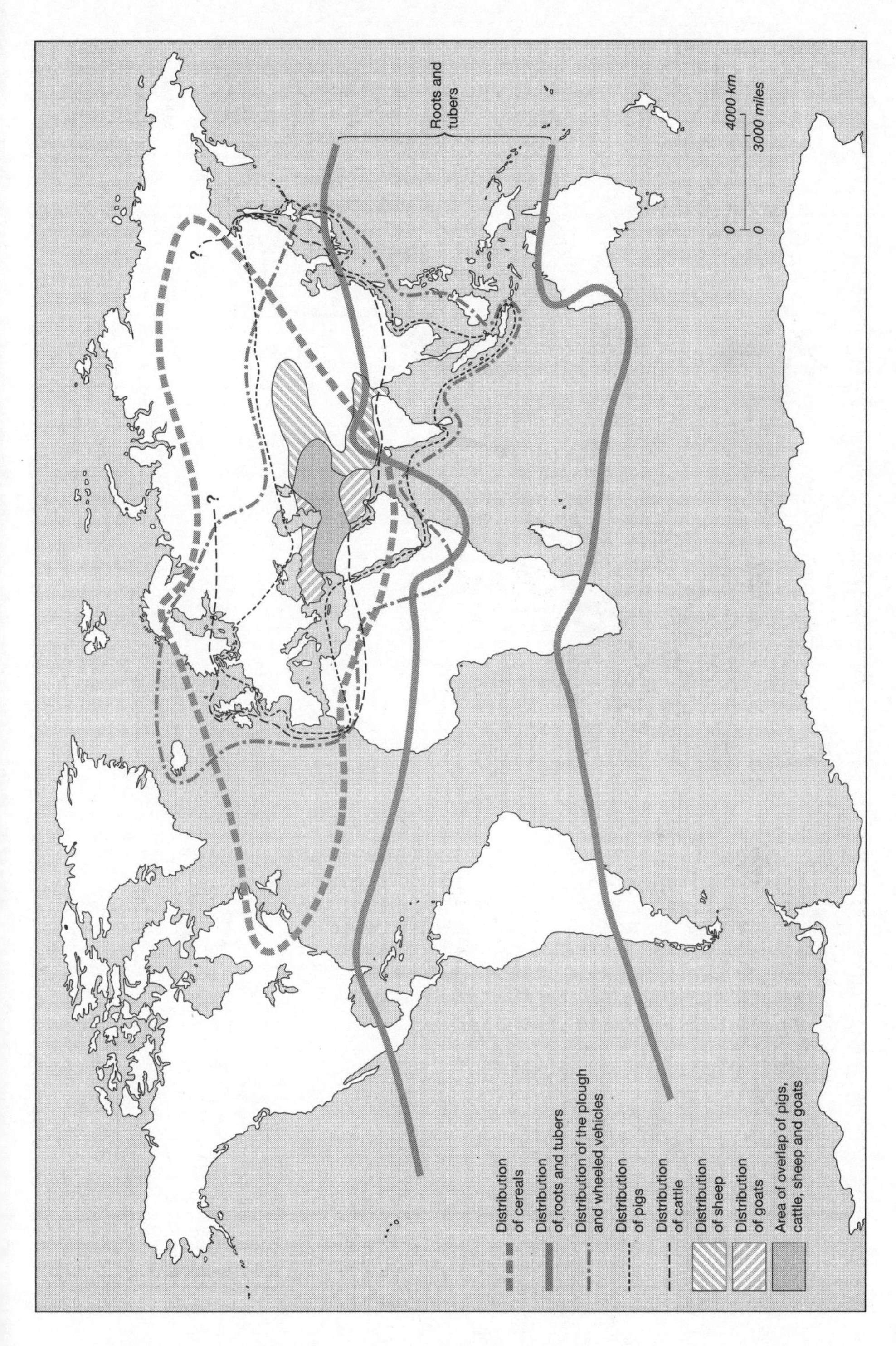

Map 3: The natural distribution of the plough, wheeled transport and major food products before AD 1500. Note the minimal overlap between the spread of tubers and roots on the one hand, and cereals on the other

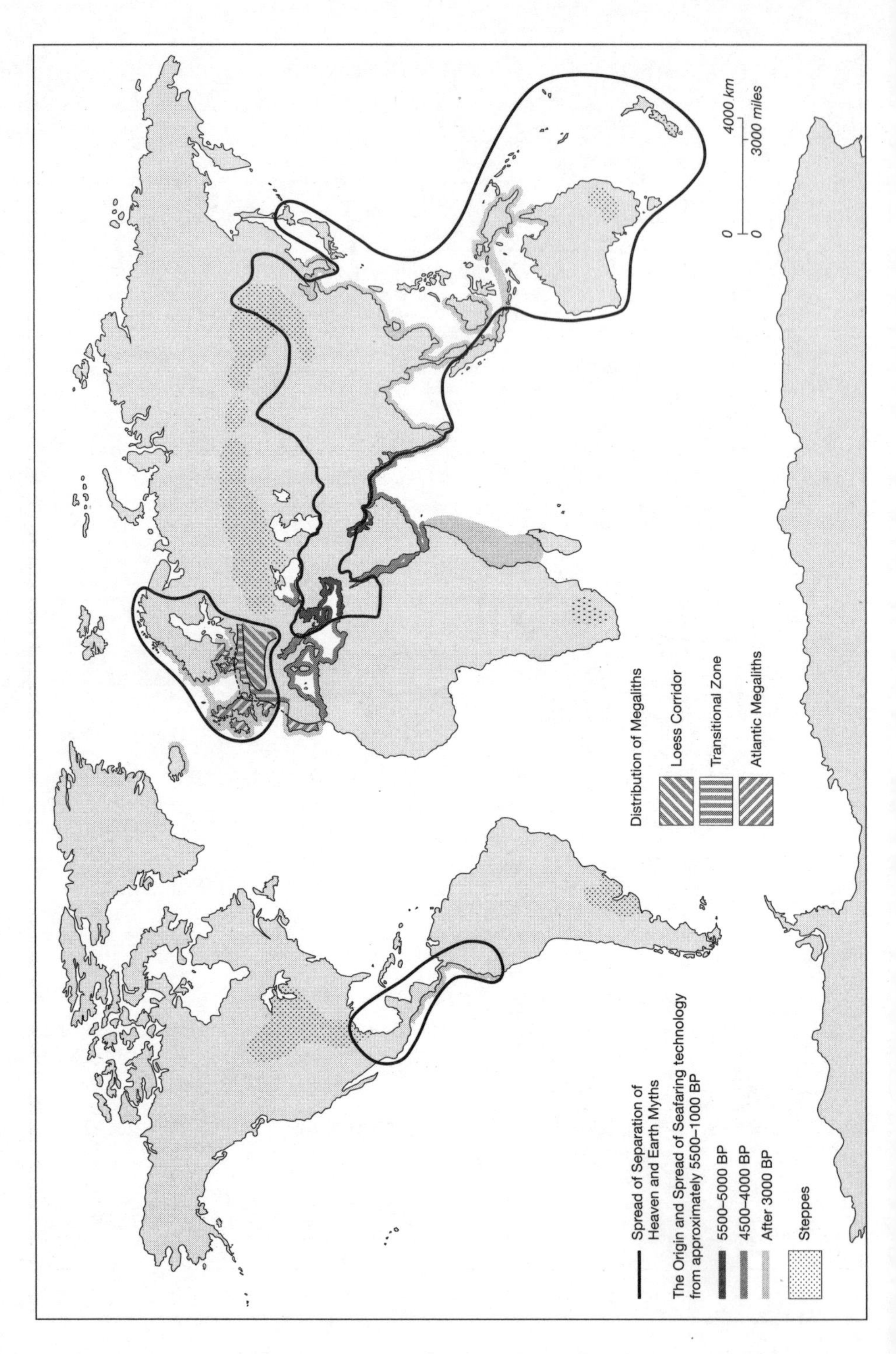

Map 4: The distribution of certain natural and cultural features discussed in the text

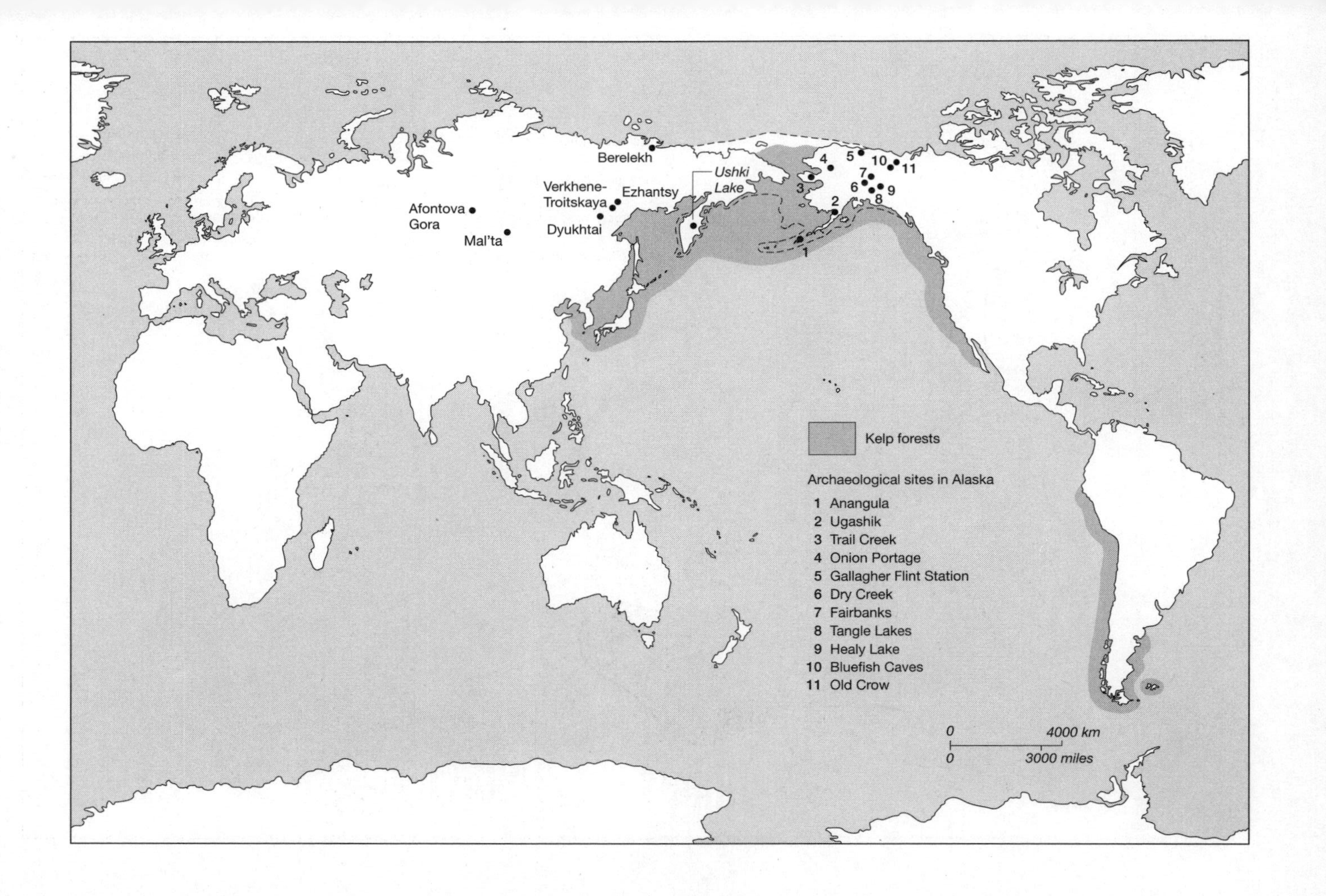

Map 5: Siberian/Alaskan settlements, the outline of the Bering Landbridge, and the distribution of the Kelp Forests around the Pacific rim

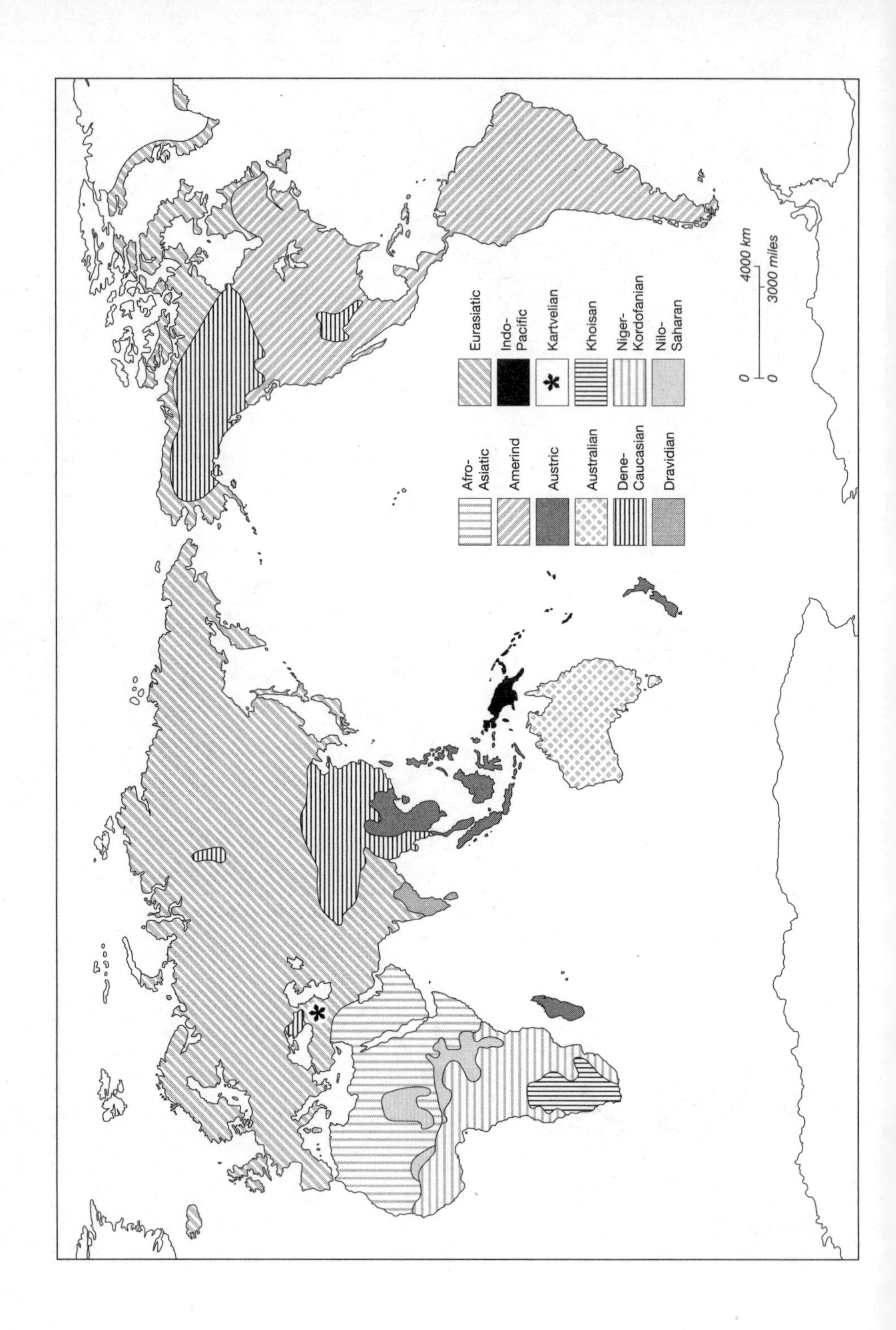

Map 6: The distribution of the world's major language families

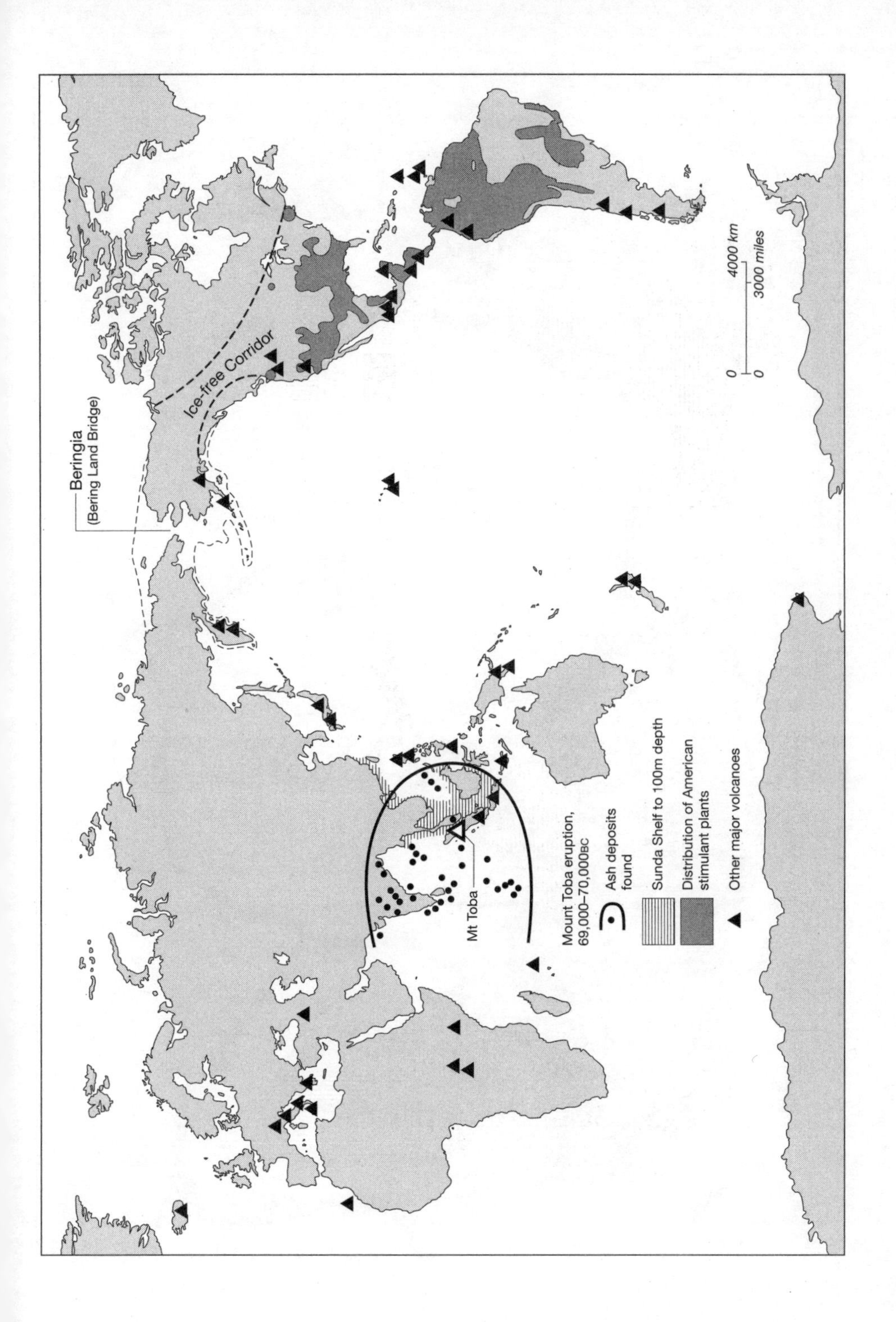

Map 7: Natural features of the Pacific rim and South Asia, discussed in the text

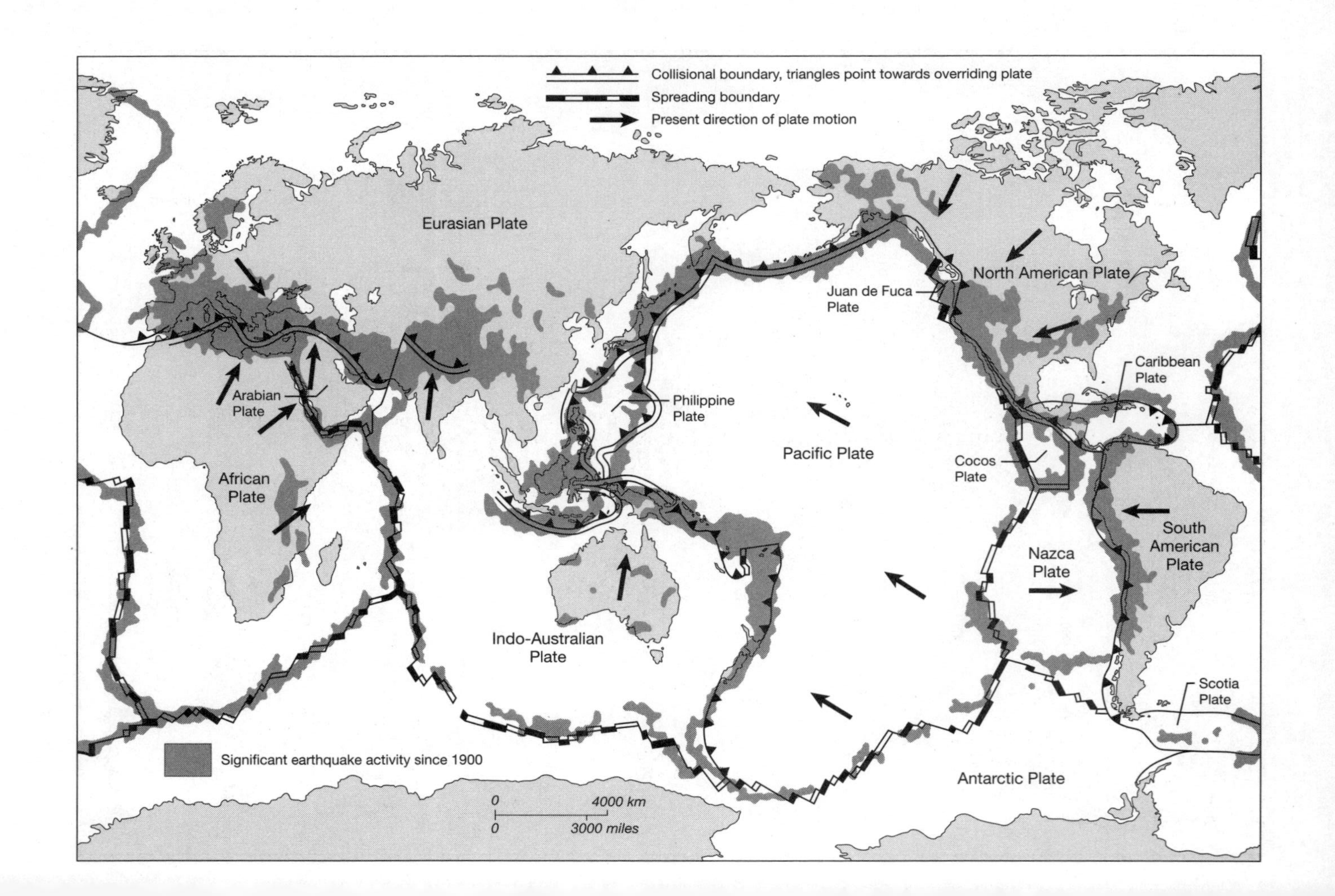

Map 8: Worldwide distribution of tectonic plates and earthquake activity

Map 9: The maximum wind speed (in mph) achievable by hurricanes over the course of an average year

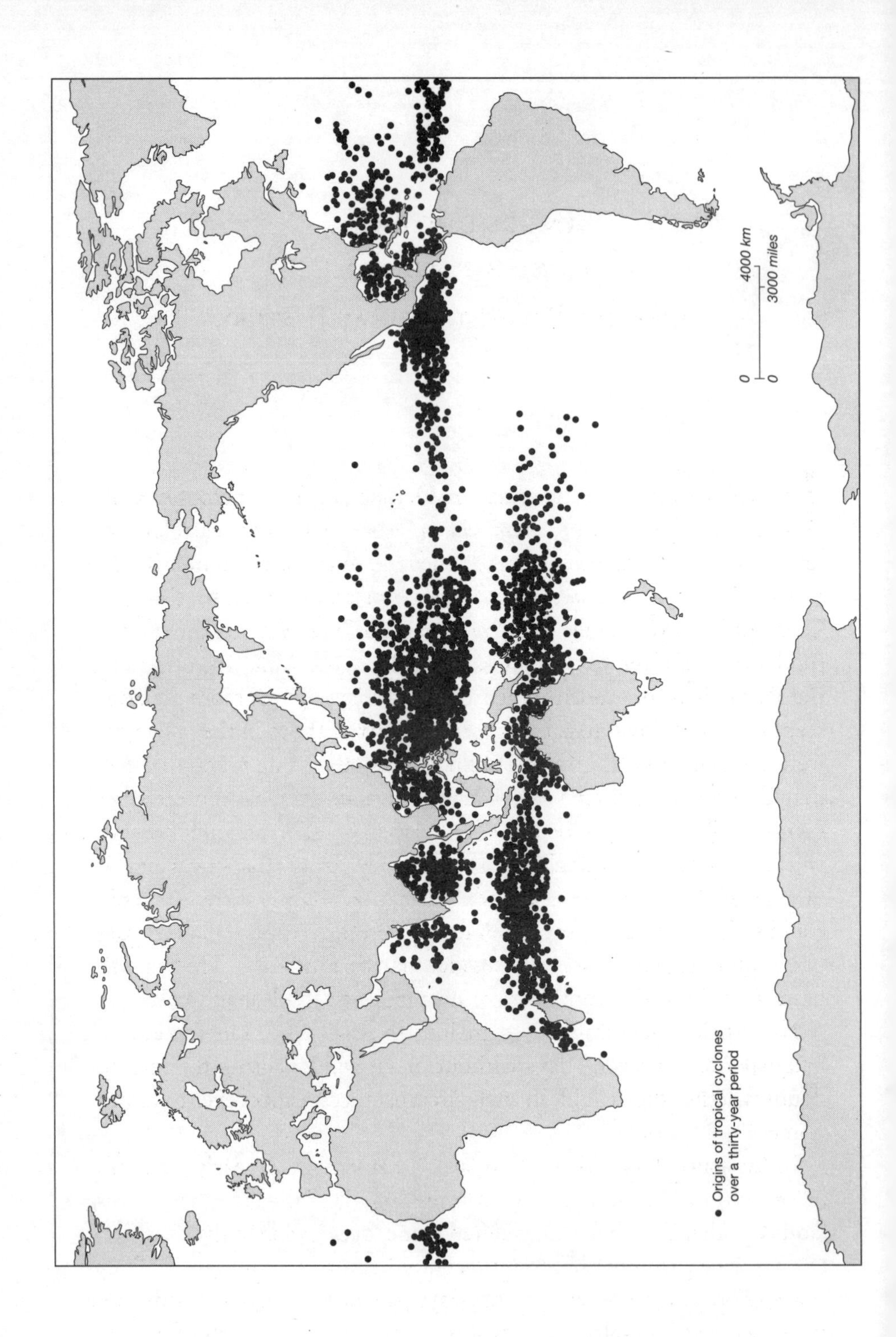

Map 10: Origin points of tropical cyclones over a thirty-year period

Introduction
15000 BC–AD 1500
A Unique Period in Human History

Just after sunset, on Thursday, 11 October 1492, Christopher Columbus, in his ship the *Santa María*, was – by his own calculations, set out in his journal – some 896 leagues (or, roughly, 3,000 miles) west of the Canary Islands and on the verge of reaching, as he thought, Cipangu, or Japan. So far it had not been a difficult voyage, though the rudder of one of the two accompanying ships, the *Niña* (the other was the *Pinta*), had come adrift twice, causing delays, which more than one sceptic has put down to sabotage on the part of those in the crew who were reluctant to sail into the unknown. By the same token, the fact that Columbus calculated one distance travelled each day, but gave the men a smaller figure, has been attributed to his need to pretend that they were less distant from Spain than in fact they were. This 'subterfuge' is now generally discounted: a medieval league was the distance a ship could sail in one hour – say seven to twelve miles – and Italian leagues (Columbus was Genoese) were smaller than Spanish ones. The Spanish figures would have meant more to his men than the Italian variety.

Nonetheless, Columbus was anxious to reach land. His expedition had experienced several days without much wind, causing the men to doubt whether they could, in such circumstances, expect ever to return home.

In his journal as early as Sunday, 16 September – nearly a month before – they had been keenly interpreting signs that they were near land. On that day they encountered some 'deep green seaweed which (so it seemed to him) had only recently been torn from land'.[1] They saw a good deal more weed as the days passed. At other times the sea water seemed less salty, as if they were nearing the mouth of a large fresh-water river; they saw large flocks of birds flying west (as if towards land), plus gannets and terns, species which 'sleep on land and in the morning fly out to sea to look for food and do not go further than 20 leagues'. At still other times they saw birds, including ducks, which they took to be river birds; or else it rained with a kind of 'drizzle without wind, which is a sure sign of land'.[2] The smaller ships were faster than Columbus's own and, during the day, they often became separated. (The Spanish monarchs, Ferdinand and Isabella, had offered a lifelong pension to the first man to spot land.) But the three ships were instructed to keep together at sunrise and sunset, 'because at those times the atmosphere was such as to allow them to see furthest'.[3] In fact, land had been 'spotted' twice before but each time it proved illusory.

On 11 October, however, the crew of the *Pinta* encountered a stalk and a twig and fished out from the sea another stick, 'carved with iron by the looks of it, and a piece of cane and other vegetation that grows on land, and a small plank'.[4]

They sailed on as the sun set that day and, at around ten in the evening, Columbus himself claimed to have seen a light. According to Bartolomé de las Casas, the sympathetic historian whose father travelled with Columbus on his second voyage, Rodrigo Sánches de Segovia, whom the king and queen had sent as comptroller, did not agree with Columbus though other crew members did. Later scholars have calculated that if Columbus *did* see a light, it must have been some sort of fire but it would have to have been a very large fire, because the *Santa María*, we now know, was then some 50 miles off land.

In fact, the first undisputed sighting of land took place in darkness, at two o'clock the next morning, Friday, 12 October, the identification being made by a sailor whose name was given by Las Casas, in his summary of Columbus's *Journal*, as Rodrigo de Triana. But since this name does not appear in the crew lists, scholars have concluded he must have been Juan Rodríguez Bermejo, a native of the town of Molinos.[5] The land was about two leagues – fifteen to twenty miles – away.

Columbus ordered his men to 'lay to' that night, taking down some of the sails, waiting for dawn. The next morning the captains of the three ships – Columbus, Martin Alonso Pinzón and his brother, Vicente Yáñez – went ashore in a small armed boat, accompanied by the comptroller, and they together witnessed Columbus claim the island in the name of the King and Queen of Spain. He called it San Salvador.

Soon, however, many islanders gathered round. 'In order to win their good will,' Columbus wrote in his journal that night, 'because I could see that they were a people who could more easily be won over and converted to our holy faith by kindness than by force, I gave some of them red hats and glass beads that they put round their necks, and many other things of little value, with which they were very pleased and became so friendly that it was a wonder to see. Afterwards they swam out to the ships' boats where we were and brought parrots and balls of cotton thread and spears and many other things, and they bartered with us for other things which we gave them, like glass beads and hawks' bells. In fact they took and gave everything they had with good will, but it seemed to me that they were a people who were very poor in everything. They go as naked as their mothers bore them, even the women, though I only saw one girl, and she was very young.' After describing the people (now known to be Tainos) physically, and how they painted their bodies, Columbus went on, 'They do not carry arms and do not know of them because I showed them some swords and they grasped them by the blade and cut themselves out of ignorance. They have no iron: their spears are just shafts without a metal tip, and some have a fish tooth at the end.'

This date – 12 October 1492, and this encounter, between an Italian acting under royal Spanish auspices, and a people now known to have spread north from South America, near the Orinoco River in Venezuela – comprise an event of almost unrivalled importance in world history: the first meeting between the Old World and the New. Yet Columbus's *Journal*, this part of it certainly, is a relatively tame document and even allowing for the fact that Spanish wasn't his first (or even his second) language, it is not hard to see why. Columbus himself had no real idea of what he had discovered, or its significance. This is underlined by the fact that, even today, we don't know where, exactly, this island was, or is. We know that it was in what are now called the Bahamas, and we know that the native name for the Bahamas was Lucayas. We also know that the native name for that particular island was Guanahaní but nothing more. Many islands in the Bahamas and the Turks and Caicos group fit Columbus's description and, in all, some nine locations have been suggested. The most likely, according to modern scholars, are Watlings Island or Samana Cay.

Columbus and his crew were relieved to reach land, not least for the fact that they could take on fresh water. But he moved on quickly, the following day, on the afternoon of Sunday, 14 October. At that time it was not yet a legal requirement in Spain for ships' captains to keep a log (that only happened in 1575), so we are perhaps fortunate in having anything at all in Columbus's hand. But his style is very repetitive, his observations are very general and, as Barry Ife has observed, the admiral's first aim seems to have been to make everything familiar – he constantly compared the terrain he had discovered with rivers and landscapes in Seville or Andalusia, rather than specifying what was new or exotic (though he did this later). 'Columbus's response to the natural beauty of the islands is undoubtedly genuine, but it is also strategic. Each island is the most beautiful that eyes have ever seen. The trees are green, straight and tall, fragrant and full of singing birds. The rivers are deep, the harbours wide, wide enough to embrace all the ships of Christendom ... what Columbus describes is not so much what he saw, as the sense of wonder with which he saw it.'[6]

There is, therefore, hidden in Columbus's account, a sense in which he was disappointed in what he found on the far side of the Atlantic. This may seem strange to us, who are the beneficiaries or the victims of his achievement, but Columbus's disappointment relates – of course – to the well-known fact that, to the end of his life, he maintained 'that he had reached the "Indies" he had set out to find. He had landed on islands close to Cipangu (Japan), and on the mainland of Cathay (China).'[7]

This insistence shows that all manner of historical forces were represented by Columbus, whether he knew it or not. In the first place, his voyages were the culmination in a mammoth series of navigational triumphs that had begun centuries earlier. Some of these voyages had been much longer than Columbus's, and no less hazardous. In some ways, they collectively represent humankind's most astounding characteristic: intellectual curiosity. Man's medieval ventures into the unknown are, save for space travel, simply impossible for us to share and therefore separate us from Columbus's time in a fundamental way.

Low-key as Columbus's landfall was, however, as the world now knows it eventually sparked a stampede across the Atlantic, a tide that still continues to an extent and changed forever the very shape of our world, with momentous consequences, both brilliant and catastrophic. It is not always recognised, however, that Columbus's discovery, whatever else it did, also marked the end – or the beginning of the end – of a particular phase in history, a singular set of circumstances that have only been appreciated – can only have been fully understood – in recent years, owing to discoveries in several fields of learning. That unique period in history had begun, very roughly speaking, more than 16,000 years previously.

The Greatest Natural Experiment in History

Until Columbus's landfall on San Salvador, from approximately 15000 BC, when ancient peoples first entered the Americas, until roughly AD 1500, to speak in round numbers, there were two *entirely separate populations* on earth, one in the New World, one in the Old, each unaware of the other. This is a period of history that has never been regarded as an epoch in its own right before, but a moment's thought will show how unusual it was, and how deserving of inquiry.

These separate populations were faced with different environments, different weather, different landscapes, different vegetation, different animals. *Nature* in the two hemispheres, as we shall show, was very different. For more than sixteen thousand years – between 600 and 800 generations – these two populations, originally so similar, adapted to their environments, developing different survival strategies, different customs, different languages, different religions and, ultimately, different civilisations. For all of this time the world was divided in a unique – an unprecedented – way, but when Columbus set foot on Guanahaní, therefore, and, without knowing it, he began a process whereby this unique parallel development was eventually brought to an end.

And this is the purpose of *The Great Divide*, to resurrect and recreate, to examine and investigate that parallel development, to observe the similarities and contrast the differences between the Old World populations and the New World populations, to see where the comparisons and the contrasts lead.

In a sense the parallel development of these two populations was the greatest natural experiment the world has seen. Not a tidy experiment of course, in the laboratory sense, but a fascinating exercise in comparison nonetheless, a unique opportunity to see how nature and human nature interact, to explain ourselves to ourselves. It is a project not carried out before.

The territories under scrutiny – entire hemispheres – are the greatest entities on Earth, along with the oceans, and some purists may be sceptical that the comparison involves too many variables to be meaningful. But I think the evidence is plentiful enough to enable us to draw some very fruitful conclusions about the important and long-term differences between the Old World and the New which *explain* – as well as describe – the markedly different trajectories of civilisation in the two hemispheres.

For perfectly understandable reasons, archaeologists and anthropologists have in general looked at the *similarities* between different civilisations across the world, sharing the view that such comparisons will, more than anything else, reveal fundamentals about human nature, human society, and the way humankind has developed over the last 10,000 years, since the end of the Ice Age. While not denying that those parallels exist, or that they are important, this book takes the obverse approach and looks at the *differences* between the two hemispheres, on the grounds that these are just as instructive, perhaps more so, and that they have been relatively neglected. These differences also throw an important sidelight on what, ultimately, it means to be human.

The book is divided into three. Part One describes how the first Americans reached the New World, what was special about their journey, how their experiences distinguished them from the peoples they left behind in Eurasia. Part Two describes the important and systematic ways in which the two hemispheres differed (and differ) – in geography, climate, their flora and fauna and the *interaction* between these separate elements. In some ways, this is the most surprising section of the book, that something as fundamental as *Nature* should vary so much between the two hemispheres. Part Three is a narrative, two narratives in fact, entwined around each other, as we follow the different trajectories as people in the two hemispheres both evolved great – but in some ways very different – civilisations.

Broadly speaking, *The Great Divide* seeks to show that the physical world which early people inhabited – the landscape, the vegetation, the non-human animal life, plus the dominant features of the climate, of latitude and the relation of the land to the sea – determined the *ideology* of humans, their beliefs, their religious practices, their social structure, their commercial and industrial activities, and that, in turn, ideology, once it had emerged and cohered, determined the further characteristic interaction between humans and the environment. It may be true, as the socio-biologists and geneticists say, that there is only one human nature. But the very different environments across the world created some very different *ideas* that early peoples had *about* human nature. And, as this book seeks to show, that was in many ways more important.

Our story will show that each hemisphere depended on, and was determined by, three very different phenomena. Vast stretches of the Old World fell under the influence of the Asian monsoon, the seasonal rainy period that extended from the eastern Mediterranean to China, which supported two-thirds of the world's farmers and which has, for reasons we shall explore, been gradually weakening in strength for the past 8,000 years. This meant that *fertility* was the main preoccupation of religion in the Old World. Second, the existence of *domesticated* mammals exerted another all-important influence over the *course* of ancient history in the Old World, in particular the nature and extent of inter-societal competition and warfare. In the New World in contrast, the dominant influences were extreme – violent – weather, and, third, the much wider availability, variety and greater abundance of hallucinogenic plants. Together, these factors meant that religion, ideology, took on a far more vivid, intense and apocalyptic tone in the Americas.

The Great Divide attempts quite an ambitious synthesis of what, to begin with, may appear to be very different disciplines: cosmology and climatology, geology and palaeontology, mythology and botany, archaeology and volcanology.

It also makes use of one of the great breakthroughs in modern scholarship that has occurred since the Second World War – namely, the understanding of the newly deciphered scripts of the four main Mesoamerican civilisations – the Aztec, Mixtec, Zapotec and Mayan. Although only four Mayan books escaped the Spanish flames during the Conquest, other books or codices, produced jointly by Spanish clerics and Native Americans, and now countless inscriptions on stone stelae, altars and flights of stone steps, and other monuments and sculptures, are understood to such effect that the last thirty years have seen a massive explosion in knowledge and understanding of pre-Columbian life in the New World.

The Great Divide uses this recent scholarship to formulate a systematic comparison between ancient history in the two hemispheres, and in doing so we shall see that there were very different trajectories as between Eurasia and the Americas. Despite these different paths, both hemispheres developed similar features, but it is the *differences* that are our concern here, in this book. They tell us just as much about human nature – and maybe more – as the similarities do.

In examining these trajectories jointly, we shall not only see what happened – when and where the civilisations began to diverge – but *why*.

• Part One •

How the First Americans Differed from Old World Peoples

• 1 •

From Africa to Alaska: The Great Journey as Revealed in the Genes, Language and the Stones

If our 'experiment' of comparing developments in the New World with those in the Old is to have as much meaning as possible, then we need to be as clear in our minds as we can be as to what extent the people in the two hemispheres were similar in the beginning. Or, failing that, we need to know how they differed. Clearly, this is not an easy task – we are talking of a time of at least 15,000 years ago and, in much of the material in this and the next chapter, a lot longer ago than that. But, although the time depth is daunting, and the material of such a nature that we need to be cautious at all times – since so much is speculation, albeit informed speculation – that should not deter us. A few years ago, it would have been impossible to answer these sorts of question but now, thanks to developments in biology (in particular genetics), in geology, in cosmology, climatology, linguistics and mythology we understand far more about our deep history than ever before. The conclusions we *are* able to draw, however tentative, are worth the effort.

Out of Africa

Because of the discovery of DNA, genes, and in particular mitochondrial DNA (normally written as mtDNA and inherited only through the mother), and the Y-chromosome (which determines male sexuality), and because we know the rate at which DNA mutates, it has become possible – through the comparative analysis of the DNA of modern peoples right across the globe – to assess who is related to whom, both now and at various times in the past.* In effect mtDNA gives us, as one expert put it, 'a cumulative history of our own maternal prehistory', while the Y-chromosome does the same for our paternal history. For our purposes, the main elements in this theoretical picture (much of which is still to be confirmed archaeologically) are as follows:

- Modern humans evolved in Africa around 150,000 years ago.
- Perhaps as early as 125,000 years ago, a group of humans left Africa, most likely across the Bab al-Mandab Strait, at the southern end of the Red Sea (when that sea was some 230 feet lower than it is now) and travelled across the southern Arabian peninsula at a time when the region was much wetter than now, occupied by lakes and rivers. No human remains have been found, but primitive stone tools, similar to those produced in Africa at much the same time by *Homo sapiens*, have been excavated at Jebel Faya, a rock shelter near the Strait of Hormuz. Genetic evidence, of individuals across the world, alive now, shows that *all* non-African people are descended from one small group that must have passed through the Arabian peninsula. During very dry periods, the Jebel Faya population may have been isolated for hundreds or even thousands of years, before moving on, eastwards, along river routes that are now submerged in the Gulf. In this way they would have avoided the arid inland deserts of the region, eventually reaching India, by way of the Iranian and Pakistan coasts. This 'beachcombing' theory about the peopling of the world is still only that, a theory, but it is supported by the genetic evidence and by the presence of ancient shell middens on many coastal sites. Furthermore, we now know that, for much of human existence,

* The details of mtDNA and the Y-chromosome research will not be given here. This research is quite well known and accessible accounts are given in many sources – for example, see Oppenheimer (1998), Wade (2007) and Wells (2007) in the Notes. Mutation rates vary according to species and sex (some studies show that males mutate four times as fast as females). One mutation in every 33 20-year generations is evolutionarily significant. Out of every 200 mutations, ~190 are neutral, ~5 are fatal and ~1 beneficial. The order of magnitude of human mutation rates is from 0.12–0.57 per base pair (of nucleotides connected by a hydrogen bond, making up the double-helix) per million years, but there is evidence that the rate has decreased since 20,000–15,000 years ago, when there was a demographic change in living patterns, when early peoples ceased to live in small, mobile, hunter-gatherer groups that may have been subject to fluctuations in populaton size, which caused bottlenecks in genetic diversity.[1]

before 6,000 years ago, sea levels were lower than now and as a consequence there was at that time perhaps as much as 16 million square kilometres more dry land in the world than there is now, ten per cent of the inhabited areas of the globe, a significant and attractive resource. We also know that, in general, marine/littoral environments provide a richer nutritional environment and support higher population densities and more sedentary settlements than do inland sites. Hunter-gatherers in ancient coastal and landbridge areas have so far been peripheral to human prehistory but that looks as though it is in the process of changing.

- This group that left Africa may not have been very large: Y-chromosome studies suggest it perhaps comprised only about 1,000 men of reproductive age and the same number of women. Add on children and older people and this represents a population of perhaps 5,000. They may not all have gone together, either. Studies of foragers show that they like to live in groups of about 150, though when they ceased beachcombing, in Australia for example, they formed tribes of between 500 and 1,000 people (which is what the European colonisers found when they arrived in Australia at the end of the eighteenth century).
- After 70,000 years ago, humans crossed into Australia.
- At 50,000–46,000 years ago, in what is now Iran/Afghanistan, a group left the coast and travelled north and west, to populate Europe.
- About 40,000 years ago, a second bifurcation took place, this time in Pakistan/north India, with another group travelling inland into central Asia.
- At about the same time, the 'beachcombers' had reached China, travelling around the 'corner' of South East Asia, and then moving inland, back west, along what would become much later the Silk Route.
- Roughly 30,000–20,000 years ago, the groups that had headed inland from Pakistan/India bifurcated, with one group travelling west, towards Europe, while the other travelled deep into Siberia, perhaps meeting up with the people moving inland from China.
- Some time around 25,000–22,000 years ago, humans reached the Bering Land Bridge which connected Siberia to Alaska, though there is no archaeological evidence for them in Chukotka, or Alaska,

until after 15,000 years ago. At that time, the world was in the grip of the last Ice Age, which endured from 110,000 years ago to about 14,000 years ago and as a result of which much of the world's water was locked away in the great glaciers – many kilometres thick – which mantled the Earth. As a consequence, the world's sea levels had fallen to some 400 feet below where they are now. In turn, this meant that the geography of the world was substantially different from what it is today. One important – crucial, fascinating – effect of this was that the Bering Strait did not then exist. It was comprised of dry land, or at least scrub land with lots of ponds and lakes but even so very passable for early humans. And so, some time between, roughly speaking, 20,000 and 14,000 years ago, early humans migrated into what would be called, later, variously the New World, the Americas, or the western hemisphere. Then, and this is no less crucial, after 14,000 years ago, when the world warmed up and the latest Ice Age came to an end, the Bering Strait refilled with water, Alaska and Siberia became parts of different landmasses and the western hemisphere – the Americas, the New World – was separated from the Old.

- Some of this evidence is shown on maps 1–10, between pages xi and xx. These maps summarise visually several of the arguments to be found throughout the text of this book.

As the crow flies (or a 747), it is about 7,500 miles from the southern end of the Red Sea to Uelen on the eastern-most tip of Siberia, but beachcombing around India and South East Asia would have more than doubled – and even tripled – that distance, and cutting across the landmass of Central Asia would not have been much shorter, and could have been more arduous, given the mountain ranges and lakes and rivers that needed to be circumvented without much in the way of technology. The journey of, say, 20,000 miles, took 50,000 years (though until early people reached the regions of intense cold, they may have spread quite quickly).

But eventually, early peoples arrived in what is now known as the Chukotskiy Poluostrov, or Chukotskiy Peninsula, overlooking what would become the Bering Strait. It is not only the close proximity of Siberia to Alaska that suggests early peoples entered the New World in this way (the strait – which was then a landbridge – is barely sixty miles wide at its narrowest point). There are three pieces of genetic evidence that, taken together, paint a coherent and convincing picture of early humankind's entry into the Americas.

The Chukchi people of eastern Siberia who, though they might be said to live at the edge of the world – the edge of the modern world at any rate – are nonetheless central from our point of view. Even today they live by herding reindeer and fishing through small holes in the ice-covered rivers.[2] No one really knows why early peoples chose to live in this hard part of the world. Perhaps they followed mammoths and other big game; perhaps they didn't *choose* to live there at all but were forced there by population pressures from the west and south. Just how hard life there was is confirmed by archaeological studies which show that there is a complete *lack* of sites in this part of Siberia between 19,000 and 18,000 years ago, suggesting that the amount of ice at its most extensive caused the area to be abandoned for a time before being re-colonised, by highly mobile hunter-gatherers who frequently moved their camps to where important animal resources were available – most sites have the remains of just one type of large-bodied prey species: reindeer, red deer or bison. Whatever happened, eastern Siberia (still a good distance from Chukotski) has been occupied – at sites such as Dyukhtai and Mal'ta – since 20,000 years ago (see map 5). This date is important, as is the location.

The date is important because no archaeological sites earlier than 20,000 years ago have ever been found anywhere in Siberia. However, there is both genetic and linguistic evidence for an earlier entry into the New World. This evidence is controversial and is not universally accepted. The location of the sites is important because agriculture has never been successfully practised this far north and so early man could not have entered the New World knowing anything of agriculture. This is in itself not surprising for agriculture did not emerge anywhere on earth until about 10,000 years ago but at least that means this is one area where the picture is clear: the Old World and the New both lacked agriculture when the Great Divide took place (though they had dogs).

The DNA evidence shows that the Chukchi are genetically distinctive. According to the Genographic Project (see below), they have a marker, a distinctive pattern of genes, technically known as M242, and other characteristics, which show that they originated in a single man living about 20,000 years ago in southern Siberia or Central Asia. These markers are also shared with Native Americans as far south as Tierra del Fuego and therefore confirm – for geneticists – that early man entered the New World from Siberia some time after 20,000 years ago.[3]

This picture was supported – and amplified – when the first results came in from the Genographic Project, set up in 2005, sponsored by National Geographic but making use of IBM's massive computational skills. This very large study examined the DNA of around 150,000 individuals on five continents, to draw up the most thorough picture of our genetic history ever mounted. The most basic technique of the Genographic Project was to examine what are termed haplogroups, distinctive and characteristic patterns of genetic mutation, which comprise 'markers' on mtDNA or Y-chromosomes and which show how people are related and *were* related in the past (M242 is a haplogroup).

This research shows two things that concern us. First, that today's Native American peoples are very similar to one another genetically and that most of the distinctive markers are somewhere between 20,000 and 10,000 years old, clustering around the 16,000–15,000-year mark. The significance of this timing is that it was during what is called the Last Glacial Maximum (LGM), the era – between 20,000 and 14,000 years ago – when the vast glaciers of the last Ice Age reached their greatest extent, when sea levels were 400 feet below where they are now, and when the Bering Strait would have comprised a land bridge between Siberia and Alaska.

One haplogroup located on the Y-chromosome is found in Native American men living all the way from Alaska to Argentina and, together with another haplogroup descended from it, is almost the only Y-chromosome lineage found in South America.[4] In western North America there is another lineage, which appears to have arrived in the New World later and never to have got as far as South America. But between them, these markers account for 99 per cent of Native American Y-chromosomes. On top of that, there are only five mtDNA haplogroups in Native Americans, in marked contrast to the *dozens* of mtDNA and Y-chromosome lineages found in Eurasia and Africa.[5] An important point about the second lineage, the one found in western North America and known as haplogroup M130, is that it is also found in South East Asia and in Australia, suggesting that this *second*, later migration into the Americas comprised people who travelled up the Pacific rim, the east coast of Asia and entered the New World around 8,000 years ago, when the Bering Strait was again submerged. They therefore must have migrated by boat. This lineage typically appears in Indians speaking Na-Dene languages, the second major linguistic family of North America (see below).

The third piece of evidence was another large inquiry, published in 2007, by a team of twenty-seven geneticists from nine countries coordinated by Sijia Wang from Harvard.[6] This team examined the genetic markers in 422 individuals representing 24 Native American populations in North, Central and South America and compared them with 54 other indigenous populations world-wide. The main results of this study were as follows:

- they found that Native American populations have lower genetic diversity and greater differentiation than populations from other continental regions (overlapping with the findings of the Genographic Project referred to earlier);
- there was decreasing genetic diversity as a function of geographic distance from the Bering Strait and decreasing genetic similarity to Siberians; the groups most similar to Siberians were the Chipewyan population (Na-Dene/Athabaskan) from northern Canada and the least similar were those in eastern South America;
- there was a relative lack of genetic differentiation between Mesoamerican and Andean populations;
- they found a scenario in which coastal routes were easier for migrating peoples to traverse in comparison with inland routes;
- they found some overlap between genetic similarity and linguistic classification;
- the study showed there was a particular allele (genetic variable) 'private' to the Americas (i.e., which exists only in the DNA of indigenous Americans), supporting the view that much of New World ancestry 'may derive from a single wave of migration'.

This picture is again overall broadly consistent with the evidence from the Genographic Project, and from the analysis of Chukchi DNA, in showing a single entry into the Americas, from Siberia, by a group that is roughly 550 generations old – in other words, who arrived in the New World between, say, (550 × 30 =) 16,500 years ago and (550 × 20 =) 11,000 years ago, probably using a coastal route rather then the inland, inter-glacier route (again, see below). Since the study by Sijia Wang and his team, other surveys, not quite so large, have nonetheless produced fairly similar results regarding the time of entry of ancient peoples into the Americas, but have suggested the entry occurred in two waves not one, the first at ~18,700 years ago, the second at ~16,200 years ago. As will be seen, this fits in with the linguistic evidence presented below.[7]

It is only fair to emphasise at this point that there are a handful of DNA studies that suggest a much earlier entry of peoples into the New World – some at 29,500 years ago, some even at 43,000 years ago.[8] However, the latest and largest studies – the Genographic Project and that by the Sijia Wang team – not only agree with each other, broadly speaking, but they also agree with the archaeological evidence discovered all across North America, from Alaska to New Mexico. Some of this evidence is outlined immediately below but there is a more extended discussion in chapter three.

A second kind of biological evidence comes from the work of Christy Turner, at Arizona State University, who is an expert on the evolutionary development of human teeth.[9] In particular, Turner has looked at the crowns and roots of 200,000 teeth of prehistoric Americans, Siberians, Africans and Europeans because (a) they show well how populations adapted to different environments, and (b) they are, he says, more stable than other evolutionary traits, and tend not to vary so much between males and females or between old and young. For our purposes, his work is most interesting where it distinguishes between what Turner calls 'sinodonty' and 'sundadonty'. Sinodont teeth, found mainly among northern Chinese and north Asian (Siberian) populations in general, are characterised by 'incisor shovelling' (scooped-out shapes on the inside of the tooth), double-shovelling (scooping out on both sides), single-rooted upper first premolars, and three-rooted lower first molars. Turner has found sinodonty in the excavated remains of northern Chinese skeletons that go back at least 20,000 years.

Sinodonty, he finds, is confined to northern Chinese and northern Asian populations *and* in ancient Alaskan and other northern American populations. In contrast, such Upper Palaeolithic skeletons as have been found further west – in the Lake Baikal area, for example – do not display sinodonty, nor do teeth from ancient burials in European Russia. The same is true too of ancient remains found among South East Asians. (Turner calls this group 'sundadonts' because in Palaeolithic times South East Asia, like Beringia, was above sea level, the continental shelf there being known as the Sunda Shelf, a phenomenon about which we shall have much more to say.) From the spread of sinodonty in northern Asia and North America, Christy Turner believes that the first Americans developed from people who migrated slowly through eastern Mongolia, the Upper Lena Basin, eastern Siberia, and from there across the Bering Strait into Alaska.

Still further biological support for this scenario is found in the fact that the infants of some American indigenous tribes are born with the so-called 'Mongol spot', a bluish birthmark at the base of the spine that soon disappears and is also found among children in Tibet and Mongolia.[10]

Putting all this genetic evidence together, therefore, we may say that the early people from whom virtually all Native Americans alive today are descended, arrived in the Americas – very roughly speaking – about 16,500–15,000 years ago, from somewhere in north-eastern Asia, the area that is now Siberia, just possibly as far south as Mongolia. There may have been small groups of people who found their way to the Americas earlier than that but their effect on later populations was negligible. And there may have been later migrations, the evidence for which will be considered shortly.

Timothy Flannery makes the point that, although the Aleuts and Inuits in Alaska share many cultural features with north-east Asians (including a form of Eskimo spoken on the Kamchatka Peninsula in Russia, and a variety of acupuncture practised in the Aleutian Islands similar to that used in China), there is very little evidence of people or ideas going *back* to Asia from the Americas. The only genetic study that throws any light on this is that produced by Nicholas Ray and colleagues who cautiously concluded that some Native Americans crossed back into Asia 390 generations (or 9,750 years) ago. A well-publicised study of the only surviving Yeneseian-speakers, the Kets in central Siberia (several thousand miles from the Bering Strait), showed a linguistic link between the Yeneseian and Na-Dene languages but, genetically, the Kets were very similar to the other Siberian groups around them, and not at all related to the Na-Dene speakers in North America. At the moment there is no satisfactory explanation for this anomaly. But we may conclude, therefore, that the main migration across the Bering Land Bridge went from Siberia to Alaska and occurred – crucially – towards the end of the Ice Age.[11]

There is one other piece of genetic evidence we need to consider before moving on. This is the work of Bruce Lahn, at the University of Chicago, who has discovered two genes which are involved in the construction and enlargement of the human brain. Each gene has several alternative forms, or alleles, but in each case one version has become far more common than others among certain populations. This disparity must mean that the effect of this allele was of great evolutionary significance, providing a selective advantage. One of the alleles is a version of a gene known as microcephalin. This first appeared some 37,000 years ago and is carried by 70 per cent of populations in Europe and Asia but is much less common in sub-Saharan populations, where it is carried by between 0 and 25 per cent of people. The second allele is known as ASPM (for Abnormal Spindle-like Microcephaly-associated) and appeared and then spread rapidly in the Middle East and Europe around 6,000 years ago. This allele is absent in sub-Saharan Africa and only weakly represented in East Asia.[12]

For these two alleles to have spread so quickly, they must each have conferred some cognitive advantage. For obvious reasons this is material that should be interpreted with the utmost caution, as Bruce Lahn himself has counselled. There is at the moment no evidence these alleles are associated with increased intelligence; set alongside the other results mentioned above, however, these discoveries may have two implications that concern us. First, so far as the mutation that occurred around 37,000 years ago is concerned, one might ask whether this allele had anything to do with the 'cultural explosion' that occurred in the palaeontological record beginning around 33,000 years ago, with the notable florescence of cave art in certain areas of Europe. And, by the same token, was the mutation that occurred some 6,000–5,000 years ago in any way related to the development of civilisation that appeared roughly 5,500 years ago? Are we seeing here a link between genes and culture that has not been suspected before, because such results were not available?

If so, then the second implication may become relevant for the arguments in this book. The first mutation, at around 37,000 years ago, would, if it was so adaptive, presumably have spread quickly throughout Eurasia and included those early people who eventually migrated into the New World. Native Americans should, in other words, have possessed this allele. This is what research shows: microcephalin is virtually universal in New World populations.

On the other hand, the second mutation, occurring roughly 6,000–5,000 years ago, would appear to have evolved *after* early men and women had crossed the Bering Strait, meaning that, in all probability, Native Americans should lack this development. And this too is what research shows: ASPM is completely absent in New World populations.

It is too soon to say whether microcephalin or ASPM conferred some sort of cognitive advantage on those who possessed it, though its rapid spread suggests that is likely, although simple brain size appears to have remained stable. Nonetheless, this is clearly an area of potentially important genetic difference between Old World and New World peoples. We know from evidence in Iceland, which was inhabited only about a thousand years ago, that substantial genetic differences can arise in such a relatively short time frame, so it is not out of the question that *some* genetic variation accounts for the differences between the Old World and the New.

That said, this area of science is still in its infancy, so no more will be made of it here, other than to draw attention to what is a tantalising possibility.

One final thought on genetics. The relative lack of diversity among Native Americans, compared with the rest of the world, as shown by the Genographic Project and the Sijia Wang team study, may imply one of three scenarios. First, that there was at some stage a genetic 'bottleneck' in Beringia, where a small genetically limited group lived for a while, perhaps in a refuge surrounded by the ice, when they were forced to breed within their small community. Second, there was much polygamy later, with some – the more successful – men having several wives and others none (just such a pattern has been observed among the Dani population in Papua New Guinea, for example, where 29 per cent of the men had between two and nine wives, while 38 per cent had none).* Or again, a third possibility, it could be the result of widespread warfare, the burden of which was borne by men, leaving the remainder to father the children (among the Dani, again, 29 per cent of men were observed to be killed by warfare).[13]

Among the consequences of such limited genetic diversity would have been the fact that the pace of evolution in the New World would have been slowed in comparison with that in the Old; and it would also have made New World peoples more susceptible to diseases introduced from outside.

Sleds and Seaweed

The archaeological evidence for an entry into the New World from Siberia is supported by the great similarity of sites either side of the Bering Strait. A group of locations nearest to the Strait in Siberia was christened (in 1967) by Yuri Mochanov, a Russian archaeologist from the Scientific Research Institute at Yakutsk, as the 'Dyukhtai culture', after a site on the Aldan River, which flows north into the Laptev Sea, on the fringes of the Arctic Ocean. Here, mammoth and musk-ox remains were excavated, associated with spear and arrow points flaked on both sides, as well as blades and wedge- and disc-shaped cores – in other words, a distinctive upper Palaeolithic culture, dated to between 14,000 and 12,000 years ago. Other sites, with bifacial tools and blades and even knives, have since been found in the area, together with bone and ivory artefacts. Nothing older than 18,000 years has been unearthed, and the bulk of remains are later. The northern-most site of the Dyukhtai culture is found at Berelekh, near the mouth of the Indigirka River, on the northern shore of Siberia.

Just as early people appear to have 'beachcombed' their way around the south-east coast of Eurasia, to reach China, so they may have beachcombed east from Berelekh along the Arctic Ocean coast of Siberia until they reached the Bering Strait – except that it was then land. Some palaeontologists, like Dale Guthrie, emeritus professor at the Institute of Arctic Biology at the University of Alaska, believe that

* We can't be certain, of course, that men without wives never sired children. There is some evidence that this happened in Australia.

the Dyukhtai microblades were intended to be slotted into antler points as weapons. If so, this could complicate matters, suggesting that this technique, which is also found in North America, was not so much learned or copied by 'New World' people from 'Old World' people, as a rational adaptation to an environment where reindeer were abundant. In other words, it is not in itself evidence of migration.

But the fact remains that there are several other cultural similarities between the Dyukhtai complex in Siberia and sites found in Alaska. Both cultures, it should be said, are terrestrial cultures, which do not feature sailing among their skills, suggesting that these early peoples at least crossed Beringia on foot, rather than by canoe or something similar. (One interesting observation that may be more than a side-effect is that the burial of a domesticated dog was recorded at a site in Ushki, on the Kamchatka Peninsula, dated to 11,000 years ago. Given that, even today, it is easier to move around in the Arctic Circle on foot during winter, with its hard frozen surfaces, than in summer, with its soggy, marshy landscape, this discovery takes on a significance it might otherwise lack.)

The several prehistoric sites that have been discovered in Alaska show a complicated picture but one that does not necessarily negate the scenario given above. As Brian Fagan says, in his *The Great Journey: The Peopling of Ancient America*, 'Despite years of patient endeavour, no one has yet found an archaeological site in Alaska and the Yukon that can be securely dated to earlier than about 15,000 years ago.'[14] A caribou tibia was found at the Old Crow site, close to the Alaskan-Canadian border, which had undoubtedly been fashioned by human hands into a 'fleshing tool', for removing flesh from hide. To begin with, this and related bones were dated to about 27,000 years ago, but were later revised to only 1,300 years ago. It has also since been discovered that certain other bone 'tools' found at Old Crow were actually naturally occurring artefacts as more became known about how predators break the bones of animals they are in the process of killing.

The Bluefish Caves sites, about forty miles south-west of Old Crow, provided butchered animals, dated by associated pollen to between 15,550 and 12,950 years ago, together with stone tools at much the same date – stone tools moreover that, as Fagan says, would not be out of place in Dyukhtai.[15] Later, similar finds were made at Trail Creek, Tangle Lakes, Donnelly Ridge, Fairbanks, Onion Portage and Denali, with most dates in the 11,000 to 8,000 years ago range. At first, this tradition was known either as the Dyukhtai or Denali or Nenana complex, but Palaeo-Arctic is now the preferred term for these and slightly later artefacts. The diminutive size of the stone work is its most striking feature, and may stem from the fact that pollen analysis in the area shows that there was a rapid vegetational change beginning about 14,000 years ago, when the herbaceous tundra (grasses, mosses) gave way to a shrub tundra (woody thickets), which would have caused the mammal population to dwindle and may well have forced early man out of Beringia. As he moved on, smaller tools would have been preferable.

Not all the sites in eastern Beringia contained microblades. Others contain large core and flake tools, including simple projectile points and large blades. And at Anangula, on the coast out along the Aleutian island chain, blade tools were made, but not the diminutive microblades as at Denali. So there was quite a bit of cultural diversity in Beringia around 11,000 years ago. We simply to not know if this represents distinct cultural traditions that existed side-by-side, or alternative adaptation strategies designed to cope with different forms of wildlife.

The evidence, such as it is, suggests that there was no 'crossing' of the Bering Strait, in any modern sense. The early peoples spread into eastern Siberia, which then extended as far east as what is now the Yukon and Alaska. Then, when the seas rose, after ~14,000 years ago, the peoples of eastern Beringia were forced even further east, where the huge glaciers were themselves melting, allowing passage south, as we shall see. The seas rose behind them and they were isolated in the New World.

An alternative view, supported by some of the genetic evidence already reported, is that early man penetrated the New World along the coast. This makes sense, not only in view of the genetics, but – it will be recalled – because early mankind, after he and she left Africa, is considered to have followed a 'beachcombing' route (though as we have seen there is as yet no direct evidence for this). It also finds support in the discovery that, at Monte Verde, an early site in southern Chile, the remains of several kinds of seaweed were found in ancient hearths, while other remains appear to represent ancient clumps of kelp which had been chewed into 'cuds', according to Tom Dillehay, one of the archaeologists involved in the excavation.[16] Several other scientists have pointed out that there are virtually uninterrupted beds of seaweed right around the northern rim of the Pacific Ocean and have proposed that, with seaweed being so useful as a source of nutrition and for its medicinal properties, it would make sense for early coastal peoples to have followed this distribution (see map 5).

Mother Tongues, Lumpers and Splitters

In the genetic study considered earlier, carried out by Sijia Wang and his team, it was observed that there was an overlap between genetics and linguistic similarity. A second study, by Nelson Fagundes and colleagues, also showed a strong link between genetics and language among the Tupian-speakers of Brazil. Such results have to be understood against the background of the well-documented consensus which now accepts that some languages have evolved from others. This was formally first set out in the late-eighteenth century by a British civil servant and judge in Colonial India, William Jones, who observed the similarities between Sanskrit and several modern European languages.* And we know, for example, that Spanish and French are derived from Latin, which itself developed out of proto-Italic.[17] In fact, all but a handful of European languages have evolved from a proto-Indo-European root, meaning that thousands of years ago, many of the languages from the Atlantic to the Himalayas had a common source. A very similar exercise has been carried out with the languages of North America. Some of the scenarios constructed by linguists fit neatly with what we may call the LGM consensus. For example, Robert Dixon, an Australian linguist, has calculated that a dozen separate groups speaking different languages entered the Americas between about 20,000 and 12,000 years ago. Daniel Nettle, an English linguist, on the other hand, argues that the diversity of languages spoken in the New World today began in the last 12,000 years – i.e., *after* they arrived in America.

It is fair to say that linguistic research is on less secure grounds than the genetic or archaeological evidence, for the very good reason that we have no real way of knowing what languages people spoke in the

* Johan Reinholt Forster, on Cook's second voyage, made the same observation for what we now call Austronesian in 1774.

past, especially before the invention of writing. The only evidence we have for non-literate societies are the languages spoken today, their geographical spread across the world, and some idea of how, and at what rate, languages change or evolve. This is better than nothing but it still means that our reconstructions of past languages are at best theoretical and at worst speculative. This is why the field of 'chronolinguistics', or 'glottochronology', has been so controversial. In all that follows, it is as well to keep the above observations in mind.

In principle, the operation of comparative linguistics is simple. For example, the word for 'two' in Sanskrit is *duvá*, in classical Greek it is *duo*, in Old Irish it is *dô*, and in Latin it is *duo*. Thousands of similar examples could be given, to underline the point that specific languages are related. The controversy arises over just how similar languages have to be in order for them to be regarded as stemming from a common origin. This is a field divided – notoriously – into 'lumpers' and 'splitters', where the former favour a relatively small number of language families spread across the world, and the latter play down these linkages. If we note here, prominently, that the splitters are every bit as eminent as the lumpers, and that the splitters' central message is that very few conclusions may be drawn about the spread of languages around the world, and that this should be borne in mind in what follows, we may then proceed to examine what the lumpers say. (It is also worth reminding ourselves that, in the genetic studies reported above, overlaps were found between genetics and language, suggesting that the lumpers have at least a case.)

Map 6 shows the major language families of the world, according to Joseph Greenberg, an American linguist and one of the major (and most controversial) 'lumpers'. This reveals that there are three major language families in the New World – Eskimo-Aleut, Na-Dene and Amerind. On the face of it, this would suggest three waves of migration. Merritt Ruhlen, a linguist/anthropologist from Stanford University (and also a director of the Santa Fe Institute), in a re-analysis of Greenberg's material, suggests that Amerind is a form of the Eurasiatic family, but whereas Eskimo-Aleut is a *branch* of the Eurasiatic family, 'Amerind is related to Eurasiatic as a whole', and is no closer to Eskimo-Aleut than is any other Eurasiatic language. Many features of Amerind (for example, kinship terms) are unique to the Americas and several features are common to North, Central and South America, suggesting to Ruhlen that this language expanded rapidly across the New World at a time when it was unoccupied by humans speaking any other languages.

The second group, Na-Dene, is evidence for a second migration, later than that of the proto-Amerind speakers, Na-Dene being related to the Dene-Caucasian family whose homeland would appear to be in South East/Central Asia and includes Sino-Tibetan (again, see map 6). It also overlaps with the genetic marker known as M130, which originated in northern China and is not found in South America (see pages 8–9).[18]

Finally, Eskimo-Aleut is a third language family, a branch of Eurasiatic, which would make it evidence for the most recent migration. This theory is supported further by the thin spread of this language family around the edges of northern Canada.

So far then, the linguistic evidence is broadly in agreement with the genetic evidence, that the main migration into the New World took place between 20,000 and 12,000 years ago, by a group of people speaking Amerind, a branch of Eurasiatic, and that there was a second migration, much later, around 8,000 years ago, by a group of people speaking a language, Na-Dene, a form of Dene-Caucasian that originated in South East/Central Asia. The linguistic evidence also suggests there was a third migration – even more recent – of the people who speak Eskimo-Aleut, around the northern rim of Canada. This need not concern us too much as the Eskimo-Aleut people will play only a small role in our story.[19]

So far, so good then. However, just as there are a small number of genetic studies that show an earlier entry into the New World, earlier than the 20,000–12,000 period (the LGM consensus), so there is one linguistic analysis that shows much the same. Johanna Nichols, at the University of California, at Berkeley, has estimated that there are in the world 167 language 'stocks' (groups of languages that can be related back to a common branching point). She does this on the basis of such features as word order (subject-object-verb, or subject-verb-object), the form of the personal pronouns, whether verbs are more 'inflected' than nouns (whether they change their endings according to sense and context), how number is treated, how singularity and plurality are represented in verbs, and so on.[20] Using this approach, she looked at 174 languages spread around the world, and from this interrelationship she was able to conclude three things that interest us.

One, there are only four large linguistic areas across the globe: the Old World, Australia, New Guinea (with Melanesia), and the New World. Two, in a region such as a continent or subcontinent which is isolated from outside influence (such as South America or Australia), the number of language stocks increases as a simple function of time.[21]

But it is Nichols' third conclusion that is the most interesting. In her own words: 'A historical interpretation [of language diversity] would posit an ancient split between the linguistic populations of the Old World and the Pacific, with the Pacific then functioning as a secondary centre of spread and source of circum-Pacific colonisation. It is circum-Pacific colonisation rather than spread from the Old World that has populated most of the world, given rise to most of the genetic lineages of human language, and colonised the New World. The entry point to the New World was of course Beringia; but linguistic typology shows that the colonisers entering through Beringia were predominantly *coastal* people involved in the circum-Pacific colonisation pattern rather than *inland* Siberian people impelled ultimately by spreads out of central Eurasia.' (Italics added.) A final gloss on this picture is that 'the first colonisation of the New World was under way by about 35,000 years ago'.[22]

On the face of it, of course, this appears to throw much that we have been discussing so far into disarray. The LGM consensus, the genetic evidence, Christy Turner's dental evidence, together with the archaeological evidence from either side of the Bering Strait, and the linguistic evidence of Greenberg and Ruhlen, cohere in showing that early humankind reached the Bering Land Bridge roughly 16,500–15,000 years ago, via an inland route through central and northern Eurasia, with a second later group crossing the strait at about 8,000 years ago, originating in South East Asia. Johanna Nichols' linguistic evidence says early peoples reached Beringia 35,000 years ago via the west coast of the Pacific rim, travelling north from island South East Asia, China and into Siberia. Can these two scenarios be reconciled?

Nichols' linguistic evidence is not like the genetic evidence for early entry into the Americas. As was referred to earlier, we may allow that one or two more or less genetically distinct but isolated groups of people entered America much earlier than the main group of migrants without seriously jeopardising the main thrust of the overall picture. But Nichols' linguistic evidence by definition applies to large groups of people, not isolated pockets.

The answer to the discrepancy must surely lie in the uncertain nature of the methodology of chronolinguistics. Many of Nichols' colleagues, while accepting her division of languages into four 'families', do not take seriously her arguments about time depth; and she does not herself use glottochronology. We shall see in chapter four that, archaeologically speaking, there is next to no evidence for the presence of early peoples in the Americas beyond Alaska before 14,500 years ago but we shall also see, in chapter two, that there is good geological, cosmological and mythological evidence for *why* there would have been a *second* wave of migrants who entered the New World much later than the first, at around 8,000 years ago, after leaving island South East Asia and travelling around the Pacific rim. In other words Johanna Nichols is right about the origin of at least some of the New World languages, but wrong about the time depth. (Remember that it is the calculation of time depth that is so controversial and unreliable in comparative linguistics.) The clue to the disparity, as we shall also see, lies precisely in Nichols' insistence that there was an ancient split between Old World and Pacific languages. Why should that be? What happened, deep in the past, to cause this split?

The next chapter will go a long way to explaining that split and we shall also see that, on their way to the New World, some of the people who populated the Americas underwent a series of unique events that produced in them some *psychological* or *experiential* characteristics that distinguished them from those they left behind in Eurasia and which could have affected their later development. We shall see that some of these special events did in fact take place about 8,000 years ago, which agrees well with the genetic evidence, referred to earlier, concerning haplogroup M130, which is associated with the Na-Dene speakers who entered the New World at precisely that time.

At one stage, it would have been difficult if not impossible to assemble such an argument about deep history, but not any more. In addition to advances in genetics and linguistics, we can say that, thanks to developments in geology and cosmology, we now know far more about our remote history than ever before and, moreover, these studies have shown surprising and consistent links with mythology.

As a result, we now know that myths are less the fanciful, woolly accounts they have traditionally been dismissed as, and much more closely based on fact than anyone had previously imagined. Once we learn to decipher them – as is now happening – they tell us quite a bit about deep time.

• 2 •

From Africa to Alaska: The Disasters of Deep Time as Revealed by Myths, Religion and the Rocks

The evidence of the previous chapter told us that early peoples, like the Chukchi, finally reached the Bering Strait from Africa, which their ancestors had left tens of thousands of years before, by one or both of two great routes, the Central Asian route or the Pacific rim route. In this chapter, instead of genetics, we shall be looking at myths and using a relatively new scientific synthesis which seeks to put the latest findings of cosmology, geology, palaeontology and archaeology together with mythology, to reconstruct distant occurrences in deep time, occurrences that were so catastrophic, traumatic and bewildering that ancient people brought all their intellectual firepower to bear on them, to make sense of their disastrous experiences. In the main, we now know, this is what myths, most of them, *are* – memories and, at the same time, warnings that disasters could well recur. By the time we are done, we shall have some idea of the early psychological differences between early Old World peoples and the first Americans.

It has been known for more than a century that the most widespread myth across the world – as well as the best known – is that referring to a vast flood, whose exact size was not calibrated, but which was reported not just in the Christian bible, of course, but in the ancient legends of India, China, South East Asia, north Australia, and the Americas. We shall be considering the flood myth(s) in some detail in just a moment but for reasons that will become clear, it suits us here to consider first the second-most common myth on earth, that of the 'watery creation' of the world.

The chief theme of this myth is the separation, usually of the sky from the Earth. This story is found in a band stretching from New

Zealand to Greece (a significant distribution, as we shall see) and it invariably has a small number of common features. The first is the appearance of light. As it says in Genesis, 1:3: 'And God said, Let there be light: and there was light.' Nearly all cosmogonies have this theme, where it is notable that *neither the sun nor the moon is the source of the first light at Creation.* Rather, the first light is associated with the separation of heaven and Earth. Only after heaven and Earth have separated does the sun appear. In some traditions in the east the light is let in because the heavy substance of the clouds that envelop the Earth sinks down to the ground, and the light, clearing the clouds, rises to become heaven. A common metaphor for this is an egg splitting. In other myths, the darkness is described as a 'thick night'.

Recent geological studies have identified a phenomenon known to scientists as the Toba Volcanic explosion. Cores drilled in the Arabian seabed have shown that there was a volcanic eruption at Toba in Sumatra between 74,000 and 71,000 years ago. This is known to have been the biggest eruption on earth during the last two million years, a massive conflagration that would have released a vast plume of ash thirty kilometres high (an estimated 670 cubic miles, twice the volume of Mount Everest), spreading north and west, to cover Sri Lanka, India, Pakistan and large areas of the gulf region with a blanket six inches deep though at one site in central India the ash layer is still twenty *feet* thick.[1] Toba ash has recently been found in the Arabian Sea and in the South China Sea, 2,400 kilometres from Toba itself.[2] The eruption left an immense caldera that now holds Indonesia's largest lake, Lake Toba, 85 kilometres long, up to 25 kilometres wide, with cliffs 1,200 metres high and water 580 metres deep.[3] A prolonged volcanic winter would have followed this eruption. (Sea temperatures, according to geologist Michael Rampino, dropped by ten degrees Fahrenheit and a total darkness would have existed over large areas for weeks or months.)[4] The aerosol clouds of minute globules of sulphuric acid, now known to be produced by massive eruptions, could have reduced photosynthesis by 90 per cent, or even shut it down completely, having a major effect on forest cover.[5]

Now if early humankind did leave Africa at about or some time after 125,000 years ago, and if the people followed a beachcombing route that took them around what is now Yemen and Aden, and on to the Iranian, Afghanistan and Pakistan coast, and if they were isolated and held up from time to time by adverse climatic variations, they could have arrived in South Asia more or less on schedule to meet the Toba eruption. This is in fact confirmed by excavations in India and Malaysia, which have found Palaeolithic tools embedded both above and below volcanic ash at this date. According to some calculations, the population of this vast area could have been reduced from an estimated 100,000 to between 2,000 and 8,000 (a similar population crash is known to have occurred among chimpanzees). But we cannot overlook the fact that population estimates for so long ago are very speculative.[6]

At a conference in Oxford in February 2010 the 'catastrophic' nature of the Toba eruption was queried, and new evidence presented to suggest that the temperature dropped by only 2.5°C. But no one is suggesting that Toba's effects were other than far-reaching and the conference also heard fresh evidence that tools made by *Homo sapiens* straddled the ash layer.[7] So two things may be inferred from this. The volcanic winter may have all but wiped out the early humans living in a wide swathe centred on India, meaning that certain specific survival strategies would have been devised and which may have been memorised in myth form; and second, that the area would have been *re-colonised* later, both from the west and the east.*

The 'separation' myth is a not-inaccurate description of what would have happened over large areas of the globe, in South East Asia, after the Toba eruption and the volcanic winter that would have followed (see map 7 for the spread of the Toba explosion). Sunlight would have been cut out, the darkness would have been 'thick' with ash, the ash would gradually have sunk to the ground and, after a long, long time, the sky would gradually have got brighter, lighter and clearer, *but there would have been no sun or moon visible perhaps for generations*. There would have been light but no sun, not for years, not until a magical day when, finally, the sun at last became visible. We take the sun for granted but for early humankind it (and the moon, eventually) would have been a *new* entity in the ever-lightening sky. Mythologically, it makes sense for this event to be regarded as the beginning of time.

* Curiously, the map of the Toba explosion overlaps the course of the tsunami of Christmas time 2004, which occurred due to an earthquake at Simeulue, off the west coast of northern Sumatra, which affected most of all Sri Lanka and southern India (see map 7).

*

The discovery of the Toba eruption, therefore, was almost as important a breakthrough for mythology as it was for geology. And as we are about to see, there are grounds for believing that many other ancient myths and legends, far from being the products of our deep unconscious – as Carl Jung or Claude Lévi-Strauss insisted – are in fact based on real events.

While myths were interesting to anthropologists, they were treated to begin with as mainly fictional accounts, revealing more about early man's primitive beliefs than anything else. Sir James Frazer, the late-nineteenth-century anthropologist and author of *The Golden Bough*, recorded many of these myths in his book, *Folk-Lore in the Old Testament*, published in London in 1918, and where he had this to say: 'How are we to explain the numerous and striking similarities which obtain between the beliefs and customs of races inhabiting different parts of the world? Are such resemblances due to the transmission of the customs and beliefs from one race to another, either by immediate contact or through the medium of intervening peoples? Or have they arisen independently in many different races through the similar workings of the human mind under similar circumstances?'[8]

Attitudes evolved somewhat when, a few years later, in 1927, the British archaeologist Leonard Woolley began to dig at the biblical Ur of Chaldea, in Iraq, the alleged home of Abraham, founder of the Jews. Woolley was to make several important discoveries at Ur, two of them momentous. In the first place he found the royal tombs, in which the king and queen were buried, together with a company of soldiers and nine ladies of the royal court, still wearing their elaborate headdresses. However, no text had ever hinted at this collective sacrifice, from which he drew the important conclusion that the ceremony had taken place *before writing had been invented* to record this extraordinary event, an inference that was subsequently substantiated. And second, when Woolley dug down as far as the forty-feet level he came upon nothing, nothing at all. For more than eight feet there was just clay, completely free from remains of any kind. For a deposit of clay eight feet thick to be laid down, he concluded that a tremendous flood must at some time have inundated the land of Sumer. Was this then the flood referred to in the bible?

Many people – then and now – thought that it was. But just as many didn't. They didn't because the bible text says that the flood covered mountain tops – i.e., it was rather more than eight feet deep – and because the flood was supposed to extend right across the world. An eight-feet flood of the Tigris and Euphrates in Mesopotamia did not suggest anything more than a local event. Or, had the ancients exaggerated? Since in those days hardly anyone travelled far, perhaps a reference to a 'world-wide flood' was just a manner of speaking.

That is more or less where matters remained for several decades. In recent years, however, new light has been cast on three events – or rather, three sets of events – deep in our past. The history of the years covering the transition from the Pleistocene to the Holocene – from the Ice Age to modern times – has undergone a major revision recently and, to put the matter briefly, the latest scholarship of the period shows three things. It shows that the world suffered not one but *three* major floods, at (roughly speaking) ~14,000, ~11,500 and ~8,000 years ago, and that the last of these was especially catastrophic, changing life drastically for many of the people then on Earth. This has produced a sudden surge of interest among archaeologists in the relatively shallow land bridges and offshore continental shelves of the world, as areas which may have been dry at various times in the remote past, and therefore locations where early peoples lived. Thousands of radiocarbon dates from ~300 sites have been obtained (some by diving), in certain cases going back more than 45,000 years, but little of substance has been discovered before 13,000 years ago. Walls, clay floors, hearths and stone tools have been found down to depths of 145 metres, at distances of up to 50 kilometres off such disparate locations as Sweden and California, in the Red Sea, in Beringia, and in the Mediterranean stretching from off Gibraltar to off Israel.

Second, this new understanding shows that the area of the world most affected by the floods was not Mesopotamia but South East Asia, where a whole continent was drowned. If these floods did have most effect in South East Asia, it would mean that the inhabitants of that sunken continent would have been forced to migrate all over the world – north to China and then to the New World, east to the Pacific islands and Australia, and back west to India, and possibly as far as Asia Minor, Africa and Europe, taking their skills with them. The third aspect of this new chronology is that many of the early skills of civilisation, such as agriculture – which have always been understood as being invented in the Middle East – were actually first developed much further east, in South East Asia and in India.

It is a contentious theory. Critics point out that when sea levels rose, they would have risen everywhere, so coastal migration would have been less likely than movement inland; these critics also insist that north Asian stone tools are very different from South East Asian ones, casting doubt on the (otherwise seemingly firm) genetic findings that there was a migration *up* the coast of East Asia. But even if the theory is only partly true, it has a major consequence for the ideas behind this book, not least because it may help to explain Johanna Nichols' conclusion, that there was a great linguistic split between the Old World and the Pacific peoples, which played an important role in the peopling of the New World.

The evidence is now substantial to suggest that the rise in sea levels after the last Ice Age was neither slow nor uniform. Instead, three sudden ice melts, the last only 8,000 years ago (6000 BC) had a devastating effect on certain tropical coastlines, which had extensive flat continental shelves. These changes were accompanied by massive earthquakes, caused as the weight of the great ice sheets was removed from the land and transferred to the seas.[9] These giant earthquakes would have generated super-waves, tsunamis. Geologically, the Earth was much more violent then than it is now.

The overall oceanographic record between 20,000 and 5,000 years ago reveals that sea levels rose at least 120 metres (~400 feet) and affected human activity in three ways. In the first place, in South East Asia and China, which have a large flat continental shelf, all examples of coastal and lowland settlement were inundated and for all time. Those settlements have been underwater for thousands of years and will most likely remain so. Second, during the final rise in sea level, at 8,000 years ago, the water did not retreat for about 2,500 years, with the result that many areas there that are now above water are nevertheless covered with a layer of silt that is many feet thick. Third, as already mentioned, the floods that devastated South East Asia required the inhabitants to move out.[10]

This picture is supported by the curious dating pattern of the Neolithic Revolution in eastern Eurasia. According to such sites as have been found (admittedly few), the Pacific rim cultures seem to have begun their development well *before* those in the West but then, apparently, stopped. For example, pottery appeared for the first time in southern Japan around 12,500 years ago; 1,500 years later it had spread to both China and Indo-China. It is important to say that these examples pre-date any of the sites in Mesopotamia, India or the Mediterranean region by as much as 3,500–2,500 years.[11] In other words, these early signs of proto-civilisation occurred much, much earlier in South East Asia than anywhere else.

(Although these sites are further away from Africa than many other Mediterranean, central Asian and Mesopotamian locations, their chronological primacy makes sense if early peoples were beachcombers. Early migrants would have realised that rivers – sources of fresh water – occurred relatively frequently along the coasts, flowing into oceans, but that, when following a river inland to its source, fresh water ran out and there was no guarantee where the next river would be. If the migrants were forced out from any one area, by population pressure, moving further along the coast was therefore less risky than moving inland.)

In addition to the early beginnings of pottery in Japan and Indo-China, around 12,000–11,000 years ago, a wide range of Neolithic tools has been found in East Asia – choppers, scrapers, awls and grinding stones, as well as hearths and kitchen waste – but these finds tend to be found in *inland* caves. There are almost no Neolithic sites in *lowland* areas dating to between 10,000 and 5,000 BC.

Two explanations have been put forward to account for this anomaly. One view has it that in island South East Asia the Neolithic period only started 4,000 years ago, with migrants coming down through Taiwan and the Philippines and introducing new skills and artefacts. This is why, these scholars say, most South East Asian caves are empty; there were few people around. Such is the view of the eminent Australian archaeologist, Peter Bellwood, who says that nowhere in South East Asia is there currently good evidence for any form of food production before 3500 BC. At this time, too, the early Neolithic, he observes a shift in cave use, from habitation to burial, which he thinks must have accompanied the beginnings of village life. The other view is more ambitious: people were living in South East Asia at the end of the Ice Age and had developed their agricultural (and sailing) skills much earlier than people elsewhere (in the Near East, for example) but were forced into long-distance migration, both east, north and west as a result of flooding brought about by the melting glaciers.[12] And, as well as forcing these people out, the associated silt covered up many sites.[13]

Wobbles, Tilts and Perfect Storms

These are clearly important assertions and so the floods need to be fully described if we are to be able to judge the merit of these new theories. This may seem like a large detour from our main story but the reader is asked to be patient: a consistent picture will emerge which suggests that the people who entered the New World first did so after a distinct set of experiences that separates them from many of those they left behind in the Old World.

We now know that the three catastrophic floods referred to above occurred because of three interlocking astronomical cycles, each different and each affecting the warmth transmitted by the sun to various parts of the Earth. Stephen Oppenheimer calls these the 100,000-year 'stretch', the 41,000-year 'tilt' and the 23,000-year 'wobble'.[14] The first arises from the Earth's orbit around the sun, which is elliptical and means that the distance from Earth to sun varies by as much as 18.26 million miles, producing marked variations in the force of gravity. The second cycle relates to the tilt that the Earth presents to the sun as it rotates. This varies – over 41,000 years – between 21.5 and 24.5 degrees and affects the seasonal imbalance in heat delivered from the sun. Third, the Earth rotates on its own axis, in a so-called 'axial precession', every 22,000–23,000 years. These three cycles perform an elaborate dance that produces an infinite array of combinations but which, when they come together in a 'perfect storm', can provoke very dramatic and very *sudden* climate change on Earth. And it is these complex rhythms which triggered not one but three floods in the ancient world.

The glaciers which melted to cause these floods were massive, the largest covering huge areas such as Canada and were several *miles* thick. One has been estimated as being 84,000 cubic kilometres. They could take hundreds of years to melt completely but eventually raised sea levels by as much as forty-four feet.

One of the interesting effects of the changes that followed the second catastrophe (after 11,000 years ago) was that, as sea levels rose, river gradients were lowered and, after 9,500 years ago, river deltas began to form all around the world. The importance of this lay in the fact that these deltas formed very fertile alluvial plains – in Mesopotamia, the Ganges, the Chao Phraya in Thailand, the Mahakam in Borneo and the Chiang Jiang (Yangtze) in China; overall more than forty such deltas have been identified as forming at that time on all continents. Many of these alluvial plains/deltas played a role in the growth of agriculture and the subsequent birth of civilisation.[15] Deltas are suitable for certain kinds of plants and not others, as we shall see.

But it was the most recent flood, at 8,000 years ago, that had the greatest effect.[16] Its dimensions were truly awe-inspiring and the reason for its sensationally catastrophic nature had quite a lot to do with the geological structure of Canada which, around the Hudson Bay area, is shaped not unlike a huge saucer that, in places, is hundreds of feet above sea level. Added to that, the Hudson Strait (leading north between Baffin Island and the Labrador Sea) acts like a spout or narrow channel into the ocean.

What seems to have happened is that the Laurentide glacier, stretching for thousands of miles right across Canada, began to melt at the edges but the water couldn't escape into the sea; it was instead trapped in the massive saucer, above sea level, and was also kept in place by the ice blocking the Hudson Strait, which acted as a giant plug. The main body of ice then began to crack and melt until, eventually, the plug gave way – and a truly massive body of water and cracked ice sluiced through the Hudson Strait out into the ocean. The glacier was a third the size of Canada and 1.5 kilometres thick.[17] It raised global sea levels by 20–40 centimetres (eight to sixteen inches) more or less instantaneously and the remaining ice would have eventually melted as it was swept out to sea, adding another 5–10 metres (sixteen to thirty-two feet) to the sea level.

The sudden removal of the ice sheets from the North American and European continents, releasing massive amounts of ice and water into the world's great ocean basins, meant that a sudden change in the spread of weight across the Earth occurred, and this would have caused great earthquakes, increased volcanism and massive tsunamis crashing ashore on all continents, an epic period of natural disasters that, as we shall see, had a profound effect on the mental life of ancient men and women. The fact that the Earth's crust is soft and springy – not at all as brittle as it might at times seem – also meant that the effects of the earthquakes and tsunamis were not uniform across the world. (The Earth is, in a way, not unlike an enormous tennis ball. It is a firm sphere but if enough pressure is applied to a certain point, it can be dented or flattened.)

The importance of this flood, so recently established, cannot be exaggerated for our purposes, as it had several important consequences. One was that a flood and tsunamis of such dimensions would have deposited layer upon layer of silt many feet thick across huge areas, layers that must have covered crucial examples of early human development between, say, 8,000 years ago and when the seas receded again many hundreds if not thousands of years later. This 'silt curtain', as Stephen Oppenheimer calls it, must in turn affect our understanding of world chronology.[18] A second consequence arises from the natural geography of the world, where the largest landmass that was inundated by the flood was almost certainly South East Asia, where there could be found the largest shallow continental shelf, stretching out into the South China Sea for 160 kilometres (see map 7). Crucially, for an understanding of early chronology, and perhaps for a full grasp of the emergence of civilisation, these two consequences can be put together.

The starting point for this synthesis comes from the fact that this area has the highest concentration of flood myths in the world.[19] Does this prove that the flood had the most devastating impact here? No, but the inference is tantalising and it fits exactly with what William Meacham, a Hong Kong-based prehistorian, noted in 1985, what was referred to earlier, that the most important gap in the Neolithic record now 'is the total absence of open sites in lowland areas [in South East Asia] dating from 10,000 BC to 5000 BC'. Moreover, after sea levels started to fall again, from 6,000 years ago, pot-making maritime settlements began to occupy sites all the way down from Taiwan to central Vietnam. Charles Higham, a New Zealand archaeologist based mainly in Thailand, argues that these settlements were actually *re*locations of maritime people who had always lived in these areas but had been flooded out much earlier on.[20] His argument is based on two factors: one, the complete lack of evidence of people moving *into* these areas from anywhere else; and two, certain cultural resemblances between these coastal peoples and a much older pre-Neolithic culture in inland Vietnam known as 'Hoabinhian'. Since Higham made his inferences, more direct evidence has been found. At several sites (near Hong Kong, for example) two cultural phases have been unearthed, separated by a layer of silt up to six feet thick. Inland sites, on the other hand, were continuously inhabited from more or less the end of the Ice Age.

More extraordinary and tantalising still, many of the artefacts found on the south China coast, at Middle Neolithic levels *below* the silt, have counterparts to those found at Ur under Woolley's silt. These artefacts include perforated clay discs (tied into nets, to help them sink), painted bowls, shell beads and polished stone adzes and stone hoes.[21] It is also the case that tattooed female figurines were found in both places – slim naked women, often with exaggerated genitalia and sometimes carrying children.[22] The heads of these figurines are distinctive – they have black bitumen hair or wigs and slanted eyes, with heavy folds under the eyelids. Paint marks and embedded clay pellets on the shoulders of the figurines hint at tattooing and/or skin scarification. A few men with similar features were also excavated in Eridu in Mesopotamia. What accounts for this similarity? Coincidence, or early contact?

These links have in fact been dismissed as 'stretching credulity beyond the limits', and it is certainly true that, between these two extremes – Mesopotamia and the south China coasts – there are several intermediate communities, in Oman and the Persian Gulf for example, which are known from their shell middens and do not share these burial practices. But the hair, wigs and tattooing features are not their only similarities. There is, for instance, the added fact that Woolley observed that the graves in which these figurines were found were of a rectangular shape, their bottoms covered in pottery that had been deliberately broken up. The cadavers had been stretched and the remains powdered in red haematite, iron ore. There is a similar practice of painting extended bodies with haematite, in wooden box burials, in the Niah Cave in Borneo. They are dated to 3800 BC.[23]

Then there is the matter of skin scarification. Tattooing is widespread in Austronesia but skin scarification is limited to Oceania, notably the north coast of New Guinea. Scarification is performed on the shoulders and torso as part of an initiation rite and is intended to imitate the teeth marks of crocodiles. (Note the reference to crocodiles.) The patterned scars so produced, according to Stephen Oppenheimer, 'resemble those of the Ubaid figurines'. Ubaid is west of Ur in southern Iraq, the Ubaid period dating to ~5300–4000 BC.

Geoffrey Bailey makes the point that while the continental shelves close to many of the Old World centres of human evolution and early civilisation are relatively narrow, the major exceptions are 'in the extensive shelf that skirts mainland China and the peninsulas and archipelagos of South East Asia, and more localised pockets around the Arabian Penisula, part of the Indian coastline and northern Australia'. Is early contact between these areas unthinkable? The remains of the early technology of sailing – boats, fishhooks, harpoons – are only found in post-glacial times but when they are found, their chronological spread is as shown on map 4: seafaring developed along what will be identified in just a moment as 'the East-West Corridor', linking Mesopotamia with South East Asia, at least 4,000 years ago (map 1). And of course it is well known that the South East Asians colonised Madagascar, thousands of miles across the Indian Ocean, albeit only in the first millennium AD.[24]

If these practices are more than coincidence, then the Indian Ocean – and Pacific rim – cultures carried on a long-term trading network and were in very early contact with Mesopotamia. Intuitively, this seems exceptionally early but we shall see shortly that it is reflected in mythology.

Added to all this, rice grains associated with pottery have been found in the Malay peninsula apparently dated to 9250 BC, while in India it is now known that two *different* forms of agriculture were begun: six-row barley with cattle, sheep and goats was introduced in *west* India in the seventh millennium BC; and rice agriculture in the Vindha Hills in the sixth-fifth millennium, where the practice overlaps heavily with the distribution of the Mundaic tribes in the north-central and north-eastern areas.[25] The Mundaic tribes speak Austro-Asiatic languages, as found predominantly in South East Asia.

The new chronology – insofar as it concerns us, and if confirmed – is therefore as follows. Rice-growing, with pottery and the exploitation of seafoods, went hand-in-hand in South East Asia between 10,000 and 7,000 years ago. Sea levels rose, thanks to massive glacier melts, covering many early sites with silt when they receded, but stimulating changes in lifestyle, including an improvement in sailing and navigational skills, causing the people of the Sunda Shelf to spread out in all directions, perhaps as far as the Middle East. (The effects of flooding in the Middle East are considered in more detail in chapter fifteen, but for now it is enough to say that, at one stage, between 15,000 and 8,500 years ago, vast areas of the Persian Gulf, 900 kilometres between the Strait of Hormuz and what is now Basra, were dry land.) But the largest amount of low-lying land in the world, which would have been most affected by any rise in sea levels, where a flood would have been most catastrophic, was the Sunda Shelf, on the south-east 'corner' of South East Asia, and stretching 5,400 kilometres east to west and 2,700 kilometres north to south. This may well account for why flood myths are more prevalent in that region than anywhere else. Such a flood would have provoked large-scale migration – east, west and north.

Is this the 'great split' that Johanna Nichols identified in the linguistic record between Old World peoples and Pacific rim peoples? Though the hard evidence is meagre and by no means universally accepted, the picture it paints fits in every way but for the chronology; and dating, we know, is the weakest point in chronolinguistics.

Collective Warnings

We now need to consider, briefly, one other relatively new finding about ancient Asian history before moving on to compare Old World and New World myths. This too is not the detour it might at first appear. It concerns a whole constellation of stories known as the Vedas. The Vedas, the sacred writings of the Hindus, envisage a 'Yuga' theory of historical and cosmic development – great cycles of humanity and of nature, disrupted by enormous natural cataclysms. One of these cycles is said to last 24,000 years, not so very different from the 23,000-year 'wobble', as Stephen Oppenheimer calls it, but more relevant, perhaps, is the newly discovered fact of three great floods in recent geological history, at ~14,000, ~11,500 and ~8,000 years ago. Is this not in effect cyclical history, broken by great catastrophes? Is that why Hinduism lays such store by cyclical history?

More specifically for our purposes, however, the Vedantic literature refers to a land of seven rivers – identified as the Indus, Ravi, Sutlej, Sarasvati, Yamuna, Ganga and Saryu – in which the Sarasvati was the most important for Vedic people, both spiritually and culturally, irrigating their central land and place of origin and supporting a large population.[26] One verse of the Vedas describes the Sarasvati as 'the best of mothers, the best river, the best goddess' and another places it between the Sutlej and the Yamuna.[27]

The problem is – or was – that today there is no major river flowing between the Yamuna and the Sutlej, and the area is well known as the Punjab (*panca-ap* in Sanskrit), or the Land of *Five* Rivers or Waters. This discrepancy led some scholars for many years to dismiss the Sarasvati as a 'celestial' river, or an imaginary construct, or in fact to be a small river in Afghanistan, whose name, Haraquti or Harahvaiti, is cognate with Sarasvati.

Beginning just after World War Two, however, archaeological excavations began to uncover more and more settlements which *seemed* related to the well-known Mohenjo-Daro and Harappa Indus civilisations but paradoxically were up to 140 kilometres distant from the Indus River itself, at sites where there is today no obvious source of water. It was only in 1978 that a number of satellite images from the spacecraft launched by NASA and the Indian Space Research Organisation began to identify traces of ancient river courses that lay along where the Vedas said the Sarasvati had been. Gradually, these images revealed more details about the channel including the fact that it had been six to eight kilometres wide for much of its course, and no fewer than *fourteen* kilometres wide at one point. It also had a major tributary and between them the channel and its tributary converted the Land of Five Rivers (of today) into the Land of Seven Rivers (*sapta-saindhava*) in the Vedas. Moreover, the Rig Veda describes the Sarasvati as flowing from the 'mountains to the sea', which geology shows it would have done only between 10,000 and 7,000 years ago, as the Himalayan glaciers were melting. Over the years, the rivers feeding the Sarasvati changed their course four times as a result of earthquakes, feeding the Ganges instead, and the Sarasvati dried up.[28]

So the Veda myths were right all along – there *was* a Sarasvati River and it was just as mighty as the scriptures said.* But it also throws into context the fact that the sacred texts show that the Vedic culture at that

* It also helps date the Vedas: since they describe the river as 'mighty', when it in fact began to dwindle some 5,000 years ago, this shows that the Vedas must be a good bit older than many people thought.

time was a maritime culture (there are 150 references in the Vedas to the ocean, rivers flowing into the sea, and travel by sea).

The rediscovery of the Sarasvati therefore underlines two things that concern us. One, the basic skills of civilisation – notably domestication, pottery, long-distance trade, sailing – were in place in South Asia (India) and island South East Asia by 5000 BC; and second, the great myths which we find spread right across the world are almost certainly based on real catastrophic events that actually devastated early humankind and form a genuine collective memory to warn us that such terrible events may one day recur.

Myths as Memories

Now that we have done our groundwork and have a strong suspicion that myths – the important ones, the original ones – are founded on fact, on real historical events, then the ways in which myths vary across the world takes on a new and tantalising significance. What, we may ask, do they tell us about the early experiences of humans in different parts of the globe? In particular, how do myths in the New World differ from those in the Old and, where they do, what does this mean? Do they help us to reconstruct the experiences of early peoples?

The genetic evidence shows that the Chukchi in Siberia and the first human groups to enter the Americas reached Beringia by central Eurasia and had arrived some time between 20,000 and 16,500 years ago at the latest. The linguistic evidence in particular suggests that a second, later group of ancient peoples travelled up the western coast of the Pacific Ocean – Malaysia, China, Russia. If the earliest peoples reached the New World at any time between 43,000 and 29,500 years ago, as some of the genetic evidence suggests, they may well have had a memory of the Toba earthquake, but none of the great floods had yet occurred. On the other hand, the second group – the Na-Dene speakers with the M130 genetic marker, whose bearers migrated up the Pacific rim and into the Americas at about 8,000–6,000 years ago – should have had fairly recent experience of flood. What do we find?

In the first place, and by way of generalisation, we may say that there is an extensive constellation of myths that occur in both the Old World *and* the New, far too many for them to have all been jointly conceived by coincidence. Allied to this, there are some important myths that occur only in the Old World and in Oceania but do *not* appear in the New World. At the same time, there are a few myths – of origin, creation – that appear in the New World and not in the Old. This is all what you would expect if early humankind originated in the Old World and migrated to the New.

A second thing to assimilate is that a number of myths have a very wide spread indeed, right across the world. For example, there is a creation myth among the Diegueno Indians of south-west California that is closely paralleled in the creation myth of the Mundaic aboriginal tribe of Bengal. In the Californian myth, two brothers who were under the sea at the beginning of time go out looking for land. After some fruitless searching, the elder brother creates land from tightly packed red ants. However, the birds which he made later could not find this land because it was still dark. He therefore made the sun and the moon. In the Bengal myth, after a watery start to the world, two birds were created in mistake for men. They then flew around the world looking for land but after twelve years hadn't found any. The creator then sent various animals to dive for earth and after several abortive attempts, the turtle brought up land, in the form of an island, which became the source of all life on Earth. In both these cases there is a watery start to the world, two brothers, or birds, go looking for land but are unsuccessful.

The wide geographic distribution of this myth is not so odd if we accept that its origin was the flooding of the Sunda Shelf in South East Asia, a flood which generations later receded, to reveal (create) more land. Then the idea migrated out from there, both to the west and to the north. We have seen that the Mundaic peoples were rice growers, who speak Austro-Asiatic, both of which traits originated in South East Asia. The Diegueno Indians of California are Na-Dene speakers which, as we saw in a previous chapter, is a language that overlaps with Dene-Caucasian in South East Asia. The Diegueno have the M130 haplogroup so most likely left island South East Asia at around 8,000 years ago. Whatever accretions have been made in, say, Bengal and California (or on the way there), the similarities of these myths suggest their common origins. The Mundaic tribes and the Diegueno originated in island South East

Asia and, following the flood, one tribe went west, the other north.[29]

Next we may consider the myths found on both sides of the Bering Strait, where we can examine the detailed – and systematic – ways in which they vary.

As has been said, many myths describe a 'watery chaos' flood, out of which land gradually emerges. In the sub-arctic regions of North America, however, the most common myth is that of the 'land diver'. In these myths, following the flood, land doesn't emerge gradually but is created by raising it up from the floor of the ocean bed. A common procedure to ensure this happens is the use of what have become known as 'land divers'. These are animals, often diving birds, who are sent down to the bottom of the ocean (by either the creator or Earth's first inhabitants) to pick up a scrap of earth on the ocean bed. Typically, after a few unsuccessful attempts, one diver returns with earth or clay in its claws or beak, and this small amount is transformed into the growing Earth. In one form or another, this myth can be found from Romania to central Asia, to Siberia.

But the land diver and land raiser myths are most typical in sub-arctic North America and among the Algonquin tribes of the eastern woodlands. The Huron, of Ontario, for example, have a myth in which a turtle sends various animals diving for earth, all of whom drown except the toad, who returns with a few scraps of land in its mouth. These are placed on the back of the turtle by the female creatrix, who has descended from heaven for this purpose, and the scraps of earth grow into the land. The Iroquois (on the north-western Pacific coast of what is now the United States) and the Athabaskan tribes also have this myth, which is in fact confined to two linguistic groups, Amerind speakers and Na-Dene speakers. The motif is not found in Eskimo flood myths or in Central or South America.[30]

Two other aspects of this set of myths claim our attention. First, the distribution of the land diver stories overlaps with a characteristic genetic marker in sub-arctic North America. Certain population groups (not Eskimos, Aleuts or Athabaskans) have what is known as the 'Asian 9-base-pair deletion' – nine pairs of proteins are missing from their DNA. This marker, this characteristic pattern of absence, is shared with certain clans in New Guinea, and also with peoples in Vietnam and Taiwan. Not only does this further confirm the South East Asian origin of at least some Americans (and underlines the distinction between

Eskimos and Na-Dene speakers), but the sheer size and diversity of the 9-bp deletion on both sides of the Pacific suggests a very old origin. One suggestion is that they represent an expansion of circum-polar Asian populations around the time of the Younger Dryas event over 11,000 years ago (i.e., at the time of the second flood). The Younger Dryas was a bitter cold snap that preceded the flooding at 11,000 years ago and this event might explain the period of extreme cold and famine preceding the flood, that is also described in the Algonquin myths.[31]

Second, these land raiser myths, which after all are fairly spectacular, lend themselves to one or more of three phenomena recognised by geographers and oceanographers. The first is 'coastline emergence'. This is something which happened on a grand scale, especially in North America. It is a phenomenon which occurs because, after the Ice Age, as the glaciers melted, they grew lighter and, with less weight on it, the continental crust lifted up. Moreover, the change in weight brought about a rise in the land that was *more* than the rise in sea level. Since the land had hitherto been crushed below sea level, it would at that time *have risen out of the sea*. Photographs of Bear Lake in Canada show several shore lines that have risen hundreds if not a few thousand feet above sea level.[32] People alive at the time would, over the generations, have noticed that the shore line had moved and, we may assume, incorporated this strange phenomenon into their myths, explaining it as best they could.

The second phenomenon involves the well-established fact that the Pacific rim is known as the 'Ring of Fire', because that is where the world's most active volcanoes are located. Volcanoes are discussed in more detail in chapter five; here, all we need to point out is that many volcanoes in the Ring of Fire are offshore, underwater volcanoes, forming part of the seabed. During underwater offshore eruptions (of which there were more than 50 in 2001–2002) solid matter – 'land' – would have been propelled forcefully to the surface.

A third possibility is that the myths dimly reflect the experience of ancient people who had lived among earlier inundations (say, following the Younger Dryas episode), and had then witnessed the raised sea levels decline, revealing more and more land as the waters receded.

Here, too, then, myths appear to follow history and suggest that, for some types of North American Indian at least, they did 'remember' in their legends an early flood, and observed the land rising as it was released from the weight of ice, as the glaciers melted, or offshore volcanoes erupted, or higher sea levels receded. The rest of the Americas lack the 'land diver' and 'land raiser' myths, though they have a rich stock of flood myths, including those with birds who fly out to seek land.

Other ancient legends would appear to offer still more tantalising glimpses into the past. For example, in his analysis of world flood myths, Stephen Oppenheimer found systematic variation in the New World. The dominant flood myth of North American Indians was the land raiser and land diver myths, as we have just seen, and these they shared with Siberians. In Central America, on the other hand, the dominant features involved super-waves, mountain-high floods, with the survivor(s) landing on the mountainside before the flood receded. These themes were shared with myths in Tibet-Burma, Taiwan, island South East Asia and Polynesia. As we shall see in more detail in chapter five this geographical distribution coincides exactly with the pattern of hurricane activity across the Pacific Ocean (pages 95–7).

In South America the dominant flood themes stress overpopulation (the gods decided to make a flood because there were too many people on earth), drought and/or famine before the flood and the use of an ark or boat of some kind. As again we shall see in chapter five, this area is known to be among the most volcanically active regions on Earth, and is also susceptible to El Niño events, which create violent winds, associated tsunamis and can trigger earthquakes, causing great loss of life. Early peoples may well have concluded that such disasters, which killed so many people, meant that the gods thought there were too many people on Earth.

Tricksters and Totems

A somewhat different myth of origin has less to do with floods than with what has come to be called 'the trickster creator'. This character is found in Norse myths, in Africa, in New Guinea but above all in North America. The trickster creator is usually an animal, like a fox, a raven or coyote, or is half-animal/half-human, and usually creates people by some sort of subterfuge or deception. This has generally been taken to refer to shamanistic behaviour, primitive religious leaders exercising power or influence through some sort of magic, who were often thought capable of turning themselves into animals, at least temporarily.

The world-wide distribution of these myths suggests that shamanism had evolved by the time early humans reached the Americas (see chapter three). It is certainly possible that religious practitioners could have evolved several times throughout history, in different places, but it is much less likely that the 'trickster' theme would have evolved several times. Could the theme of creating humans by subterfuge have evolved before early humans grasped the male role in procreation? There is some evidence that humans in the Old World discovered the male role in reproduction only when the dog was domesticated, because it was domesticated first and because among the large mammals it is the one with the shortest gestation time. The dog, the fox and the coyote are all members of the *Canidae* family, whose gestation times range from 52 days (fox) to 63 days (coyotes, dogs and dingoes). This possibility is discussed in more detail in chapter seven.

Paul Radin, in his study of North American Indian trickster myths (among the Winnebago, Tlingit and Assiniboine) concludes that the trickster has three primary traits – his voracious appetite, his wandering and his unbridled sexuality. Given that he is also on occasions a mix of being a god, or almost a god, or an ex-god, and at the same time a buffoon, Radin concluded that the trickster represents the spirit, or the threat, of *disorder*, indicating from where dangerous disorder is most likely to arise (conflicts over food and sex), suggesting that this too is a myth that stems from an experience in deep time, perhaps when the population thinned and was under threat from limited or nearly non-existent food supplies and where the act of procreation required discipline if the tribe were to survive. It may also represent an ambivalent attitude to the gods who, in the past, had let mankind down, scarcely behaving in god-like ways. Is this a folk memory of the New World inhabitants' precarious and constricted time in Beringia, cut off between water to the west and ice to the east?[33] Is the unbridled sexuality of the dog an unconscious reference to its role in revealing the male principle in reproduction?

Of the many other myths that exist in both the New World and the Old, the most important generalisation, from our point of view, is that they support the idea that the early inhabitants of the Americas come from both inland central Asia *and* island South East Asia.

According to Stephen Oppenheimer, there are very few motifs that are totally absent in the New World, but there *are* systematic variations and these form a consistent and coherent picture. The most substantial systematic difference is a constellation of ten linked motifs not generally found in Africa, the Americas, or Central and North East Asia. Instead, this constellation occurs in a distinctive swathe (referred to earlier) from Polynesia across China, South Asia, and then the Middle East, ending in northern Europe (as far as Finland).

For example, in the Americas there is a relative dearth of 'watery chaos myths' beyond the north-west Pacific coast. Another difference is that New World myths lack almost any references to sea monsters or dragons (the one exception is an Aztec myth). In any one of the three floods referred to above, far more areas, and far more populated areas, would have been accessible to crocodiles, whose main range of activity was in Indo-China. This is confirmed by the fact that, in the myths, most dragons and serpents attack coastal peoples, not fishermen. On this reading, the dragon and sea monster stories are perhaps a deep folk memory of a plague of crocodiles that occurred when shallow coastal areas were flooded at one stage in the distant past.[34] (Recall the skin scarification practices mentioned earlier among the Austronesians, which are supposed to represent the teeth marks of crocodiles.)

Besides 'watery chaos', first light and separation, other elements in this constellation of myths are the use of the 'word' by the gods to create light, descriptions of incest, parricide and use of the deity's body parts and fluids as building materials for the cosmos. None of this occurs in New World myths, though it is common across Eurasia, where we also often find a divine couple who are bound together, and separate to create heaven and Earth, who are then mutilated and torn apart by their offspring, who use the parts of the parent deity to create the landscape (blood is used for rivers, for example, or the skull for the dome of the sky). Many myths in this constellation contain episodes of post-flood incest, usually between a brother and a sister. Sometimes the participants are aware of the taboo, at others it isn't mentioned. This would appear to be a forceful way for primitive peoples to reinforce the memory that, after the flood and/or the Toba eruption, the race almost died out and/or was isolated (from other islands?), the population reduced to such an extent that brothers were forced to mate with sisters.

Again such myths are not in general found in the New World. Nor are there many myths of 'land-splitting heroes' in the Americas. In land-splitting myths, the floods are caused by sea creatures or monsters when they chipped bits of islands from continents or from larger islands, when creating the geography of the region (this motif is widespread in Indonesia). It would appear to be a remnant of an earthquake (or, again, a flood) which may have created offshore islands or rearranged them, the resulting floods being associated with the above-mentioned plague of crocodiles.

Nor are there myths in the New World, as there are in the Old, where the world is created by the word of god. This is well known in the West from the bible, of course: 'And God *said*, Let there be Light: And there was light' (Genesis 1:3). Similar motifs are known in Babylon, Egypt, India, Polynesia and other areas of the Indo-Pacific. The emphasis on the 'word' may indicate the importance of language in early forms of identity.

This overall pattern is amplified by a second group of myths which are also notable by their absence in the New World. These include the 'dying and rising tree god', the myth of the warring brothers, and the so-called 'moon/lake tryst'. The dying and rising tree god or spirit had a very wide distribution across the Earth, from the Norse myth of Odin to the Egyptian myth of Osiris, to the Christian story of Jesus, to the Moluccan myth of Maapitz, to the New Britain myth of To Kabinana.* Moreover, this myth overlaps in certain locations with the theme of warring brothers, or sibling rivalry: Set/Isis in Egypt; Bangor/Sisi in Papua New Guinea; Wangki/Sky in Sulawesi; and, of course, Cain and Abel in the bible. This conflict is generally taken to reflect the different lifestyles of agriculture and either foraging or nomadism – in other words, it is *post*-agriculture. In the moon/lake tryst the hero falls in love with the reflection of the moon in a lake.[35]

What matters for us with these myths is less their meaning (for the moment) than their distribution which, as mentioned above, is all but

* The distribution of deciduous forests around the world is confined to north and south of the tropics so does not really parallel these myths.

identical to the other group of ten motifs just considered. This spread is shown in map 4 and, broadly speaking, once again extends from Indonesia and Borneo, up through the Malay peninsula, India, the Arabian Gulf, Mesopotamia, the Mediterranean civilisations to western and northern Europe. This range of locations has in common that it occupies and overlaps with the great 'East-West Corridor' (as shown in map 1 and discussed in more detail in chapter five), a broad swathe of coastline running from the tip of Malaysia at Singapore as far west as Pointe St Mathieu, near Brest in Brittany, in France.* Is this evidence for *very* ancient contact between these regions? We shall see in a later chapter how east-west movement across the globe has been much easier than north-south movement.

The new synthesis of cosmology, geology, genetics and mythology is exciting but we have taken it about as far as it will go. From it we may conclude (and repeating the proviso that this is all very speculative) that one group of early humans who first peopled the Americas arrived no later than 14,000 years ago and very probably at 16,500–15,000 years ago. They shared with everyone else an experience of a global seaborne flood and creation myths in which man is fashioned out of clay. But they showed no awareness of either agriculture or navigation, having reached the cold and limiting region of Siberia, and then Beringia, before these skills were invented (or needed). Similarly, they showed a very rudimentary awareness of a great global catastrophe, other than flood, in which the skies darkened for generations and only slowly cleared, with light preceding the sun and the moon by a very long time.

These myths concur with the genetic and linguistic evidence, that there was a later migration, possibly at 11,000 but more likely at 8,000–6,000 years ago. It follows from this that early Americans had no awareness of the cultural conflict that gave rise to the 'warring brothers' myths, or the dying and rising tree god myth, which originated in South East Asia too late for it to be incorporated into New World mythology. This too suggests these people left the Sunda Shelf before agriculture was invented. The two constellations of myth motifs were arguably the most important ideas of ancient times in the Old World,

* The general geography of the Earth, and its effect on prehistory and history, is considered in chapters five and six.

shaping – as we shall see – most of the religions and traditional histories from Europe to South East Asia.

This all suggests (and still speculating) that the period between 11,000 and 8,000 years ago on the Sunda Shelf was very problematical, entailing several catastrophes which both expelled many people and gave rise to powerful myths among those who remained. As Johanna Nichols has said, there was a major rupture between the people who headed north, eventually to colonise the New World, and those who remained, or headed back west, to form part of the civilisations of Eurasia.

In fact, what the distribution of myths enables us to say is that, just as Johanna Nichols identified four large language phyla across the world (see above, page 20), so there are four large 'phyla' of myths, though with a somewhat different distribution. These four areas encompass, first, Africa. For our purposes, Africa can be largely set aside: the continent features as the starting point of a journey for humankind where the main episodes of interest take place elsewhere. And this is reflected in Africa's myths of origin about which, as Stephen Belcher says, 'No generalisations are possible.' Tricksters are found in myths right across the continent, as are giants and ogres; snakes are common and most myths take place in a rural, village environment, and concern hunting or cattle herding (the latter, therefore, are of rather recent origin). Baboons and chimpanzees feature, often as early forms of human being; sky gods and the moon are also quite common but none of the patterns we shall be attaching importance to exists in African myths. The second area is the long swathe of cultures from northern Europe through the Mediterranean and Middle East, India and South East Asia and on to island South East Asia (the great East-West Corridor); the third area encompasses northern Asia (China, Siberia, Japan, Korea); and the fourth comprises the New World.[36]

In general, two broad conclusions may be drawn from this brief survey of myths, one concerning the New World, the other the Old World. In the New World the very ancient myths (such as the watery creation of the world) tended to be superseded by those that forced themselves on people by their experiences in the American continent itself – land divers, land raisers, tricksters, violent tsunamis. This is an early indication of a trend or theme we shall see more of as the book proceeds: the role that extreme weather – storms, hurricanes, volcanoes and earthquakes – plays in New World ideology. In the Old World what draws our attention is the distinctive distribution of the watery creation myth, the use of the 'word' to create light and the dying and rising tree god set of myths. Dying and rising refers to fertility, a dominant issue in Old World ideology that, as we shall see, did not have quite the same importance in the New World. The spread of these myths, from the 'corner' of South East Asia, up through China and across India into the Middle East and west and north Europe, following what we shall be calling the great East-West Corridor, shows that this route was in use very early on. It would have a profound effect on the development of Eurasia in all senses – ideologically, commercially, technologically. There was nothing like it in the New World.

• 3 •

Siberia and the Sources of Shamanism

The final matter to consider before following early humans *into* the New World is a specific region of the Old World that the migrants had to pass through before completing their Great Journey. This was the region immediately to the west of the Bering Strait: Siberia. As well as being remote, cold and empty, Siberia is the home of shamanism, a phenomenon that will play a large role in our story.

Although it has been called into question lately, most anthropologists accept that shamanism is a hunter's religion, and probably the earliest evolved manifestation of religious activity, spiritual discipline and medical practice, and which seems to have been in existence in prehistoric times across a wide swathe of the Eurasian landmass. Shamanism is even depicted in cave art – for example, at Trois Frères in Ariège, in the French Pyrenees, south of Toulouse, where a figure seventy centimetres high displays antlers, owl's eyes, lion or bear's paws, a fox's tail, and is wrapped in an animal skin which almost conceals an erect penis (see figure 1).[1]

Shamanism is primarily concerned with the need to take the life (of animals) in order to live oneself and reflects an early cosmology, an equilibrium achieved by the idea that one has to pay for the souls of the animals one needs to kill in order to survive, and where the shaman flies to the owner of the animals to negotiate a price.[2]

In hunting societies, the main belief is – and was – that 'All that exists lives'.[3] We are surrounded by enemies, 'invisible spirits with gaping mouths'. A second belief is that the cosmos exists on a series of levels – six, seven or nine, as the case may be – all linked by a 'world tree' or pillar or mountain, and where the shaman's primary ability is –

Fig. 1 'Animal-shaman' dancing. Les Trois Frères cave, Ariège, France.

and was – 'soul flight' which enabled him, in a trance (an important element), to travel between the different levels of the cosmos in order to perform feats that would benefit the community. A third belief was that people had multiple souls and that dreams were evidence of one or more of these souls leaving the body, and going on journeys. Illness resulted when these souls could not for some reason rejoin their bodies (possibly because ill people found it difficult to sleep, or alternatively were so exhausted by their illness, and slept so deeply, that they couldn't remember their dreams). And it was part of the shaman's function, again in trance, to go on a soul journey, find the missing or lost soul and lead it back to the ill person's body, restoring that body to health.[4] These soul journeys often involved crossing menacing landscapes, in which the shaman was either dismembered or reduced to a mere skeleton, and for that reason shamans underwent special initiation procedures, having to spend a considerable period in the wilderness, surviving on their own and making the intimate acquaintance of the landscape and the wild animals, developing his (or, more rarely, her) survival skills. The activities of the shaman tend to concentrate on matters that, though important, are erratic: illness, weather, predators, prey.[5]

The word 'shaman' comes from the language of the Evenk, a small Tungus-speaking group of hunters and reindeer herders in Siberia. Its literal meaning is 'one who knows' and it was first used, according to Piers Vitebsky, only to describe religious specialists from this region. Other tribes in the arctic north have other words, which haven't caught on to the same extent.[6] Shamans exhibit various forms of behaviour – which we shall come to – but the aspect of shamanism that is of obvious interest to us at this point is its geographical spread. In Vitebsky's words, 'There are astonishing similarities, which are not easy to explain, between shamanistic ideas and practices as far apart as the Arctic, Amazonia and Borneo.'[7] Although shamanism is found in Borneo, the strongest links are in fact laid out like a giant figure '7', along the northern rim of Eurasia, from the Lapps in arctic Russia to Siberia, and then on down the Americas to the Amazon.

It is fair to say that more would be known about the phenomenon of shamanism – its nature and distribution about the world – but for twentieth-century history, in particular the Cold War. Several hundred photographs of prehistoric cave art, and ancient sculptures, have been published in the West (in France, for example), but, according to one account, there are 20,000 illustrations published in Russian (Soviet) journals but not readily available elsewhere in accessible journals or books. This has inhibited the growth of our knowledge while at the same time confirming the distribution of shamanism across Eurasia.

Siberia, even today, is an inhospitable place. An area equal in size to Europe and the United States put together, it contains, in the Kolyma basin, the coldest inhabited location on Earth, where winter temperatures can sink to -70° Centigrade and where, when summer arrives in May, the frozen rivers can shatter 'with the sound of cannon fire'. Siberia contains the world's largest forest, covering roughly 2,000 million acres and, of all the inhabitable parts of the globe, it remains the most thinly populated. According to Soviet sources, there are some 120 linguistic groupings in the area though much of the *taiga* (boreal forest, birch, poplars, conifers) and tundra (shrubs, sedges, mosses, scattered trees) of central and eastern Siberia was roamed by reindeer-hunting and herding tribes who spoke the language known as Tungus. On the Kamchatka Peninsula and its offshore islands – the eastern-most tip of Asia – the language spoken was and is Eskimo: Yupik and Unangan.[8]

At the time the Russians moved into Siberia, in the late sixteenth century, the Chukchi, a tribe we have already met, were still a Stone Age people, living in tents with bone-tipped arrows and spears and recognising no authority higher than family networks.[9] In this area, Christianity was never fully imposed on the people, who continued to offer blood sacrifices (often of dogs) to the saints, usually in secret. In Soviet times, shamanism was officially abolished by decree, aided by collectivisation, which moved Russians into Siberia from other regions of what was then the USSR. Great cruelty was sometimes used to stamp out traditional practices, with shamans allegedly being thrown from helicopters and challenged to 'show their powers of spirit-flight'. By 1980 the Soviet authorities claimed to have suppressed shamanism completely, though this is doubtful.[10]

Shamanism may have begun as the world's oldest religion, when and where it did, partly because of the psychopathology of the arctic region, and partly because of man's relationship with the deer. There is some evidence, for example, that early shamans were psychologically out of the ordinary – either epileptics or neurotics, whose abnormal behaviour was regarded by primitive peoples as 'an altered state of consciousness'. In the arctic region, with its intense cold and long periods of darkness, recognised forms of mental illness, such as *meryak* or *menerik* (arctic hysteria), could appear similar to the shaman's trance.[11] It is notable that in some areas of Siberia and the Americas the initiation rites for shamans include a period of sickness. What we understand as abnormal behaviour may not always have been regarded as pathological in ancient times but rather as 'spirit possession'.

Reindeer are magnificent animals. They are large; the males have their extraordinary crown of antlers, giving them an easy, and easily identifiable, nobility; they are skittish but not fierce, unlike bears and the big cats; and they are natural herding animals. Because of their configuration, they were and are useful sources not just of meat, but of hides, bone and sinew. This configuration may have provoked the traditional attitude of hunting societies towards game, which has been described as 'a complex of worship and brutality'.[12] Early hunting societies (the 'reindeer civilisations' as Fernand Braudel called them) usually had some belief in a 'Master' or (less often) 'Mistress' of the Animals, a guardian of the animal species which played an important role in the life of the tribe or clan, and represented their collective soul or essence. On this understanding, the Master of the Animals 'releases' animals to human hunters, so they can kill certain creatures and obtain food and other necessities, but in return the hunters owe certain obligations – they must make agreed sacrifices and observe particular rules. This is where the shaman comes in.

One of the defining attributes of a shaman is the phenomenon known as 'soul flight'. By means, usually, of an 'altered state of consciousness' (to use a modern western phrase not necessarily recognised by the shamans themselves), and often a form of trance, the shaman flies – either around the landscape, to locate the animals to be killed, or to the upper world, or the netherworld, or the bottom of the ocean, to locate the Master of the Animals, to negotiate the price to be paid for the creatures.

But there is more to shamanism than this. Often a link is made between hunting and seduction. The penetration of the animal's body in hunting is sometimes seen as analogous to sexual union.[13] For example, in the Upper Amazon, among the people known as the Desana, the word 'to hunt' also means 'to make love to the animals'. The analogy is elaborated in the careful preparations that are made, meaning that the prey is 'courted' and even sexually excited, so that it will be attracted towards the hunter and 'allow' itself to be shot. To this end, the hunter must be himself in a state of heightened sexual tension, which is achieved by sexual abstinence immediately before the hunt, and by making himself more appealing to his prey through ritual purity and body decoration of which face paint is an important ingredient. In Siberia, the shaman is often felt to have power over the animals because he has once been an animal himself. In preparatory ritual dances, the rutting and mating of animals is often simulated, with explicit sexual gestures.[14]

According to Piers Vitebsky, Siberia and Mongolia are the 'classic shamanistic areas'. A conference at Harvard in the spring of 2011 heard other recent evidence suggesting that the elk, a species of large deer, migrated across the Bering Land Bridge at more or less the same time as did humans. If this is confirmed, it would underline the idea that shamanism crossed the continental divide at that time too. Soul flight does occur in the Americas but mainly in the arctic and sub-arctic, with North American Eskimos very similar in this regard to Siberian peoples.[15] For Eskimo shamans, 'dismemberment, dramatic flights through the air and journeys to the bottom of the sea are common'. As the distance from the Bering Strait increases, soul journeys become rarer, as does the occurrence of deep trance. Instead, the shaman is initiated into his role through isolation and fasting. Especially in the Great Plains region, trance and journeying are replaced by dreaming and by the 'vision-quest'. This latter is a procedure in which young men – and less often young women – remove themselves to the wilderness for days at a time, to prove their hardiness but also to seek a vision from the spirits of the natural world. In many tribes this has become a widespread initiation rite all by itself – to teach young people basic survival skills – but shamans develop the visions obtained in this way in more detail.[16]

In Central and South American societies, the shaman is the dominant figure in the tribe. Their view of the cosmos is essentially the same as it is in Siberia – a layered heaven linked to Earth by a world tree, or pillar, with the shaman having the ability to fly to the upper and lower worlds. The initiation procedures, also, are much the same: an initial sickness, the experience of being dismembered or reduced to a skeleton, and marriage to a 'spirit-spouse'. Chanting is a particularly distinctive technique of entering trance in South America, and so are two other devices not found elsewhere to anywhere near the same extent – the widespread use of hallucinogens and the close identification of the shaman with the jaguar. These aspects are considered in more detail in later chapters but here we may note that, recently, David Lewis-Williams and Thomas Dowson have brought new evidence to bear which confirms the great antiquity of shamanism by showing that shamanic practices were 'a significant component' of Palaeolithic rock art.

In the first place they noted evidence of hallucinogens being used in these confined spaces; at the same time the images themselves, often abstract designs, are, they say, similar to those produced in 'entoptic' (trance-like or drug-induced) states. This links in with the well-known account by Peter Furst of fly agaric use among the Koryak, another Siberian tribe. According to Furst, fly agaric, a psychoactive fungus, has been used as a sacred inebriant of the shamanistic religions of the northern Eurasiatic forest belt, especially those of Siberian reindeer hunters and herders, from the Baltic to Kamchatka, where it was known as the 'mushroom of immortality'.[17] Perhaps the most remarkable aspect of the Koryaks was that they would drink the urine of intoxicated people – intoxicated with fly agaric – and even the urine of reindeer who had eaten the mushroom and became affected. The urine of people and deer who had eaten the mushroom was apparently even more potent than the mushroom itself and in this way the Koryaks could 'prolong their ecstasy for days'.[18]

There is, too, an overlap between the role of the shaman and the myth of the trickster, referred to earlier. Being a shaman often involved trickster-type behaviour. The shaman must change form to 'fight and outwit' obstructive spirits, and in myths the earliest shamans employed trickery to capture the sun, to give light to the people, or else stole the secret of fire, or of agriculture, from the jealous gods.

One final cultural characteristic that crosses the Bering Strait is the fact that shamanism is closely associated with transvestism. In Siberia the costume of the male shaman is invariably decorated with female symbols, while among the Chukchi shamans may dress as women, do women's work, and use a special language spoken only by women (they are referred to as 'soft men'). Among North American Indians there is a widespread tradition of transvestism known as *berdache*. These make especially powerful shamans – the Mohave, for example, rank their shamans, rating females as more powerful than males but *berdaches* as more powerful than either. The Navajo, Lakota and Cheyenne Indians all believed *berdaches* could cure insanity and were powerful aids in childbirth.[19]

Ronald Hutton concludes that the idea of the shaman is a more diffuse concept than many other scholars believe but all are agreed that, at the most fundamental level, shamanism is based on the hunt – the primary role of the shaman being to ensure success in hunting. After that, his (or her) functions fall into two: to perform feats that benefit the community; and to become 'expert' on the workings and experiences of the individual soul, in particular conducting the soul of the sacrificed animal to the deity to whom the offering was made, showing families where their dead relative should be propitiously buried, and then conducting the soul of the dead relative to the world beyond.

That the Bering Strait is where it is – in the remote far north – was to prove of great importance in the long run. Whether the early peoples who populated the New World came from inner Asia or island South East Asia, they had to pass through Siberia on their way. And whether they practised shamanism to begin with, and whatever other genetic or psychological or mythological characteristics they had, Siberian shamanism was imported wholesale into the New World as its basic ideology. It was to have an influence that, as we shall see, outweighed everything else.

· 4 ·

Into a Land Without People

Beringia: Where the Great Divide Took Place

The question as to whether America was part of Asia, or a landmass in its own right, was only settled in 1732, nearly a quarter of a millennium after Christopher Columbus had first set foot on Guanahaní, when Ivan Fedorov and Mikhail Grozdev finally discovered Alaska. In 1778 Captain James Cook sailed through the Bering Strait, noting that only a short reach of sea sixty miles wide separated the continents, convincing many that this was the point of entry for the first Americans.

The man who first put forward the idea that there was once a land bridge between Russia and America was Fray José de Acosta, a Jesuit missionary, in 1590.[1] By then he had lived in Mexico and Peru for nearly twenty years and he took it as an article of faith that, since Adam and Eve had begun life in the Old World, man must have migrated to the Americas. Moreover, he thought transoceanic travel unlikely, preferring the idea that 'the upper reaches' of North America were joined or 'approach near' Russia, with a narrow enough water gap that migration would not be inhibited.

He noticed too the spread of smaller animals, considering it unlikely that they had swum across even a short stretch of water: overland travel was much more likely. He also suggested that it wasn't a deliberate migration, rather a 'gradual expansion ... without consideration in changing by little and little their lands and habitations. Some people the lands they found, and others seeking for newe, in time they came to inhabit the Indies.'[2]

Angelo Heilprin (1853–1907), a geologist, also drew inferences in 1887 from animal distribution, observing that Old and New World animal species were relatively dissimilar at southern latitudes, more similar in mid-latitudes, and 'nearly identical' in the north. To him it was clear that, 'if species diversification was a function of distance from the north, then the species must have dispersed from that direction'.[3] Not long afterwards, another geologist, the Canadian George Dawson (1858–1901), noted that the seas separating Alaska from Siberia were shallow and 'must be considered physiographically as belonging to the continental plateau region as distinct from that of the ocean basins proper'. Dawson added that 'more than once and perhaps during prolonged periods [there existed] a wide terrestrial plain connecting North America and Asia'. He had no idea of the Ice Ages but accepted that continental uplift had from time to time raised the seabed above the water level.[4] In 1892 great excitement was caused when some mammoth bones were discovered on the Pribilof Islands, three hundred miles west of Alaska. 'Either these giant hairy elephants were awfully good swimmers, or the islands were once high spots in a broad plain, conjunct ... with the entire Alaskan and Siberian landmasses.'[5] W.A. Johnson, another Canadian geologist, added a final gloss. In 1934, he made a link between sea level changes and the Ice Ages, the existence of which had been affirmed only in 1837. 'During the Wisconsin stage of glaciation [~110,000–11,600 years ago],' he wrote, 'the general level of the sea must have been lower owing to the accumulation of ice on the land. The amount of lowering is generally accepted to be at least 180 feet, so that a land bridge existed during the height of the last glaciation.' This agreed well with the argument that the Swedish botanist Eric Hultén made about the same time, that the area of the Bering Strait had been a refuge for plants and animals during the Ice Age. It was Hultén who named the region Beringia, and argued that it provided the terrestrial route by which ancient humans reached the New World.[6]

The evolution of scholarship surrounding the Bering Land Bridge is a fascinating story in its own right. It falls into three aspects – first, the attempts to prove that there really *was* a land bridge, at various remote times in the past; second, an exploration of what the land bridge was like physically, what its geography comprised, what plants and animals it could support; and third, an inquiry into what sort of people travelled across, and when.

The Pleistocene era, more popularly known as the Ice Age, began roughly 1.65 million years ago. Most scientists think that it ended some ten thousand years ago, though others argue we are still in it, 'merely enjoying an interglacial reprieve'.[7] During the Pleistocene, warming trends followed cooling ones in great cycles that could last hundreds of thousands of years. The most recent cold cycle began about 28,000 years ago, with temperatures falling relentlessly until roughly 14,000 years ago. Conditions were far harsher than anything known today, especially at the polar regions, and particularly so in the northern hemisphere owing to the direction of the Earth's rotation, which made ocean currents and the weather they affected worse there than anywhere else, and because there is more (heavy, dry) land in the northern hemisphere than in the southern one, producing irregularities in the Earth's orbit. All of which meant that more snow fell in winter than melted in summer, building up in layers that melted slightly in the short summers, then re-crystallised. Each year's snowfall pressed down on the layer of the year before, creating great masses of ice.

As a result, year by year, the glaciers' edges spread out, so that they gradually merged, one with another, their dimensions becoming truly awesome. For instance, the Laurentide Ice Sheet, the biggest in North America, built up to a height of nearly two miles. It was centred on what is now Hudson Bay, but it eventually smothered all of what would come to be called Canada, 4,000 miles across. To the north it merged with the Greenland Ice Sheet and in the south it eventually extended as far as what is now known as the state of Kentucky.[8] To the west, the Laurentide merged with North America's other great ice sheet, the Cordilleran, which stretched 3,000 miles down the coastal mountains of western North America, from Puget Sound to the Aleutian Archipelago. Northern Europe was blanketed in much the same way, in England as far down as what would become Oxford, while ice covered most of the world's principal mountain ranges, not to mention Antarctica. One surprising exception was the interior of Alaska, where the arid conditions meant there was little moisture to make ice or snow. It was a phenomenon that would prove all-important.

Over this 14,000-year period (28,000–14,000 years ago), as had happened several times before, water was removed from the seas in enormous quantities by evaporation, with the clouds formed being blown over the land by winds, there to fall as precipitation, usually snow, contributing further to the build-up. Eventually, the glaciers held around one-twentieth of the world's water, and half of that was to be found in the Laurentide Ice Sheet. As a result, the sea eventually dropped by about 125 metres, or 400 feet (not 180, as Johnson said) below where they are now.[9] As the waters receded, Asia and North America 'began to reach for each other like the outstretched arms of God and Adam on the ceiling of the Sistine Chapel. When the finger tips touched, a charge of new life streamed into the Americas.'[10] An imaginative analogy but the contact was achieved with geological slowness. Gradually, the shelf widened until, at 18,000–14,000 years ago – the height of glacial activity, or LGM, for Late Glacial Maximum – the shelf between Alaska and Siberia was exposed as dry land for over nine hundred miles, north to south. 'North America and Asia were joined at the head like great, sprawling Siamese twins [see maps 1, 5 and 7]'.

There was more to it than a simple land bridge forming. Every time the sea levels fell and rose, Alaska effectively switched continents. During the cold cycles, when the land bridge was exposed, glaciers formed in Canada, and Alaska was cut off from the New World – it became, in effect, the eastern end of Siberia. When the glaciers melted, and sea levels rose, dividing Alaska from Siberia, she became what she is today, the north-western tip of North America.[11]

Geological studies, involving drilling cores of both land and ice, show that there have been some sixteen Ice Ages in the last million years alone, separated by 'inter-glacials'. The Bering Land Bridge would have connected the continents during most of these glacials, when animals – if not humans, who had yet to evolve or reach Siberia – could have switched between the New and Old Worlds.[12] The last land bridge is of most interest precisely because humans were around at that time. This period endured from, roughly speaking, twenty-five thousand to fourteen thousand years ago.

Even without ice, it was a harsh landscape, dry and windy. Loess (windblown glacial silt) built up in great grey dunes. Vegetation was thin, the land little more than a polar desert, a drier version of today's tundra. Despite this, Beringia – as it came to be called – was home to a variety of animals. Or that is what research shows. Woolly mammoths were probably the biggest of the beasts that roamed the region. Their six-inch-thick hairy coats, hanging in ragged skirts, offered sufficient protection against the cold and the bitter winds. Ground sloths, weighing as much as six thousand pounds, and the long-horned steppe bison, were almost as big. Horses, which evolved in North America, migrated across the land bridge going the other way, *into* Asia, but they almost certainly had much thicker coats than today's horses. Various forms of antelope, moose, caribou and sheep – all these inhabited the land bridge during the Ice Age. And so too did huge sabre-toothed tigers with their six-inch-long canines capable of piercing the thick hides of the mammoth and bison, plus giant lions and packs of timber wolves. An ancient form of bear, bigger even than today's Alaskan grizzly, completed this exotic bestiary (see figure 2).[13]

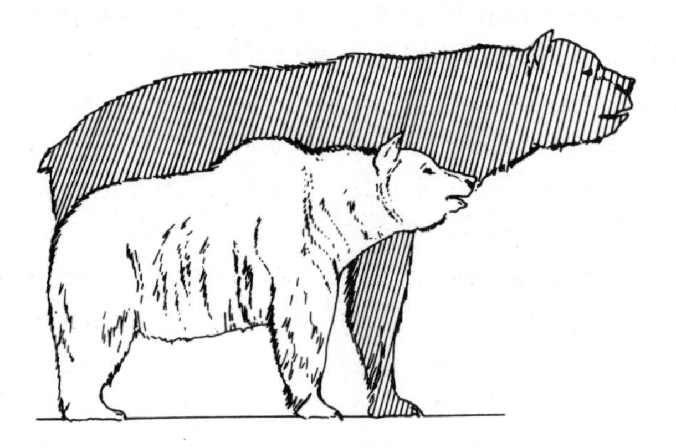

Fig. 2 A comparison of body sizes of the prehistoric bear, *Arctodus simus* (shaded) and the modern grizzly, *Ursus arctos horribilis*.

The identification of this plant and animal wildlife was itself an achievement of scholarship. Eric Hultén, the Swedish botanist referred to earlier, studied the plants of Siberia and Alaska in the 1930s, to produce his *Flora of the Aleutian Islands*. Hultén was statistically minded and, as well as describing the plants, he noted their distribution, in particular their spread around Canada's Mackenzie River and Siberia's Lena River.[14] Plotting these distributions on a map, he saw that they spread out in a series of ovals, stretching east-west. Moreover, these ovals, as well as being symmetrical, were also concentric, and the axis of symmetry was always located in a line drawn *through* the Bering Strait. 'If they spread a little bit to the east of this line, they also spread a short way to the west of it. Should they spread far east, they also spread far west.' A consistent picture therefore suggested itself to Hultén. He imagined there must have been at one time a dry landmass 'stretching from mostly unglaciated Siberia into mostly unglaciated Alaska'. This landmass was isolated from the much colder areas around it and therefore acted as 'a great biological refugium, a place where northern plants and animals survived extinction and from whence they evidently spread when the glaciers receded'.[15] The area was for him 'a broad highway of biological exchange', with animal and plant traffic travelling in both directions. Wanting to emphasise the importance of the area, he gave it the name Beringia, after Vitus Bering who was actually Danish-born and therefore a fellow Scandinavian. (Bering had sailed through the strait in 1728 but had not himself reached Alaska until after Fedorov and Grozdev.)

Hultén's work was soon built on by Louis Giddings, a Texan archaeologist who, in the 1940s, at Cape Krusenstern, south-east of Point Hope, identified 'no fewer than 114 beach ridges each parallel to the shore line and extending more than three miles inland'. Not only that, each ridge furnished a series of archaeological discoveries in which the outer ridges (closer to today's coastline) were older than those further inland. The natural conclusion was that these seafaring cultures had transferred inland by stages as the seas rose and encroached on their dwellings, showing neatly how human habitation and sea level were intimately linked.[16]

But the individual who did more than anyone else to advance the scientific understanding of Beringia was David M. Hopkins, a graduate of the University of New Hampshire. Together with William Oquilluk, a famous Iñupiat historian from north Alaska, Hopkins' first project was a study of fossil mollusc shells, because he understood that their distribution and sedimentation would reveal when the Bering Strait was open and when it was not. The basis for this study was the natural history of the giant snail *Neptunea*. This has existed in the north Pacific all the way back to the Tertiary geological epoch, some sixty-five million years ago. But there was no evidence of *Neptunea* in the Atlantic sediments until the early Pleistocene, about a million years ago. What this suggested was that a land bridge blocked marine migration for most of the Tertiary period and was only flooded out at the start of the

Pleistocene, allowing *Neptunea* to move north and east and to reach the Atlantic.[17]

Oquilluk introduced Hopkins to many shell deposits, and as a result they collaborated on a seminal paper, published in *Science* in 1959, dedicated to the Bering Land Bridge. Their conclusion was that the bridge had been in existence throughout most of the Tertiary era (from sixty-five to two million years ago), that the evidence from fossil fishes suggested that a waterway cut through the bridge in the middle of the Eocene (about fifty million years ago), and that it was fully submerged around a million years ago. Since then, the bridge had appeared and been submerged numerous times as the Ice Ages had come and gone. The land bridge disappeared for the last time, they said, some 9,500 years ago.[18]

Details of this broad picture have been fleshed out by others. Around the turn of the century, the vertebrate palaeontologist, W.D. Matthews, from the American Museum of Natural History in New York, observed that though mammals migrated across the land bridge in both directions, far more travelled from the Old World to the New than vice versa. Presumably, he said, this had to do with the fact that Eurasia is a much larger landmass than the Americas and extends much further in an east-to-west direction, in north temperate latitudes. This would mean that more populations of Old World animals could evolve separately. Their relatively similar environmental conditions would provoke more varied adaptive strategies, meaning there were more discrete species to take advantage of the land bridge. In North America, heavily glaciated during the Ice Ages, only the relatively small area of Alaska was available for species to evolve.[19]*

Charles Repenning, of the US Geological Survey, took Matthews' reasoning further. Based on a detailed study of fossil mammals, he argued that there were three, and possibly four, 'surges' of animal migration across the land bridge into the New World. The first influx, he said, occurred at about twenty million years ago, the types of fossil suggesting that the land bridge then was 'warm-temperate, humid and

* Recent research supports this, showing that most large mammals evolved in the Central Asian steppes shortly after 55 million years ago, during a short spell of global warming. Mammals such as horses (then the size of cats), deer, cows and dog-like predators appeared in the fossil records in China's Hunan province and then spread into Wyoming across the land bridges that then existed.[20]

forested'. A further surge took place in the middle Pleistocene, roughly 750,000–500,000 years ago, when the bridge consisted mainly of temperate grasslands, though with some forest. Not until the late Pleistocene (125,000–12,000 years ago) did mammal remains suggest arctic conditions, 'with a flora dominated by tundra, steppe, and the scrubby northern forest called taiga'.

A very different kind of evidence comes from two colleagues of Hopkins, Joe Creager and Dean McManus, who found the vestiges of an ancient river at the bottom of the Bering Strait. They drilled the seabed south of Cape Thompson, near Point Hope, and recovered core sediments that indicated a 'brackish-deltaic' river. Apparently they had struck an ancient estuary formed by a river that linked in with today's Kobuk and Noatak Rivers of north-west Alaska. Radiocarbon dating put the estuarine remains at twelve to fourteen thousand years ago. This was named the 'Hope Seavalley'.[21]

Although the picture that has been gradually revealed has by and large proved consistent, one important area of disagreement has concerned the vegetation on the land bridge. There are two views. According to one side, the analysis of pollen shows that ancient Beringia was only sparsely populated, by herbaceous tundra at the higher levels and sedge-grass meadows in the lower locations. If so, this would have been sufficient to support only primarily rodents, and does not square at all with the fact that, at the same time, elsewhere in Alaska that was unglaciated during the last Ice Age, fossils of large mammals, especially bison, horse and mammoth, have been found.[22] On top of it all, R. Dale Guthrie, now an emeritus professor at the University of Alaska Fairbanks, has observed that the three most common species – bison, horse and mammoth – were gregarious, existing in herds. According to this version, then, Beringia was populated by many large mammals, not just rodents.

Matters began to be resolved in the 1970s, when Siberian gold miners, digging underground near the Selerikan River, stumbled upon the remains of an ancient horse, still frozen in the earth. The miners informed the Siberian Academy of Sciences who excavated the animal and found it to be 'a beautifully mummified specimen of a stallion' which, unlike today's stallions, had a two-inch thick coat, 'light yellowish colour underneath and a coffee brown above, with a black mane and a dark streak running all along its spine to its black tail'. Dated to

37,000 years ago, the most interesting thing about the remains, from our point of view, was that its preserved stomach contents were found to be 90 per cent herbaceous material, two-thirds grasses and one-third sedges. Other mummified mammals were discovered subsequently (baby mammoths, for instance), and fossil insects, and the Siberian findings were confirmed, indicating a landscape of dry substrates, but with no trees, generally suitable for ungulates (hoofed mammals).[23] On top of this, the hooves of such fossil ungulates as have been found indicate that they lived on firm, dry ground, enabling them to escape their predators by running.[24] Even the remains of plant fragments stuck between the teeth of fossil ungulates show that grasses dominated their diet by a wide margin. Beringia at the LGM was essentially a 'steppe' environment.

The final confirmation of the Beringia landscape was discovered by Hopkins himself. In 1974, near a lake at Cape Espenberg on the northern Seward Peninsula, he came across a layer of tephra, or volcanic ash, that was fully one metre deep and had congealed within it masses of twigs and tufts of grass. The nearby lake was in fact a *maar* or circular lake of the kind produced when a volcano erupts at ground level. Hopkins was aware of previous research which had established that this maar, Devil Mountain, erupted 18,000 years ago, at a time when the land bridge existed. It therefore followed logically that the twigs, roots and grasses congealed in the tephra constituted *land bridge vegetation*. Examination of this plant material showed it to be a dry meadow and herb-rich tundra, a mix of herbs and grasses, in particular the sedge *Kobresia*, plus the occasional willow, and a carpet of mosses. This amalgam of *Kobresia*-dominated vegetation has since been discovered in the stomachs of several preserved mammals and confirms the steppe-type landscape. Hopkins, who died in 2001, held firmly to the view that the Bering Land Bridge could support whole herds of grazing animals, and the predators who sought them out.

Computer models carried out more recently confirm Hopkins' claim that areas of Beringia that were submerged by rising sea levels were more productive biologically/nutritionally than the higher areas that remained above the water, that there may have been a climate refuge in central Beringia for a time, perhaps causing a migration 'bottleneck', and that climate varied between east and west Beringia, changing the vegetation and animal life dependent on it and forcing people eventually to migrate. In general the climate change would have favoured the wapiti or elk, and contributed to the demise of the mammoth and horse.[25]

Importantly, this environment meant that the land bridge would have supported humans.

Animals Who Had Never Met Humans

When early man entered Beringia, it was, as we have seen, the eastern extension of Siberia. As sea levels rose around 14500 BC, and Siberia was cut off, Beringia became part of what is now Alaska. And sea levels rose, of course, because the world was warming up, the Ice Age was coming to an end, and glaciers all over the world were beginning to melt. For early man in Alaska, this had two direct consequences. In the first place, as map 7 shows, an ice-free corridor opened up between two enormous glaciers – the Laurentide and the Cordilleran – which covered most of Canada and Northern America down beyond the Great Lakes. Second, as the glaciers retreated, and the weight of ice over the land lessened, the North American continent (as did others) rose and beaches appeared along the coast. In this way, two routes southwards opened up by which means early men and women could move into new territory.

Opinions differ as to what was the actual route taken. As was mentioned in an earlier chapter, the mtDNA study by Sijia Wang and colleagues showed that genetic diversity is greatest among the Pacific coast Indians, indicating that the seashore was the probable route. On the other hand, Gary Haynes, professor of anthropology at the University of Nevada at Reno, favours the ice-free corridor, he says, partly because, although this passageway would have been cold and sparsely vegetated at first, 'with ice-blocked lakes and wetlands covering most of it', migratory birds would have acted as 'scouts', 'beckoning' humans to follow their movements to the south since the birds returned every year to where they had been. Haynes argues that traversing the coast in small boats would not have worked, since the populations would have been too small to be viable. C. Vance Haynes, at the Department of Geochronology at the University of Arizona at Tucson, imagines as plausible a trek of 5,700 kilometres from the Tanana Valley of Alaska to Anzick in Montana, where the biotic environment would have been suitably rich, a journey, he says, which could have been completed in six to twelve years, or maybe less.[26] The new genetic evidence, alongside the continuous existence of kelp beds around the northern Pacific coastal rim, remains persuasive however.

Then, after a few tentative forays, Gary Haynes says a further move south would have been attempted. The landscape then was very different from now. It was dominated by the massive lakes that developed along the edges of the ice sheets.[27] These were expanses quite unlike anything anywhere else, both then and now. Lake Missoula, at the southern edge of the Cordilleran ice sheet, was the size of Lake Ontario today, but was dwarfed by Lake Agassiz, to the west, which lasted from 12000 BC to 8000 BC and was four times the size of what Lake Superior is today, equivalent in area to Ireland or Hungary.

The geography was unstable, too, in a way that is much less true today. As the ice melted, many shrinking glaciers, or 'calves' of glaciers, formed blockages in valleys, keeping lakes dammed behind them. But as they shrank, there came a point where they simply gave way, and the water was released in unpredictable massive floods, yet another litany of catastrophes to be remembered, possibly, in myth.

So the northern reaches of North America had their dangers, as between the unstable and inhospitable landscape and the predatory animals. Valerius Geist finds an inverse relation between human sites in North America and the extinction of megafauna, leading him to ask if the great carnivores, especially the prehistoric bear, *Arctodus simus*, kept humans out of the New World for thousands of years before dying off (see figure 2). Humans are known to have met *Arctodus* at only one site but are also known to have entered North America along with moose, caribou, timber wolf and glutton, so maybe there is something to this (and confirming that the ice did block humans and other animals for a time). There is evidence that *Arctodus* was more aggressive than modern bears – it was more often caught in traps whereas modern bears tend to avoid humans. And Native North American tribes traditionally did not like hunting bears.[28] They thought they were the animal that most resembled humans, that they were wiser, and could overhear human conversations across vast distances and were, perhaps, shamans themselves inhabiting bear form.[29]

Despite this, many people would have pressed on to the south, when the prairies, steppes and plains would have been reached and where humans first came across many other mega-mammals, who preferred warmer weather and all of whom had never been hunted before, and some of whom did not exist in the Old World (thirty-five original genera of mammals had shrunk to eight at this point).[30]

Geological studies show that, about fifty million years ago, in the era known as the Eocene, Europe and America were joined by a land bridge, probably through Greenland. As a result several large mammals, including an ancestor of the horse and a primitive hoofed mammal, the Coryphodon, existed on both continents. After that, the continents separated and their fauna began to evolve in different ways. At the end of the Eocene, Asia and America were contiguous and animals started to arrive in North America via Beringia which, as we have seen, from time to time was comprised of land. This is when elephants, mammoth and mastodons migrated into what would become the New World.[31]

South America was a separate continent from North America until about two million years ago, when the Panama isthmus was formed, and although the two landmasses were close, and some species island-hopped between the two, South America developed its own very distinctive fauna, which included the edentates (sloths, glyptodonts – with bone body-armour, like massive tortoises – and armadillos) and liptoterns (horse-like creatures). At two million years ago, the continents were joined and gave rise to the so-called 'Great Interchange', when animals from the north spread south, and vice versa. Thanks to the La Brea tar pit excavations (a cluster of asphalt tar pits, stretching over more than a hundred acres, which formed tens of thousands of years ago in what is now downtown Los Angeles), we know what animals were around when man arrived in the New World. The tar pits were often covered in water, and up to sixty species of ancient animal came to drink. Some fell in and sank into the tar, which preserved them. They date to somewhere between 40,000 and 25,000 years ago and comprise 'a quite breathtaking disproportionate over-abundance' of carnivores, which make up more than 90 per cent of the La Brea contents, mostly wolves but closely followed by sabre-toothed cats, lion, sloths, tapirs, bison and bobcats.[32] There was also a large number of day-flying birds of prey, attracted by the carcasses in the asphalt, the most spectacular of which was *Teratornis meeriami*, which had a wingspan of nearly four metres.

The sheer variety, novelty and size of these mammals and carnivores was one unusual aspect of the landscape into which early man was drawn on his arrival in the New World. None of these creatures had seen man – or been hunted by him – before. Another unusual phenomenon was the relative speed with which, at around this time, the megafauna disappeared.

This disappearance, and the reasons for it, comprise another of those controversies which beset palaeontology, archaeology and anthropology and, perhaps, are destined always to do so, since the phenomenon under consideration occurred so long ago and where, as a result, the hard evidence is invariably less than adequate. In what follows, it is as well to bear in mind the remarks of Thomas Dillehay, professor of anthropology at Vanderbilt University, in Nashville, Tennessee, who reminded us in 2001 that 'only one reliable skeleton of late Pleistocene age has been excavated in South America, and only two have been recovered in North America', making all conclusions and generalisations unusually tentative.[33]*

The paucity of early skeletal remains in the New World is puzzling because it is in marked contrast to both the Old World and the Australian experience. We know of only one deliberately meaningful burial before the era known as the Early Archaic period (10000–9000 BP). A few other bones have been found but they are fragmentary, their contexts are ambiguous and they have rarely been carbon-dated adequately: although some of the early dating produced sensational results (e.g., 70000 BP), these have not been confirmed and, according to later studies, only one is earlier than 10600 BP. The rest come in at between 10200 and 7000 BP.[36]

In the Old World in strong contrast, well-defined burial practices and rituals are common after ~50000 BP. Even among the Neanderthal people in Eurasia (~60000 BP), ritual burials were common, it being not unknown for archaeologists to find dedicatory offerings of flowers,

* The site of Monte Verde in South America remains an anomaly. Dated to 14500 BP, it boasted hearths, a store of firewood, wooden mortars with grinding stones, even three human footprints 'where someone had walked across the soft, wet clay brought to the site for refurbishing the fire pits'. Medicinal herbs, molluscs, salt, bitumen and the remains of mastodon and palaeo-llama were also found.[34] The tools there, says Thomas Dillehay, are more reminiscent of late Pleistocene technologies in Australia and parts of Asia than North America. Does this mean South America has more cultural affinity with those regions, or is it simple a sampling bias in a field where the evidence is thin?[35]

exotic flints, red ochre and other materials in what were clearly carefully prepared graves. Moreover, similar burial patterns are seen among the early peoples of Australia. Archaeological remains suggest that this continent was peopled some time between ~70000 and ~40000 BP, with individual and group burials, associated with ritual and grave offerings dated to between ~30000 and 10000 BP. Timothy Flannery makes the point that, despite being settled earlier than the New World, Australia has provided much more evidence of that settlement than have the Americas. He interprets this as vitiating the ideas that America was settled much earlier than the Holocene period.[37]

Since ritual burials are so common, so early, throughout Europe, Asia and in Australia – certainly from around 40000 BP onwards – it is more than a little puzzling that we do not find more Ice Age remains in the New World. Are palaeontologists looking in the wrong places? Has there been interference with the sites? Are the radiocarbon dates *still* unreliable? Was early man in America so mobile that the *lack* of burial sites became an adaptive trait – it identified a homeland and that consumed too much social energy? As Dillehay puts it, 'None of these reasons feels substantial. If [early Americans] were cannibalistic, or cremated bodies, we should still have found something.'[38] And even if the early humans in the New World were very mobile, some human remains should have been found along known migratory tracks – but they haven't been. The Neanderthals were just as mobile as the early Americans and yet they left remains. The best Dillehay can offer is that palaeontologists and archaeologists are unfamiliar with early New World burial patterns and so they haven't recognised that what was left behind *are* remains. Whatever the reason, the mystery endures.

Life Beyond the Glaciers

Gary Haynes has given one of the most thoughtful accounts of the first humans to occupy the New World.[39] He begins by making the point that they would most likely have been hunters because, with the environment being relatively unstable and constantly changing – certainly in the region of the melting glaciers – it would have been easier to adapt the hunting techniques of one animal species to another than to adapt to different plant species in different environments, plant behaviour (the annual cycle) taking much longer to reveal itself and which would have required a more sedentary lifestyle.

At one time the evidence for early humans in the New World before ~13000 BP was thin and ambiguous, the picture changing at that date with the appearance of the culture known as Clovis, named for the Blackwater location, near Clovis in New Mexico. The defining characteristic of the Clovis culture was a 'fluted' hand-axe, a 'flute' in this context being a concave groove running along the centre of the blade, a phenomenon that doesn't exist in Old World stone hand-axes and the function of which will be discussed shortly.

In recent years, however, evidence has begun to accumulate showing that the Clovis culture, though it may well have been distinctive, did not represent the oldest human presence in the Americas south of the glaciers. Michael R. Waters and a whole raft of colleagues have identified evidence for the occupation of sites much older than Clovis, from Texas to Florida to California to Pennsylvania, some of which physically underlie the Clovis evidence stratigraphically and extend as far back as 15,500 years ago. At the same time Waters and his team have revised the dates for the Clovis era as ranging from ~13,250 to ~12,800 years ago, a span of ~450 years, possibly less. Several sites in South America have yielded radiocarbon dates that are synchronous with Clovis – in Argentina and Chile for example. Waters says it would probably have taken 600–1,000 years for people to travel this far from the ice-free corridor.

In another study, Jon Erlandson and a team from the University of Oregon found a cache of finely crafted spearheads on the Californian Channel Islands (more than seven kilometres off the mainland) alongside more than fifty shell middens dating back mostly 12,000 years and in some cases to 13,000 ± 200 years. These delicate barbed points, possibly used only in water, are very different from those found at Clovis sites, though Clovis-like flutes were also found on the islands, implying that different groups coexisted and traded at that early time. Also found were pieces of ochre and obsidian, the latter mined in quarries in eastern California, more than 300 kilometres away. Besides showing that these ancient Californians were capable of seafaring, Erlandson says the discoveries confirm that Clovis could not have been the first New World presence, and may be evidence for a coastal migration of early peoples following a 'kelp highway'.

Waters says the evidence is consistent with two models. In one, because the Clovis technology seems to have appeared synchronously across the United States, there could have been a rapid spread of Clovis people over an empty continent, with migrants exiting the ice-free corridor and spreading across the contiguous United States in as little as 100 years. The ice-free corridor was navigable at least 200 years before the oldest known Clovis sites and the stone technology of the Nenana culture was well established at the Broken Mammoth site, in Alaska, at least a hundred years before that (Waters finds that the Nenana complex shows 'strong similarities' to the Clovis assemblage). The alternative model, supported by the most recent evidence, is that Clovis *technology* spread rapidly across North America through a pre-existing but culturally and genetically undefined human population.

The earliest identifiable cultures in Alaska, as mentioned, are not Clovis. The first arrivals made blades (not microblades) and unifacial points (i.e., polished on one side only, with one flat surface) that did not have flutes. We know from their remains that these peoples hunted large mammals such as wapiti (or elk, a species of deer) and bison and they also ate small mammals and birds, such as swans, geese, duck, ptarmigan and fish. But they seem to have given rise to two traditions, one being the Denali (microblade) complex, referred to earlier, which derives from Dyukhtai in north-eastern Asia, and could be a specific adaptation to conditions in the Denali area; the second is the Nenana tradition of macroblades, which boasted various forms of stone tools, unfluted but similar overall, as noted above, to what would be elaborated later among the Clovis hunters and foragers.

Haynes says that 'thousands of fluted points' have been found in every region of North America. Fluting, he says, was developed to solve a hafting problem and was so successful that it spread rapidly throughout the continent. He further says that the people who made these fluted points also shared other significant traits: (1) they did not establish year-round camps; (2) they did not accumulate debris-middens as seen in the Old World sites of the same time period; (3) they did not create rock art or in most cases leave artwork in camps (some art has been found but it is 'negligible'); (4) they were highly mobile; (5) they did not manufacture artefacts needed to process plant foods or fish or smaller mammals; (6) they may have actively killed – or scavenged – mammoths or mastodons more than any other large mammal.[40] Timothy Flannery says that flutes may have served another purpose, namely to allow animals to bleed more freely and therefore die more quickly.[41] Thomas Dillehay says fluting appeared in South America around 13000 BP.[42]

Clovis people preferred larger animals, rather than plants or small-game foods, because big game is more readily located, is not so susceptible to environmental change – say drought – as are plants, for example, and mega-mammals can be tracked more easily without what Haynes called the 'arcane' special knowledge needed to find and process non-toxic plants.[43] Moreover, he says, the Clovis period is unique in New World prehistory in that 'never again is the archaeological record for so many different parts of the continent so similar'.[44] The dates of the many sites fall within the 15000 BP–10000 BP range and this may be one 'isolable' range, or several – the evidence is not sufficient to arrive at a final conclusion. Nonetheless, Haynes believes, unlike Waters, that the 'Clovis florescence' began after 13000 BP and that it was over by 11600 BP.[45]

The pre-Clovis sites, on both sides of the Bering Strait, are very sparse, very few and contain very small amounts of recognisable artefacts, so that the evidence for mammoth-hunting in North East Asia and Siberia is 'at best very scarce and at worst non-existent' (although it has been so poorly studied that evidence supporting large-mammal hunting could still turn up). Against this background, the relatively rapid appearance of fluted points all over unglaciated North America at around 13000 BP suggests therefore that this was a highly successful adaptation to a new situation that early humans encountered as they emerged beyond the southern edges of the great melting glaciers. At that date, says Haynes, the population of North America stood at 'not much more than' 25,000, possibly a thousand groups of twenty-five to thirty each.[46]

According to David G. Anderson and colleagues at the University of Tennessee, Knoxville, in a 1998 study, some 12,000 fluted points had so far been reported across the continent, somewhat less than have been found for post-Clovis dates. Judging by what has been excavated where, they say that not more than about 30,000 points of this kind will ever be found, which would give an estimated population for the entire continent of barely one million people, split into 30,000+ separate groups of about twenty-five people each.[47] From the distribution of the fluted points, other scholars have suggested that a group of, say, a hundred people (macrobands) may have foraged across areas of approximately 20,000 square kilometres, and may have had 'staging centres', where people congregated from time to time after dispersing along the main rivers. Some areas have more fluted points than others, especially in eastern (more wooded) locations (rather than the plains), suggesting that they passed through certain areas without settling them. One estimate is that the early settlers moved on by 250 kilometres each generation and that the morphology of the points changed every ~250 years. If these figures are correct, the rise in population, from 25,000 to 1 million, is a measure of the success of the 'fluting' adaptation.[48]

Other rhythms that have been discerned in the remains of Clovis and post-Clovis people are that great cyclical movements overlapped the serial patterns, as people returned to quarries and refuges, meaning that quality tools were carried long distances and selectively 'cached'. Quality stone points have been found as far as 1,800 kilometres from where they were mined, and sometimes attached to them are substances interpreted as 'hafting cement'.[49]

Since the invention of fluting by Clovis people was so successful, and unique to the New World, the question naturally arises as to *why* this technique was adopted. Here too we meet yet another controversy. It is a notable feature of the Clovis era that, during this time, the large mammals that existed in North America at about 13000 BP disappeared 'for ever' within a few hundred years. We know this because a mega-mammal landscape is set across a number of familiar trails and fixed resource points (such as watering holes, coppices), because areas of vegetation are affected by feeding and trampling, because the water holes themselves are enlarged and deepened by wallowing, and because dung beetles feed on the droppings of mega-mammals and spread with them. Within a few hundred years, this type of landscape disappears. Michael Waters and his team confirm that 'The extinction of mammoths and mastodons coincides with the main florescence of Clovis'.

Furthermore, as S.A. Ahler and P.R. Geib, at the University of North Arizona, Flagstaff, demonstrated in 2000, the culture which followed Clovis, known as Folsom, named after a site of that name, also in New Mexico, also has fluting but it is subtly different, with more stable hafting and a more uniform configuration, suggesting that these later points were used for only one species of big game – the bison – whereas the Clovis flutes, with their greater variety of shapes, meant these people 'had fewer worries about access to raw materials', their points being used with a much larger range of creatures.[50]

Although it would be natural to expect the Clovis culture to begin either *in* the ice-free corridor, or immediately south of the glaciers, Haynes argues to the contrary, for it beginning near the present-day Mexican border, because it was a prime mammoth habitat, and where the intense cold would have been a distant memory. He says that Clovis people would have been small – among modern foragers small stature and slow growth are 'adaptive' because 'small babies are easier to carry' when people have to walk a lot. And others have noted how, among the !Kung foragers in Africa today, shorter men are more successful hunters than are taller men.[51] The lack of archaeological evidence for houses meant that the Clovis people moved on from any one site within a month, more or less. And there is no evidence for the use of fire among them. Fire was used extensively in later times – clearing brush makes animals easier to find, removes dead vegetation and encourages new growth, which attracts animals. It can also be used to encircle creatures, it can make travel easier and is a form of communication. But in all these cases, we should see signs of 'fire-driven mosaics' of vegetation overlaying the climate-driven and mega-mammal-driven patterns, together with abundant charcoal. But none of this has been observed.

Were the Mega-mammals Slaughtered to Extinction?

After ~11600 BP the configuration of points breaks up into several different forms – not just Folsom but 'Midland', 'Agate Basin', 'Plainview' and others. Haynes therefore agrees with Paul Martin, of the University of Arizona, who was the first to argue, in 1984, that Clovis hunters had used their exceptionally efficient fluted points to hunt the North American mammals to extinction, mammals which had allowed themselves to be outmanoeuvred because they had evolved in a landscape without people, and were therefore exceptionally tame.[52]

This theory has been criticised, not least by R. Dale Guthrie at the University of Alaska at Fairbanks who, in a 2006 paper in *Nature*, provided new radiocarbon dates for many ancient remains, showing that some species of large mammal, including bison, wapiti and to a lesser extent moose, actually showed an *increase* in numbers before and during human colonisation of the New World. As Guthrie put it, there was no 'blitzkrieg' of mega-mammals; the impact of humans on their new habitat was more subtle.

Nonetheless, Haynes still concurs with Paul Martin. He says that the continental United States 'contains more mega-mammal kill-sites' than there are elephant kill-sites in all of Africa, a much larger landmass.[53] Furthermore, not only is Africa bigger but early humans have been around far longer there than the ~13,000+ years that humans have been in the New World (south of the glaciers), at least 100 times in fact, according to him.* Judging by the quantity of fluted points found associated with mammoths and mastodonts, mammoths were more often hunted and scavenged, though this may be because mastodonts occupied wetter territory and the fluted points used to kill them sank deeper into the ground and have not been found.[54] Another important observation is that mammoths killed by Clovis hunters were notoriously under-utilised – they were not fully butchered – and this may indicate that they were opportunistically attacked, people reacting to the climatic stresses that were affecting the mega-mammals.

This is one of the rival theories, that climatic change – in particular that the world was warming up and drying out just then – is what killed off the mega-mammals in an entirely natural progression (though Jared Diamond points out that America's megafauna had already survived the ends of twenty-two previous Ice Ages – what made this one so different?).[55] Another theory, published only in 2007, by a team of twenty-five geophysicists, is that a wayward comet hurtled into Earth's atmosphere around 12900 BP, fractured into pieces and exploded in giant fireballs, producing immense wildfires, killing off both the mega-mammals *and* the Clovis people, with the population of the New World being reduced by 70 per cent. At eight well-documented Clovis sites, the geophysicists say they found evidence of extraterrestrial debris – including nanodiamonds and carbon molecules containing the rare isotope, helium-3, the former only ever found on Earth in meteorites

* Africa has far fewer archaeologists than North America available to carry out investigations, but even allowing for this, Haynes says, the kill-sites in Africa are 'unexpectedly poor by comparison'.

and the latter far more abundant in the cosmos than on Earth. The chemical signature of wildfire – polycyclic aromatic hydrocarbons – were also found in three sites. Moreover, at no fewer than 70 sites in the United States and Canada a black sedimentary layer (suggesting widespread burning) has been found, dating to 12,900 years ago, with Clovis remains below. This corroborates the idea that there *was* an extraterrestrial impact at that time and that it had a marked effect on the landscape and on the population of early humans and mega-mammals in the Americas.[56]

The idea of a population decline has been queried by other studies and Haynes argues that the under-utilisation of the carcasses of mammoths points to the fact that Clovis foragers found it easy to procure them, either by scavenging or by killing, and they did not make full use of meat and other by-products in the way they would have needed to if hunting had been more difficult. The two theories can be put together, of course, in that the mega-mammals may have been slowly starving, progressively weakened by the climatic perturbation that was brought about by the end of the Ice Age. And of course the Clovis people may have chosen weakened or vulnerable animals – a rational decision. Even so, the under-utilisation of the mega-mammals that were killed certainly suggests a source of food in plentiful supply. The later use of totems by certain New World tribes – totems which govern the hunting practices of people in regard to a specific animal, where those practices help preserve the animal in question, to maintain future supplies – may well be a folk memory (and a warning) of circumstances where primary food supplies were wiped out.*

Gary Haynes has also compared the Clovis culture with its contemporaries in the Old World. He makes the fundamental point that the difference in sheer size of Eurasia and the Americas, and the antiquity of the human presence in the Old World, could account for the most important differences, though Eurasia was not as big then as it is now, because ice covered large areas of the north. Even accounting for this, however, there were some instructive differences between Clovis and the late-Pleistocene/early-Holocene cultures of Eurasia.

* Mammoth survived until about 4000 BP on Wrangel Island, in the Arctic Ocean, off the northern tip of Siberia.

Of the three most important differences, the first was the widespread presence of artworks, the most striking of which were the so-called 'Venus figurines' found in sites distributed from France to the Ukraine. Venus figurines are so called because they exhibit exaggerated anatomical features (these are discussed in more detail in chapter seven). A second difference is the size of settlements. At Dolni Věstonice, in the Czech Republic, and dating to 29000–25000 BC, the site contains the bones of at least 150 mammoths, with signs of shelters made from mammoth bones, wood, rocks and dirt, hearths and ashy deposits, indicating fire, more than 2,000 fragments of clay figurines and many other artefacts, both stone and organic.[57]

Both Gary Haynes and Olga Soffer, the latter at the University of Illinois, argue that there is some overlap between Clovis and the culture known in Europe as Gravettian (named for La Gravette in the Dordogne in France, at 28,000–22,000 years ago). Gravettian was different from Clovis in that the former spread in parts of Europe that had been occupied before, whereas the Clovis people were entering land unoccupied by humans. But both Gravettian culture and Clovis spread over huge geographical regions, meaning they were both very successful adaptations, though there is still a lively disagreement as to whether, in the Gravettian case, this was the spread of people or of ideas. Both Gravettian and Clovis people hunted large mammals such as the mammoth but there is disagreement about the importance of these large animals in the respective diets. In western Europe, the Gravettian faunal record is dominated by reindeer and fur-bearing carnivores (fox and wolf, mainly) but mammoth 'has a commanding presence' in Moravia, Poland and the central Russian plain. Clovis faunal lists, as we have seen, are most often dominated by mammoth, with bison and a few other types more rarely represented. Gravettian sites sometimes show elaborate living features 'and items of personal adornment are relatively abundant'.[58]

There are also similarities between Clovis and the Solutrean culture (also French, 21,000–17,000 years ago), notably in lithic technology, but Solutrean tools – and Eurasian tools more generally – are comparatively varied and contain forms not found in Clovis sites. However, the most intriguing comparison is with the Magdalénian, named for La Madeleine, in the Vezere Valley, again in the Dordogne part of France, 18,000–10,000 years ago, where flint is carried over long distances, few if any sites are occupied year-round, and there is almost no artwork. Large lithic assemblages are rare, there are next to no houses with hearths and 'all signs point to very mobile and transient foraging groups'.[59]

These factors may be put together to conclude that, with the possible exception of a few sites (for example, Shoop in Pennsylvania and Bull Brook in Massachusetts), there are no large Clovis sites (of twenty acres or more) with debris middens, and, again with the possible exception of Gault, Texas, no art or decorative work. Recent studies by Genevieve von Petzinger and April Nowell, at the University of Victoria in British Columbia, created a stir at a Paleoanthropology Society meeting in Chicago in April 2009, when they showed that, alongside the great cave paintings of the Ice Age, some 26 signs had been overlooked but were common at sites stretching from France to South Africa to China to Australia to North and South America. These figures may have had a spiritual significance but, at the moment, their meaning – if they had meaning (and many occur in similar clusters, suggesting that they did) – escapes us.[60]

This all tends to confirm that the founding population of the New World was very small and very mobile, that it was not repeated, on any scale, and that to begin with it became reliant on a fairly narrow range of mega-mammals – presumably because they were relatively easy to hunt – and that artwork was not developed because there was no need to establish either dedicated territories or tribal identities.[61] And/or that food was in such plentiful supply that they had no need to keep records that assisted their memory of animal habits.

• Part Two •

How Nature Differs in the Old World and the New

• 5 •

Rings of Fire and Thermal Trumpets

The fact that the New World was devoid of people when the first Americans arrived, was – obviously enough – a matter of crucial importance in accounting for differences between the two hemispheres. But it was not the only one, and perhaps not the most important. We now need to consider the phenomenon of meta-history, the ways in which the broad trajectory of human development has been and is governed by deep forces affecting the entire globe or large areas of it. We have already seen how the orbits of the Earth in relation to the sun determines Ice Ages and interglacials, and the floods which characterise them, and in turn how that has influenced the spread of people across the Earth. And, as was discussed immediately above, the spread and movement of the continents helped to determine the types of animal that evolved around the world. We now need to consider several other of these deep forces.

It has been known for some time, for example, that heat affects life in predictable and important ways. It encourages the proliferation of life forms hostile to man, in particular insects, so that there is a faster transmission of disease in hot countries. In some cases, this is so virulent as to make urban life unviable.[1] The unusual spread of temperature and even rainfall patterns mean that farmers in Europe can grow crops all year round.[2] In Spain, Portugal, Greece and southern Italy, olive trees and grapes do better than cereals and pasture pays more than agriculture, two factors that may have played a part in hampering those countries' developments as industrial nations.[3] And not until the advent of iron tools could the dense forests north of the Alps be cleared for farming land, which is one reason why northern Europe lagged behind the Mediterranean countries for thousands of years, and then overtook them.[4]

Broadly speaking, there were – and are – three important ways in which the New World differs from the Old World on such matters. Some are essentially climatological factors, some are geographical factors, and some are the biological results of climate and geography. They are interrelated, and they have influenced the broad sweep of developments in the two hemispheres, everything from food production, to systems of government, to modes of warfare and religious belief. The first and most basic was the weather.

The Origins of Weather

There are three areas of the world which for the most part govern weather patterns. These are the Pacific Ocean, the North Atlantic and the Himalayan mountain range and the associated Tibetan plateau. Their configuration not only helps shape weather world-wide but has exercised an effect on history by influencing how civilisations developed, where and when.

In a sense, the world's weather begins in the Pacific. Because of the prevailing trade winds, which blow east-to-west at equatorial latitudes, and because there is a chain of islands running from South East Asia to Australia, at the south-western corner of the Pacific, a 'pool' of warm water, many thousands of kilometres across, collects off the Philippines, where it evaporates to create great, dark nimbus clouds which produce the torrential rains known as the monsoon. Monsoon comes from an Arabic word, *mausem* (season) and traditionally refers to a season of rains moving north in the northern summer, southward in winter. In western India and Pakistan rain showers fall from June to September, sometimes until November, with millions of tropical farmers – in fact, two-thirds of the world's population of farmers – depending on this circulation.

The north Atlantic is important because, as a look at a map of the world will show, while the Pacific Ocean is effectively cut off from the Arctic region – the Bering Strait is barely sixty miles wide – the Atlantic is thousands of miles wide at its northern extremity. This means that, from time to time, glaciers break off from the polar region and float south, gradually melting. This water evaporates and is born eastward on the prevailing winds, to fall as snow on the Himalayas and the Tibetan plateau. This is the greatest mountain range on earth and there is no other body of land this high, with this extent. The effect of so much snow over such a large area, in central Asia, has the effect of consuming much of the sun's energy, because the snow has to be melted before the land can be warmed up. It is known that the monsoon strength varies on the 21,000, 43,000 and 100,000-year orbital cycles that were discussed in an earlier chapter but, more to the point, this process has been going on since the last great flood, 8,000 years ago, which released glaciers into the Atlantic in larger numbers than before, and which has had the overall effect that, over that time, the monsoon in Asia has been decreasing in abundance. As we shall see shortly, this decline in the strength of the Asian monsoon – whose effects extend to the eastern end of the Mediterranean and northern/eastern Africa – and the increasing aridity which has ensued, have determined both the rise and fall of several civilisations across the Old World. Using pollen data from western Indian lakes, Reid Bryson and A.M. Swain, from the Center for Climate Research at the University of Wisconsin at Madison, concluded in 1981 that the summer monsoon was at its peak in the early Holocene (10,000–8,000 years ago) but is reduced by as much as two-thirds now.[5]

The warm pool of water in the south-western corner of the Pacific, off the Philippines, exists where it does in normal years and is complemented by an equally large pool of cold water in the eastern Pacific, off the coast of Chile, Peru and California. The eastern Pacific is normally very cold, even close to the shore, with the result that there is next to no evaporation, and rain clouds rarely form. The Peruvian coast receives virtually no rainfall, and both Mexico's Baja Peninsula and California have years of almost total drought.[6]

This dryness had a profound effect on the development of early man in the New World, as we shall see, but, every so often, for reasons that are not entirely clear, this 'normal' state of affairs is reversed. The east-west trade winds in the Pacific weaken, or even cease entirely, and are replaced by winds in the opposite direction, from the south-west, which now generate in the Pacific what are known as Kelvin waves, giant undulations below the surface of the ocean, which drive warm water back east towards the Americas. This water flows over the cooler water and warms up the sea surface in striking fashion. Back in the western Pacific, where the waters are now cooler than usual, cloud formation is inhibited and there is drought across South East Asia and in Australia. On the Pacific coast of South America and California, on the other hand, there are severe storms – very severe storms – and a hundred years' rain can fall in a few days.[7]

This pattern was first spotted in 1892, when a Peruvian sea captain, Camillo Carrillo, published a short paper in the *Bulletin* of the Lima Geographical Society, in which he noted that the change in water flow and sea temperature disrupted the anchovy fisheries close inshore (anchovies like cold water). He further noted that the Peruvian sailors, who were aware of this periodic fluctuation, also observed that, when it occurred, it did so just after Christmas-time, and for that reason they had nicknamed the occurrence El Niño, the Christ Child. The phenomenon is now known to scientists as the El Niño Southern Oscillation, or ENSO ('oscillation' because the predominant water movement in the southern Pacific oscillates between east→west and west→east).

El Niño's effect extends far beyond the Pacific coast of the Americas. The great bulk of warm, moist air that collects over South America during an ENSO is large enough to disrupt the normal air flows across the Earth, bringing heavy rains to much of the North American west coast, keeping arctic air out of other parts of North America, which as a result enjoy unusually mild winters. ENSO also brings drought to Brazil and parts of Africa.

So the effects of an El Niño event are felt more or less everywhere on Earth. No less important, especially for the subject of this book, just as the Asian monsoon has been getting weaker over the past 8,000 years, so there is evidence that El Niño has been occurring more and more frequently over the past 6,000 years. The two aspects may be linked, in that the chain of islands off South East Asia that separates the Pacific from the Indian Ocean, does allow some water to pass through and, as sea levels have risen over the past 8,000 years, so these islands have let through relatively more water. In an ENSO event, therefore, when the western Pacific is relatively cold, more cold water is let through into the Indian Ocean and this too helps decrease the effect of the monsoon. Studies have shown that droughts occur in India only in El Niño years though not all El Niños bring drought: the link is there but the relationship is still not completely understood.

The evidence tends to show that before 5800 BP, the El Niño phenomenon was very rare. For about 3,000 years after that, they occurred, probably, only once or twice a century. Until the last quarter of the twentieth century, ENSOS occurred every seven to fifteen years and, at the present time, they happen every two to seven years. This is a remarkable change that is of central importance to our story.

Climate and Civilisation

The evidence that we have at the moment shows most clearly how the monsoon has affected the development of civilisations in Mesopotamia and, moving west to east, the Indus Valley in Pakistan/India, and in China. We also know now to some extent how El Niño has affected the development of civilisation in South America.

The flood plains of the Tigris-Euphrates Rivers are often referred to as the 'Fertile Crescent' and are the site of some of the most ancient urban cultures on Earth. The earliest of these, the Ubaid culture, emerged ~7000 BP. However, the Fertile Crescent lies just west of the Asian monsoon system and, after ~5000 BP, when studies show that the climate changed from relatively moist to drought-prone, as affected by the weakening and more variable monsoon conditions, the small farming villages of the early Ubaid culture consolidated into larger settlements.[8] There was now a need for more large-scale and integrated irrigation projects (at both local and state levels) to cope with increasingly harsh climatic conditions. Ur was one of the (earliest) cities that emerged at this time – it was thriving around 4600 BP – but its first phase was ended by an attack from the Akkadian empire at around 4340 BP. This too was a time of particularly dry climate, which may have already weakened Ur (though it received much of its wealth from taxation) *and* may have similarly affected the Akkadians, provoking their attack. We know that Ur continued in existence, as a smaller city-state, but regenerated after about 4110 BP, when the great ziggurat was built, as the temple to Nanna, the moon deity. However, Ur was attacked again in 3950 BP and never recovered. This attack was mounted by the Elam people, based around the city of Susa in what is now south-west Iran, these people themselves having migrated from further east as increasing aridity made life harder on the higher plateaus.

Around 4200 BP all of the ancient civilisations in Mesopotamia – even the Akkadian – either contracted markedly or collapsed entirely and excavations confirm that this collapse was triggered, at least partly, by drought. These studies have shown, say Peter Clift and Alan Plumb, that, at around 4200 BP, there was a marked rise in the levels of eolian dust (silt-sized dust particles, deposited on the surface of the Earth by wind, and a common way of dating sites) in the Gulf of Oman, which geochemical and mineralogical analysis shows came from the west, from Sumeria and Akkadia. Carbon-14 analysis likewise shows that this excessively dry spell, with strong dust storms and low water levels in the Dead Sea, lasted for some 300 years, accounting for the societal collapse.[9] The overall picture, therefore, is that, in this region, at around 5000 BP, city-states formed to cope with the increasing desiccation – cities being where elaborate collaborative irrigation systems were formed – but then, at ~4200 BP, the desiccation had reached a point where even city-states could not cope with the strains this imposed, as a result of which the society collapsed, and the population dispersed to survive as best it could.

A complicating factor, but still weather-related, is that northern Mesopotamia was a rain-fed society, which tended to mean that administrative areas, laid out like a patchwork quilt, developed earlier and were equal (and therefore more competitive and warlike), whereas southern Mesopotamia developed later, being irrigated by rivers and therefore was geographically more linear, the irrigation systems implying a more hierarchical (and stable) administration. This latter arrangement tended to cope better with the later desiccation.[10]

This is a pattern paralleled in Egypt and the Indus Valley, where – in the second case – the earliest human settlements on the Indian subcontinent are to be found in what is now western Pakistan and eastern Afghanistan.[11] These small-scale farming communities are known as the Mehrgarh Culture, so called after a site near modern Quetta in Pakistan. The earliest Mehrgarh remains have been dated to 9000 BP which, though they show no evidence of pottery, do appear to have been in contact with the Fertile Crescent, with the earliest farming being developed by semi-nomadic people growing wheat and barley and herding sheep, goats and cattle.[12] Judging by the development of pottery, and its elaboration, this culture was flourishing at 7500 BP but, towards 5500 BP, less and less material was being buried alongside the inhabitants, indicating a society in decline, and suggesting that the weakening of the summer monsoon was putting stress on the communities, which seem to have been largely abandoned between 4600 and 4000 BP. This decline corresponded with the rise of the Indus Valley civilisation on the flood plains of the Indus/Sarasvati Rivers to the east, with the inhabitants of the Mehrgarh culture migrating because their own homeland became more arid as the summer monsoon continued to weaken.

The Harappan (and Mohenjo-Daro) civilisation that grew out of these displacements was very advanced for its time, with a brilliantly developed water, sanitation and waste disposal system.[13] Water was obtained from wells, houses had rooms set aside for bathing, and covered drains ran down the main streets – there was at this time little apparent problem with water. It was this culture which developed the Indus script (more or less simultaneously with other scripts in Mesopotamia and Egypt), and by 4600 BP, Harappa-Mohenjo-Daro was a complex civilisation. But its brilliance, the surplus it generated, enabling technological achievement – including the plough – and its non-food-producing artists, scribes and artisans, concealed the environmental stresses with which the brilliant organisation was designed to cope and, again around 4200 BP, the civilisation underwent a profound change, with the population moving southward into a post-urban phase of smaller settlements. Studies of the sedimentation in and around Lake Lunkaransar, for example, show that, at about 4230 BP, the lake was only temporarily filled with water, indicating a 'rapid weakening' of the summer monsoon. Analysis of the plankton in the mouth of the Indus River shows that its water discharge slowed rapidly at 4200 BP, for the same, monsoon-related reason.[14]

Which brings us back to the Sarasvati River. Peter Clift and Alan Plumb remind us that this river is mentioned in the Rig Veda no fewer than seventy-two times, and was comparable to the Indus in size, but does not exist today. However, the Rig Veda, written down at about 5500 BP, describes it as being a major river and, as was alluded to earlier, nearly 2,000 of the 2,600 Harappan sites that have been discovered lie along the palaeo-channel of the Sarasvati. This major change in riverine layout may well have been related to the weakening of the monsoon.

Much the same picture emerges in China – more humid conditions in the early Holocene, followed by drawn-out drying after ~8000 BP. Stone tools found in China and dated to ~11000 BP, around the end of the Younger Dryas, show a wide variability, suggesting a transition by early man to broad-spectrum foraging and seed-processing by hunter-gatherers. This developed into agriculture, with the evolution of permanent settlements associated (at least in time) with the drying of the climate and the weakening of the monsoon. The earliest human settlements in China are found in the Yellow River Valley between the modern cities of Xian and Lanzhou, where cores drilled into the earth show swamp and river sediments that contained abundant aquatic mollusc shells, indicating wet conditions between ~8000 and ~6000 BP, followed by drying at ~6000–5000 BP.

The oldest cultures known in the Yellow River area are the Dadiwan (7800–7350 BP), followed by the Yangshao (6800–4900 BP, better known perhaps after Banpo, its best-excavated site) and the Dawenkou (6100–4600 BP). In most cases, settlements were far more numerous after 6000 BP, indicating a weakening of the monsoon and a drying of the climate which drove populations to work together in larger groupings to maximise agricultural production.[15]

The evidence is also plentiful that the Yangshao people cultivated wheat, millet and rice, using the plough, from 6000 BP. They kept domestic animals but also hunted and fished, and had specialised, highly polished stone tools. They may have cultivated the silkworm and had painted pottery. They were succeeded, around 5300 BP, by the Majiayao in the upper Yellow River region, which developed copper and bronze technology. In turn, they were followed by the Qijia culture at 4400–4000 BP and the Longshan culture on the Shandung Peninsula.[16]

Clift and Plumb observe that the bone and dental remains in these cultures show that there was an onset of a cooler and drier climate at around 5900 BP, and that this marked a decline of more egalitarian societies, such as the Yangshao, to be replaced by the more stratified, chiefdom-like society of the Longshan, who employed pottery wheels. The Longshan culture also marked a transition to city life, with earth walls and defensive moats, and rice cultivation. This, they say, was an adaptation to cope with a weaker monsoon.

Like the Qijia culture, the Longshan died out at about 4000 BP, accompanied by the disappearance of high-quality pottery, and all over the Yellow River Valley the density of settlements after 4000 BP decreased, with conditions becoming less hospitable.[17]

The picture which emerges, therefore, is fairly consistent. There was, across the Old World – or those civilisations that we know most about, in Arabia, northern Africa, in India – a slow weakening of the monsoon after ~8000 BP, when the last great flood occurred. The great glaciers of the polar arctic, melting in the north Atlantic, would have evaporated, producing increased snowfall over the Himalayas and the Tibetan plateau, which in turn consumed more and more of the sun's energy, weakening evaporation in the south-west Pacific, sapping the strength of the monsoon. This provoked people to live closer together, in urban contexts.

Then there was a further rapid shift to even drier conditions after ~4200 BP, which appears to be slightly before the Chinese cultures collapsed. This may be due to faulty dating, or may indicate that these cultures were able to hang on for a while – for as long as two- to three-hundred years – before finally succumbing. This would also fit with the emergence of the oldest known imperial dynasty to control large areas of China – the Xia dynasty – which dates from ~4000 BP. Indeed, say Clift and Plumb, maybe this is what 'dynasty' implies – the ability to cope well with a deteriorating climate with greater centralised government, spread over larger areas, more varied and therefore better able to sustain food production, after other, smaller entities have succumbed to severe droughts and floods.[18]

The El Niño Timetable

On the other side of the Pacific, in South America, there was no monsoon but there was El Niño, in some ways the other 'end' of the world's weather system. The climate of the Pacific coast is of interest to us in two ways. In normal times, in the past, before say 5800 BP, archaeological evidence shows that the El Niño phenomenon was very rare, occurring much less than it does today or in historical times.* We

* It was not totally absent, as some have suggested, at least according to David K. Keefer *et al.*, who examined the site of Quebrada Tacahuay, in Peru, which dates to 12700–12500 BP, and which contains some of the oldest evidence of maritime-based economic activity in the New World. Sediments above and below the hearth, lithic tools and processed maritime fauna, were probably generated by El Niño events.[19]

know this from the analysis of fish and shellfish remains in the middens of Peruvian and Chilean archaeological sites, which contain bones and mollusc shells of species that prefer colder waters than are brought about by El Niño events.

After ~5,800 years ago, however, two things happened. In the first place, and for almost 3,000 years, El Niños began to occur more often than before but still relatively infrequently, perhaps only once or twice a century. Second, the coastal cultures began living in large villages and building temples. These infrequent El Niño events would have brought with them widespread devastation – the plentiful anchovies in coastal waters would have disappeared, along with the seabirds and sea mammals which fed on them, tidal waves would have flooded inshore, terrible winds would have devastated buildings and trees. With such catastrophes occurring, say, every other generation, or every three generations, they would be fresh in people's memories and very much part of the folklore and myths of these early peoples. Is that why they huddled together in villages at this time, and constructed temples, so they could appeal to their gods to protect them from these devastations? (More will be made of this later on.)

This situation persisted for nearly 3,000 years. Then, according to the mollusc and other evidence, at about 3,000 years ago there was another change. Now the El Niños started to occur even more frequently, closer to the modern rate. This rate is itself interesting. Currently, El Niño events occur every three to seven years but in the earlier part of the twentieth century they occurred every seven to fifteen years. The question naturally arises therefore as to whether El Niño has been getting progressively more common as the years have passed since 3,000 years ago, and if so why? Is there a link here between the weakening monsoon over the past 8,000 years and the increase in frequency of El Niño? Or is that the wrong way round? We know that there is a link – albeit imperfectly understood – between El Niño events and drought in India.[20]

We shall come back to this matter time and again but for now the important point is that, at 3,000 years ago, when El Niño started to become much more common, the first civilisations of Peru collapsed and it was several centuries before urban structures reappeared, and even then under very different political systems.

Here, therefore, we see both similarities and differences between the

Old World and the New. Urban cultures – primitive civilisations – appeared and disappeared in both hemispheres, most likely as a result of grand-scale changes in the weather, but the climatic phenomena were different and the time-scale, though it may have been linked, was also different. In the Old World the weakening monsoon at first caused communities to form, to develop irrigation techniques to cope with desiccation. But then, later, desiccation got so bad that not even city-states could cope with the situation. In the New World, the more frequent onset of ENSO events was an important factor in causing villages and temples to be built, to deal with the phenomenon by communal worship, but then got so frequent that the villages and temples could not cope with *that* situation. The perturbation became so common that people concluded it was not worth the trouble to rebuild after an ENSO-caused devastation. This divergence will be echoed in later developments.*

The Un-pacific Pacific

There were two other geographical factors that affected the New World civilisations disproportionately. As is now well known, the Earth's surface consists of a number of tectonic plates, vast slabs on which the continents gradually slide over the mantle below the crust at a rate of

* These ideas are supported, albeit indirectly, by the seminal work of Paul Wheatley (1921–1999), chairman of the Committee on Social Thought at the University of Chicago, and chairman also of the Centre for South East Asian Studies at the University of California, Berkeley. In his book, *The Pivot of the Four Quarters* (1971), Professor Wheatley was at pains to point out that, in antiquity, urbanisation was not what it is today. Across the seven civilisations that he studied – Mesopotamia, Egypt, the Indus Valley, the north China plain, Mesoamerica, the central Andes and the Yoruba territories of south-west Nigeria – the first cities were invariably *religious ceremonial centres* before, and sometimes long before, they were anything else. They were established only after an array of geomantic considerations had been satisfied and were laid out as *axes mundi*, projecting symbols of cosmic order and focused sacredness 'on to the plane of human experience, where they could provide a framework for social action.'[21]

Cities began, in effect, he maintained, as tribal shrines, with the priests in Sumeria being probably 'the first persons to be released from the stultifying routine of direct subsistence labour'.[22] Craftsmen were very few until 3500 BC, and writing and early calendrical systems were used to preserve elite cohesion but proved brittle as systems of government, easily susceptible to threats from outside – either by other groups or climate change.

a few centimetres a year. Where these plates meet we find almost all the Earth's volcanoes and this is also where the majority of earthquakes occur, as one plate jostles another. About 75 per cent of the Earth's seismic energy is released along the boundaries of the Pacific Plate and another 23 per cent comes from a zone extending eastwards from the Mediterranean. The rest of the world accounts for just 2 per cent.[23] Map 8 shows the layout of these plates, plus the range of the active volcanoes and the spread of recent important earthquakes.

Collectively, this shows that, for the most part, the active volcanoes line the edge of the Pacific (the so-called 'Ring of Fire'), and this does not include underwater volcanoes, which are also disproportionately common around the Pacific rim. It is known that earthquakes and volcanic activity are related to tidal activity, in itself related to the gravitational effects exerted by the moon, and that they occur on a roughly regular cycle of just over four years, occurring mainly in (the northern) winter.[24] There is also a relationship between volcanic activity and El Niño, in that underwater volcanoes, when they erupt, release heat into the ocean in massive amounts and this can trigger an ENSO event. As an indication of the level of activity, there were 56 confirmed underwater eruptions in 2001–2002.

We can be more specific than this, however. In 1981 the Smithsonian Institution in Washington DC published a directory, gazetteer and chronology of volcanism during the last 10,000 years, under the lead editorship of Tom Simkin. This identified 5,564 eruptions from 1,343 volcanoes all over the world.[25] Volcano lists were not new. The first volume of the Catalog of Active Volcanoes of the World (CAVW) was published in 1951 and updated via the *Bulletin of Volcano Eruptions* since 1960, but the Smithsonian publication had 400 more eruptions than any previous publication. That chronology tells us that in the years 1400–1500 there were only three recorded eruptions in Meso- and South America, two in Guatemala and one in Peru. In the years 1500–1600, on the other hand, 139 eruptions are reported for exactly the same area. Geological phenomena being the slow-moving entities that they are, this startling disparity cannot reflect a real sudden upsurge in seismic activity of some 4,600 per cent. Instead, the arrival of the Europeans must account for the greater *reporting* of erruptions. This is reinforced by the records in Italy. That country does not have an especially large number of volcanoes (18 compared with 35 in Mexico and 57 in north Chile and Bolivia) yet it has by far the greatest number of reported eruptions, pre-1492.

Beyond the Ring of Fire, two areas of less dramatic but still extensive activity are the Rift Valley of north-east Africa and the north and east coasts of the Mediterranean. Central and northern Europe, Russia and mainland Asia (China and India), together with Australia and the eastern parts of the Americas are relatively free of earthquake activity.

A closer look at the so-called Ring of Fire shows further that the volcanoes of the western rim lie not along the Asian mainland but on the peninsulas and offshore islands: Kamchatka, Japan, the Philippines, Indonesia, Sumatra, the Sundas and New Guinea. This means that the only areas where volcanoes occur on continental mainlands, and where early civilisations formed, are around the north and east Mediterranean and in Meso- and South America.

If we accept the slow pace of geological change, and that records for Meso- and South America were fuller – and therefore more accurate and reliable – after the Europeans arrived, then we may say that the sixteenth-century figure – 139 eruptions – was more typical. This translates into one eruption every thirty-seven weeks. Put another way, Popacatépetl (whose name in Nahuatl means 'Smoking Mountain'), which is 43 miles south-east of where Tenochtitlán, the Aztec capital was, and was easily visible, erupted in 1519, in 1521 and 1523, while Santa Ana in El Salvador erupted in 1520 and 1524, and volcanoes in Nicaragua erupted in 1523 and 1524. In fact, it was not until 1554 that a year passed *without* a volcanic eruption in Meso- or South America. Only Italy could keep pace but that was thanks to one volcano, Etna, which erupted 39 times between 1500 and 1541 but then quietened down for nearly a decade.

Undoubtedly, then, what we now call Latin America, at the time of the Conquest, had – and had had, in the previous centuries – far more active volcanoes than anywhere else on the world's mainland continents (as opposed to offshore islands) where high civilisations developed. It may also be worth pointing out that earthquakes and eruptions are more serious for sedentary peoples, who can't move away. South America, with its many small, isolated language groups, dependent on potatoes, manioc and maize, would have been especially vulnerable to quakes and eruptions.

Even extinct volcanoes can have dangers, collapsing inwards and causing landslides which, near the sea, can provoke tsunamis. Evidence has also emerged recently that, at the end of the Ice Age, thinning glaciers on volcanoes could destabilise vast chunks of their summit cones, triggering 'mega-landslides'. One is now known to have occurred around 11,000 years ago on Planchón-Peteroa, a glaciated volcano in Chile, when around one-third of the volcanic cone collapsed, according to Daniel Tormey, of ENTRIX, an environmental consultancy based in Los Angeles. On this occasion ten billion cubic metres of rock crashed down and smothered 370 square kilometres of land, travelling 95 kilometres in total.

The point is well made, therefore, that, overall, Latin America is and has been *by far* the most volcanically active mainland area of the world where ancient civilisations have existed. For human beings in antiquity, these events were mysterious and potentially very dangerous and fearsome, obscuring the sun and bringing darkness, showering wide areas with hot ash, triggering earthquakes, sending vast plumes of hot molten lava over wide areas, devastating the landscape, destroying crops and houses and at times killing many people.

Volcanoes are not all bad. Eruptions contain potassium and phosphorus, needed by plants, and the weathering of volcanic rocks releases other nutrients.[26] Obsidian is a volcanic glass valued in many early cultures, partly for its properties – it is shiny and produces very sharp blades – but also for the fact that it is associated with the mysterious and threatening behaviour of volcanoes. Yet it is not hard to see how such phenomena could affect the beliefs and behaviour of people living nearby. (In February 1943 a Mexican farmer, Dionisius Palido, watched the birth of a volcano in his cornfield, as a slight depression rose thirty feet *in one day*, 550 feet in a week, and over a thousand feet in a year. Inside nine years it grew to 6,800 feet and its lava flow destroyed several towns.)[27]

Ancient peoples would have been hyper-conscious, anxious even, about living in a potentially hostile landscape, where the gods were never quiet for very long, where they seemed to be often angry, insofar as volcanic activity and its associated phenomena, such as earthquakes, seemed like punishment, where material was periodically thrown up from the underworld by enormous unknown forces, or the ground shook and trembled without warning. Today there are between 200 and 300 active volcanoes in the Andes. In the roughly 500 eruptions since 1532 an estimated 25,000 people have been killed. Both Mexico and Peru are prone to earthquakes, with Arequipa in Peru averaging one quake per century and Mexico's southern coast being most at risk: in 1985 a quake off Acapulco killed more than 4,000 people in Mexico City, some 300 kilometres away. Popacatépetl showed renewed activity in 1995 and 1996.

These phenomena comprised a central ideological/psychological predicament in pre-Columbia Latin America, the consequences of which are explored throughout the rest of the book.

Skies on Fire

The second climatological factor that affected disproportionately the areas of the New World where civilisations developed was the hurricane. These violent windstorms, which may precipitate dense rain, can comprise cyclones with diameters extending anywhere from 50 to 1,000 miles and generate winds of from 80 to 130 miles per hour. The very name, hurricane – which occur in the east Pacific, the north Atlantic, the Caribbean Sea and the Gulf of Mexico – is derived from *huracán*, a Tain or Carib god, and/or from *hunraken*, the Mayan storm god. According to Taino legend, Guacar was one of two sons of the creator god who was jealous of his brother's success in creating plants and animals, and so changed his name to Juracan and became the evil god of destruction.[28]

Early post-Conquest accounts described how much native Indians feared these winds, and the associated rains, in which lightning flashes came so quickly 'the sky seemed to be completely full of fire' and the next minute 'a thick and dreadful darkness descended'. The 'strong and frightful' wind ripped large trees out of the ground and collapsed cliffs.[29]

Map 9 shows that hurricanes originate over open water, occur only in the tropics and subtropics, and mainly die out once they reach land. The greatest wind speeds are achieved in Mesoamerica, north Australia and south India. Mesoamerica, being a narrow isthmus, suffers more than most *and* gets hit from both sides (as do Australia and India but they are much bigger landmasses). In some areas of Mesoamerica rainforest hurricanes are so common that the trees have adapted in an evolutionary sense: they snap off their crowns at about thirty feet above the ground. Once the storm has died down, the trees begin to re-sprout from the remaining part of the trunk.[30]

A final factor on the subject of winds is that North America is shaped, as Timothy Flannery puts it, like 'a great thermal trumpet'. What he means by this is that North America 'is a unique place and that one of its main distinctive features is its climate', determined by its shape, 'a great inverted wedge' with a 6,500-kilometre-wide base deep in the sub-Arctic, narrowing to the south until it is just a peninsula sixty kilometres wide, eight degrees north of the equator in a narrow isthmus abutting South America. On its eastern side the wedge is limited by the Appalachian Mountains, on the west by the Rockies. To this unique configuration may be added the fact that land cools and heats more rapidly than water, meaning that temperatures there are more varied than at sea. In winter, super-chilled air that forms over the continent's northern expanses surges south, funnelled by the north-south mountain ranges. In the summer a huge pool of air that is warmed over the Gulf of Mexico surges in the opposite direction 'bringing the tropics to the far north'. This has had a 'prodigious' effect on North America, Flannery says, and means that temperatures in the continent can vary incredibly over a brief period and where the turbulent cool air from the north encounters the breezes of the hot south tornadoes are spawned. 'Ninety per cent of the world's tornadoes occur in North America – most originating between the Rockies and the Mississippi River.' This has two effects that concern us. One is that the 'climatic trumpet' adds to the violence and unpredictability of the winds abutting the Gulf of Mexico: coming on top of the volcanoes, hurricanes and El Niño phenomenon, this makes Mesoamerica one of the most challenging areas in the world, climatologically. Second, the very varied conditions of North America favour the evolution of cactuses. Cactuses are unique to the New World and we shall see in a later chapter what the significance of this is.[31]

Then there is the configuration of the Pacific Ocean itself. As map 8 confirms, the Pacific (tectonic) Plate is nearly entirely covered by the Pacific Ocean. While the plate itself is only a little more than six miles thick, it has an average of 2.5 miles average depth of water resting on it. As the sun and moon appear to rise in the east, large volumes of water are pulled by gravitation against the American coast. As the sun and moon appear to set in the west, large amounts of water are likewise pulled against the south Pacific islands and Asian countries, this regular movement of a massive body of water causing a repetitious pulsing and stretching of the Pacific Plate. More than that, the western Pacific Plate is being forced under the Asian continent, and the Nazca Plate (in the eastern Pacific) is being forced under the South American continent by these daily movements. The Nazca and Pacific Plates are in fact the fastest-spreading sea floors on Earth.

This area is thus the most unstable configuration anywhere in the world, with more seismic and volcanic activity than any other mainland area.

We may conclude this section, therefore, by emphasising that the area where the New World civilisations developed occupied the region in which, to a 'primitive' mind, the gods were more violent, more destructive, far *angrier* than anywhere else on Earth. Although the early inhabitants of South and Central America were in no position to make this comparison, their experiences of natural calamities and disasters were quite enough to colour their religion and give it a unique stamp, as we shall see in due course.

Monsoon and Religion

There is more to say about climate and the gods around the world. For instance, worshipping gods to bring rain, to sustain crops and livestock, is by no means unique to Asia, but it is only in Asia (and one small region of the Euphrates) that the rains and the great rivers have *become* gods.[32] This is especially true now of the Ganges, a monsoon-fed river which the Hindus consider to be holy, a place of pilgrimage, where their ashes are scattered. The Ganges is personified by the Goddess Ganga, a beautiful woman often shown with a fish's tail instead of legs, and riding the Makara, a crocodile-like water monster (a folk memory, perhaps, of the plague of crocodiles referred to in an earlier chapter).[33] There are several legends about the origin of Ganga, including one that she was the daughter of Himavan, king of the mountains. In general the great monsoon-fed rivers (which, of course, have their origins in the mountains) are regarded as purifying streams for the souls of the departed. The greatest religious ceremony on the Ganges is made during the Kumbh Mela, a ritual bath held every three years or so and attended by between 50 and 70 million people each time, even today.

Another river goddess is Sarasvati, also mentioned in a previous chapter. She was and is regarded as the goddess of learning, education, arts and skill. She is the consort of Brahma, the source of all knowledge, her name meaning 'the one who flows'.[34]

The Indian year is divided into the cold months, the hot months and the rainy months, Hindus using the monsoon (the *caturmasa* or 'four months') to explain various aspects of their belief-system. For example, the monsoon season is associated with Vishnu's Sleep (Vishnu, from the verb '*Vis*', 'to pervade', being the Hindus' supreme deity). During the monsoon, Vishnu is believed to have retired below the ocean for a four months' sleep, so that the *caturmasa* is a time when the Earth is without its protector and at the mercy of demonic powers. This legend clearly reflects the fact that the monsoon represents a threat as well as a life-giving opportunity for so many farmers. For that reason, the end of the monsoon season is also celebrated (the *Rama-Lilas*).

This may not be all there is to the link between climate and religion. The greater variability of the monsoon after ~8000 BP may have meant there was more need for worship. And various authors have suggested that the (rain)forested areas, which have evolved under the monsoon system, where life is everywhere and abundant, should favour polytheism. 'Tropical forests teem with life and the cycle of birth, life and death are endlessly replayed.' This contrasts with the monotheisms of Judaism, Christianity and Islam which evolved in the deserts of the Middle East. In the wilderness of the desert, where life struggles to survive, it is only logical for a deity to create life out of nothing. In the tropical forest, humans form a small part of a teeming world, giving a deeper *respect* for nature, while in the desert the *mastery* over nature is what counts. (In *Genesis*, in the bible, God gives man 'dominion' over the animals.)[35]

If there is something to this, it may be that the gradual drying of the Eurasian climate since ~8000 BP has also had religious consequences, in that there has been a long-term shift from shamanism to polytheism to monotheism as forests receded. This deforestation was especially true of the Mediterranean region, say Clift and Plumb, where, to begin with, the rain gods or storm gods were the pre-eminent deities. Hadad and Ba'al were important Semitic storm gods, related to the Akkadian god, Adad. The Anatolian storm god Teshub, the Egyptian god Set, the Greek god Zeus and the Roman god Jupiter were all important sky gods and/or rain/storm gods. The progression from an all-important sky god to monotheism seems a shorter route than from the full-blown polytheism of, say, Hinduism, though even there Varuna is lord of the sky, ocean and rain, both all-powerful and, should he so wish, destructive.[36] This confused trajectory will be clarified somewhat later on.

East-West is Best

Besides the very great importance of weather, there were several other systematic differences between the Old World and the New that had a profound effect on the development of their inhabitants. Some were geographical, others biological.

In order to make a start, even if it means jumping ahead of ourselves, let us first consider other features displayed in the maps. These show first of all, as José de Acosta, Wilhelm Hegel and, most recently and in most detail, Jared Diamond have pointed out, that the Americas are essentially a north-south landmass, as is true, though to a lesser extent, of Africa, whereas Eurasia in marked contrast is essentially east-west. Now, jumping even further ahead of ourselves, let us examine the geographical spread of the major civilisations that developed across the world before modern times, before 1492. Consulting the maps, the following observations stand out:

The spread of these major civilisations was far from random. The impressive civilisations of the Old World all extended, roughly speaking, between latitudes seven degrees north and fifty degrees north, both tropical and temperate zones. Against that, the main ancient American civilisations – Chavín, Moche, Olmec, Maya, Toltec, Inca, Aztec and so on – all occurred between eighteen degrees south and twenty-five degrees north. The overall north-south spread of early civilisations is pretty much the same in both hemispheres – about forty-three degrees of latitude in each case – but the American civilisations are *entirely within* the tropics. We shall see what this implies shortly. Furthermore, the overall extent of the Old World civilisations is much, much greater than those in the New World – 8,500 miles from east to west, and 3,000 miles north to south, or 25.5 million square miles, compared with 1,250 miles east-west and 3,000 miles north-south in the Americas, or just 3.75 million square miles. The sheer size of Eurasia gave it an inbuilt 'advantage' over the New World.

When we use the word 'civilisation' in this particular context, we are using it to refer not so much to the emergence of great public buildings, monumental arts, or advanced irrigation systems so much as the fact that these were, essentially, the areas of primary food production, which arose independently across the world. Food production soon spread but these are the areas where domestication, of plants and animals, was independently conceived. The main spreads of food production were: from South West Asia to Europe, Egypt and North Africa and to Central Asia and the Indus Valley; from the Sahel and West Africa to East and Southern Africa; from China to tropical South East Asia, Indonesia, Korea and Japan; and from Mesoamerica to North America.[37]

This spread around the world was not, however, uniform. To put the matter plainly, north-south spread is far more difficult, far slower, than east-west spread. This is because, obviously enough, as you travel north to south, or vice versa, there is a much greater difference in the weather, in mean temperatures, in hours of daylight, and soil conditions, as compared with travelling east-west. The most conspicuous failures in this regard, according to Jared Diamond, are: the failure of both farming and herding to reach Native North American California from the US south-west; the failure of farming and herding to reach Australia from New Guinea and Indonesia; the failure of farming to spread from South Africa's Natal Province to the Cape.

These are by no means the only examples. There is no shortage of evidence showing that, because of its basic north-south configuration, there have been greater difficulties as regards both agricultural and cultural diffusion in the New World. For example, because the llama, guinea pig and potato of the Andean highlands never reached the Mexican highlands, Mesoamerica and North America went without domesticated mammals save for dogs. The domestic turkey of Mesoamerica never reached South America, or the eastern United States. While alphabets of Middle Eastern origin eventually spread out across the Old World, as far as Indonesia, the writing systems of Mesoamerica never reached the Andes. Most important of all, perhaps, 'The wheels invented in Mesoamerica as parts of toys never met the llamas domesticated in the Andes.'[38] The Romans grew peach and citrus fruits from China, cucumbers and sesame from India, hemp and onions from Central Asia, while in the New World the sunflowers of North America never spread to the Andes.[39]

Other examples include the habit of tobacco smoking, which was first developed in Mexico and carried across the Mississippi and the Appalachians in the first millennium AD (and then on of course to Europe), but never spread back to Peru, whose inhabitants were still using tobacco as snuff in 1532. The invention of hieroglyphic writing and numerals (including the zero) and a marvellously complex and precise calendar conceived by the Maya priesthood, had still not reached the Peruvian empire a thousand years later.[40]

Jared Diamond and his colleagues have gone so far as to calculate the rate of spread of some of the ancient food production techniques in different areas of the globe. From south-west Asia west to Europe and east to the Indus Valley, domesticated crops spread at the rate of about 0.7 miles a year. From the Philippines east to Polynesia the spread appears to have been much faster, at 3.2 miles a year (perhaps because so much of the distance was open sea). The earliest dates for cultivated maize in the New World are in Mesoamerica and in the Andes and Amazonia at ~5500 BP. Maize travelled to the south-western United States at less than 0.5 miles a year, while the llama spread from Peru to Ecuador at an even slower 0.2 miles a year.[41]

The profusion of so many separate languages in the Americas supports the same general picture. In 1492, the New World peoples were speaking an estimated 2,000 languages, grouped into perhaps thirty families. This very great diversity (there are approximately 6,000 languages in the world today), which must have evolved in a mere 15,000 years from a restricted racial stock that populated the New World from somewhere in eastern Asia, reflects the fact that there was not, in the development of New World societies, as there were in the Old World, great 'linguistic expansions', as we know occurred with the Aryan and Semitic, the Bantu and the Chinese languages, which in the course of time extinguished many smaller, less powerful language families in the Old World. We shall see why this happened in a later chapter, and why it didn't happen in the New World, and that it is consistent with other developments.[42]

The very great language diversity of the New World in itself supports the idea of many small, relatively isolated societies, presumably each with a restricted geographical terrain. It also implies relatively little history of warfare and imperial-style campaigns of conquest which would have changed this linguistic mosaic (though the Incas are an exception). In a remarkably short time, these isolated societies will have developed different dialects, then different, mutually incomprehensible tongues, further reinforcing their isolation.

A second observation that we may make from the maps is that the major Old World civilisations – the Assyrians, the Romans, Mohenjo-Daro, the Gupta, the Han and so on – all developed along the great 'East-West Corridor', a geographical configuration that takes in, west-to-east, the Mediterranean Sea, the Suez-Red Sea Isthmus, the Rivers Tigris and Euphrates in Mesopotamia, the coastline of the Arabian Gulf, of India, Burma and Thailand, and the island chain of South East Asia, all the way round to China. No great daring or stamina would have been needed to navigate these coasts, rivers and isthmuses: the 'Corridor' in effect forms a continuous and not inconvenient travel link between the Strait of Gibraltar (latitude 36° N) with the Strait of Singapore (latitude 1° N) and beyond, a distance of just over 10,000 miles. The prevailing winds would have aided sea travel as it developed and there was no shortage of great rivers draining into this corridor. It was a natural asset aiding trade and contact and had no real equivalent in the New World. The layout of the East-West Corridor matches both the pattern of the spread of sailing technology and the distribution of certain myths – these are shown on maps 1 and 4.

The great steppes of Eurasia, where the horse was first domesticated, was also an east-west configuration. This has had a profound influence on world history, far more than is generally realised. The dimensions of the steppes of the world also feature on map 4.

We need to make special mention of the Mediterranean Sea. This body of water, besides being totally surrounded by land, being east-west in configuration and located in temperate latitudes, has a number of other features which mark it as one of the most favoured spots on Earth, so far as the development of humankind is concerned. It consists, for example, of many peninsulas, bays and inlets, ecological niches which vary biologically and give the inhabitants a strong sense of place, encouraging competition and exchange. It contains a large number of islands, some of them quite large islands. Being geographically distinct, and distinctive, these islands developed strong local cultural traditions – the obsidian of Melos, the alum of the Lipari Islands, the sulphur of Sicily, the healing earth of Lemos.[43] As Peregrine Horden and Nicholas Purcell say, this encouraged 'connectivity', an important quality, which helped account for the pre-eminence of the Mediterranean.

At a time when sea travel was faster and much cheaper than overland travel, people accompanied their goods being traded, with the result that islands in antiquity had much bigger populations than they do now, relatively speaking: which meant that, with so many islands, the Mediterranean was relatively rich in people.[44] Connectivity was also aided by the fact that the sea was bordered by so many mountains that much of the Mediterranean was within sight of land.

Other advantages of the Mediterranean were that it produced so many *non-perishable* foodstuffs (cereals, cheese, oil, wine) that could be traded, that the absence of large tides (because of the narrowness of the Strait of Gibraltar) meant that relatively more salt was produced in the Mediterranean area, helping to preserve meat products, and that the many rivers draining into the sea provided, in their estuaries, plentiful alluvial wetlands where crops could be grown.[45]

Deltas and Diseases

A still further observation is that at least four of the Old World river systems seem to have helped to spawn great civilisations: Nile → Pharaonic Egypt; Tigris/Euphrates → Assyria/Babylon; Indus/Sarasvati → Mohenjo-Daro/Harappa; Yangtse → Han. In Egypt, for instance, the Nile flows from south to north but the prevailing wind blows from north to south, facilitating two-way traffic on the river. And again, in marked contrast, none of the major New World rivers – the St Lawrence, the Mississippi, the Magdalena, the Orinoco, the Amazon, the Paraguay or the Plata – was associated with major ancient civilisations. The inhabitants of these regions did of course build villages and in one case, Cahokia, on the Mississippi, began to embark on monumental architecture; but more generally the great rivers of the

New World were not associated with great civilisations. The significance of this extraordinary difference may have something to do with the fact that, between 9,500 and 8,000 years ago, river deltas formed throughout the world as a result of a rise in sea levels over the continental shelves. This rise reduced the gradients of rivers, meaning that water flow towards the sea slowed, and rivers naturally meandered more, or silted up, causing occasionally catastrophically abrupt channel shifts. Despite the risks of such abrupt channel shifts, these deltas nonetheless formed the basis of fertile alluvial plains, a process that occurred in Mesopotamia, the Ganges in India, the Nile in Egypt, the Chao Phraya in Thailand, the Mahakam in Borneo and the Yangtze in China. As was mentioned earlier, over forty of these deltas have been dated throughout the world.[46] These deltas, or flood plains, created conditions that favoured the kind of plants that tended to grow in the Old World but not those more common in the New World. This configuration had a profound effect on the development of agriculture and of civilisation.

These geographical/climatic configurations are basic. As we shall see in more detail as the book proceeds, the great East-West Corridor facilitated the transmission of peoples, goods, diseases and ideas between the various regions of the Old World; the cold deep water off the Pacific coast of the New World attracted a unique mix of fish and marine mammals, a world of abundance, which in turn promoted a unique developmental trajectory for early peoples. We must be careful not to oversimplify a complex picture but the different configurations of water in the two hemispheres did ultimately affect everything from agriculture to religion. The development of deltas – alluvial floodplains – promoted civilisations on the coasts or near seas and oceans, facilitating the transmission of trade, humans and ideas.

• 6 •

Roots *v.* Seeds and the Anomalous Distribution of Domesticable Mammals

These fundamental geographical differences were built on by two other factors. There was, first, a substantial difference between the Old World and the New in regard to plant life. There developed, at the most basic level, two radically different types of agriculture across the globe, each with a notably separate distribution. These two types are seed culture and vegeculture.

As its name implies, seed culture refers to crops which reproduce sexually, by means of seeds – in particular, grasses/cereals such as wheat, oats, barley, rye, millet, rice and maize and which, with the exception of the last mentioned, *were native only in the Old World*. Vegeculture, on the other hand, refers to crops such as roots and tubers, manioc, potatoes and yams which reproduce vegetatively. Roots and tubers are not confined to the New World but the Americas *are* home to many varieties of root and tuber crops, not only yams, canna, manihot and sweet potatoes, but also crops adapted to cool mountainous areas, such as potato, oca, ulluco, arracacha and añu. Map 3 shows the markedly different distribution of these two types of crop around the world.

From all this it will be clear that seed culture occurs mainly in the drier tropics, subtropics and temperate zones, whereas vegeculture occurs entirely within the tropics. (Roots and tubers were of negligible importance in the Middle East and China.)[1]

These two types of plant life – cereals and roots/tubers – also give rise to two types of cultivation, the so-called *milpa* cultivation and *conuco* cultivation. Milpa cultivation refers to seeds and is associated with swidden agriculture – that is, where forests are slashed and then burned to create cleared areas – fields – in which the seeds may be planted. Conuco refers to root crop cultivation, often involving the preparation of earth mounds (*montones*) in which stem cuttings or other vegetative parts are planted. Although the conuco system may employ swidden techniques, many plots of tubers and roots are maintained for years on end in one place.

Besides their very different distribution across the world, and their different modes of cultivation, there are three other all-important differences between roots and tubers on the one hand, and seed vegetables on the other. First, because roots and tubers grow underground (and are tropical) other vegetation grows above them to form a canopy. This means that the soil in which they grow is far less likely to be eroded (by wind, say) – such soil is, therefore, more likely to remain richer in nutrients. At the same time, the natural growing conditions of roots and tubers mean they can be harvested continuously throughout the year, not all at once like cereals, and that too keeps the soil richer, since there is no annual harvest cycle to denude the ground of its fertility all at once. As we shall see later, this different rhythm also has ceremonial/religious implications.

The second difference between tubers/roots and cereals is that the former, being almost exclusively tropical plants, are adapted to an ecological system in which there are well-marked wet and dry seasons, requiring the plant to store up sufficient reserves for it to survive the dry months. This means that roots and tubers build up starch to tide them over these times and in fact they are mainly eaten for their starch content, whereas cereals are much richer in protein. The various forms of wheat and barley, for example, consist of 8–14 per cent protein, pulses contain 20–25 per cent, whereas rice and corn and tubers/roots are much lower (taro, for example, contains just 1 per cent protein). It follows that people living off roots and tubers have to go elsewhere for their protein – mainly by hunting animals. Some scholars have even attributed cannibalism to the need of certain tribes to find protein where they can. (This process is known as 'protein capture'.)[2]

The consequences of all this are profound. We should not forget that in some areas of the world – California, Australia, the Argentine pampas, western Europe – plant food production never developed indigenously *at all*.[3] Where it did occur, roots and tubers were, essentially, a much more stable crop system than cereals. The ground is not exhausted anywhere near as quickly as it is under cereal agriculture, which required early cereal cultivators to shift to new terrain every year until earlier fields, lying fallow, had regained their fertility. All this meant that seed cultures (both large seeds like wheat and rice and small seeds like millet) were much more likely to expand into new areas, in contrast to root cultures which remained more or less static. And in fact, in some areas of the world – for example, South East Asia and the West Indies – seed agriculture did spread and take over from vegeculture.[4] Vegecultures, on the other hand, tended to settle – and then remain – near river banks, seashores or the edge of savannah areas, where early peoples could supplement their starchy roots with hunting expeditions (or fishing) for protein. It also meant that fertility, or at least the fertility *cycle*, was much less of an issue in root cultures, and so their religions reflected this principle far less.

A further aspect of early plant life is the way in which agriculture tended to develop not just with one plant but with 'packages' of a small number of plants. These 'packages' are known as 'founder crops'. In the Middle East, for example, the founder crops comprised wheat along with barley, peas and lentils. Agriculture developed in the so-called Fertile Crescent (parts of Palestine, Jordan, Israel, Turkey and Iraq) because its latitude, Mediterranean climate – the largest of its kind in the world – and its mountains meant that harvests of wild grasses could be staggered throughout the autumn at different heights: people could move up and down the slopes as the plants became ripe. This package came about because many cereals, while they do contain protein, are not especially rich in it, but that deficiency is made up for by pulses (peas and lentils) which, as we have seen, can be up to 25 per cent protein (flax was also domesticated as a founder crop, acting – with raffia in Africa and cotton in India – as a source of fibre). As, again, Jared Diamond has pointed out, cereals and pulses together, therefore, 'provide many of the ingredients of a balanced diet'.[5] In the same way, the 'package' that developed in China consisted of rice, millet and soybeans (the latter with 38 per cent protein), with hemp for fibre; and in Mesoamerica, the package was corn, beans and squash, with cotton, yucca and agave for fibre.

Three other crucial differences between cereals and roots/tubers should be noted. One is that cereals grow quickly and, because they quickly denude the land of its fertility, this is another reason why they spread rapidly. The earliest dates known for cultivated cereals is around

10,500 years ago, in the Fertile Crescent in South West Asia. Thus the crops and animals of the Fertile Crescent came to meet early man's basic needs surprisingly well – carbohydrate, protein, fat, clothing, traction and transport.[6] They spread west to Greece by 8500 BP and to Germany around 7000 BP. Dates for China are at about 9500 BP and the Indus Valley at 9000 BP, with Egypt at 8000 BP. Contrast that with the founder crops in the eastern US, where four plants were domesticated between 4500–3500 BP, a full 6,000 years after wheat and barley were domesticated in the Fertile Crescent. But they were not enough to serve as a complete diet and Native Americans needed to supplement these founder crops by wild foods, mainly wild mammals and water birds. Farming did not become their main activity until 2500–2200 BP, when more seed crops – knotweed and maygrass – were added to their diet, though these were at least high in protein – 17–32 per cent, compared with 8–14 per cent for wheat, 9 per cent for corn and even lower for barley and rice.[7] There was nothing wrong with the general environment of the eastern United States so far as growing crops was concerned – the soils were rich and the rainfall reliable. The florescence of culture that developed there did so, however, only at about 200 BC–AD 400, 9,000 years after village life first emerged in the Fertile Crescent and it was not until AD 900 that the Mexican crop trinity (maize, beans, squash) triggered enough surplus to create the towns of the Mississippian florescence, as we shall see in more detail later. The Native Americans of the eastern United States never domesticated local wild pulses, fibre crops or fruit or nut trees.

There were a number of reasons for this delay in New World domestication. The trajectory of maize was one factor, which is explored in chapter ten. On top of that, a further series of interlocking factors hastened the spread and efficiency of cereals. In the new, drier conditions on Earth that we have been discussing, cereals grew naturally and well where surface water periodically inundated the ground – i.e., the flood plains of river basins (often near or in the relatively new deltas). Roots and tubers, on the other hand, dislike waterlogged ground. This may be a more important difference than it looks. Cereals could grow on the flood plain/deltas, meaning that early peoples in the Old World often developed their societies at the mouths of rivers – near the seas and oceans, which facilitated contact with others. Root and tuber peoples, on the other hand, established their societies *above* the flood plains, because roots and tubers cannot grow in waterlogged soil. So these societies tended to be confined to the higher river valleys (away from seas and oceans) where contact with other peoples was naturally more limited by the terrain. This isolation, as was mentioned earlier, further encouraged the emergence of dialects, and then mutually incomprehensible languages, in a vicious circle. In the Old World the agricultural developments encouraged movement and contact – and therefore competition – whereas in the New World early agriculture, much of it anyway, discouraged contact.

Second, the rewards of cereal cultivation are greater than that for any other form of agriculture.[8] The gathering and threshing of grain is less laborious than the digging of roots and, for any given volume of plant, the food value is greater. Emmer wheat and barley have big seeds and can be easily harvested. Also, because annual plants (which is what cereals are) put little energy into making woody or fibrous stems (because they don't need to last through the winter), they put all their effort into producing large seeds. It has been calculated, therefore, that cereals can produce a ton of seeds per hectare, yielding 50 kilocalories for every one kilocalorie of work. Another, and perhaps more interesting reason, was that the *organisation* needed to harvest cereals, which matured only once a year, was much greater than with roots and tubers. This would have encouraged communal life and had religious consequences. Perhaps most important of all, cereals can be stored easily and remain in a near-perfect state for months on end. Surpluses may therefore be more easily accumulated, in the first place to protect increasingly dense populations against crop failure, and then to be used as the basis of exchange/trade. Surpluses are also eventually necessary for supporting non-food-producing specialists, such as scribes, priests, kings, artists, which are such a feature of civilisations. Roots and tubers are dug up as needed – under such circumstances the concept of 'surplus' is harder to imagine.

This development also interlocked with the invention/introduction of 'broadcast' seeding, fields with one crop only, where the farmer throws handfuls of seeds across a wide expanse. This really only works with the advent of ploughing – otherwise broadcast seeding would merely be a way of feeding the birds. So this too is an interlocking matter with the domestication of animals: men could handle simple ploughs but cattle could pull much heavier loads, deepening the furrows of the land (see chapter seven). This meant in turn that fields could be given over to one crop, further improving yield and efficiency. In the New World, however, where there were few animals to domesticate, and where, to begin with certainly, roots and tubers were the founder crops, even such seeds as there were, were planted individually by hand, with the help of a hoe. As a consequence, most New World fields became 'mixed gardens', and this too, of course, obviated any need for a plough, in another vicious circle.

One other factor concerning plant life needs to be singled out. The New World, in particular Meso- and South America, contains 85 per cent of the known psychoactive plants that grow on Earth. This too is happenstance but, combined with the distribution of shamanism across the world, and the route by which early humans entered the Americas, via Beringia, it was a coincidence that was to have such profound consequences that an entire chapter will be devoted to the subject (see chapter twelve).

Cakes and Ale

A final point relates to the way the domestication of plants links in with the domestication of animals and which completed the 'package' effect. In the Old World these two processes followed relatively closely upon one another. In the Middle East, for example, sheep, goats and pigs had been domesticated by 10000 BP, and possibly a little earlier, but in any case within a thousand years of wheat. It is possible that, when the first fields were created, and discovered to yield less well in subsequent years, early man also noticed that where mammals grazed – and more importantly defecated – the plants grew back there more quickly. So was conceived the notion of manure and its fertilising properties. But early man quickly found that these animals could also provide meat, milk (and then cheese), that their hides could be manipulated to make leather and thongs and much else.

And here too there was a major difference between the Old World and the New. In some areas – North America, Australia, sub-Saharan Africa – there were no animals at all who were candidates for domestication. On the other hand, in Eurasia there were close to a dozen types of mammal which were capable of being (and were) domesticated: sheep, goats, various forms of cattle, pigs, horses, two types of camel, donkeys, reindeer, water buffalo, yaks. Moreover, seven of these were located in south-west Asia, or the Middle East, the area where wheat and barley were common. In South America, there were just three mammals suitable for domestication: the llama, the alpaca, and the guinea pig, plus the turkey. In North America, as noted earlier, there were none at all. Another major difference between the hemispheres was that, as well as the much larger variety of animals that were domesticated in the Old World, so those animals – by chance – had many more uses. These differences profoundly affected differential development between the Old World and the New.

One of the reasons for this remarkable disparity between the hemispheres is probably the very great size of Eurasia, which determined that it had, overall, no fewer than seventy-two species of large mammal which could be candidates for domestication. Recent research has shown that most of these large mammals evolved in the great steppes of Eurasia.[9] This compares favourably with fifty-one species of large mammal in sub-Saharan Africa, none of which have been domesticated, with twenty-four in the Americas and just one in Australia.

Domestication is not easy and is not to be confused with keeping animals as pets. Most big mammals appear to have been domesticated in the 'era of domestication' between 10,000 years ago and 4,500 years ago. The dog appears to have come first (at 12,000 years ago), then sheep, goats and pigs at 10,000 years ago, cattle at 8,000 years ago, the horse and water-buffalo at around 6,000 years ago, the llama/alpaca at 5,500, the donkey in north Africa about 5,000 years ago, and the various forms of camel at 4,500 years ago (though why the camel was not adapted to pull the plough much earlier is a mystery).

The main characteristics that domesticated mammals share is that they are herd animals with a strong dominance hierarchy and show not much in the way of territorial behaviour. This social structure is ideal because, in domestication, humans take over the ranking.[10] Also, goats, cows, sheep, camels and other mammals can be milked, producing milk products – butter, cheese and yoghurt – meaning they can be a *continuous* source of protein over their lifetime, not just when they are slaughtered. Most solitary, territorial animals cannot be herded.

As with plants, so the spread of animals around the New World was much slower than in the other hemisphere. None of the Andes' domestic mammals (llama, alpaca, guinea pig) ever reached Mesoamerica in pre-Columbian times, meaning that the Olmecs, Mayans and Aztecs had to do without pack animals. Without pack animals, which could become draft animals, there was no real need for the wheel (though wheelbarrows, invented in China to cope with terrace farming, would have been helpful).

The growth of sedentism, and the introduction of agriculture, allowed a marked rise in population and a corresponding increase in population *density*. This stimulated a virtuous cycle, or autocatalytic process, in which, as population densities rose, food *production* became increasingly favoured because it provided the increased outputs needed. Sedentism also meant that early mankind could shorten the interval between offspring (hunter-gatherers and nomads typically wean their children later to prevent early re-conception). The birth interval for farm people (sedentary communities) is around two and a half years, compared with twice that for hunter-gatherers. It was only natural, therefore, for population to grow much more under the regime of seed/package agriculture than under vegeculture, with the result that, eventually, the Old World in general grew far more crowded than the New.

Moreover, the need to *store* grain, from one harvest to the next, produced a number of innovations that seem never to have occurred to people living off roots and tubers, who were 'harvesting' them throughout the year as needed and therefore had much less need of storage. One innovation was pottery, no doubt discovered when some clay fell into the fire and hardened. Pots were the ideal rat-resistant vessels, and their existence seems to have promoted the idea that, if a certain amount of grain was being harvested to store for the next year, just in case the following harvest should fail, why should the farmer not produce *more* than he needed and *trade* the surplus.

Pottery is useful as a storage vessel, of course, only in sedentary societies. Hunter-gatherers, and people living off tubers/roots and hunting – on the move a lot – preferred baskets and woven bags and their bottle gourds, which were lighter, more supple, no less durable and much less likely to shatter. However, once again we see a process of interlocking developments, for the invention of pottery had several knock-on effects. One was that it enabled oats, rice and millet to be boiled in water, facilitating the making of gruel and/or bread (though it was not inevitable – in the Sahel millet communities, for example). Another was the invention of smelting – metallurgy, which presumably was also discovered accidentally, when stone was heated along with clay, and early man got the shock of his life when stone of one colour (probably the green of malachite) yielded a bright red malleable substance we know as copper. A third knock-on effect was the invention of beer. The combination of pottery and cereals (especially barley) produced fermentation – what must have seemed another magical process was presumably discovered accidentally as well, as people observed that an old gruel left to stand 'did not spoil, but instead tasted sweet and had distinct effects on the mind and emotions'.[11] In turn that led to the invention of wine with *its* psychological and emotional effects. We shall see in later chapters that in fact the development of ceramics may have had much more to do with the advent of alcohol and other psychoactivre substances than hitherto thought (and this is true in the New World also – see chapter eleven). Barley was linked with malting and, in brewing, yeast was produced which also facilitated baking. And third, as trade flourished, and more and more pots were used for the transport of goods, and had to be sealed to prevent theft or damage en route, their contents were identified by a series of tokens which, as we shall also see in a later chapter, led to the first signs or symbols which in turn would evolve into script, writing.

And so the domestication of cereal grasses was a fortuitously momentous event, which fundamentally shaped the lives – and then the civilisations – of early man, governing ultimately the linked emergence of pottery, metallurgy (the plough in particular but also weapons), population density, baking, brewing and writing. The difference between cereal cultivation on the one hand, and root/tuber cultivation on the other, was, together with the weakening monsoon/increasing frequency of ENSO events, probably the most basic difference in the history of the world, and which gave the western hemisphere an historical trajectory very different from that of the Old World. Even in the New World, the great civilisations all arose in association with seed agricultural (maize-growing) areas. Root/tuber farmers, who were also hunters, never developed civilisations – they remained confined, as we shall see, to chiefdoms.*

In the previous chapter we saw that early man probably arrived in the New World with a stock of myths that was subtly different from those understood by many Old World people. To this we may add the difference in climate, the difference between seed culture and vegeculture, the presence/absence of domesticated mammals, and say further that what this also means is that, essentially, the world may be divided into three huge landmasses or continents – one 'quick' east-west continent, Eurasia, and two 'slower', north-south continents, the Americas and Africa (map 1, inset). Australia is considered separately later, albeit briefly.

In Eurasia (which, for our purposes, includes the north coast of Africa, along the Mediterranean shoreline), the years 12000 BP to, roughly speaking, 5500 BP, saw the development of farming across the whole landmass, from the Atlantic coast to the Pacific, the development of pottery and metallurgy (the 'cultures of fire'), culminating in the rise of cities and the first great civilisations. There were no real barriers to this expansion – the great East-West Corridor has already been identified, and the Sahara did not become the barrier it is now until about 2500 BP. In addition, there were no bottlenecks or deserts to block the east-west passage of peoples and ideas, and the main mountain ranges – the Alps, for example, the Himalayas, the Tien Shan and the Qilian Shan in China and Krygystan, and the Altai in Russia/Mongolia – were themselves essentially east-west configurations, and again did not act as substantial barriers to the diffusion of ideas, new practices and peoples. Indeed, in the existence of the Eurasian steppes (shown in broad outline on map 4) the natural features of the Old World actually facilitated movement and, as we shall see, had a major effect on the course of Old World history. At the end of this period, some time between 5500 and 5000 BP, major civilisations arose in the Tigris-Euphrates Valley, the Nile Valley, the Indus Valley and the Yellow River and Yangtze Valleys.

We have already discussed the (very effective) barriers to diffusion in the Americas, one consequence of which was that no major civil-

* The African evidence, such as it is, modifies this 'pure' picture but doesn't undermine it.

isation ever emerged in the continent of North America before the arrival of the Europeans, though there were some large and sophisticated chiefdoms. A not dissimilar picture emerges in Africa, where civilisation did not emerge south of the Sahara until 800 BC at the earliest, in West Africa. There, the main crops to begin with were yams and then millet with bananas and taro introduced from Oceania later on; and there was only one animal available for domestication, the guinea fowl. It was, therefore, in a situation not dissimilar to South America.

At 5000 BP, the Bantu in their original homeland in West Africa, had cattle and yams but lacked metal and were still hunting, fishing and gathering. Smelting appears to have emerged in Africa around 4000 BP. Again, as in South America, there were barriers to north-south diffusion. For example, Egypt's wheat and barley never reached the Mediterranean climate in the Cape of Good Hope until the Europeans arrived with it. The Khoisan (the Kalahari bushmen) never developed agriculture at all. The Sahel crops (the savannah at the edge of the Sahara), adapted to summer rain, could not grow in the Cape. The Sahara, after 2500 BP, and the tsetse fly together impeded the spread of livestock, particularly cattle and the horse. It took more than 2,000 years for cattle, sheep and goats to cross the Serengeti; and pottery – recorded in the Sudan and the Sahara as early as 10000 BP – did not reach the Cape until AD 1.[12]

Put this way, it seems that the Old World has had all the advantages, and to some extent that is true. Its great size, its basic east-west configuration, its location largely away from the tropics, were all vitally important. And we are not done yet. If the geographical configuration of the Earth, and the distribution of plants about it, comprised the most basic division between the hemispheres to begin with, it was to be the role of domesticated mammals which – more than any other single factor – shaped history thereafter.

• 7 •

Fatherhood, Fertility, Farming: 'The Fall'

The backbone of this book is the differential development of ideology – the understanding of human nature and what it implies for the organisation of society – in the two hemispheres and this and the next chapter explain how those different trajectories began to diverge, thanks to a whole complex of interlinked changes that occurred together, beginning first around the tenth millennium BC.

The very presence of grave goods, of whatever kind, suggests that ancient people believed at least in the possibility of an afterlife, and this in turn would have implied a belief in supernatural beings. Anthropologists distinguish three elements to a religion: that a non-physical component of an individual can survive after death (the 'soul'); that certain individuals within a society are particularly likely to receive direct inspiration from supernatural agencies; and that certain rituals can bring about changes in the present world. The beads found at the burial site in Sungir, 150 miles east of Moscow and dated variously to 28000–25000 BP, strongly suggest that people did believe in an afterlife at that time, though we have no way of knowing how this 'soul' was configured. The remote caves spread across central and eastern Europe and decorated with so many splendid paintings, were surely centres of ritual (they were lit by primitive lamps, several examples of which have been found, burning moss wicks in animal fat). We have already encountered the caves of Les Trois Frères in Ariège in southern France, near the Spanish border, where there is what appears to be an upright human figure wearing a herbivore skin on its back, a horse's tail and a set of antlers – in other words, a shaman (see figure 1 above, page 49). At the end of 2003 it was announced that several similar figures carved in mammoth ivory had been found in a cave near Shelklingen in the Jura mountains in Bavaria. These included a *Löwenmensch*, a 'lion-person', half-man, half-animal, dating to 33,000–31,000 years ago, and would appear to confirm that shamanism was the earliest form of magical or religious belief system. An image of a shaman at Les Trois Frères shows a horned creature, in what appears to be a dancing pose, with two lines coming from his nose (figure 3). We shall see in a later chapter that shamans who take hallucinogenic snuff often excrete substances through their noses, substances which may be regarded as sacred. (This applies both to ancient peoples in the New World rainforests and the paintings of the present-day San people of South Africa.)[1]

Fig. 3 'Dancing shaman' with effluent extruding from his nose, *c.* 14000 BC. Les Trois Frères cave, Ariège, France. Compare figure 10, page 316.

David Lewis-Williams, emeritus professor of cognitive archaeology at the University of Witwatersrand in South Africa, is convinced of the shamanistic nature of the first religions and believes they explain the layout of cave art. He puts together the idea that, with the emergence of language, early humans would have been able to share the experience of two and possibly three altered states of consciousness: dreams, drug-induced hallucinations, and trance. These, he says, would have convinced early humans that there was a 'spirit world' elsewhere, with caves – leading to a mysterious underworld – as the only practical location/entrance for this other world. He thinks that some of the lines and squiggles associated with cave art are what he calls 'entoptic', caused by people actually 'seeing' the structures of their brains (between the retina and the visual cortex) under the influence of drugs. No less important, he notes that many paintings and engravings in the caves make use of naturally occurring forms or features, suggesting, say, a horse's head or a bison. This art, he suggests, was designed to 'release' the forms which were 'imprisoned' in the rock. By the same token, the 'finger flutings', marks made on the soft rock, and the famous hand prints, where early artisans used bone pipes to spray pigment around their outstretched hands, leaving a silhouette, were a kind of primitive 'laying on of hands', designed again to release the forms 'locked' in the rock.[2]

He also notes a system of organisation in the caves. Probably, he thinks, the general population would have gathered at the mouth of the cave, the entrance to the underworld, perhaps using methods of symbolic representation that have been lost. Only a select few would have been allowed into the caves proper. In these main chambers Lewis-Williams reports that the 'resonant' ones have more images than the non-resonant ones, so there may have been a 'musical' element, either by tapping stalactites, or by means of primitive 'flutes', remains of which have been found, or drums.[3]

Finally, the most inaccessible regions of the caves would have been accessed only by the shamans. Some of these areas have been shown to contain high concentrations of CO_2, carbon dioxide, an atmosphere which may, in itself, have produced an altered state of consciousness. Either way, in these confined spaces, shamans would have sought their visions. Some drugs induce a sensation of pricking, or being stabbed, which fits with some of the images found in caves, where figures are covered in short lines. This, combined with the shamans' need for a new persona every so often (as is confirmed today, among 'Stone Age' tribes) could be one of the origins of the idea of death and rebirth, and of sacrifice which, as we shall see, looms large in later religious beliefs.[4]

The Promise and Dangers of Birth

The widespread depiction of the female form in very early Palaeolithic art also needs some explanation and comment. These so-called 'Venus figurines' are found in a shallow arc stretching from France to Siberia, the majority of which belong to the Gravettian period – around 25,000 years ago. There has been, inevitably perhaps, much controversy about these figures. Many of them (but by no means all) are buxom, with large breasts and bellies, possibly indicating they are pregnant. Many (but not all) have distended vulvas, indicating they are about to give birth. Many (but not all) are naked. Many (but not all) lack faces but show elaborate coiffures. Many (but not all) are incomplete, lacking feet or arms, as if the creator had been intent on rendering only the sexual characteristics of these figures. Some (but not all) were originally covered in red ochre – was that meant to symbolise (menstrual) blood? Some figures have lines scored down the back of their thighs, perhaps indicating the breaking of the waters during the birth process.

Some critics, such as the archaeologist Paul Bahn, have argued that we should be careful in reading too much sex into these figures, that it tells us more about modern palaeontologists than it does about ancient humans. His point is well made. Nevertheless, other early artworks do suggest sexual themes. There is a natural cavity in the Cougnac cave at Quercy in the Cahors region of France which suggests (to the modern eye) the shape of a vulva, a similarity which appears to have been apparent also to ancient people, for they stained the cave with red ochre 'to symbolise the menstrual flow'. Among the images found in 1980 in the Ignateva cave in the southern Urals of Russia is a female figure with twenty-eight red dots between her legs, very possibly a reference to the menstrual cycle. At Mal'ta, in Siberia, Soviet archaeologists discovered houses divided into two halves. In one half only objects of masculine use were found, in the other half female statuettes were located. Does this mean the homes were ritually divided according to gender?[5]

Whether some of these early 'sexual images' have been over-interpreted, it nonetheless remains true that sex *is* one of the main images in early art, and that the depiction of female sex organs is far more widespread than the depiction of male organs. In fact, there are *no* depictions of males in the Gravettian period and this would therefore seem to support the claims of the distinguished Lithuanian archaeologist, Marija Gimbutas, that early humans worshipped a 'Great Goddess', rather than a male god. The development of such beliefs possibly had something to do with what at that time would have been the great mystery of birth, the wonder of breast-feeding, and the disturbing occurrence of menstruation. Randall White, professor of anthropology at New York University, adds the intriguing thought that these figures date from a time (and such a time must surely have existed) when early man had yet to make the link between sexual intercourse and birth. At that time, birth would have been truly miraculous, and early man may have thought that, in order to give birth, women received some spirit, say from animals. Until the link was made between sexual intercourse and birth, women would have seemed mysterious and miraculous creatures, far more so than men. Is this why there are *no* images of males, or of the male function, in Gravettian art?

Anne Baring and Jules Cashford, in their book, *The Myth of the Goddess: Evolution of an Image*, describe these early figurines as the Palaeolithic Mother Goddess and this goes straight to the heart of the matter, so far as *The Great Divide* is concerned.[6] For a number of things need to be explained about these figurines and the principles/ideas they represent. First, why do they occur when and where they do? Second, what is their meaning? Third, why do they not occur in the New World? Enrique Florescano, in his book, *The Myth of Quetzalcoatl*, records two images of a buxom fertility goddess in the New World but this is the only evidence for such a deity in the Americas that I have been able to find.[7]

Randall White is surely correct in his argument that there must have been a time when ancient humans had not made the link between sexual intercourse and birth, 280 days of human pregnancy (on average, from the end of menstruation to parturition) being just too long a delay for such a link to be observed. This delay in understanding would help explain why the Venus figurines were only ever carved – in ivory or stone – and hardly ever painted on cave walls; and why, often, they lacked heads and feet, or these features were highly stylised. The figurines were carved because they had to be portable: the clan or tribe took the statuettes with them as they followed the herds, and the figurines were carved in such a way that only the important, practical features were included. Haim Ofek has shown how such figurines as have been found are spread in a broad east-west sweep stretching between the glaciers of northern Europe and those of the more southerly range of mountains running from the Pyrenees and the Alps to the Taurus, Caucasus and Zagros uplifts, and on to the Pamirs and the northern edges of the Himalayas, in general the space between these two ranges being the habitat of migrating deer. The southern band of ice-covered mountains, he says, would have deterred or barred lion and other predators from warmer latitudes. The figurines were carried by the hunter-gatherers as they followed the herds.

Elizabeth Wayland Barber and Paul Barber, in their book, *When They Severed Earth from Sky: How the Human Mind Shapes Myth*, have convincingly shown how many common myths are based on fairly accurate observations by ancient peoples of phenomena they thought important and yet used mental devices still familiar to us to remember these phenomena and warn their descendants.[8] The Barbers show how giants are to be understood as volcanoes, or the remains of mammoth bones, how volcanic eruptions may be understood as gigantic 'pillars of stone', how their eruptions are impressive evidence of an underworld with powerful forces, how storm gods became winged horses and tsunamis 'bulls from the sea'. On this basis, we may infer that ancient people were perfectly well aware of the miraculous nature of birth but they were no less aware, too, that it could be a sensitive, even perilous time, and so they recorded on the figurines the elaborate and detailed configurations of the female body which indicated that birth was imminent. Without being aware of the link between sexual union and birth, and given that not all women swell up equally during pregnancy, they could have had no conception of the biological rhythms governing gestation and so the bodily signs that birth was imminent would have been the only practical way they could know when to organise their lives so that the chances of a danger-free birth were maximised, perhaps by secluding the soon-to-be-mother in a favourite or familiar cave where predators could be kept at bay.

The discovery of a small Venus statuette, carved from mammoth ivory, was announced in May 2009, having been excavated in the Hohler Fels caves in Germany and dated to 35,000 years ago. Nicholas Conard, part of the excavating team at the University of Tübingen, noted that the figure had exaggerated sexual features and a small loop where the head would be.[9] He thought the statuette was 'hung on a string and worn as a pendant'. This fits the general picture but tends to confirm that the statuettes were carried around, supporting the interpretation offered here.

The earliest Venus figurines are recorded at 40,000–35,000 years ago, and they die out (the Venus of Monruz, Switzerland) about 11,000 years ago. This dating is suggestive too, being close to the time when mammals were first domesticated. The gestation period of cows is much the same as for humans (285 days) and the horse is even longer (340–342 days), but the dog is a mere 63 days, barely two months, so it may have been through observation of the (newly) domesticated dog's behaviour that early peoples first spotted the all-important link between coitus and gestation. Findings reported only in March 2010, by Bridgett M. Von Holdt and Robert K. Wayne, of the University of California at Los Angeles, using DNA evidence, put the domestication of dogs somewhere in the Middle East at 12,000 years ago.[10]

Intuitively, this seems very late for humankind to have made the discovery of the link between sexual union and birth. And yet, according to Malcolm Potts and Roger Short, Australian aborigines did not associate sexual intercourse with pregnancy until they domesticated the dingo – a form of dog with a similarly short gestation period (64 days).[11] Nor is the idea inconsistent with the fact that male gods do not appear to have evolved until the seventh millennium BC, unless we include the (predominantly male) shaman himself, as was suggested in an earlier chapter. If we do allow the shaman, then this would mean that humankind's earliest ideology would have involved the worship of two principles: the Great Goddess and the mystery of fertility, and the drama of the hunt, highlighting the problem of survival. An image of a wounded bear at Les Trois Frères cave in France shows it covered in darts and spears, with blood pouring from its mouth and nostrils. Anyone who has been to a bullfight will recognise this configuration.

As we shall see, this theme of the Great Goddess was to become the dominant ideological motif in the Old World, in contrast to the New, where different images prevailed. We need therefore to ask ourselves why this image, this motif, occurred where it did, when it did, and why its range was limited.

This is not a question that can be answered totally satisfactorily but we can go some way towards an understanding of the phenomenon. The first thing to say is that, as the Ice Age came to an end, between 40000 and 20000 BC, say, when the glaciers and permafrost retreated, and as grassland spread, the woolly mammoth, woolly rhinoceros and reindeer gave way to great herds of bison, horse and cattle.[12] Later still, the grassland itself gave way to thick forests and so the herds moved east, with the hunters following them. A third of cave paintings are of horses, while bison and wild ox make up another third. Reindeer and mammoth hardly appear, though many bones of such animals have been found.[13]

We saw in an earlier chapter that shamanism was the ideology of hunters, especially in relation to the reindeer, which is more at home in cold and frozen conditions than other large mammals. Are we seeing here then, in cave art, a sort of half-way stage between what we might call 'raw shamanism' and man's first close encounters with mammals other than reindeer – horses, cows, sheep and goats – who are relatively new as prey, less well understood and whose habits, therefore, need to be recorded? While many animals were painted in profile, so far as their bodies were concerned, their hooves were painted full on, which suggests that the profiles and the shapes of the hooves were being memorised for later, or being used for the instruction of children.

If the 'new' animals were being followed across the 'new' grassland habitat, between the two glaciated ranges, as identified by Haim Ofek, by people carrying the figurines with them, the distribution of the steppe would also explain the spread of Venus statuettes, which end where the steppe ends, around Lake Baikal. Lake Baikal and the Lena River, to the north, mark a kind of natural boundary, beyond which the horse and cattle would not have gone. This is confirmed by the maps of the distribution of cows and sheep as also shown on map 3.

Therefore, if, as was outlined in a previous chapter, early humans reached the New World via the Bering Strait from Mongolia and/or beachcombed their way north from South East Asia, deriving their protein chiefly from fish, they would not have incorporated the Eurasian post-glacial ideology – essentially large-mammal hunting – into their psychological make-up. They would have presumably discovered independently the link between coitus and birth but, for one reason or another – perhaps the relative absence of herding mammals – did not elevate it into a general principle to be worshipped. Maritime peoples and beachcombers are unlikely to have had any conception of reproduction among fish or sea mammals. Similarly, rainforest inhabitants, sharing their habitat with only wild animals, would have had much less opportunity to observe mating and almost no opportunity to witness births, since the process is so dangerous in the wild. This is all speculative, of course, but plausible and intellectually consistent.

Once the link between sexual union and birth was understood, and that it applied *generally*, in regard to other mammals that might serve as food, the idea of *controlling* fertility – a crucial aspect of domesticating animals – would have become a possibility. This too is therefore consistent with the discovery of the link between coitus and birth at around only 12,000 years ago.

Shrinking Land and Wild Gardens

The domestication of plants and animals took place across the Old World some time between 14,000 and 6,500 years ago and it is one of the most heavily studied topics in prehistory. It is safe to say that while we are now fairly clear about where agriculture began, how it began, and with what plants and animals, there is no general agreement, even today, about *why* this momentous change occurred. The theories, as we shall see, fall into two types. On the one hand, there are the environmental/economic theories, of which there are several; and there are the religious theories, of which at the moment there is only one.

The domestication of plants and animals (in that order) occurred independently in two areas of the world that we can be certain about, and perhaps in seven. These areas are: first, south-west Asia – the Middle East – in particular the 'Fertile Crescent' that stretches from the Jordan Valley in Israel, up into Lebanon and Syria, taking in a corner of south-east Turkey, and round via the Zagros mountains into modern Iraq and Iran, the area known in antiquity as Mesopotamia. The second area of undoubted independent domestication lies in Mesoamerica, between what is now Panama and the northern reaches of Mexico. In addition, there are five other areas of the world where domestication also occurred but where we cannot be certain if it was independent, or derived from earlier developments in the Middle East and Mesoamerica. These areas are the highlands of New Guinea, China, where the domestication of rice seems to have had its own history, a narrow band of sub-Saharan Africa running from what is now the Ivory Coast, Ghana and Nigeria across to the Sudan and Ethiopia, the Andes/Amazon region, where the unusual geography may have prompted domestication independently, and the eastern United States, where it very likely derived from Mesoamerica.

One reason for the distribution about the globe of these areas has been provided by Andrew Sherratt. His theory is that three of these areas – the Middle East, Mesoamerica and the South East Asian island chain – are what he calls 'hot spots': geologically and geographically they have been regions of constant change, where incredible pressures generated by tectonic plates moving over the surface of the Earth created in these three places narrow isthmuses, producing a conjunction of special characteristics that are not seen elsewhere on Earth. These special characteristics were, first, a sharp juxtaposition of hills, desert and alluvium (deposits of sand or mud formed by flowing water) and, second, narrow strips of land which caused a build-up of population so that the isthmus could not support traditional hunter-gathering. These 'hot spots' therefore became 'nuclear areas' where the prevailing conditions made it more urgent for early man in those regions to develop a different mode of subsistence.[14]

Whatever the truth of this attractively simple theory, or in regard to the number of times agriculture was 'invented', there is little doubt that the very first time, chronologically speaking, that plants and animals were domesticated, was in the 'Fertile Crescent' of South West Asia. First, there were three cereals which formed the principal 'founder crops' of neolithic agriculture. In order of importance, these were: emmer wheat (*Triticum turgidum*, subspecies *dicoccum*), barley (*Hordeum vulgare*) and einkorn wheat (*Triticum monococcum*). They first appeared in the tenth and ninth millennia BP. Second, the domestication of these fast-growing, high-yielding cereals was accompanied by the cultivation of several 'companion plants', in particular the pea (*Pisum sativum*), the lentil (*Lens culaniris*), the chickpea (*Cicer arietinum*), Bitter vetch (*Vicia ervilia*) and Flax (*Linum usitatissimum*). In each case, the original wild variety, from which the domestic crop evolved, has now been identified; this enables us to see what advantages the domestic variants had over their wild cousins. In the case of einkorn wheat, for example, the main distinguishing trait between wild and cultivated varieties lies in the biology of seed dispersal. Wild einkorn has brittle ears, and the individual spikelets break up at maturity to disperse the seed. In the cultivated wheat, on the other hand, the mature ear is less brittle, stays intact, and will only break when threshed. In other words, to survive it needs to be reaped, and then sown. The same is true for the other crops: the domesticated varieties were less brittle than the wild types, so that the seeds are only spread once the plant has been reaped, thereby putting it under man's control. Comparison of the DNA of the various wheats all over the Fertile Crescent shows that they are fundamentally identical, much less varied than the DNA of wild wheats. This suggests that in each case domestication occurred only once. A number of specific sites have been identified where domestication may have first occurred. Among these are Tell Abu Hureyra and Tell Aswad in Syria, which date back to 10,000 years ago, Karacadağ in Turkey, Netiv Hagdud, Gilgal and Jericho, in the Jordan Valley, and Aswan in the Damascus Basin also in Syria, which date back even further, to 12000–10500 BP.[15]

In the case of animal domestication the type of evidence is somewhat different. In the first place we should note that the general history of the Earth helped somewhat: after the last Ice Age most species of mammal were smaller than hitherto. One or more of three criteria are generally taken as evidence of domestication: a change in species abundance – a sudden increase in the proportion of a species within the sequence of one site; a change in size – most wild species are larger than their domestic relatives, because humans found it easier to control smaller animals; and a change in population structure – in a domestic herd or flock, the age and sex structure is manipulated by its owners to maximise outputs, usually by the conservation of females and the selection of sub-adult males. Using these criteria, the chronology of animal domestication appears to begin shortly after 9000 BP – that is, about 1,000 years after plant domestication. The sites where these processes occurred are all in the Middle East, indeed in the Fertile Crescent, at locations which are not identical to, but overlap with those for plant domestication. In most cases, the sequence of domestication is generally taken to be: goats then sheep, to be closely followed by pigs and cattle (i.e., the smaller species before the bigger ones). There was no radical break; for many years people simply tended 'wild gardens' rather than neat smallholdings or farms as we would recognise them. Pigs do not adjust to the nomadic way of life, so their domestication implies sedentism.[16] Animals required fodder so the plants they preferred would also have been cultivated.

For the Old World, then, the location and timing of agriculture is understood, as are the plants and animals on which it was based. Further, there is a general agreement among palaeobiologists that domestication was invented only once and then spread to western Europe and India. Whether it also spread as far afield as South East Asia and central Africa is still a moot point.

Much more controversial, however, are the reasons for *why* agriculture developed, why it developed then, and why it developed where it did. It is also a more interesting question than it looks when you consider the fact that the hunter-gathering mode is actually quite an efficient way of leading one's life. Ethnographic evidence among hunter-gatherer tribes still in existence shows that they typically need to 'work' only three or four or five hours a day in order to provide for themselves and their kin. Why, therefore, would one change such a set of circumstances for something different where one has to work far harder? In addition, reliance on grain imposed a far more monotonous diet on early humans than they had been used to in the time of hunting and gathering. Again, why the change?[17]

The most basic of the economic arguments stems from the fact that, as has already been mentioned, some time between 14000 and 10000 BP, the world suffered a major climatic change. This was partly a result of the end of the Ice Age which had the twin effects of raising sea levels and, in the warmer climate, encouraging the spread of forests. These two factors ensured that, in a world with as yet no metal tools, the amount of open land shrank quite dramatically, segmenting formerly open ranges into smaller units. The reduction of open ranges encouraged territoriality and people began to protect and propagate local fields and herds. A further aspect of this set of changes was that the climate became increasingly arid, and the seasons became more pronounced, a circumstance which encouraged the spread of wild cereal grasses and the movement of peoples from one environment to the next, in search of both plants and animal flesh. There was more climatic variety in areas which had mountains, coastal plains, higher plains and rivers. This accounts for the importance of the Fertile Crescent.

Mark Nathan Cohen is the most prominent advocate of the theory that there was a population crisis in prehistory and that it was this which precipitated the evolution of agriculture. Among the evidence he marshals to support his argument is the fact that agriculture is not easier than hunter-gathering, that there is a 'global coincidence' in the simultaneous extinction of megafauna, the big mammals which provided so much protein for early humans, a further coincidence that domestication emerged at the end of the Pleistocene age, when the world warmed up and people became much more mobile, and that the cultivation of wild species, before agriculture proper, encouraged the birth of more children. It is well known, as has been pointed out, that nomads and hunter-gatherers control the number of children by not weaning them for two years. This limits the size of a group that is continually on the move. After the development of sedentism, however, this was no longer necessary, and resulted, says Cohen, in a major population explosion. Evidence for a population crisis in antiquity can be inferred, he says, from the number of new zones exploited for food, the change in diet, from plants which need less preparation to those which need more, the change in diet from larger animals to smaller (because larger ones were extinct), the increasing proportion of remains of people who are malnourished, and smaller than hunter-gatherers, with shorter life-spans, the specialisation of artefacts which had evolved to deal with rare animals and plants, the increased use of fire, for cooking otherwise inedible foodstuffs, the increased use of aquatic resources, the fact that many plants, though available as food in deep antiquity, were not harvested until around 12000 BP, that grass (cereals) is a low priority in food terms, and so on and so on.[18] For Cohen, therefore, the agricultural revolution was not, in and of itself, a liberation for early humans. It was instead a holding action to cope with the crisis of overpopulation. Far from being an inferior form of life, the hunter-gatherers had been *so* successful they had filled up the world, insofar as their lifestyle allowed, and there was no place to turn.[19]

It is another attractively simple hypothesis but there are problems with it. One of the strongest criticisms comes from Les Groube, who is the advocate of a rival theory. According to Groube, who is based in France, it is simply not true that the world of deep antiquity was in a population crisis, or certainly not a crisis of *over*population. His argument is the opposite, that the relatively late colonisation of Europe and the Americas argues for a fairly thinly populated Earth. For Groube, as people moved out of Africa into colder environments, there would have been fewer problems with disease, simply because, from a microbial point of view, the colder regions were safer, healthier. For many thousands of years, therefore, early people would have suffered fewer diseases in such places as Europe and Siberia, as compared with Africa. But then, around 25,000–15,000 years ago, an important coincidence took place. The world started to warm up, and humans reached the end of the Old World – meaning that, in effect, the *known world* was 'full' of people. There was still plenty of food but, as the world warmed up, many of the parasites on humans were also able to move out of Africa. In short, what had previously been tropical diseases became temperate diseases as well. The diseases Groube mentions include malaria, schistosomiasis and hookworm, 'a terrible trinity'. A second coincidence then occurred. This was the hunting to extinction of the megafauna which were all mammals, and therefore to a large extent biologically similar to humans. All of a sudden (sudden in evolutionary terms), there were far fewer mammals for the microbial predators to feast on – and they were driven on to people.[20]

In other words, some time after 20,000 years ago, there was a health crisis in the world, an explosion of disease that threatened humankind's very existence. According to Groube's admittedly slightly quirky theory, early humans, faced with this onslaught of disease, realised that the migrant pattern of life, which limited childbirth to once every three years or so, was insufficient to maintain population levels. The change to sedentism, therefore, was made because it allowed people to breed more often, increase numbers, and avoid extinction.

One thing that recommends Groube's theory is that it divorces sedentism from agriculture. This discovery is one of the more important insights to have been gained since the Second World War. In 1941, when the archaeologist V. Gordon Childe coined the phrase 'The Neolithic Revolution', he argued that the invention of agriculture had brought about the development of the first villages and that this new sedentary way of life had in turn led to the invention of pottery, metallurgy and, in the course of only a few thousand years, the blossoming of the first civilisations. This neat idea has now been overturned, for it is quite clear that sedentism, the transfer from a hunter-gathering lifestyle to villages, was already well under way by the time the agricultural revolution took place. This has transformed our understanding of early humans and their thinking.

Of Dogs, Dingoes and Domestication: Genesis, The Fall and the Meaning of Monogamy

The fact that sedentism *preceded* agriculture stimulated the French archaeologist Jacques Cauvin to produce a wide-ranging review of the archaeology of the Middle East, which enabled him to reconcile many developments, most notably the origins of religion and the idea of the home, with far-reaching implications for the development of our more speculative/philosophical innovations. In other words, he explained how, in the Old World, there was a development in ideology to move beyond shamanism.

Cauvin (who died in 2001) was director of research emeritus at the Institut de Préhistoire Orientale at Jalés in Ardèche, France (between Lyons and Marseilles). He started from a detailed examination of the pre-agricultural villages of the Near East. Between 12500 BC and 10000 BC, the so-called Natufian culture extended over almost all of the Levant, from the Euphrates to Sinai (the Natufian takes its name from a site at Wadi an-Natuf in Israel). Excavations at Eynan-Mallaha, in the Jordan Valley, north of the Sea of Galilee, identified the presence of storage pits, suggesting 'that these villages should be defined not only as the first sedentary communities in the Levant, but as "harvesters of cereals"'.[21]

The Natufian culture boasted houses, grouped together (about six in number), as villages, and were semi-subterranean, built in shallow circular pits 'whose sides were supported by dry-stone retaining walls; they had one or two hearths and traces of concentric circles of posts – evidence of substantial construction'. Their stone tools were not just for hunting but for grinding and pounding, and there were many bone implements too. Single or collective burials were interred under the houses or grouped in communal cemeteries. Some burials, including those of dogs, may have been ceremonial, since they were decorated with shells and polished stones (note the early presence of dogs). Mainly bone artworks were found in these villages, usually depicting animals (an aspect of shamanism?). At Abu Hureyra, between 11000 BC and 10000 BC, the Natufians intensively harvested wild cereals but towards the end of that period the cereals became much rarer (the world was becoming drier) and they switched to knot grass and vetch. In other words, there was as yet no phenomenon of deliberate specialisation.[22]

Cauvin next turned to the so-called Khiamian phase. This, named after the Khiam site, west of the northern end of the Dead Sea, was significant for three reasons: for the fact that there were new forms of weapons; for the fact that the round houses came completely out of the ground for the first time, implying the use of clay as a building material; and, most important of all, for a 'revolution in symbols'. Natufian art was essentially zoomorphic, whereas in the Khiamian period female human figurines (but not 'Venus'-shaped) begin to appear. They were schematic initially, but became increasingly realistic. Around 10000 BC the skulls and horns of aurochs (a now-extinct form of wild ox or bison) are found buried in houses, with the horns sometimes embedded in the walls, an arrangement which suggests they already have some symbolic function. Then, around 9500 BC, according to Cauvin, we see dawning in the Levant '*in a still unchanged economic context of hunting and gathering*' (italics added), the development of two dominant symbolic figures, the Woman and the Bull. The Woman was the supreme figure, he says, often shown as giving birth to a Bull.[23]

Cauvin sees in this development the true origin of non-shamanistic religion. His main point is that this is the first time humans have been represented as gods, that the female and male principle are both represented, and that this marked a change in mentality *before* the domestication of plants and animals took place.

James Mellart, the original excavator of Çatalhöyük – see below – agreed with this reasoning, but we now need to amend part of Cauvin's theory.

It is easier to see why the female figure in Khiamian art should be chosen than the male. The mystery of birth had conferred on the female form a sacred aura, easily adapted by analogy as a symbol of general fertility. For Cauvin, therefore, the bull conveniently symbolised the un-tameability of nature, the cosmic forces unleashed in storms, for example. Moreover, Cauvin discerns in the Middle East a clear-cut evolution. 'The first bucrania of the Khiamian ... remained buried within the thickness of the walls of buildings, not visible therefore to their occupants.' Was this because they wanted to incorporate the power of the bull into the very fabric of their buildings, so that these structures would withstand the hostile forces of nature? Perhaps they only wanted metaphorically to ensure the resistance of the building to all forms of destruction by appealing to this new symbolism for an initial consecration (i.e., when the houses were built). After that, however, bovine symbolism diffused throughout the Levant and Anatolia and at 'Ain Ghazal, north-west Jordan, we see the first explicit allusions, around 8000 BC, to the bull-fighting act, in which man himself features. Man's virility is being celebrated here, says Cauvin, and it is this concern with virility that links the agricultural revolution and the religious revolution: they were both attempts to satisfy 'the desire for domination over the animal kingdom'. This, he argues, was a psychological change, a shift in 'mentality' rather than an economic change, as has been the conventional wisdom. As people acquired more familiarity with herding, with non-territorial mammals, their interest and appetite for control over such creatures would have grown.[24]

But there is surely more to the story – the transition – than this, especially if we return to the idea that early peoples were later in understanding the link between coitus and gestation than has been generally thought. Animals, as we have seen, were domesticated on the whole about a thousand years after plants. Why the delay? Was it because people had only just discovered how infant humans – and therefore infant cows and sheep – were 'made'?

As many prehistorians have noted, the moon at certain stages in its cycle resembles the horns of a bull. This would have been recognised by ancient peoples, who would also have noted the menstrual cycle, linked to the phases of the moon and therefore, by implication, to the bull (recall the Venus figurine with 28 red dots painted on it, mentioned above). Early peoples would have observed that menstruation stopped immediately before gestation and may therefore have linked the 'bull-shaped' moon with human gestation and birth. Is this why so many early Neolithic images show women giving birth to bulls? It is important to say that bulls are represented by their bucrania far more than by their penises, a discrepancy that is difficult to understand on straightforward 'fertility symbol' reasoning. Furthermore, no one can actually have *seen* a woman giving birth to a bull, so what did this imagery mean? In shamanistic religions, shamans undergo soul flight to other realms and can take the form of animals. On such a system of understanding, the bull/shaman of the heavens could visit Earth and enter inside a woman. Did early peoples think that women were made pregnant by one or other of the mysterious forces of nature?

That is a plausible theory made more so by the fact that, in the Middle Eastern sites of Çatalhöyük and Jericho, which followed the Natufian and Khiamian cultures, at ~11,000–9,500 years ago, yet more changes are observable – two in particular. First, as Brian Fagan and Michael Balter have pointed out, at both Çatalhöyük, in Turkey, and Jericho in Palestine, 'there was a new preoccupation with ancestors and with the fertility of animal and human life'.[25] And there was, secondly, another change, with people being buried not in communal graves (as Cauvin noted) but under the houses where they had lived. Sometimes they were decapitated, with the ancestors' skulls being plastered over and given new features. These are the very houses where, previously, bucrania had been sequestered, in the walls and under the floors of houses.

Now recall that the dog is present in these Middle Eastern sites – this is precisely the area where, Von Holdt and Wayne say, domestication of the dog first occurred. If people had only recently discovered the link between coitus and birth, as is here being suggested, a number of things would have followed. For example, not only would the discovery have revealed new relationships in Neolithic society between men and women, and between parents and children, but it would have transformed ideas about ancestors as well. Until that point, the understanding of 'ancestors' would have been general, communal, tribal – ancestors were 'the people who had gone before'. After the breakthrough, however, 'ancestorship' would have become a much more individual, *personal* phenomenon, which is why ancestors at Çatalhöyük and Jericho were now buried under the floors of the houses where they had lived, replacing bucrania. They were decapitated precisely because of this new, more individualistic understanding. Decapitation and plastering with human facial features was a way of preserving the power of *specific* ancestors, *specific* relations, rather than a more general, communal force that had been the practice earlier. Plastering the skulls with human features was another way of claiming specific ancestors, a way of remembering, conserving, preserving the forces of particular forebears.

Is the disappearance of the Venus figurines across Eurasia, at about 12,000 years ago, and the appearance of the female figures at these Middle Eastern sites, about 11,000 years ago, *giving* birth, rather than *nearing* birth, coincidental, or yet more evidence of the change in understanding that we are considering?[26]

And there is one other factor that has only recently come to light. We now know that, as people transferred from hunter-foraging to a cereal-based, more sedentary diet, changes occurred in the pelvic canal of females. The pelvic canal, recent science shows, is very susceptible to nutrition and the change in diet caused it to narrow – even today, the canal has not regained its Palaeolithic dimensions.[27] And so, we may conclude that, at the very time sedentism and a change in diet were taking place, and a new understanding of reproduction – and what it meant for family/religious life – was occurring, *the act of birth itself was getting more traumatic and dangerous.*

At this point we may well ask ourselves whether, with all that was happening, more or less at once, and with some of the changes being shocking – even moving – then, like other powerful events, would they have been remembered in myth form?

Did this happen? There is one piece of evidence, one myth, that suggests there *was* just such a powerful change in human consciousness.

Could it be that this development, this all-important change in mentality, is in fact contained in the very first book of the bible? Is this what the beginning of the bible is all about? Is this why the bible begins as it does? Genesis is known partly for its account of the Creation of the world, and of humankind, but also for the otherwise rather strange episode of the expulsion of Adam and Eve from the Garden of Eden because they had eaten from the Tree of Knowledge against God's explicit instructions.

Some parts of this story are easier to understand/decipher than others. The expulsion itself, for instance, would seem to represent the end of horticulture, or the end of humankind's hunter-gatherer lifestyle and its transfer to agriculture, and the recognition, discussed earlier, that the hunter-gatherer lifestyle was easier, more enjoyable, more harmonious, than farming. The bible is not alone in making this observation. In more or less contemporary traditions (Elysian Fields,

Isles of the Blessed, in Hesiod or Plato) human beings are understood to have hitherto lived free from toil, in a fruitful Earth, 'without help from agriculture', and 'untouched by hoe or ploughshare'.[28]

The central drama of Genesis, however, is of course that Eve acts on the serpent's advice and induces Adam so that they both eat from the Tree of Knowledge, after which they discover that they are naked. This is incomprehensible unless we acknowledge that both 'knowledge' and nakedness here refer to sexuality, or sexual awareness in some form. And indeed, as the biblical scholar Elaine Pagels tells us, the Hebrew verb 'to know' ('*yada*) 'connotes sexual intercourse'. (As in 'He knew his wife.')[29] What can this link between knowledge and nakedness mean other than that they become aware of their bodies, how they differ, how and why that difference matters, and that the knowledge they now have is of how sexual reproduction works? This knowledge is shocking and moving because it shows that reproduction is 'natural'; humans are not made by some miraculous divine force but by sexual intercourse. This is why it is felt as a Fall.

There are other clues to this change once we look for them. As Potts and Short again observe, hunter-gatherers are polygynous,[30] but it is now, Elaine Pagels says, that marriage becomes monogamous and 'indissoluble'.[31] People understood the nature of paternity for the first time and it became important to them. It also becomes relevant, as again Elaine Pagels points out, that in Genesis III:16, the text reads: 'To the woman he [God] said, "I will greatly multiply your pain in childbearing; in pain you shall bring forth children."'* Elsewhere, Adam and Eve are given 'dominion' over the animals.[32] Their naming of the animals was clear proof of the authority God had given them over other creatures.[33]

Are we not seeing here a mythical account of the transition from hunter-gathering to farming and some of its associated implications? More than that, is it not accompanied by an *attitude* that is not entirely at variance with modern scholarship: that the transition was not wholly good, that harmony with nature had been lost and that even childbirth had become more painful and dangerous? (In addition to the changes in the dimensions of the birth canal, other contemporary research has

* This is the wording of the Revised Standard Version (1952). The King James Version (1611) of the same passage reads: 'I will greatly multiply thy sorrow and thy conception; in sorrow thou shalt bring forth children.'

shown that birth intervals of fewer than four years are more perilous than longer intervals. Although people at that time did not have the advantage of modern science, they were a lot nearer the event itself, and in a better position to note the changes that were occurring.)[34]

Genesis does not 'date' when humanity linked sex and birth, not directly, but it does associate the link with the transition to farming. Therefore, is Genesis, and the Fall it records, really reporting a very great, shocking breakthrough that humanity made around 12,000–10,000 years ago, that coitus and gestation are related? Is that why animal domestication *followed* plant domestication by about a thousand years, because the link was only then made? Timothy Taylor, in *The Prehistory of Sex*, reports that around that time, among the Inuit of Alaska, they had what archaeologist Lewis Binford recognised as 'lovers' camps' – places where new couples could 'get away from it all' to cement their relationships.[35] Does this reflect a new understanding? Taylor also makes the point that, around 10,000 years ago, the caves where the Ice Age art proliferated, also seem to have been forgotten.[36] Whatever was going on was pretty important. Venus figurines were no longer needed, cave art was no longer needed. Mammal reproduction was understood and domestication was in place.

These are tentative arguments but their main strength lies in the consistent picture they paint. Around 12,000–10,000 years ago, as well as a transition to sedentism, urbanism and domestication, people discovered the link between coitus and birth and this produced a seminal change in attitudes to ancestry, the male role, monogamy, children, privacy, property – it was above all a momentous *psychological* change as much as anything else and that is why it was recorded, in coded form, in Genesis.

All this modifies but in effect amplifies Cauvin's second general point – over and above the fact that recognisable religion as we know it emerged in the Levant around 9500 BC – namely, that these changes took place *after* cultivation and sedentism had begun, but *before* domestication/agriculture proper. The radical change in mentality and lifestyle took a while to complete itself.

Cauvin's central point, then (and there are others who share his general view), is that the development of domestication was not a sudden event owing to penury, or some other economic threat. Instead, sedentism long preceded domestication, with bricks and symbolic artefacts already being produced. From this, he says, we may infer that early humans, roughly 12000–8000 BC, underwent a profound psychological/ideological change, essentially a religious revolution, and that this accompanied the domestication of animals and plants. This religious revolution, Cauvin says, is essentially the change from animal or spirit worship, shamanism, to the worship of something that is essentially what we recognise today. That is to say, the human female goddess is worshipped as a supreme being. He points to carvings of this period in which the 'faithful' have their arms raised, as if in prayer or supplication. For the first time, he says, there is 'an entirely new relationship of subordination between god and man'. From now on, says Cauvin, there is a divine force in the Old World, with the gods 'above' and everyday humanity 'below'.[37]

On this reading, the all-important innovation was the *cultivation* of wild species of cereals that grew in abundance in the Levant and allowed sedentism to occur. It was sedentism which allowed the interval between births to be reduced, boosting population, as a result of which villages grew, social organisation became more complicated and, perhaps, a new concept of religion was invented, which in some ways reflected the village situation, where leaders and subordinates would have emerged. Once these changes were set in train, domesticated plants at least would have developed almost unconsciously as people 'selected' wild cereals which were amenable to this new lifestyle.

It was sedentism that allowed closer observation of the behaviour of the dog, leading to an understanding of mammal reproduction, which changed attitudes to marriage, property, ancestorship, and brought with it the understanding to domesticate other herding animals.

These early cultures, with the newly domesticated plants and animals, are generally known as Neolithic and this practice spread steadily, first throughout the Fertile Crescent, then further, to Anatolia and then Europe in the west, and to Iran and the Caucasus in the east, gradually, as we shall see, extending across all of the Old World.

The shift to sedentism, and then to agriculture, and to worship of the Goddess and the Bull, were momentous changes, and they were, of course, linked. The Goddess, as a fertility symbol – fertility in its human, animal and vegetal contexts – emerged not only with the change to sedentism but also as the monsoon began to weaken and the fertility of the land became ever more problematic and people and cattle lived in closer contiguity. Despite this, it is as well to remember that, in the Old World, the symbols of fertility were not just plants but animals, mammals, the Goddess and the Bull. Just as the relationship between people and cattle, sheep, goats and horses, was a move beyond their links to mammoth, rhinoceros and reindeer, so the worship of the Goddess and the Bull was a half-way house, a move beyond 'raw shamanism' that reflected the developing understanding of fertility and that animal fertility was easier to understand than vegetal fertility. Nor should we forget Cauvin's other point, that bull worship was worship of his power, symbolic of the hostile forces of nature. The hostile forces of nature – earthquakes, storms, eruptions, tsunamis – are in fact more destructive of sedentary communities, who have more to lose, than they were of hunter-gatherer communities, who by definition are always moving on.

Cauvin, then, was right by and large. Sedentism and the domestication of plants and animals brought with them a significant move beyond shamanism – divinities involving the human form were a conceptual breakthrough that would have momentous consequences. The bull interests us too, of course, because it didn't exist in the New World and was therefore not available to be worshipped there as a god.

More than that, since neither the bull nor most of the other domesticated large mammals of the Old World existed in the Americas, the *interaction* of domesticated mammals and plants could not exist in the New World either. And it was *this* interaction which produced what was to be the single most important difference between the two hemispheres: the form of existence known as pastoral nomadism.

· 8 ·

Ploughing, Driving, Milking, Riding – Four Things That Never Happened in the New World

The four activities that make up the main title of this chapter are chosen precisely because they are all activities common enough in the ancient Old World but which did not exist in the New. Together, they comprise a way of life that shaped developments in Eurasia – both practical, technological and economic developments and ideological and political developments as well.

Beyond Human Muscle: the 'secondary products revolution'

As, again, Andrew Sherratt has commented, cultivation alone, without the use of domestic mammals, was able to sustain complex urban societies (as in the New World) but he goes on to say this: 'It is not without significance that the next threshold, that of industrialisation, was attained only in the Old World, for the employment of animal power as the first stage in the successive harnessing of increasingly powerful sources of energy beyond that of human muscle was only possible where these animals were domestic, not wild.'[1] In other words, the real difference made by these (usually large) domesticated mammals lay less in them being sources of meat, as they had been to begin with, and more in their 'emergent properties'.

Sherratt's most cited paper is 'Plough and Pastoralism', in which he identified what he called the 'secondary products revolution', following the invention of agriculture. For him, the next most important development was the ways in which domesticated animals interacted with domesticated plants. The world has 148 species of large herbivores or omnivorous mammals, candidates for domestication. Seventy-two of these are in Eurasia, of which thirteen were domesticated; fifty-one are in sub-Saharan Africa, of which none was domesticated; twenty-four are in the Americas, of which one, the llama, was domesticated; and there is one in Australia, which remained wild.[2]

Of particular importance here was the plough, representing the first application of animal power to agriculture. Not long after came the cart, which allowed agriculture to be intensified and aided the transport of its product. The cart, even more than the plough, allowed surplus goods (milk, plus wool and cheese which did not deteriorate anywhere near as quickly as milk) to be traded far and wide. Not only did the Old World have the *power* of domesticated animals that the New World lacked, but it had *products* from those domesticated mammals that did not require them to be slaughtered. Live animals were valuable, more valuable than animals that were kept in order to be sacrificed. This too would, in time, have significant ideological/religious consequences.

These developments didn't happen straight away, with the birth of agriculture, but took several thousand years to emerge. In fact, the secondary products revolution took place, generally speaking, in the fourth millennium BC, at much the same time, significantly, as the emergence of civilisation. But the New World went without all this.

The plough, for instance, made possible the cultivation of what had hitherto been regarded as poor-quality soils, resulting in the colonisation of wider areas that had not been cultivable before. The bulk transport offered by carts helped here too for it meant that ever more difficult land, hilly land for example, could be grazed by sheep, and the wool that was produced more easily transported to markets than before. More and more marginal land could be exploited, encouraging the pastoral sector, transhumance and even nomadism. None of these forms of subsistence – in particular plough-using agriculturalists and pastoralists – was available in the New World. In the Old World these two systems often went side-by-side, sometimes symbiotically, and sometimes in open conflict – they constitute an element in Old World history that is absent in the Americas. As we shall see, repeatedly, the conflict between pastoral nomads and sedentary peoples was one of the main motors shaping Old World history. Not the least of the differences between plough agriculturalists and the pastoralists is their systems of heredity.

Besides allowing more marginal land to be cultivated, the plough encouraged 'broadcast' seeding, whole fields given over to one foodstuff, as against the New World gardens, cultivated by small digging sticks. Broadcast seeding allowed greater surpluses to be built up, more wealth to be amassed, facilitating the emergence of civilisations benefiting from non-food-producing specialists of one kind and another.

The 'secondary products revolution', Sherratt concluded, separated two stages in the development of Old World agriculture: an initial stage of hoe cultivation, based on human muscle-power, in which animals were kept purely for meat; and a second stage of plough agriculture and pastoralism, using animal sources of energy. It was these secondary products, he says, that marked the birth of the kinds of society that were to develop in Eurasia.

The Traction Complex and the Rise of Milk and Wool

Cattle were fully domesticated by the sixth millennium BC but were not used as traction animals until the later fourth millennium. After that, cattle and carts spread as a closely related 'traction complex'.[3] (This spread is deduced partly from the dating of models made in clay, in the Harappan culture in the third millennium BC, and in China in the second, though the spread of carts across Asia was also associated with the domestication of the horse – see below.) Sherratt further concluded that the occurrence of model wheels in New World contexts shows that the principle was known but also that, since they were not used in transport, the (un)availability of draught animals was the critical factor in this technology.[4]

The spread of the plough closely mirrored that of the cart, which both seem to have been first used 'somewhere in northern Mesopotamia' in the early fourth millennium BC. It was the plough which created an interest among early peoples in metallurgy.

It was also in the fourth millennium that another four or five species of animals were domesticated which, though hunted in earlier times, had, says Sherratt, not been economical to domesticate. Furthermore, this group could be used not only as pack animals, or draught animals, but also could be *ridden* – these were the equids and the camels. There were/are four main groups of equid: zebras in sub-Saharan Africa, asses in North Africa, half-asses (hemiones) or onagers in the Near East and true horses in Eurasia.[5] The horse's main habitat was the steppe belt from the Ukraine to Mongolia and it was first domesticated by the sedentary cattle-keeping but non-agricultural communities of the middle Dnepr. The keeping of horses began there and spread out in the middle of the fourth millennium, once again when the traction complex was spreading. Domesticated horses spread slowly into forested Europe but also south where riders on horseback begin to be shown on terracottas of the Old Babylonian period.[6] The onager was less easy to domesticate and is often shown muzzled in illustrations of this period. The horse gradually replaced the onager as a traction animal and propelled further changes in technology such as the horse bit and spoked wheels, around 2000 BC. This would eventually lead to the chariot, the very great significance of which is considered later.[7]

The ass or donkey was domesticated at much the same time, its natural range being from Algeria to Sinai. Its advantage over other equids is its docility and its low dietary intake, making it a far more economical animal. Another advantage here was the discovery of what became known as 'hybrid vigour', that asses crossed with horses produce the mule, an infertile intra-specific hybrid that was, however, more vigorous than either parent species, and was used for the longer, more arduous overland trade networks.[8]

The camel, as is well known, falls into two regional populations which were probably domesticated independently. The woolly, two-humped 'Bactrian' camel is adapted to the cooler steppe and mountain fringes of Eurasia, while the single-humped 'Arabian' camel is better suited to the more arid regions of Arabia and North Africa. The camel can carry twice as much as a donkey, is faster, and needs less frequent feeding and watering. The two-humped variety was part of the traction complex from the third millennium on.

These five independent but parallel episodes of domestication took place in the fourth millennium in adjoining zones, so that a 'transportation complex' arose across a wide area, with varied conditions stretching from western Europe to Mongolia – Sherratt calls this a revolution in transport.[9] The Old World, therefore, was connected –

in trade, in the movement of goods, people and ideas – in a way that the New World never was.

But more than transport was involved. First there came milk (or possibly, before that, blood). The breakthrough here was that the products of domesticated animals could nourish populations without the necessity of killing the creatures, conserving resources. The attraction of milk is that it contains an amino acid, lysine, which is absent in cereal diets, and it also contains fat, protein and sugar, not to mention calcium. On top of that it can be converted into a number of storable products, such as cheese, butter and yoghurt.

However, it is now known that most of the world (Mongoloid peoples, New World peoples, Melanesian, Australoid, Khoisan, many Negroid peoples, and about half the inhabitants of Mediterranean countries) cannot digest lactose, the disaccharide that is synthesised only in the mammary gland, and the theory has lately been advanced that 'lactose tolerance', as it is known, evolved only recently in northern Europe as a way for those peoples to assimilate vitamin D, otherwise available to more southerly peoples in sunlight, and which, via calcium, helped to prevent rickets. The adoption of milk-drinking is believed to be shown from the changing shapes of pottery, in which open bowls are gradually replaced by pouring vessels. They may have had some sort of ritual use too, since the production of koumish, an intoxicating fermentation milk product, also emerged at this time.[10]

The fact that domesticated mammals were kept for their milk or other products, and not just killed for their meat, meant that the relationships between humans and large mammals was closer in the Old World than in the New. This had a marked effect on disease transmission, on immunity to disease, and on ideas about animals – what they were capable of, physically and mentally, what the ethics were of animal husbandry – and this all influenced ideas about *killing* animals – see, for example, chapter 19 below.

Early on in the Old World, vegetable fibres formed the basis of the earliest textiles, as happened in the New World. In the Americas cotton replaced skins and leathers whereas in the Old World flax was the most used vegetable fibre. Elaborate weaving techniques were already developed long before wool came into use. Wool, which would in time become the most important Old World fibre, emerged during the urbanisation of Mesopotamia (or at least the *record* of it does). We know from archaeological discoveries that flax was still being used in Europe in the mid-third millennium BC, while in the earliest proto-literate period in Mesopotamia over thirty signs representing sheep are known, including 'hair sheep', 'woolly sheep' and 'fat-tailed sheep'.[11] Shearing lists show that there was already a spring shearing period in the calendar.

In Europe in the third millennium, there is a marked increase in the proportion of sheep bones found in excavated sites, 40 per cent as against 10 per cent earlier. From Greece to Switzerland, wool gradually replaced leather and linen (flax) towards the end of the third millennium, and we also see a change in fastenings, which become suitable for the looser weaves of wool. Wool products were among the first goods to be exchanged on a large scale – the traction complex and the properties of wool had a major impact on the trading development of the Old World – and this had something to do with the change to a predominant male role in agriculture, with women being left at home to spin and weave. The emerging dominance of the male role is something we shall return to.[12]

As Sherratt also points out, when animals are raised for their meat, it is most economical to slaughter them relatively early, as soon as they reach their adult size. Until that point they are growing quickly but once they are fully grown, the cost of keeping them alive does not diminish but no more meat, so to speak, is 'manufactured'. Milking, however, is very efficient in terms of the amount of energy that is recovered for the amount of food that is consumed. This also means that, relatively speaking, female animals are preferable to male ones, who are more likely to be slaughtered, though with sheep males also generate wool. In general it seems that sheep were kept for six to eight years before being slaughtered. Either way, keeping animals alive, for their secondary products, was a new way of amassing and conserving wealth. As we shall see, much of the history of the wealth/industry of the Old World concerns wool.

Animals used for traction are for the most part castrated males which, at three or four years, are virtually fully grown. Castration, which has a marked effect on mammals, and the idea for which may have originated accidentally with humans, turned out to be very useful for transport/energy purposes and was, of course, not available in the New World where all large mammals save for the llama, were wild.

Overall, Sherratt's review of the evidence concludes that before the fourth and third millennia BC, domestic animals were exploited only for their meat, but that the secondary products revolution, in the mid-fourth millennium, 'brought about major changes in animal husbandry', changes that were to have marked sociological effects.

Ploughs, Pastoralism, and Population Patterns

The changes which took place in the third millennium came about as a result of the expansion of early mankind into more marginal areas, itself stimulated by the changes that occurred during the early domestication of crops and animals. But the increased scale of investment now in animal husbandry brought about a new phase in man-animal relationships, of which the basic feature was an increase in the scale of animal-keeping, something which never occurred in the New World.

The practice of milking allowed the population of domestic animals to contain breeding stock, working and production stock, and creatures kept for meat.[13] Milk likewise allowed the emergence of pastoralism. It was continuously obtained, far less risky than hunting (the other alternative on the semi-arid steppes) and was helped immeasurably by the invention of riding. The evolution of lactose tolerance was an important ingredient here. Most likely, according to Sherratt, this all began in northern Iraq, Syria and Palestine where the Ghassulian culture (~3800–3350 BC) shows a ceramic 'inventory' which includes vessels for manipulating liquids, including a 'butter churn' and a mortality pattern of its livestock that suggests secondary products were in place.*

Overall, the mechanical plough increased the farmer's ability to prepare his land by a factor of four, which not only increased the productivity of good land but enabled for the first time the exploitation of poorer land. At the same time, the increased production of wool created circumstances for an exponential increase in trade. Pastoralism became more and more attractive.

* A study published in 2009 suggested that the gene for lactose tolerance in fact emerged in central Europe and the Balkans, among cattle herders living 7,500 years ago. Milk residue has also been found in pottery remains in Anatolia, dated to the seventh millennium. Parts of the secondary products revolution therefore must have begun earlier than Sherratt thought. We also know now that lactose tolerance evolved separately in Africa.[14]

Pastoral peoples continued to spread where grain-growing was unprofitable and this dichotomy would prove crucial in the years ahead. Between pastoral nomads and sedentary farmers there was, alternately, the opportunity for symbiosis, trade, but also for conflict, in the occasional need for equal access to land, for grazing or cultivation. This conflict, already introduced, was to have a major effect on the course of Old World history.

In more arid regions – such as Africa – pastoralism spread without the plough. Use of the plough stretched from Europe to India but in Africa it was used no further south than the upper Nile. Below that, pastoralism, involving milking, penetrated most of the Sahara and East Africa, without either the cart or the plough. (The tropical African evidence has taxed archaeologists. In the most-heavily studied ancient sub-Saharan societies, those of West Africa, the inhabitants must have included individuals who had seen metal ploughs on the northern reaches of the continent, and their own societies had cattle for thousands of years. Yet they never added the plough over that time. One answer seems to be that, in at least some areas, such as the Middle Senegal Valley and the Inland Niger Delta, the annual floods were of such a size, and so chaotic and unpredictable, that available agricultural areas could not be predicted, even from year to year, and so such areas never emerged and stabilised. Certainly, evidence shows that large proportions of the population switched from time to time from seed agriculture to fishing. This diversity probably hindered the development of political hierarchies, conceivably for millennia.)

On the Eurasian steppes, the plough was used in oases but the lactose intolerance prevalent among the Chinese inhibited pastoralism and therefore draught animals, which is one reason why China had ploughs pulled by humans for so long.[15] Some scientists attribute the prohibition on pork to the need to concentrate on milking and traction, for which the pig is unsuitable.

But the overall effect of plough cultivation was to change fundamentally the pattern of settlements. The greater efficiency of the plough both enabled fresh land to be exploited and at the same time *used up* the land more quickly, so it had to be left fallow for a number of years. This is shown in the appearance of mines and quarries in a broad swathe across northern Europe, from Britain to Russia, outfits which were producing flint and stone axes for forest clearance. People in northern Europe grew crops on the land for as long as they could (often prolonged somewhat because animal manure helped), then moved on, without returning. In southern Europe they occupied higher ground, as milk-based pastoralists, preferring that to lowland cereal cultivation.[16]

In the Middle East the advent of the plough brought about a five-fold increase in the number of sites during the Uruk period (~4000–3100 BC), followed by increasing concentration of the population in a few well-defended locations.[17] It is not known whether the cities were defending themselves against other cities or against pastoralists. Trade networks were growing.

'While the Mesoamerican evidence shows that animal-traction is not a necessary precondition for the development of urban communities, the rarity and much later appearance of towns in the New World suggests that while a variety of settlement systems may eventually reach an urban form, the higher energy of the Old World systems greatly accelerated movement along this trajectory.'[18]

The cart and traction complex had a major effect on the development of the steppes of Asia. This dovetailed with the opening up of the forests in northern Europe and, again according to Sherratt, was an important influence on, among other things, language. It is this opening up, and the similar spread of the traction complex into North Africa, that accounts for the dispersal of the Indo-European and Semito-Hamitic languages across the Old World. 'The eastward movements of population on to the steppes in the third millennium BC linked east Europe with the Pontic and Trans-Caspian region as far as the Tarim Basin and the Iranian plateau, and there seems no doubt that it was such relatively rapid movements in the semi-arid zone which gave Indo-European its geographical range.' (The distribution of wheeled transport is shown on map 3.)[19]

The changes we are discussing also implied alterations in the relations between the sexes and in the rules of inheritance. In simple hoe agriculture, it is said – and this applies world-wide – the major contribution comes from female labour: sowing, weeding, harvesting, and that brings with it matrilineal inheritance, and is characteristic of aboriginal societies in the woodlands of the American south-east, which are generally regarded as the closest parallel to pre-plough agriculturalists.[20]

In contrast, both plough agriculture and pastoralism are associated with male dominance and patrilineal descent. According to one world-wide survey, two-thirds of plough agriculturalists, and the same proportion of pastoralists are patrilineal. This change is shown to be associated with increased use of the loom, and with traces of weaving equipment that show up in greater numbers in the third millennium.[21] The main point is that, as more land is brought under control, marital alliances become more important in consolidating holdings so that arranged marriages multiply. Competition for land intensifies and, in this way, inequalities in wealth begin to emerge. This was more true in the Mediterranean region than northern Europe, where land was more plentiful, at least to begin with.

All these systems penetrated only slowly into east Asia. 'There an alternative system of protein capture based on fish (especially in rice paddies) and on the pig was already supporting a relatively dense population, and the expanded pastoral sector which the secondary products revolution required could not easily be brought about. For this reason, the civilisation of China was in many respects comparable to those civilisations of the New World where domestic animals played a minor role.'[22] This patterns fits neatly with the point made in an earlier chapter about the limited distribution of the images of the Great Goddess.

The secondary products revolution, even in its amended version, was an acceleration and an expansion, containing within it sources of wealth creation and conflict on a grand scale. With the vicuña and the llama, the New World had sources of wool and beasts of burden. But these mammals could carry little more than a man could so they did not represent the marked expansion of energy that cattle and horses did in the Old World. The absence of milk meant that cheese in particular, as a long-lasting source of protein, never featured in New World life.

• 9 •

Catastrophe and the (All-Important) Origins of Sacrifice

The Woman and the Bull, identified by Cauvin as the first true gods (i.e., as abstract entities rather than animal spirits, a move beyond shamanism), found echoes elsewhere, at least in Europe in the Neolithic period. These echoes occurred in very different contexts and cultures, together with a symbolism that itself differed from place to place. But this widespread evidence confirms that sedentism and the discovery of agriculture *did* alter early humans' way of thinking about religion.

This is shown, first of all, in the development, between, roughly speaking, 5000 BC and 3500 BC, of megaliths. Megaliths – the word means 'large stones' – have been found all over the Old World but they are most concentrated, and most studied, in Europe, where they appear to be associated with the extreme western end of the continent – Spain, Portugal, France, Ireland, Britain and Denmark, though the Mediterranean island of Malta also has some of the best megalithic monuments. Invariably associated with (occasionally vast) underground burial chambers, some of these stones are sixty feet high and weigh as much as 280 tons. They comprise three categories of structure. The original terms for these were, first, the menhir (from the Breton *men* = stone and *hir* = long), usually a large stone set vertically into the ground. The cromlech (*crom* = circle, curve, and *lech* = place) describes a group of menhirs set in a circle or half-circle (for example, Stonehenge, near Salisbury in England). And third, the dolmen (*dol* = table and *men* = stone), where there is usually an immense capstone supported by several upright stones arranged to form an enclosure or chamber. The practice now is to use plain terms such as 'circular alignment' rather than cromlech.

Most of the graves were originally under enormous mounds and could contain hundreds of dead. They were used for collective burial, on successive occasions, and the grave goods were in general unimpressive. Very rarely the chambers have a central pillar and traces of painting can be seen. As Mircea Eliade has said, all this 'testifies to a very important cult of the dead': the houses where the peasants of this culture lived have not stood the test of time, whereas the chamber tombs are the longest-surviving structures in the history of the world and this has given Andrew Sherratt the basis of his explanation for megaliths. His theory is not the only one available and two accounts will be given here because they are not necessarily mutually exclusive and because the other account offers an explanation for some of the most enduring (and otherwise puzzling) aspects of ideology/religion where there are marked differences between the Old World and the New.

The Meaning of Megaliths

The spread of farming to Europe occurred in two trajectories. In the Balkans and central Europe, the 'west Asiatic package' was imported wholesale, which involved cereal farming, livestock keeping, pottery and the construction of villages with substantial houses. This pattern occupied the central European loess corridor, whereas further south along the north Mediterranean coast, cereals were less important (loess is very fine silt, many feet thick, left behind by retreating glaciers). Farming villages therefore spread throughout Europe along the loess corridor but where the loess ended the village pattern broke down and the megaliths began. 'Whereas the village was the basic settlement unit and primary community of Neolithic central Europe, early settlement in western Neolithic Europe was insubstantial and dispersed. The element of permanence seems to have been provided not by the settlements themselves, but by monumental tombs and enclosures.'[1]

In central Europe, farming villages – cereal cultivation plus the keeping of small amounts of livestock – would have produced sufficient structure for stable lineages to emerge, controlling inheritance and ritual life. Beyond the loess belt, however, where the land was poorer, and unsuited to cereal cultivation, pastoralism became the preferred form of farming, with larger herds and with the population correspondingly spread more thinly over larger areas (a form of subsistence unavailable in the New World). In such circumstances, villages did not meet the requirements of the new situation but megalithic structures – great ceremonial centres unifying these larger areas – did. People would have come together at these centres several times a year, for marriage, burial and other ritual ceremonies that helped unify these otherwise more dispersed communities.

These monuments spread across western and northern Europe over many centuries, their long-term survival showing they met a general need. Their earliest forms were often long mounds of earth and timber.[2] But then, further out from the loess corridor and later, stone replaced timber for revetments and internal structures, still often in long mounds. Later still, round forms became more frequent, and the chambers increased in size. To begin with, these structures were very similar in design to the contemporary houses in the loess belt, but they progressively diverged from this pattern. What seems to have happened is that, in a dispersed society where labour was the most important commodity, moving large stones symbolised the strength of that community, what could be achieved by coordinated effort. Monumentality began in France ~4600 BC, only reached Denmark around ~3800 BC, but many megalithic structures continued to be used as centres of ritual long after they had ceased being used for interment. Later, passageways were introduced which, Sherratt says, may have allowed access to the tombs after burial, indicating a different attitude to the dead.[3]

This is reinforced by other ways in which the layout of the megalithic structures also changed over time. The earlier ones were built to contain closed cists and were decorated on the outside, whereas the later ones, with passages allowing re-entry, were decorated on the inside. There was also a change in orientation: east in the case of the former, southeast in the latter.[4] For Sheratt this suggests the existence of two – perhaps competing – ideologies, 'an ideological struggle between the central European tradition and that of the local French'. And this perhaps puts into context the recent work by Jean-Yves L'Helgouach and colleagues who have shown that at some time around 3800 BC, 'a massive phase of iconoclasm' occurred in the area around Locmariaquer, on the Brittany-Atlantic coast, in north-west France, at which time many sculpted standing stones were pulled down, and incorporated into a new generation of large passage graves. The largest site where this iconoclasm took place was at Gavrinis, an island off Brittany which has a profusion of intricately carved ornamentation, together with a 'sensitive' astronomical alignment. This marked change in ideology seems to have coincided with the arrival of migrants from the south and east, the Chasséen. (Is this why the orientation of the structures was changed from east to south-east, as a form of folk memory of where the Chasséen came from?)

Among the items excavated from this period were distinctive pieces of apparatus known as *vase supports*. Sherratt said it is 'tempting' to interpret these as devices of cult apparatus. They were perhaps narcotic burners, 'an interpretation which would also explain the remarkable character of the Gavrinis engravings [the ornamentation] as entoptic images produced under the influence of drugs'.[5]

Are we seeing here, then, the beginnings of yet another move away from shamanism pure and simple? (If, that is, shamanism was ever pure and simple.) In the change from exterior symbols to interior ones, and the use of stone to create an enclosed ritual space, is this an attempt to recreate, above ground, and at convenient locations, the caves that were the original shamanistic centres? Chris Scarre, at Durham University in the United Kingdom, has pointed out that many of the huge stones in megalithic structures have been taken from sacred parts of the landscape, 'places of power' – waterfalls, for example, or cliffs – which have special acoustic or sensory properties, such as unusual colours or textures. This, he says, helps explain why these stones are transported sometimes over vast distances but are otherwise not modified in any way.[6]

These 'caves', if that is what they were, would have been brought above ground so as to be associated with the ancestors.[7] (The incorporation of astronomical alignments may suggest that such ceremonies had a regular calendrical occurrence, calculated by phases of the heavenly bodies. It is a not uncommon ethnographic observation that such ceremonies may employ mind-altering drugs which are consumed by certain individuals in the context of these rituals, and whose effects are interpreted in terms of communication with an 'other world'.)[8] A further development, especially in Brittany and Britain, were the henges, which had no roofs, and in which the ceremonial aspect was more important than the funerary.

What ideas lay behind the worship in these temples? Colin Renfrew, emeritus professor of archaeology at Cambridge, has shown in his researches on the island of Arran, in Scotland, that the megalithic tombs there are closely related to the distribution of arable land and it therefore seems that these tombs/temples were somehow linked to the worship of a great fertility goddess, which was adapted as a cult as a result of the introduction of farming, and the closer inspection of nature that this would have entailed.[9]

It does seem to be true that several megalithic circular alignments were prehistoric astronomical observatories. Knowledge of the sun's cycle was clearly important for an agricultural community, in particular the midwinter solstice when the sun ceases to recede and begins to head north again. From the mound, features on the horizon could be noted where the midwinter solstice occurred (for example), and stones erected so that, in subsequent years, the moment could be anticipated, and celebrated. Sun observatories were initiated round 4000 BC but moon ones not until 2800 BC.

Moreover, there may be a further layer of meaning on top of all this. A number of carvings have been found associated with megalithic temples and observatories – in particular, spirals, whorls and what are called cup-and-ring marks, in effect a series of concentric 'C's. Elsewhere in Europe, as we shall see in just a moment, these designs are related to what some prehistorians have referred to as the Great Goddess, the symbol of fertility and regeneration (though not everyone accepts this interpretation). In Germany and Denmark, pottery found associated with megaliths is also decorated with double circles and these too are linked with the Great Goddess. Given the fact that menhirs almost by definition resemble the male organ, it is certainly possible that the megalithic cromlechs were observatory/temples celebrating sexuality/fertility. The sexual meaning of menhirs is not simply another case of archaeologists reading too much into the evidence. In the bible, for example, Jeremiah (2:27) refers to those who say to a stone, 'You have begotten me.' Belief in the fertilising virtues of menhirs was still common among some European peasants at the beginning of the twentieth century. 'In France, in order to have children, young women performed the *glissade* (letting themselves slide along a stone) and the *friction* (sitting on monoliths or rubbing their abdomen along certain rocks).'[10]

It is not difficult for us to understand the symbolism. The midwinter solstice was the point at which the sun was reborn. When it appeared that day, the standing stones were arranged so that the first shaft of light entered a slit in the centre of the circular alignment, the centre of the world in the sacred landscape, which helped to regenerate the whole community, gathered there to welcome it. A good example of this is Newgrange in Ireland.

In the great megalithic ceremonial centres, therefore, we perhaps see the beginnings of a new religious complex. In the enclosed ritual space, a space surrounded by the remains (and the spirits?) of the ancestors, narcotic burners were lit so that the shamans, and others perhaps, could enter altered states of consciousness and visit other worlds for divinatory purposes. However, in societies that were dispersed, where each family or clan had its own herd of cattle and/or sheep, the fertility of the land, *and* the animals, *and* the community would have been doubly, triply, important. Hence the importance of the woman and the bull, by far the most impressively fertile male and powerful living force anyone had ever experienced.

And in such a dispersed society, with each family tending its own herd or flock away from others, the role of the shaman would have been more and more confined to communal ceremonies and rituals, held only a few times a year. Under these conditions, most of the time the shaman would have been present much less often than he was in village communities and his services less available.

The Disturbed Sky of the Bronze Age

The second theory about megaliths takes us back into the realm of natural disasters and catastrophes. In an earlier chapter the phenomena of the Toba eruption, at 74,000–71,000 years ago, and the three great floods – at 14,500, 11,000 and 8,000 years ago – were explored, together with their effects on mythology, and the distribution around the globe of potentially menacing weather patterns (the monsoon and ENSO), volcanoes, hurricanes and earthquakes. We now need to consider another realm of natural catastrophes involving disturbances in the heavens – comets, meteorites and asteroids – all of which would have been visible to ancient peoples and some of which impacted the surface of the Earth.

This aspect of our deep past has been in and out of intellectual fashion since the Second World War. One reason for this was the work of Immanuel Velikovsky who, in a series of books, *Worlds in Collision* (1950), *Ages in Chaos* (1953), and *Earth in Upheaval* (1955), made a number of sensational claims, that several of the planets of the Solar System had threatened Earth in historical times, in particular that Venus had caused major catastrophes by passing close to Earth at a time corresponding to the end of the Middle Bronze Age and that Mars had done much the same a few centuries later.[11] Despite the fact that several reputable scientists, such as Albert Einstein, found Velikovsky's work convincing, and despite the fact that a number of his cosmological predictions subsequently were confirmed, many scientists were outraged by his claims and in America there was an attempt to suppress publication of Velikovsky's books.

As a result of the controversy, the Society for Interdisciplinary Studies was formed in 1975, devoted to the scientific study of catastrophes and their role (if any) in prehistory and history. The 1970s were themselves a period of change in fashion (or in intellectual climate), one reason being that the Apollo landings on the moon finally confirmed that most or all lunar craters have an impact rather than a volcanic origin and a series of other probes clearly showed that all the bodies of the solar system have been heavily bombarded. In addition, an increasing number of impact craters have been found on Earth itself.[12]

As a result of all this activity, what seems established now are two things that concern us. First, that during the Bronze Age, especially around 3000 BC and for several centuries after that, the night sky *was* disturbed, far more than it is now, for example, with one or a few comets recurring annually, coupled with epochs when, according to W.M. Napier, once of the Royal Observatory in Edinburgh, and Oxford University, 'the annual meteor storms reached prodigious levels', meteor storms being 'probably the most impressive spectacles the sky has to offer'.[13] Furthermore, 'At some intensity level beyond modern experience, they may become an ecological hazard.'[14] In addition to that, known impact events in geologically/astronomically 'recent' times have included the Rio Cuarto string of craters in

Argentina 2,000–4,000 years ago, caused by a bouncing asteroid, and with a known impact on mythology, the Henbury impact crater, in the north-western territory of Australia, at 3000 BC, the Broken Bow crater in Nebraska at 1000 BC, the Wabar craters in Saudi Arabia at AD 500, meteorites in China in AD 1490, which killed 10,000 people, and the well-known Tunguska impact over Siberia in 1908. This list of known catastrophes is by no means complete but is sufficient to show that they do occur not infrequently and have occurred in recorded history (see figure 6 on page 157).

In fact, Bruce Masse, of the United States Air Force and the University of Hawaii, has compiled a list of 1,124 'naked eye events' recorded by Chinese, Korean, Japanese, Arab and European observers between 200 BC and AD 1800.[15] Figures 4 and 5 show, respectively, eclipse demons from the American south-west, dated to the fourteenth century AD and a series of 'frightful' celestial deities from Mesopotamia. Masse makes the point that the popularity of these deities was not constant but that they waxed and waned over the centuries according to how active the night sky was.[16]

Fig. 4 'Eclipse demons' from the American southwest, 14th century AD.

Fig. 5 'Frightful' clestial demons from Mesopotamia.

c. 3000 BC	Henbury crater, 160m diameter
c. 2350 BC	Early Bronze Age civilisations disappear – climate change data
c. 2100 BC	More Bronze Age cities disappear
c. 2000 BC	Campo Cielo, Argentina – bouncing asteroid
c. 1800 BC	Major dust event
c. 1650 BC	Destruction of Middle Bronze Age cities
c. 1365 BC	Destruction by fire in legends
c. 1200 BC	Destruction of Late Bronze Age cities
c. 1159–1140 BC	Decline in annual growth rings in Irish bog oaks
c. 1000 BC	Nebraska – Broken Bow crater
c. 800 BC	Bronze Age ends
c. 850–760 BC	Climate change data from the Netherlands
c. 200 BC	Fireball peak
c. AD 400–600	A blitz of fireballs
c. AD 500	Wabar craters, Saudi Arabia
c. AD 536	Dark Age triggered – dust veil event
c. AD 580	Gregory tells of climate chaos
c. AD 679	Gregory tells of climate chaos
c. AD 800	Vikings find west coast of Europe uninhabited
c. AD 1000	Comet and fireball peak
c. AD 1176	South Island of New Zealand burned ('fire from space')
c. AD 1490	10,000 killed by meteorites in China
c. AD 1700	Tsunamis in Japan – no earthquake cause known
c. AD 1800	5 April, event in North America
c. AD 1819	9 or 19 November, event Canada and US
c. AD 1885	24 February, event in Pacific
c. AD 1892	3 May, event in Scandinavia
c. AD 1908	30 June, event over Tunguska, Siberia
c. AD 1930	13 August, event in Brazil – 800 square miles of jungle destroyed
c. AD 1935	11 December, event in British Guyana
c. AD 1947	Sikhote – Alin impact in USSR

Fig. 6 List of known catastrophes in ancient times.

One of the most important impacts would appear to have been the so-called 'flood comet' of 2807 BC. By studying a number of myths, Masse was able to arrive at an exact date for this event, 10–12 May 2807 BC, and concluded that the collision probably occurred in the Atlantic-Indian Ocean basin, near Antarctica, producing a massive tsunami and several days of torrential rains world-wide through the injection of water vapour into and through the upper atmosphere, creating vast cyclonic storms which lasted for up to a week.[17] Masse also thinks this event coincided with the beginning date for the ENSO, and that populations world-wide suffered a sizeable die-off at about this time, giving rise, he says, to the mythological traditions of repeated creations, especially in the New World. Masse traces myths in Argentina, Nebraska, China, Mesoamerica, Mesopotamia, Palestine (Sodom and Gomorrah), Egypt and India to this event.

With this as background, we may return to the megaliths and in particular Stonehenge, in the United Kingdom, which is one of the largest, most complete and most famous sites. Duncan Steel, director of Australia's Spaceguard Project and a member of NASA's Spaceguard Committee, has a theory about Stonehenge that is controversial, but then *all* theories about Stonehenge are controversial and Steel's at least has the merit of explaining why the attempts to understand this arrangement of standing stones have so far failed. For him, there is no ancient rhythm to discover: Steel thinks that Stonehenge is, or at least started out as, a 'catastrophe predictor'.

He shares the view that the Bronze Age night sky was more disturbed than our own is and that the Great Cursus – the earthwork near the Stonehenge circle, three kilometres long and the first part of the monument to be built, and originally understood to be an ancient racetrack (*cursus* = course) – was constructed when there was a trail of comets and debris, about ten degrees long, in the night sky, in effect a *meteor storm* as the Earth passed through the trail/tail of a comet with some real physical damage being done to the Earth's surface.[18] The showers could have lasted anything from six to seven hours or even twenty-eight hours, he says, and the arrival path of the comets does seem to have an 'orientation consistent' with Stonehenge I, the Great Cursus.

Steel's view is that something 'truly exceptional' must have sparked the initial construction of Stonehenge and that as well as the layout of the stones we need also to consider the nearby existence of a number of burial mounds in which were found the bones of individuals whose dating is a good deal younger than the mounds themselves. He therefore suggests that these mounds are not in fact burial mounds at all but were instead early forms of (his words) 'air raid shelters', hideaways for people to enter when a catastrophe from the sky (a meteor shower) was imminent, this attempt at prediction being the original purpose of Stonehenge. He finds further support for this idea in the so-called Aubrey holes, which he says have dimensions that enable humans to sit in them and be protected. In other words, this is where the priests/mathematicians would have sequestered themselves, protected, while they observed the threatening features in the sky and aimed to calculate when the worst catastrophes would occur.

By the same token, the importation to Stonehenge of the famous, and famously massive, blue stones from hundreds of miles away in the south-west of Wales – long a puzzle – would be explained because of their similarity to chondrites, stones dropped by meteor storms (similar practices with stones are known elsewhere in the world).[19]

This is an ingenious solution to the mystery of at least one megalithic monument and it helps explain why Stonehenge was abandoned for a while (after the meteor showers disappeared from the sky), and was later resurrected as a solar- and lunar-linked entity. As was mentioned earlier, it does not necessarily conflict with Sherratt's ideas.

It is included here, however, because of its timing and because it fits with the general idea that catastrophic events – whether volcanic, tectonic, celestial, or climate-driven – do seem to have been more prevalent in the past, more than we have traditionally believed and more than they are now. This is crucial to understanding the history of the Old World and the New.

The Invention of Sacrifice

Gunnar Heinsohn, at the University of Bremen in Germany, goes further. He argues that *civilisation itself began as a response to catastrophe* and that this helps explain one of the most powerful, seemingly barbaric and yet mysterious aspects of early religions – the widespread existence of human sacrifice. As we shall see, the pattern and trajectory of sacrifice was very different as between the Old World and the New.

As Heinsohn points out, people alive today are used to an evolved world 'of tiny and harmless changes, imposed in particular by [Charles] Lyell and [Charles] Darwin' and so are unfamiliar with more catastrophic times; in the same way, theories about sacrifice were formed before modern geology had changed our views about the past. To Heinsohn, however, it is clear that the texts and pictures of Bronze Age antiquity 'show us lavishly decorated actors who clearly represent destructive celestial bodies and who are participating in blood sacrifice'. He finds a world-wide 'combat myth of gods and heroes who encounter and defeat dragons, monsters, demons and giants'.[20] He further argues that priest-kings and temples had the role of catalysts in the formation of cities in which sacrificial compounds emerged early. (We recall here Leonard Woolley's discoveries of a civilisation and sacrifice at Ur immediately above evidence of a flood. Many of the Mesopotamian Bronze Age gods embodied fearsome principles as well as fertility.)

In chapter five, the work of Clift and Plumb was introduced, showing that early civilisations both formed and collapsed in response to climatic changes. In chapter fifteen we shall see in more detail that, in both hemispheres, urban structures do appear to have been first erected following environmental catastrophes of one kind or another, but here, in this chapter, our purpose is to explain sacrifice.

There are several theories about sacrifice. The most well known are what we might call the psychological or anthropological theories of Walter Burkett, a philologist from Zurich, Réné Girard, a literature professor at Stanford University, and Jonathan Z. Smith, a religious historian from the University of Chicago. Their theories take violence to be at the heart of civilisation, envisaging a sort of Freudian 'primal murder', which is lodged in the collective memory and acts as a unifying and therefore a religious force in the community – indeed, on this view sacrifice *defines* the community. Other theories see sacrifice as a function of domestication: it is not found among hunter-gatherers and only domestic animals are ever sacrificed. The domesticated animal, in being half-way between man and wild animals, is also – and by analogy – half-way between humans and the gods. Some scholars think that animals were domesticated *for the purpose of sacrifice*.[21] Still other theories root sacrifice in plant domestication: existing plants have to be beheaded in order to generate new growth.

None of these theories is very convincing. Instead, it seems clear that since human sacrifice is so extreme only extreme circumstances would have initiated the practice.

In this, catastrophe plays an obvious enough role. Catastrophes, by definition, kill people, sometimes hundreds of people, sometimes tens of thousands of people. Often, the bodies of the victims disappear for all time, buried under volcanic lava, swept away by tsunamis, crushed in the fissures created by earthquakes, obliterated by asteroid collisions. Other victims would have been visible, either as dead bodies or as injured, when a lot of blood would have been spilled. In such circumstances, ancient peoples, looking upon these forces – volcanoes, hurricanes, earthquakes, tsunamis, rainstorms, asteroids – as evidence that the gods were in some sense angry or dissatisfied or disappointed with humankind, would have drawn the (for them) obvious enough conclusion that the gods 'required' human bodies, or human blood, for their own nourishment. In turn, sacrifice would have been an obvious enough propitiatory response to this predicament. And, as Heinsohn points out, myths about great men inventing sacrifice and prayers after they had survived destructive floods are not confined to Mesopotamia. 'In Chaldea, Ziusudra is such a hero sacrificing after a flood. Assyrians have Utnapishtim and Hebrews Noah in a similar role. In India, Manu invents sacrifice after the flood. In Greek traditions, Perseus, Deucalion, Megaros, Aiakos etc., started to sacrifice after a flood.' Much the same happens in Egyptian, Chinese and Algonquin myths of North America, in the latter case where the hero Nanaboush starts the practice of praying after a flood.[22]

Much earlier than this, in Çatalhöyük, between 6000 and 5000 BC, the wall paintings on shrines included depictions of volcanoes erupting and vultures attacking human bodies. In one image a twin-peaked volcano erupts, fire spouts from the summit and lava streams down the shoulders of the mountain. Dots cover the image, which may be a representation of the ash dust that rained down from the eruption and may have blocked out the sun. The only twin-peaked mountain in Turkey is Hasan Dag, which is also the source of the obsidian that the inhabitants of Çatalhöyük found so useful and magical.[23]

Later, at Knossos in Crete, excavations have revealed a number of children's bones in basement rooms, dated to 1450 BC, bones that bear knife marks that had resulted from defleshing. British archaeologist

Peter Warren believes these were remains of sacrifices designed to avert a great disaster, 'such as a series of earthquakes, or other natural catastrophes'. (A violent earthquake did indeed overthrow the building.)[24] In Japan, many a Shinto shrine has a human sacrifice legend as its foundation story, usually a mighty animal, a demon living in a mountain, often with snake-like tails – clearly a volcano.[25] Muslims, and before them pre-Islamic Arabs, worship a black meteorite that fell to Earth from the sky.[26] Who knows how many were killed in the impact?

Catastrophe would, of course, have to be explained, as the result of some action on the part of the community. This would have marked the emergence of priest-kings, rather than shamans, who would have led collective worship, prayers rather than soul voyages, to assuage widespread grief brought about by the catastrophes. Some scholars believe that the walls around early cities, which began to appear at about 2900 BC, were for defence against tsunamis rather than other peoples, at least in some cases.

With sacrifice comes the idea of redemption, that humans can in some way 'atone' for what had led to the catastrophe and this too would have been the task of priest-kings, to explain why disasters had occurred and what sacrifices would work.

Catastrophes also explain some of the more mysterious rituals of early Bronze Age religions. In some cases, for example, priests administered powerful laxatives. This, Heinsohn argues, is a re-enactment of the phenomenon that, amid the terrifying disasters, many people would have lost control over their bowels. Similar reasoning explains the use of *phalloi* or the smearing of themselves with soot or ash, or cutting off or burning hair. Catastrophes, spreading panic and fear, can lead to spontaneous erections while the soot or ash being smeared over bodies recall the ash that would have followed volcanic explosions. Fires provoked by lightning or burning magma would have set fire to the hair.

This complex of ideas is a much more credible explanation for the advent of sacrifice than those often given, for example that sacrifice may also have begun in a less cruel way, beginning at a time when grain was the main diet, and meat-eating still relatively rare. In many agricultural societies, for example, the first seeds are not sown but thrown down alongside the furrow as an offering to the gods. By the same token, the last few fruits were never taken from the tree, a few tufts of wool were always left on the sheep and the farmer, when drawing water from a well, would always put back a few drops 'so that it will not dry up'. Admittedly, we have here the concept of self-denial, of sacrificing part of one's share, in order to nourish, or propitiate, the gods. Elsewhere (and this is a practice that stretches from Norway to the Balkans) the last ears of wheat were fashioned into a human figure: sometimes this would be thrown into the next field to be harvested, sometimes it would be kept until the following year, when it would be burned and the ashes thrown on to the ground before sowing, to ensure fertility. But all these practices seem to be secondary developments of the earlier practice, and pale echoes: only extraordinary events can explain what is to us the barbarity and yet universality of human sacrifice. From here on, it will constitute a large part of our story.

The Great Goddess and the Moon

At much the same time as megalithic ideas were proliferating, but in a different part of Europe, a different form of worship of essentially the same principles was evolving. This part of the continent is generally referred to as 'Old Europe', and includes Greece and the Aegean, the Balkans, southern Italy and Sicily and the lower Danube basin and Ukraine. Here the ancient gods have been studied by the Lithuanian scholar, Marija Gimbutas.

She finds a complex iconography grouped around four main entities. These are the Great Goddess, the Bird or Snake Goddess, the Vegetation Goddess, and the Male God. The snake, bird, egg and fish gods played their part in creation myths, while the Great Goddess was the creative principle itself, the most important idea of all. As Gimbutas puts it, 'The Great Goddess emerges miraculously out of death, out of the sacrificial bull, and in her body the new life begins. She is not the Earth, but a female human, capable of transforming herself into many living shapes, a doe, dog, toad, bee, butterfly, tree or pillar.' (This transformative ability has echoes of shamanism but she has decorously changed sex.) Gimbutas goes on: '... the Great Goddess is associated with moon crescents, quadripartite designs and bull's horns, symbols of continuous creation and change ... with the inception of agriculture.'[27]

In many regions, Gimbutas says, the Great Goddess also became associated with the moon. This was partly on account of the waxing and waning of the moon, its constant alteration in shape, its apparent 'death' every month and resurrection when it reappeared in the sky three nights after its disappearance, partly on account of its temporal/cyclical association with the female menstrual cycle, and partly on account of the appearance of the new moon being similar in shape to the horns of cattle, in particular the bull.

The central theme, says Gimbutas, was the birth of an infant in a pantheon dominated by the mother. The 'birth-giving Goddess', with parted legs and pubic triangle, became a form of shorthand, as the capital letter M, as 'the ideogram of the Great Goddess'.[28]

Gimbutas's extensive survey of many figurines, shrines and early pottery produced some fascinating insights – such as the fact that the vegetation goddesses were in general nude until the sixth millennium BC and clothed thereafter, and that many inscriptions on the figurines were an early form of linear proto-writing, thousands of years before true writing, and with a religious rather than an economic meaning. By no means everyone accepts Gimbutas's ideas about proto-writing but her main point was the development of the Great Goddess, with a complicated iconography, yet at root a human form, though capable of transformation into other animals and, on occasion, trees and stones. Here too the Goddess was a sort of half-way figure between a shaman and what came later.*

Here then, in Old Europe, we have – or appear to have – a spread of ideas derived ultimately from the Middle East, from Natufian and Khiamian cultures, as outlined in the previous chapter. These ideas probably spread along with the idea and practice of farming. They confirm that fertility and the potentially hostile forces of nature – such as catastrophe – continued to be the twin motors of early religious belief. As our story proceeds, we shall see how these twin forces continued strong in Eurasia and the Americas, but nevertheless took very different paths.

* It is perhaps worth saying at this point that religious images, in all parts of the world, on all continents, are exceedingly complex, with multiple overlapping, interlinked and often contradictory meanings. This being so, there is always a danger that the analysis of religious images can be akin to numerology, where meaning is found in all manner of improbable patterns. The reader is cautioned that the patterns discussed here are only a few among many.

• 10 •

From Narcotics to Alcohol

We now need to introduce a new topic into our story, one that hardly features at all in regular histories but about which this book will have a lot to say. For it is our contention that the role of this topic, while not neglected entirely, has been very much downplayed, or overlooked. That topic is narcotics, hallucinogens, alcohol – in a word, drugs.

Neolithic temperate Europe had relatively few stimulants compared with elsewhere in the world but even so early peoples would have had an extensive knowledge of the various mood-altering substances which occur naturally and, with the emergence of writing, there is no shortage of evidence of the use of drugs in this way. Notably, there was the ritual use of *Nymphaea caerula* in the Osiris cult, known from Dynastic Egypt, and the Aryans who according to the Rig Veda drank intoxicating *soma*, regarding it as a 'divine beverage'. Herodotus outlined the Scythians' ritual practice of inhaling *Cannabis* smoke.[1]

In the Old World, two plants in particular stand out: the opium poppy (*Papaver somniferum*) and hemp (*Cannabis sativa*). Opium was native to Europe, being found mostly in Neolithic and early Bronze Age archaeological contexts in Switzerland, southern Germany and eastern France, though this may be an artificial bias of differential preservation, for the poppy appears to have survived best in the record of lakeside-dwelling peoples, where bog conditions favoured its identification in such remains as have been found. Later, the poppy is found in the south or Mediterranean region, while hemp grew further east and on the steppes.[2]

Poppies grew (as weeds) among cereals and, being nutritious and full of flavour, they would have come to man's notice quite naturally. Moreover, the narcotic alkaloids (medically active organic plant extracts usually containing a nitrogen base, such as morphine, codeine, and papaverine) are formed in the sap of the unripe seed-head or capsule, and may be extracted in weaker solution by soaking. A stronger form is obtained by puncturing the capsule. This is important because the dilute form was used medicinally in the ancient world as a painkiller or analgesic (Galen mentions it in the second century AD) but earlier literary references, and illustrations, show that the stronger form was often used in religious contexts, certainly by the second and first millennia BC.

Poppy seeds have been found among the earliest *Bandkeramik* assemblages in the Rhineland and in caves in Spain. (*Bandkeramik* refers to early pottery, dated to ~5500–5000 BC, decorated with linear bands.) These were, without question, domesticated forms of poppy because they were unable to seed themselves. Near Granada in Spain, at a burial site dated to 4200 BC, bags of esparto ('needle') grass were unearthed which contained a large number of poppy capsules, suggesting the poppy heads had a symbolic significance and were not just used as food.[3]

Similarly, we know that cannabis was used in the Iron Age from western Europe to China. It was found (in abundance) in the Hochdorf Hallstatt D wagon-burial (near Eberdingen, near Stuttgart, in Germany, 450 BC), and is mentioned in both Herodotus and Han medical texts. It was used in eastern Europe from the third millennium BC, where in one instance a 'pipe cup' was discovered, containing charred hemp-seeds.

To begin with, these narcotic substances were probably inhaled, the seeds being burned as part of a purification or communion ritual.[4] For example, different ceramics have been found in the Balkan Neolithic, becoming common in the fifth millennium BC, often described as altars and consisting of a small dish on four feet. Heavily decorated, often with animal heads, they seem to have been burners, and were sufficiently common to suggest they were used in a domestic ritual.[5] Later on, in the fourth millennium, in France, we see the emergence of *vase-support* ceramics, mentioned earlier. These are small, shallow bowls on square or round stands and again are profusely decorated, many of which also show traces of burning. Found in both caves and megaliths, in vast numbers in Brittany, within arcs of standing stones, their ritual use seems established. A specific narcotic appears to have been the focus of the central ritual.[6] Andrew Sherratt asks whether it is a coincidence that the appearance of the apparatus of a southern cult should occur in northern France at the same time as Breton megalithic art reached what he called its psychedelic climax in the entoptic forms and hallucinogenic images of the carvings at Gavrinis? Given that *Papaver somniferum* originated in the Alps, its identification so far north, in such circumstances, strongly suggests a ritual (cult) use.[7]

Evidence of other cults occurs in the later Neolithic. Bowls with a 'sunburst' design inscribed on them are found across eastern Europe in the third millennium BC, retaining a specific form of design over more than a thousand miles from the Pontic steppes, a distribution that, more or less, parallels that of cannabis. The way that ceramic design can change suddenly but then become distributed unrelated to area or time, also suggests the emergence and spread of specific cults involving psychoactive substances.

More provocative still, ceramic assemblages in all parts of Europe were transformed during the fourth and third millennia BC by the appearance of drinking cups and vessels for manipulating liquids. Sherratt argues that this development reflects, not milk drinking but the impact of alcohol on prehistoric Europe. Some of these vessels even look like poppy heads (see figure 7). Are we now dealing with opium this time in soluble form? Analysis of residues confirms that this is so in some cases.[8] Ritual equipment for the consumption of liquid opium, shaped in the form of opium plants, became the hallmark of the Funnel Beaker Culture (TRB) in northern Germany, Poland and Scandinavia (4000–2700 BC).*

This raises the question as to whether the spread of cord decoration (corded ware), from a steppe origin, was a parallel process. The cord in question was hemp, *Cannabis*, and in northern Europe it was initially infused rather than smoked, as it was in its area of origin. Does the emergence of the Corded Ware beaker mean it was subsequently combined with alcohol? (This is not so far-fetched: we still do much the same with the close relative of Cannabis, the hop, *Humulus lupulus*, in making beer.)[9]

* The Funnel Beaker Culture is generally known by its shorthand form, TRB, from the German in which it was first identified: *Trichterbecherkultur*.

The proliferation of different forms of ceremonial pottery sets makes it likely that each variation was designed for a specific type of food or drink, and that psychoactive substances played a part in each of them.

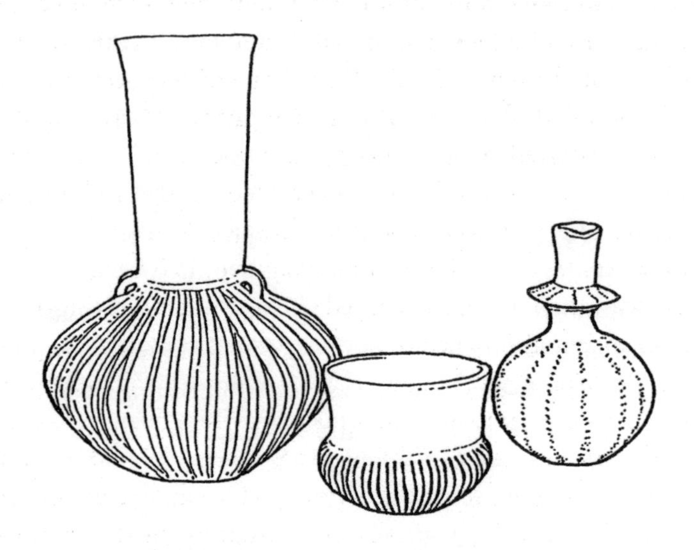

Fig. 7 Vessels for consuming opium in liquid form?

And sometimes the evidence is much stronger than that. For example, in one chamber burial in western Slovakia ten bodies, each with their hands held before their face, were interred above a collection of amphorae and drinking cups. Sherratt says these individuals must have been buried willingly though they must also have been stupefied, since they made no attempt to avoid asphyxiation. It is not too much to think that they had been poisoned. 'The stupefiant was presumably contained in the amphorae and drunk from the cup: a sacrament of death. Both the substance which it contained and the beliefs of those who drank from it, must have been extremely powerful to carry out such a ceremony.' This has all the hallmarks of a sacrifice, one moreover that was carried out on the elite who had preferential access to the sacred world through intoxication.

Later, more communal burials are encountered, still with amphorae present and sometimes including domestic animals, paired cattle in particular. Here too we are seeing a sacrifice, as is the case with other burials where two or more individuals are arranged around a table with pottery and cups, and even musical instruments.[10]

These sacrificial burials are all dated to ~3200 BC. On the north European plain, therefore, the second half of the fourth millennium BC was characterised by such communal ceremonies, frequently involving some form of drinking ritual. A further change occurred with the advent of Corded Ware, a pattern of individual burial, where adult males were interred with a specific kit: the stone battle-axe and a drinking set consisting of a cord-decorated beaker and an amphora. Continuing offerings were *not* made at such tombs. They were not ceremonial centres as earlier communal graves had been.

There was, then, a clearly marked change in ritual eating and drinking, from an initial practice of shrine-centred cult to a more personal pattern. What does this suggest? For Sherratt, and others like Ian Hodder, this development was associated with a change to more fluid settlements, as discussed in the previous chapter, based less on cereal agriculture than on more mobile livestock-husbandry, and on a change from collective graves (megaliths) to more individual, less permanent graves. 'Sacred places' were de-emphasised, stone structures *above* ground were replaced by pits *below* ground, with the bodies being buried once and for all, rather than remaining accessible. This is a major change, implying new beliefs about death, yet another new ideology.[11]

Bowls *v.* Beakers

The advent of personal drinking equipment in the new arrangement does not necessarily mean that psychoactive substances were generally available. Rather, Sherratt argues, they were a prerogative of individuals 'acting on behalf of the community, in a shamanic or priestly role'. They were probably older males, heads of families. (This practice continues today, among the Tuva Mongols, where the drinking of alcohol is permitted only to males aged over forty.)[12] Nevertheless, we *are* seeing a further change. Instead of forming part of a *communal* ceremony, ritualising the stability of long-lasting groups, drinking was domesticated or, in Sherratt's preferred term, profaned. In the more fluid societies associated with pastoralism, the sacred medium 'escaped' from its role in the established shrines, and was redeployed. Instead of there being one or a very few shamans per tribe, or village, with access to other worlds by means of trance, achieved through psychoactive substances, each family leader now fulfilled the same role, and began to come together with others of similar authority, where milder psychoactive substances were *shared* in ceremonies of hospitality and bonding, so that they could each support the others, and jointly face threats from outside. This is a process that would culminate in phenomena such as the *symposium* in classical Greece (considered later). This was in effect, as Sherratt puts it, the domestication of ecstasy. Moreover, and crucially, the milder alcohol was more suited to this bonding, communal process, rather than the stronger narcotics which isolated people from one another when they were under the influence.

He even goes so far as to say that the transition from TRB to Corded Ware 'has certain formal analogies with the Protestant Reformation in the same area 4,500 years later: the shift from a hierarchical model, with a mediating priesthood and elaborate shrines intended to last in perpetuity, to an emphasis on personal salvation, individual responsibility, architectural and liturgical simplification, often with a millenarian perspective of impending cosmic change'.[13] No less important, it was also accompanied by a shift from female imagery to an exclusively male imagery, reflecting the strengthening of patriarchal authority in dispersed pastoral societies. In such ways, as we shall see repeatedly, domesticated mammals had a profound effect on the evolution of social/ideological life.

The increasing amounts of pasture opened up by deforestation, which followed the introduction of the animal-driven plough, brought with it more flexible economic and social arrangements, the most important element of which was a larger role for animal-keeping.[14]

These changes were accompanied by others which emphasised how the cultures of the Copper Age (~6000–3500 BC) shared properties not possessed by their predecessors. These include not only the increasing appearance of metal goods in the (much richer) graves, but evidence of flints from further and further away (trade networks were proliferating with the development of the cart, especially carrying copper from the Carpathian Basin, centred on what is now Hungary), the spread of the horse from the Pontic steppes and the emergence of a certain pottery style – similar over vast differences and across cultures – associated with drinking.

In the east these drinking vessels had two handles, in the west a single handle, a cup or tankard. 'If material culture is fossilised behaviour, it may be possible to infer the appearance of conventions of hospitality or at least the social dispensing of liquids. It may be no coincidence that the funnel-necked beaker appeared among the northern farmers in an area demonstrably linked by trade to the Carpathian basin, when Atlantic Europe had no such distinctive vessels – they were "bowl cultures".'[15]

Sherratt has in fact traced the spread of drinking vessels ('a great diaspora' he calls it), beginning in the Aegean/Anatolian region and extending through Europe in a vast anti-clockwise movement. The vessels are found in graves often laid out with 'sets', as we today have sets of decanters and glasses, and they are of such a design that their prototypes must have been metal.

Thus we have a major cultural block, linked by drinking vessels, with metal prototypes. All this is accompanied by two further changes: graves have a distinctive personal weaponry kit, and the horse is present.

This is a further aspect to the domestication of ecstasy. Horses were expensive, and needed careful handling – otherwise they were potentially dangerous. In such circumstances, the use of the more intensive psychoactive substances – opium and cannabis – may have become inappropriate. Instead, the milder alcohol was substituted and, as we shall see in just a moment, after its discovery it became far more widely available.

Moreover, and perhaps more significantly, the overlap between the traction complex and new (alcohol-related) ceramics indicates a common origin and a parallel diaspora out of Anatolia. Similarly, the enlargement of the pastoral sector brought with it a concomitant enhancement of the male role in the system of production. The central fact here was that horses and wool-sheep were brought from the steppes into a Europe that was being cleared of forests. Horses, being expensive, conferred great prestige on their owners, who eventually formed a male warrior elite and it was their subculture which was reflected in the weapons and drinking vessels left in their graves.[16] An analysis of these drinking vessels tells us more.

The Cult of Alcohol

Alcohol was most likely discovered at much the same time as the horse was domesticated, the plough was invented, and the woolly sheep became part of the secondary products revolution, with its origins in the south. The naturally occurring sugars available to early man were glucose, fructose, maltose and lactose, available in honey, fruits, sprouting grain and milk, to produce respectively mead, wine, beer and koumish.

The prime candidate for producing the very first alcohol is the date-palm, since it is one of the most concentrated natural sources of sugar. Early Mediterranean cultures contain some of the earliest known drinking vessels.[17]

The religious overtones of wine in later European and Near Eastern societies suggests that its use in prehistoric times would have been accompanied by cult and by ceremony. The diaspora of drinking vessels across Europe suggests the spread of an 'alcohol cult', though in many areas it was beyond the limits of viticulture, and beyond the limits of technology for metal-working. Sherratt therefore suggests that we are seeing here a *substitution*, most probably of beer for wine.[18]

The third millennium drinking sets comprise the origin of a tradition of 'alcohol-based hospitality' that in the much later classical world we know as the *symposium*. This tradition began as a feast of merit, a gathering of warrior-companions. As kinship networks increased in size and spatial layout, owing to pastoralism and the horse, they became less linked to immediate communities. This meant that from time to time the need for armed bodies became greater and so warrior-feasting and hospitality became the preferred form of solidarity and interaction, when people came together intermittently to counter threats or mount raids on more settled communities.

There was most likely a variety of early alcoholic drinks. Koumish is unlikely – it needs too great a surplus of milk. Bee-keeping was a later technology, so mead is unlikely also. Fruit and sap in temperate zones are too low in sugar and, generally speaking, very few cereal-cultivators convert their grain into beer. (In temperate North America no alcohol was produced in pre-Conquest times, even though grain was used further south, in tropical Mexico and Peru, to make beer.)[19] Early man may have chewed certain substances, having discovered that the enzyme in spittle can ferment fruits in the mouth. But then drinking vessels would not have been needed.

Barley and emmer wheat were used in ancient Egypt and Mesopotamia, the beer being drunk through straws – as shown in illustrations dated to 3200 BC. (Herodotus found it sufficiently unfamiliar that he described it in detail.) The Latin word for beer, *cervisia* (Spanish *cerveza*), is a compound of the Latin for cereal and a Celtic element for water, suggesting a later, separate, invention.

'The best guide to the nature of Bronze Age brews is still the famous birch-bark container from an oak-coffin burial in the tumulus at Egtved in Denmark, where the residue had three components: honey, indicated by pollen of lime, meadowsweet and clover; fruits and leaves used for flavouring, including cranberry; and cereal grains, most probably emmer wheat.'[20]

The Advent of Wine

As time went by, therefore, and as societies grew more complex, especially where they handled livestock, the availability of grain, and the vine, for the production of alcohol – more moderate intoxicants than opium or cannabis – came to loom larger in the lives of populations. 'Powerful hallucinogens of "magic mushroom" type may thus be prominent in the early stages of social evolution, less so thereafter, though they may continue to have contextually appropriate uses, for instance employment in battle. Mild (and euphoriant) intoxicants, such as *kava* or alcoholic drinks (or tobacco) are more appropriate for ceremonies at which larger numbers participate and are occasions for male-bonding, entertainment, speeches and negotiation. Since the psychological experience of these substances is itself culturally constructed to a large extent, they may come to be used in a more "moderate" way, even if initially employed to induce trance-like states. This *moderation* of orthodox usage allows the wilder, ecstatic employment of psychoactive substance in the context of cult.'[21]

Certainly, wine in the Mediterranean region became practically 'synonymous with civilisation'. The spread of urban life was invariably associated with wine consumption, as we know from the widespread use of amphorae: thousands of these ordinary containers have been recovered from Mediterranean wrecks. Wine metaphors pervade Mediterranean traditions.[22]

The domestication of ecstasy continued when wine, as an elite drink, replaced beer in Mesopotamia in the early first millennium (the Neo-Assyrian period). Extensive vineyards were laid out, a practice paralleled in the Aegean and the Minoan and Mycenaean palace-centres in the second millennium, from where wine-drinking probably passed to Italy. In the works of Homer, warrior bands held bonding feasts, though wine was nowhere near as widely available as it would become by the second century BC, when Cato allowed seven amphorae of wine a year for his slaves, 'a bottle a day'. By the time that the armed aristocrat gave way to the hoplite phalanx, the *symposion* had become 'an all-male institution devoted to harmony and conviviality associated with civic life'. The symposium itself was a ritual of sorts, held under the sign of Dionysos, 'who had tamed the madness-inducing drug by teaching humankind to mix it with water'. Later these drinking clubs became notorious for political conspiracy and famous for philosophical discourse.[23]

Wine-drinking was propagated eastward when grape-wine production reached Tang China from Persia along the Silk Route in Sassanian times (~225 BC–AD 650). Silver cups of sheet metal made the same journey and revolutionised traditional Chinese metalworking practices, until then based on bronze casting.[24] This produced a range of knock-on effects, notably the 'lotus-petal' form of goblet, reflected later in Chinese ceramics; also attempts to imitate fine sheet-metal forms led eventually to the invention of white-wares and, ultimately, porcelain. The domestication of ecstasy had far-reaching consequences and we are not done yet.

What we see in the Old World, overall then, is a gradual replacement of strong, smoked or infused hallucinogens by more moderate (and mainly euphoriant and liquid) intoxicants associated, one might say, with a less hierarchical, less shamanistic form of religious experience, and marking instead a turn to a more domestic, even democratic form of worship, in which female deities are replaced by male ones, paralleling the growth of pastoral communities where animal-keeping, ploughing, driving, milking and riding were the characteristic activities and much more male-dominated than arable-farming villages. This interaction of animal-keeping activities and religious forms could not take place in the New World, where a different trajectory was followed, as we shall see.

It was of course not always quite as neat as this ideal picture implies. Mark Merlin argues that, after its early use in the Neolithic and early Bronze Age in and around Switzerland, opium travelled south and east, on the coattails of the tin, amber and gold trade, being widely used subsequently in Greece, Crete and Egypt from the late Bronze Age on (in Egypt necklaces of poppy capsules, made of faience, were very popular).[25] Opium was very valuable as a 'famine food', he says, but he also points out, for example, that the Greek classical religion owed much to the poppy and therefore retained many shamanistic features. These were reinforced by the incursions of the Scythians (see chapter sixteen), so that the caduceus or staff of Hermes had a soporific quality that made it a symbol of the shaman's trance; and the poppy had sacred associations with Nyx, goddess of the night, often shown distributing poppy capsules, with Hypnos, god of sleep, and his son, Morpheus, god of dreams. Demeter, the Greek Earth Mother Goddess was associated with the poppy – she is shown holding an ear of wheat and a poppy capsule on the Agora in Athens, and so is Persephone, who is shown holding sheaves of grain, bunches of lilies and poppy capsules.[26] Opium also plays a role in the Eleusian mysteries. Despite the general replacement of narcotics by alcohol, elements of shamanism hung on until the advent of Christianity (as we shall see) and it is possible to see shamanistic elements in later times.

Finally, on the subject of Old World intoxicants, we may note a similar process of the domestication of ecstasy further east, and for much the same reasons.

Fire Shrines and Flying Carpets

Among the oases of the Upper Oxus and other northward-flowing rivers in Turkmenistan and Tadjikistan, at the western limit of what would later be called the Silk Route, a range of fortified citadel-like sites, in mud-brick and dating to the second millennium, have been shown to enclose temple complexes with both ash repositories from sacred fires, and 'preparation rooms' where 'vats and strainers' for liquids

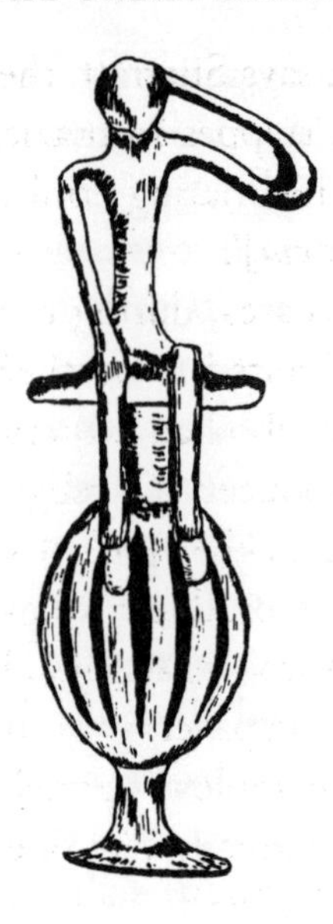

Fig. 8 Early Greek bronze artefact found at Kozani, which appears to represent an inverted opium poppy, supporting a human figure whose gesture emphasises the psychoactive effects of the narcotic.

were found. Pollen analysis shows traces of *Ephedra*, *Cannabis* and *Papaver*. Cannabis and Papaver we have already encountered. Ephedra is a shrub which produces the euphoriant norpseudoephedrine and the stimulant ephedrine. These substances were prepared by grating with stone graters, also found at the sites, and were pounded with fine imported stone pestles. These ground products were consumed as liquids, though to begin with they may have been smoked, as bone pipes were also found, some depicting 'wild-eyed faces'. Cylinder seals also found on the sites show animal-masked figures playing a drum or leaping over a pole (shamans?).

This association, of fire shrines and the consumption of psychoactive drinks, suggested to the (Russian) excavator that these citadels were where the tradition of Iranian fire rituals began, rituals that were famously reformed by the prophet Zarathustra in the early first millennium BC. 'The genesis of the tradition that gave rise to Zoroastrianism would thus have taken place in the context of interaction between oasis communities – probably familiar with alcohol, since wine had been prepared in western Iran since the fourth millennium – and steppe and desert tribes which were part of the expanding pastoralist complex'.[27] In other words, says Sherratt, the ritual plants traditionally smoked or infused on the steppes in braziers 'would now have been prepared as euphoriants or inebriating drinks – including the substance later known as *haoma* [or *soma*]'.

And so it was out of this area, during the second millennium, that Aryan-speaking groups migrated to northern India, to assimilate the Dravidian-speaking people who had hitherto comprised the collapsed Indus civilisation, and introduced the religious ideas described in the Rig Veda. On this reckoning, the drink described as *soma* comprised an infusion of several plant products known earlier on the steppes. *Soma*, then, was not a mushroom, as some have suggested, but a mix of Ephedra, Cannabis and Papaver, plus, perhaps, one further plant which fits well into this complex: harmel or Syrian rue (*Peganum harmela*). This plant was originally identified as *soma* by Sir William Jones in 1794. Archaeologically, its use has been dated to the fifth millennium BC, in the Caucasus, and it is burned in Central Asia as an intoxicant as well as being used medicinally.[28] It is in fact a more powerful hallucinogenic than Ephedra *et al.*, since it contains harmine. Not only was this used in World War Two as a 'truth' drug by the Nazis, it also forms the chief ingredient of *Banisteriopsis caapi*, the South American vine used to prepare *yagé* (see chapter 12). Given these properties and the fact that Peganum is a source of red dye used in Persian carpets, the possibility arises that (a) the entoptic images produced by harmine are reflected in the geometric designs characteristic of traditional Persian and central Asian carpets; and (b) did the hallucinogenic properties of this drug, and the flying sensation it has been known to induce, give rise to the tradition of the 'flying carpet'?*

What also follows from this analysis, if correct, is that *soma* would have been of variable strength, depending on its constituents. Here too, alcohol, after it was invented, may have formed a more modest – and therefore more appropriate – substitute in a highly mobile society

* Mott T. Greene makes a convincing case for *soma/haoma* being ergot, *Claviceps* spp., a fungus parasitic on wild grasses, in which the psychoactive ingredients are alkaloids. He bases his argument on their distribution (mainly mountains, as it says in the Rig Veda), their method of preparation, as outlined also in the Rig Veda, and on the fact that, because they are parasitic on grasses, no one plant is identifiable as the source of *soma* – hence the historical ambiguity.[29]

that depended on the potentially dangerous horse. Certainly, Ayurvedic texts recommend alcohol as an element of diet. Alcohol was only proscribed from the sixth century BC on, as Hinduism became more strict. Brahmins have turned to cannabis, widely smoked and, again, as a milder euphoriant.

A final aspect of all this is that different groups may use different intoxicants, or euphoriants, or different methods of consumption, as ways to define the group. This may be why the Chinese initially used rice and millet, then grapes, to make wine. Steppe influences there were constantly resisted as alien.[30]

Parallels between what happened in the Mediterranean and Europe, and on the steppes, and in Persia, India and China, can be found in South East Asia and Melanesia, where the most widespread and perhaps the oldest indigenous psychoactive substance is the betel nut. Traditionally chewed, this nut (*Areca catechu*) is mixed with lime (like coca) and a leaf of the vine *Piper betel*. Residues of betel nuts have been found at Spirit Cave, Thailand, dating to ~6000 BC. This plant, related to *Piper methysticum*, used in Oceania to produce kava, was probably smoked or chewed at first but then it too became a drink after alcohol was introduced and knowledge about fermentation spread. This background throws a suggestive light on the spread of the distinctively decorated pottery type known as Lapita Ware, which extended into Melanesia around 1500 BC, and is ethnographically associated with kava.

Across the Old World, psychoactive substances – poppy, henbane, belladonna, aconite and mandrake – continued to be used, as aspects of folk medicine and in fringe cults for many thousands of years. But, more generally and more forcefully, alcohol, a relative latecomer, gradually displaced earlier psychoactive substances that were infused or smoked. This was surely related to its milder properties, in more technologically complex contexts, and to its ever-wider availability. Its association had as much to do with conviviality as with the exploration of other worlds (which could sometimes be frightening), and with the increasing masculinisation of pastoral societies (it is now known that men are more tolerant of alcohol than are women).

Drinking and driving (and riding) were arguably of even greater concern in the past than they are now. Their interaction played a crucial role in the evolution of religious life, an interaction that was simply not available in the New World. The long-term effects of the replacement of narcotics by alcohol will be played out throughout the rest of this book.

· 11 ·

Maize: What People Are Made Of

If we had to sum up what has gone before and describe in a few words the main features shaping early life in the Old World, those words would be: the weakening monsoon, cereals (grain), domesticated mammals and pastoralism, the plough and the traction complex, riding, megaliths, milk, alcohol. One way to highlight the differences between the two worlds is to perform the same summing-up exercise for the Americas before we embark further on our journey. For the New World the crucial and equivalent words would be: El Niño, volcanoes, earthquakes, maize (corn), the potato, hallucinogens, tobacco, chocolate, rubber, the jaguar, and the bison.

This simple list immediately shows that, with the absence of domesticated mammals (we shall come to the llama and its cousins presently), the New World was for humans a much more *vegetal* environment than the Old World and that simple difference is profound. For a start, it meant that people in the New World were tied to the land much more than those in the Old World. For the most part (not invariably, but for the most part), they were tied to the land and the plants that they knew, and as a result their mobility was restricted. As was first mentioned in chapter six, there was little pastoralism or nomadism in the New World, and what there was did not have the historical impact that these life-ways did in the Old World. Allied to this, with *people* more rooted to the land, the movement of *ideas* was restricted, too. Not completely restricted but *more* restricted. The overall impact of this was to slow down the pace of development in the New World, as compared with the Old, and to ensure that certain features of the Old World that we take for granted never appeared in the Americas –

pastoralism and nomadism have already been referred to, so have the wheel and the plough, not to mention – as we shall see – various religious innovations. It is worth reminding ourselves that at the time of European contact with the aborigines of Australia in the late eighteenth century, when the continent had a population of 300,000, there was not a *single* domesticated plant.[1] And that was also true of the !Kung in the Kalahari. There would appear to be nothing 'inevitable' about the course of human development.

So far as the specific aspects of the New World are concerned, it is tempting to start with maize, or corn, since that plant is a grain, and was to prove as important in its way for the New World as wheat and barley and rice were for the Old and, since contact times, has proved its worth right across the globe. However, it makes more sense instead to start with the potato because, in its primary region, it was never replaced by maize as so many other cultigens were.

Rooted in the Andes

Although species of the wild tuber-bearing *Solanums* may be found as far north as Colorado, according to Redcliffe Salaman the potato was never cultivated in Central or North America in pre-Columbian times. (It *was* sometimes eaten but never cultivated.) In South America, on the other hand, there are many forms of wild potato and it has been cultivated there for at least 2,500 years.[2] Some 2,500 forms of root tuber are known in the Andes and potatoes were found at the early site of Monte Verde (mentioned above, in chapters 2 and 4), dated to 12,500 years ago. One of the initial drawbacks of the potato was that its glycoalkaloids can be poisonous unless it is properly prepared and its natural bitterness taken away.[3] This is why in some areas it was eaten with small amounts of earth – mainly clay – which removes the bitterness and counters the poisonous effects of the glycoalkaloids. (Some monkeys and birds have been observed to do the same thing.) The fact that ingenious ways needed to be found to overcome the potential poisons within the potato would have slowed down its domestication.

The significance of the potato lies in its adaptation to great heights and the associated cold. The earliest migrants in the Andes almost certainly came from the east, from the Amazon basin. The genetic evidence referred to in chapter one supports this idea, and so too does the decoration found on the earliest pottery in the Andes. If immigrants had come from anywhere else (down the coast, or even across the Pacific) he or she would not have found the jaguar and the boa constrictor so fearsome as they obviously did (because they would not have encountered them), or the potato and coca plant so useful, each of these animals and plants being shown as decoration on the earliest South American ceramics.

Maize cannot be grown above 12,000 feet, and is rarely found above 11,000 feet. At higher altitudes, only the potato, *quinoa* and *oca* can be grown but the potato far outweighs the others in importance (except around Tiwanaku, in western Bolivia, where *quinoa* takes precedence). It was thus the potato that enabled early man to settle in the highlands and escape malaria and the other diseases (and fears) of the rainforest or jungle. These highlands are not the most hospitable of environments but the Andes chain does stretch 4,000 miles from north to south, so that it is scarcely a small area. Moreover, llamas and other camelids are at home at such heights, so the Andes are not quite such an improbable area to be settled.

On the Andean highlands, from Colombia in the north to Chile in the south, several naturally occurring species of wild, tuber-bearing *Solanums* grew, among them the parent of the potato as we know it today. Some varieties can exist even at the snow line, at 15,000–16,000 feet, though they are scarcely edible. But at lower levels, when they do yield useful crops of tubers, they are still very resistant to frost. It was their great variability, and their ability to adapt to different soils and altitudes, that made them so useful to the early Andeans. The mountain chain crosses many latitudes and its valleys and high plains exist at different altitudes and face different directions. The potato had adapted to all these different niches.

The earliest way of preserving the potato, known today as *chuño*, was in use at least 2,000 years ago but probably goes back much earlier. The smallest potatoes are usually selected, laid out on flat ground and allowed to freeze overnight. This process lasts for three nights in a row but between the nights the potatoes are exposed to full sunlight, when they are trampled by foot. This action removes any remaining water and the skins. They are then exposed to two more nights of freezing and once dried they can last for months, or even years, and may be used as flour, or in soups, mixed with meat. Not much use was made of llama meat in ancient times, and none at all of llama milk. The chief source of animal protein was the guinea pig, allowed to run free in dwellings.[4]

The potato could be planted at any time of the year and in many areas more than one crop was produced, the limiting factor being frost. In general, though, the higher a potato was grown the less taste it had, and at higher altitudes most potatoes went into *chuño* manufacture.[5] *Chuño* was traded far and wide, mainly in exchange for maize, manioc and ceramics.

Recent experiments and excavations have shown that the ancient Bolivians discovered that, by building raised terraces, and leaving rainwater between them, temperatures could be manipulated so that the terraces were some six degrees warmer than the surrounding land and the effects of severe frost could be cleverly minimised. The Andeans were far more sophisticated in their growing techniques than they have usually been given credit for.

The earliest potato-growing cultures were known as Proto-Chimu and Proto-Nazca (~1100 BC), when the first representations of potatoes are found on pottery. There are signs that it was a cult object on the coast (i.e., worshipped) but the designs on the pots are also notable for several particular features.[6] One is that the images of maize and potatoes occur on the pots alongside those of jaguars and pumas, the jaguars often baring their teeth and snarling, suggesting it was worshipped as a ferocious god (we shall come back to this, more than once). Second, there are many depictions of *maimed* individuals, of people with congenital harelips, with faces deformed in other ways, such as with split noses, and still other pots depicting figures with amputated legs. The suggestion is that these people were felt to be special or sacred in some way, possibly that they were seen as half-way creatures, half-way between humans and jaguars perhaps. The figures with amputated legs could have been depictions of individuals who had been attacked by jaguars and survived. Some of the deformities are swellings, perhaps indicating (to us) infections, but which ancient peoples equated to (naturally occurring) deformities in potato shapes and therefore sacred in some way. If so, this reinforces the fact that the potato itself was also regarded as sacred. There are known to have been ceremonies with both the potato and the coca plant worshipped.[7] (The coca plant was also sacred – see chapter 13).[8]

Essentially, then, the potato, together with the llama and the vicuña, enabled early man to live in the Andes, an otherwise inhospitable environment though one that was, relatively speaking, safe and disease-free. Unlike grasses, however, and this is important, the potato kept the Andeans where they were: the early potato did not travel.

The Primacy of Beer

Since contact times, the potato and the maize (corn) plant have proved so useful to the rest of the world that we now tend to regard them as 'miracle' plants, and this is true so far as it goes. Potatoes and maize now provide staple foodstuffs across huge areas of the globe.

But this world-wide admiration for maize conceals the fact that much recent research has changed dramatically our understanding of the ancient plant – its history is nowhere near as straightforward as once was thought. The traditional view, insofar as there was one, was that maize is a grain, like wheat and barley and that, like them, it proved to be the basis of agriculture in the New World, enabling surpluses to be built up, and it was therefore directly associated with sedentism and, eventually, the emergence of urban civilisation. Its early morphology, as a wild plant, had much greater differences from the domesticated variety than did wheat or barley, say, and this, combined with the north-south configuration of the New World, conspired to produce circumstances where civilisation took longer to develop in the Americas.[9]

This picture is no longer tenable.

Maize, *Zea mays*, must now count as one of the most studied botanical entities in history. It has been studied by botanists, of course, but also by anthropologists, by geographers, by archaeologists and geoscientists, by linguists and evolutionary geneticists, by desertologists and horticulturalists, by land management specialists and Indian preservationists. And the new picture which has emerged is consistent, producing four main conclusions:

- Maize came into use *much* later than early studies reported;
- Its primary use, in most places, was as beer, used in religious ritual;
- Its spread across the New World was even slower than previously thought, and much, much slower than the spread of cereals in the Old World;
- Even when ancient peoples in the western hemisphere adopted the use of maize, there was no great rapid spread of farming in the New World as there was in the Old, nothing like the historical force that gave rise to the expansion, for instance, of the Proto-Indo-European language.

The actual origin of maize does seem to be settled. It developed from a wild annual, teosinte, the present-day range of which is centred in the Rio Balsas region of western Mexico, extending west to Jalisco and south-east to Oaxaca, with one early site there, at the Guila Naquitz cave, showing teosinte at levels dated to 3420 ± 60 BC.[10] By around 2500 BC, people were selecting the grain for larger kernel sizes, and for increased protein and starch quality and, by 1800 BC, cobs had been developed which had no shattering rachis (or stem), showing that *Zea* was already dependent on humans for dispersal.[11] There was a long period of improvement in maize, though it seems to have undergone its greatest change at about AD 200.

The central problem with maize is that many of the early – and spectacularly ancient – dates have been shown to be wrong. At one site in northern Chile, for example, where early dates for maize were put at 6000–5000 BC, later research has changed that to AD 1050 ± 32. The main reason for this has caused a new word to be coined, 'bioturbation', the process by which animals – such as crabs – can contaminate archaeological sites, driving material closer to the surface deeper into the ground, associating those later remains with earlier stratigraphic levels.[12]

Part of the problem has stemmed from the fact that, in many cases, maize pollen and maize phytoliths – what are generally referred to as 'microbotanical' remains – are at variance with 'macrobotanical' remains, the residue of actual plants, stalks or leaves. (Phytoliths are microscopic remains, usually of silica, which do not decay and allow ancient plants to be identified and dated.) In the Lake Titicaca region, for example, macrobotanical remains are consistently about a millennium later than maize phytoliths.[13]

A revealing clue to this disparity comes from the observation that, even today, many people in the central Americas chew the stalk of teosinte, the sugary juice still being used in beer production by people such as the Tarahumara in northern Mexico.[14]

So this hypothesis, that the early spread and use of *Zea* (either teosinte or incipient maize) was prompted by the value of the stalks in producing a fermentable juice, may well explain some of the anomalies in the early studies. It could certainly explain why pollen and phytoliths have been found much earlier than macrobotanical remains – in its early use it was never stored as grain, but used for its sugar and this seems to be supported by surveys recently reported from Ecuador, highland Peru, Mexico, and Mato Grosso in Brazil, that in the very earliest times maize was primarily a ritual plant used to make fermented beer or *chichi* (or *chicha*), consumed in the context of gift-giving, ritual feasts or other religious ceremonies. As happened in the Old World, ceramic bottles may well have been invented to manipulate beer – and may explain why maize was depicted on the bottles discovered at the temple of Chavín de Huántar (850–300 BC). In sites such as En Bas Saline, Loma Alta and La Centinela in Ecuador, ritual feasting contexts strongly suggest that maize was first appreciated as beer, for its intoxicating properties, and that this was the primary use for ceramics. These practices lasted from 4000 to 800 BC.

To an extent the linguistic evidence supports this interpretation. In some languages, Uto-Aztecian, for example, the terms for maize differ markedly, suggesting such peoples were familiar with it long before it became significant economically. Among the Mayan languages, however, the terms for maize are so similar that they must have coincided with its (much later) rapid adoption as an economically significant staple food.[15]

If we might call the development of *chichi* beer phase one in the use of maize, phase two occurred when it became widespread as a crop in the fourth millennium BC (it generally appears in the record around 3500 BC). However, it was *nowhere* a major dietary element before the first millennium BC – so let's call this phase, when maize achieved the status of a staple food, phase three. Sometimes the delay was even longer: among the Teenek on the Gulf coast near modern-day Veracruz, for example, maize was present at 7000 BC but permanent villages, with ceremonial centres based on maize agriculture, do not appear until 500 BC. In Soconusco, a region not far geographically from the place of origin of *Zea*, the stable isotope signatures indicate that maize does not become economically important until several thousand years after its appearance in that area. (Stable isotopes are a measure of bone chemistry derived from the main foodstuffs people eat during their lifetime.)

It is the same story in Andean South America where, in Ecuador, for example, maize does not become economically significant until after the first millennium BC and in coastal and highland Peru almost a thousand years after that.[16] Maize shows up in the Andean diet at ~2200–1850 BC but does not measure as significant in bone chemistry until two millennia later. In Peruvian sites which contain ceramics, maize is not an economic staple, but is only associated with ceremonial, including the very early site at Aspero (see chapter 15), where Ruth Shady, the lead excavator, confirms that maize had only ceremonial significance, often being left as a ritual offering, but was 'peripheral' as an economic force.

Its appearance on the altiplano of the Titicaca Basin, at ~750 BC, shows that even then maize only occurs in elite and ceremonial contexts and never (as is true today) achieved economic importance.[17] Around the Titicaca Basin itself a few grains of maize are found as early as 900 BC but it doesn't become economically important until after AD 250. The evidence suggests that Tiwanaku was an exchange (trade) centre for many different kinds of plants, the point being that, in the Andes, different plants would grow at different altitudes and so communities at one level would have regarded plants that grew at other levels (but not at theirs) as luxuries, which would have been much in demand. Among these sought-after plants were coca, peppers, hallucinogens and maize.

Some archaeologists have proposed the model of the 'vertical archipelago' in these circumstances, in which different members of the same kin group occupy different altitude/ecological zones, growing different plants and exchanging them with their kinfolk. Alternatively, this state of affairs has been used to explain the domestication of the llama, with llama caravans being employed as the major source of inter-zonal trade or exchange.[18]

The Maya elite did become maize-dependent but this was not until AD 200–900 and even then the society was not uniformly reliant on it. In fact, there is some evidence that Mayan men ate more maize than women, possible evidence of yet more religious/ceremonial significance.[19]

With the Incas, even later, as we shall again see in another chapter, there was widespread growing of maize, but once more this was mainly for the production of *chicha* beer. In fact, vast quantities of maize were grown, with whole populations being forced to move, so as to grow the plant, which was turned into beer in enormous quantities for ceremonial purposes. The growing and distribution of intoxicants was a central element in the way the Inca elite maintained its authority. John Staller, an anthropologist from the University of Kentucky, says that *chicha* beer consumption among the Inca was so common as to make a significant contribution to the diet. 'The prominence of maize as beer and food in ritual surrounding the Inca Solar cult, the Cult of the Dead, and ancestor veneration, reflects a close symbolic link to status and rank.'[20] *Chicha*, he says, had several overlapping meanings, including 'to fertilise', 'water' and 'saliva'. In Inca ceremonies, drinking alcohol took pre-eminence over eating. *Chicha* was also intimately involved in Inca sacrificial rituals as we shall see later.[21]

The truth now appears to be that maize achieved an economic importance much, much later in South and Mesoamerica than previously thought (it was never economically important in the Caribbean, while the Amazon and Orinoco are too tropical for maize). In the Andes, potatoes and *quinoa* continued throughout as the mainstays of the diet.

This picture is moreover borne out by linguistics which show that the vocabularies for maize go back for more than 4,000 years and that Proto-Otomanguean speakers (in Mexico) were involved in the initial domestication of teosinte.[22] However, of the four major linguistic families in Mesoamerica – the Uto-Aztecan, the Otomanguean, Mixe-Zoque and the Mayan – maize terms are unrelated, confirming there was no rapid spread. The Amuzgo (another Otomanguean language also spoken in Mexico) classify maize as a grass while Mayan-speakers separate out maize from other life forms, suggesting they added maize on to a pre-existing taxonomic framework.[23] A similar process is evident among the Sioux, whose word for maize is a relatively late variant of other (presumably pre-existing) cultigens, suggesting it was a late arrival, tacked on to pre-existing taxonomies.

Among North American tribes of the south-west, maize was not in use until the first millennium AD but there is no evidence for a proto-language maize term, and this is consistent with the theory that maize did not develop as a dietary staple among such groups before about AD 600. The latest evidence from the south-west is that, before maize arrived, the inhabitants were 'protecting, encouraging and cultivating' a large variety of grasses, with various forms of water management, and different races of maize were incorporated into their pre-existing systems. In other words, in North America there was no sudden 'invasion' of Mexican maize.[24] Other evidence suggests that maize moved faster to the north of Mexico, where it developed, than to the east or south. Even so it took nearly two thousand years from its first appearance until it became the main element of the diet. In the Mississippi Valley, in Ohio and in Ontario, it was the same story: maize appeared around AD 500 but the tribes there were not reliant on the grain until around AD 1000–1200.

One place where maize does appear suddenly, and in great quantities, is at Cahokia, on the Mississippi (near what is now St Louis), at ~AD 750–850, together with the equally sudden predominance of limestone tempered pottery, and a high-maize diet apparently among those individuals who were sacrificed. This too reinforces the idea that maize had a ceremonial role, even in Cahokia. A similar change is seen in the body morphology of the Cahokians excavated there – tooth decay shows a marked increase, as well as orthopaedic disorders, notably of the joints. So it seems that maize did become a sudden staple at Cahokia after AD 750, when mound plazas appear. At the same time, it is worth noting that, as T.P. Myers, of the Nebraska State Museum, has pointed out, the corn at Cahokia and other sites of Mississippian culture, is a flint corn, Northern Flint (with an especially hard outer shell), 'so distinct from its Mexican ancestors that it is almost a different species'. The difference this made, says Myers, was so marked that it 'jumped from almost no dietary importance to half of the diet in a generation or two'. His point is underlined by the fact that maize had reached New Mexico – not so far from Cahokia – by 1000 BC, but the hunter-gatherers there did not become reliant on the maize crop until AD 600, possibly for climatic reasons.

The same change, from being minimally present between AD 600–700, but becoming abundant during AD 750–850, occurred across many regions of eastern North America. 'At the same time, settlement density increased with more houses occupying less space, leaving more land for cultivation.'[25] Simultaneously, slab metates (large stones, with shallow concave grinding surfaces) decline rapidly in importance – these had probably been used for grinding chenopodium, knotweed and maygrass, on which the population had hitherto subsisted.

And so, in the New World, outside North America, we see three phases of maize: its use as beer, its slow adoption as a grain, and a much later use as a staple. Only in eastern North America is the Old World pattern paralleled, very rapidly, unlike anywhere else in the western hemisphere, and grain subsistence, and the surplus it provided, led fairly straightforwardly to urbanisation.*

Blood and Numbers

What *were* these ceremonies where maize was such a feature? There was a widespread tradition that humans were created from maize, that blood and maize were equivalent. Prayers were recited for maize, many words were created to describe its parts, its development, its role in cuisine. There was a maize calendar – many scholars believe that the 260-day calendar unique to the Americas (see chapter twenty) is related to the growing season of the corn plant. The early civilisation of the Olmec had a maize deity-ruler and the Maya believed that the first eight human beings were formed of maize (after two previous unsuccessful creations of humans, from mud and from wood).[26]

There are many legends about the secret of maize, with precious corn seeds being hidden in sacred mountains. Among the Maya there was a tradition that maize originated in the 'hearth of creation', an equilateral triangle hanging from Orion's belt in the constellation of that name, the triangle being formed by the stars known to them as Alnham, Saiph and Rigel.[27] Other traditions say that maize came from the sun's groin, or from a dwarf with golden hair. Maize plays a part in

* This is a very rough parallel, of course, and not without irony, since we now know that sedentism in the Old World preceded full-blown plant domestication.

childbirth rituals, rain petitions, medical rituals, community renewals and funerary rites. Bloodied kernels of maize are sown in honour of newly named children.[28] Maize is associated variously with the numbers 3, 7 and 8. The maize deity is represented on many Mayan glyphs.

Taking all this together, we can see that, in the Americas, the two most important food crops were very different in character from the cereals of Eurasia. The potato was specially adapted to the relatively extreme environment of the Andes mountains. This environment had its advantages but it meant that the potato, highly adapted as it was, was unable to spread, and those dependent on it could not spread either. Maize, although it is a grain, like wheat, barley and rice, was also a tropical plant, not a temperate one, and this helped to account for its high sugar content which meant that, in the very beginning, its primary use was to be chewed, for its sugar and the associated fermentation that gave it psychoactive properties.

This latter fact has also to be seen against the background of the much higher number of psychoactive plants to be found in Meso- and South America, which are the subject of the next chapter, and undoubtedly contributed to the use of maize in ceremonial contexts. All this meant that the role of maize in the New World was quite different from that of the cereals in the Old World. Aside from North America, at a very late date, there was no simple association between the domestication of maize, sedentism, the development of surplus, leading to the subsequent growth of civilisation. In other words, in the New World there was not the uniform spread of cereal farming as there was in Eurasia. Maize was valued in most locations, to begin with and for *thousands* of years, as the provider of an alcoholic drink, producing altered states of consciousness and conviviality, helping different hunter-gathering tribes maintain friendly relations with their neighbours/rivals in reciprocal gift-giving feasting ceremonies.

Valuing maize for its psychoactive properties seems to have slowed down appreciation of it as a foodstuff. No doubt this also had something to do with the relatively poor quality of rainforest soil, as discussed earlier, which took much longer to regain its fertility, and the absence of ploughs, which inhibited the innovation of broadcast sowing.

A final factor to be considered is that the protein value of maize is much increased when it is paired with beans (maize, beans and squash spread together in the New World, much as did wheat, barley and oats in the Old World). Maize, and beans, together with amaranth and sage, were very successful eventually but perhaps discovering the advantages of their combined use also took time.[29]

Among hunter-gatherers, the gift-giving ceremonies, with altered states of consciousness, would have maintained shamanistic forms of religion far longer than in the Old World. As in the Old World, however, alcohol consumption – to begin with – was associated with elite status; but in small tribal bands, and with no domesticated mammals to deal with, without driving or riding, there would not have been the same pressures to dispense with the more powerful narcotics and hallucinogens that were much more available in the New World – and especially in South and Central America – than anywhere else.

• 12 •

The Psychoactive Rainforest and the Anomalous Distribution of Hallucinogens

Just as the Old World had far more domesticated mammals than the Americas, so the situation is reversed with regard to hallucinogens. In 1970 Weston La Barre published a significant paper in the journal, *Economic Botany*, entitled 'Old and New World Narcotics: A Statistical Question and an Ethnological Reply'. In that paper La Barre attempted for the first time to account in terms of cultural history for the astonishing proliferation of sacred hallucinogens in Indian America. The 'statistical question' of his title referred to the 'striking anomaly' between the much greater number of psychoactive plants known to the original Americans, who had utilised between eighty and a hundred different species, as compared with the much smaller number – no more than eight or ten – used in the Old World. Since the Old World has a much greater landmass than the Americas, and its flora is as rich and as varied, if not more so, one would have expected there to be more hallucinogens in Eurasia. Added to which, humans have been around much longer in the Old World than in the New.

La Barre's answer was that American Indian interest in hallucinogenic plants 'is directly tied to survival in the New World of an essentially Paleo-Mesolithic Eurasiatic shamanism, which the early big game hunters carried with them out of northeastern Asia as the base religion of American Indians'. In other words, the first Americans were 'culturally programmed' to consciously explore and exploit their new environment in search of means 'by which to attain the desired "ecstatic" state'.[1] It was also La Barre's thesis that 'while profound socio-economic and religious transformations brought about the eradication of ecstatic shamanism and knowledge of intoxicating mushrooms and other plants over most of Eurasia, a very different set of historical and cultural circumstances favoured their survival and elaboration in the New World'.[2]

Peter Furst, in his book, *Hallucinogens and Culture*, argues that the basic foundations of the symbolic systems of American Indians 'must have been present already in the ideational world of the original immigrants from northeastern Asia'. These foundations, he insisted, are shamanistic and they include numerous concepts recognisable even in the highly structured cosmology and ritual of hierarchic civilisations, for example, the Aztecs, where we find such phenomena as: the skeletal soul of humans and animals; the restitution of life from the bones; the belief that all phenomena in the environment are animate; the belief in soul flight; the belief in the separability of the soul from the body during life (e.g., by soul loss, by straying during sleep, or by rape or abduction, or else the soul's deliberate projection, as by shamans in their ecstatic dreams); an initiatory ecstatic experience, especially of shamans, who must often begin their vocation by undergoing sickness; supernatural causes and cures for illness; the belief that there are different levels of the universe with their respective spirit rulers, and the requirement for feeding these on spirit food; the belief in human-animal transformation – indeed, says Furst, *transformation rather than creation* becomes the origin of all phenomena; the belief in animal spirit helpers for the shaman; supernatural masters and mistresses of animals and plants.

Furst concluded that with the concept of transformation being so prominent in these traditional systems, 'it is easy to see why plants capable of radically altering consciousness would have come to stand at the very centre of ideology'.[3]

As La Barre's original hypothesis had it, Asia and Europe formerly shared in the shamanistic world view but the Neolithic Revolution and subsequent economic and ideological developments brought about profound changes in religion 'although ancient shamanistic traditions are here and there still visible even in the institutionalised churches'. In the New World, in marked contrast, the original practice of hunting and food gathering, and the shamanistic religious rituals that went with them, persisted much longer, even into the great civilisations that arose in Mesoamerica and the Andes.[4]

Just how old these practices are has been shown from radiocarbon studies which confirm that the hallucinogenic mescal bean, *Sophora secundiflora*, was already used by Palaeo-Indians towards the end of the late Pleistocene big-game hunting period, 11,000–10,000 years ago – 'not long after the cessation of the last overland migration from Asia'. And it was still being used as part of the desert culture of the North American south-west at AD 1000.[5]

The remains of *Sophora* seeds, and associated paraphernalia and rock paintings, have been found by archaeologists in a dozen or more rock shelters in Texas and northern Mexico, as often as not grouped together with another narcotic, *Ungnadia speciosa*. The oldest were dated to 7265 BC and the latest at the time the site was abandoned.[6] 'At Fate Bell Shelter in the Amistad Reservoir area of Trans-Pecos Texas, a region rich in ancient shamanistic rock paintings, the narcotic seeds of *Sophora* and *Ungnadia* were found in every level, from 7000 BC to AD 1000, when the desert culture finally gave way to a new way of life based on maize agriculture.' Even more important for a general understanding, studies at Bonfire Shelter, a well-known rock shelter site, near Langtry in south-west Texas, showed *Sophora* seeds present at its lowest occupational stratum, dated to 8440 to 8120 BC – that is, well into the big-game hunting era. More than that, the seeds were excavated alongside Folsom and other projectile points, together with the bones of the large extinct species of the Pleistocene bison, *Bison antiquus*.

Furst found it remarkable that a single hallucinogen, the *Sophora* bean, should have enjoyed 'an uninterrupted reign' of over 10,000 years as the focus of ecstatic-visionary shamanism. It was all the more extraordinary, he said, because, of all the many hallucinogens native to the New World, only the genus *Datura* ('jimsonweed') poses such a physiological risk as does *Sophora*. 'Clearly, the individual, social and supernatural benefits ascribed to the drug must have outweighed its disadvantages.'[7]

Over enormous areas of North America many aboriginal peoples achieved much the same ends by non-chemical means, such as: fasting, thirsting, self-mutilation, exposure to the elements, sleeplessness, incessant dancing and other means of total exhaustion – bleeding themselves, plunging into ice-cold pools, laceration with thorns and animal teeth, and other painful ordeals, as well as a variety of less brutal 'triggers': different kinds of rhythmic activity, self-hypnosis, meditation, chanting, drumming and music. In one, holes were cut into men's shoulders or arms and bison skulls hung from the bleeding fissures. In another technique, some shamans used mirrors of obsidian and other materials to induce a trance state (some Indian shamans in Mexico still do). A final widespread technique was the spirit-quest ordeal practised by certain Plains Indian tribes, such as the Oglala Sioux and the Mandan.[8]

Such ordeals were not uncommon in ancient Mexico either. Self-mutilation is portrayed in the art of several pre-Hispanic cultures such as the Maya. These include bloodletting rites that must have been incredibly painful – perforation of the penis, tongue and other organs with cactus thorns, stingray barbs and other sharp instruments (see chapter 21 for an extended discussion). In one, holes were cut in the shaman's back and he was suspended with ropes passed through the folds of his skin. A well-known Maya carving, dated to ~AD 780, from the ceremonial centre of Yaxchilán in the Usumacinta region of Chiapas, shows a kneeling woman in the act of drawing through her tongue a twisted cord, set with large sharp thorns. She is richly dressed and was clearly a member of the elite (see figure 13, page 418). Such rites are often discussed in Maya literature in terms of blood sacrifice – blood being the most precious gift mankind could offer to the gods in ancient Mesoamerican thought. 'But in point of fact [it] must have constituted a violent shock to the system, sufficient to bring about alterations in consciousness to the point of visions.'[9]

Vines of the Soul

We have already (in chapter 10) explored the psychoactive substances used in the Old World. Of the more than eighty psychoactive substances known to have been employed in the New World, we have space here to mention only the ten most studied plants and even then very briefly.

The first is the hallucinogenic harmala alkaloids (harmine, harmaline, harmalol and harman), originally isolated from a substance we have already met, *Peganum harmala*, or Syrian rue. Syrian rue is one of at least eight plant families of the Old and New Worlds in which harmala alkaloids are now known to be present. The most numerous, and the most culturally interesting, again according to Peter Furst, is

Banisteriopsis, 'a malpighiaceous [liana-like, vine-like] tropical American genus that comprises no less [*sic*] than a hundred different species, of which at least two, *B. Caapi* and *B. Inebrians* ... are the basis of the potent hallucinogenic ritual beverages of the Indians of Amazonia'.[10] In Quechua, the language of the Incas, and many Peruvians today, the drink is known as *ayahuasca*, or 'vine of the souls'. In the north-west Amazon, where it was widely used, it is known as *yajé* or *yagé*, a Tukanoan word spoken by, among others, the Desana of Colombia, a tribe intensively studied by Gerardo Reichel-Dolmatoff.

The active substance is found in the bark, stem and leaves of the plant and in one experiment the alkaloids were shown to remain active for 115 years. Furst estimates that the use of *yajé* is at least 5,000 years old, Tukanoans placing the hallucinogen at the very start of tribal history, where it is said to have emerged in human form after the male sun had fertilised the female Earth with its phallic ray and the first drops of semen had become the original people. The Tukanoans equate vines with lines of descent, each phratry (kinship group) having its own distinctive vine and its own distinctive way of preparing the sacred drink. The pottery vessel that holds the liquid, for example, symbolises the maternal womb and its base is painted with a vagina and clitoris. Before the vessel can be used it must be purified with tobacco smoke.[11]

The consumption of *yajé* is highly ritualised. Its distribution is announced by sounds on musical instruments, at set intervals, preceded by ceremonial dances. The hallucinations are called '*yajé* images' and, to ensure these images are always bright and pleasant, individuals must have abstained from sex and eaten only lightly for several preceding days.

People who take *yajé* expect to awaken as a new person. They believe that, under the influence, they return to the uterus, where they see the tribal divinities and witness the creation of the first couple, of the animals, of the establishment of the social order. The origins of evil also manifest themselves – these include illness and the spirits of the jungle, in particular the jaguar. Reichel-Dolmatoff's research has shown that in Tukano society the shaman is closely associated with the jaguar, a powerful nocturnal animal, equally at home on land, in the trees or on water. Shaman-jaguar transformations are closely associated with trances, whether brought about by *yajé*, tobacco or snuff. Jaguars are regarded as avatars of deceased shamans and under the influence of *yajé* the Tukano may have a 'bad trip' when they are overcome by the menacing jaguar or huge snakes.[12]

One final 'origin': the Tukano attribute the origin of art to *yajé*. They say that the striking polychrome images that adorn the front of their houses, on their pottery and their musical instruments have all first appeared under the influence of their psychedelic drink, each design having a fixed meaning. Here again images of felines predominate but the more abstract signs have meanings too, in terms of incest, exogamy, fertility and so on. This recalls the 'entoptic' images drawn on cave walls in Pleistocene times and studied by David Lewis-Williams.[13]

Among the Aztec, their main sacred hallucinogen (they used several) was known as *Ololiuhqui*. The name means 'round thing' but is now known to refer to *Rivea corymbosa* (a species of Morning Glory) and which, together with *Ipomoea violacea* (another variety of Morning Glory), was worshipped throughout central America – among the Mazatecs, Mixtecs, Chinantecs and others.[14] The chemical make-up of these substances was not actually settled until the 1960s, when quantities of seeds were sent to a Swiss pharmaceutical company for analysis. This revealed that their main (psycho)active ingredients were the ergot alkaloids – *d*-lysergic acid amide (ergine) and its isotope, *d*-lysergic acid amide (isoergine), both closely related to *d*-lysergic acid diethylamide or LSD.[15] The curious aspect to all this is that *Rivea* is in fact mainly an Asiatic genus of woody vine, with five Old World and just one New World species. But it was only ever developed for shamanistic use in Mesoamerica, never in the Old World or, for that matter, in South America. Ergot, of course, is what Mott T. Greene believes was the main ingredient in *soma/haoma* (see above, chapter 10).[16]

In Central America, however, *Ololiuhqui* was itself regarded as a divinity. It was worshipped, the seeds prayed to, petitioned with incantations and even honoured with sacrifices and incense.[17] Both species featured in the art of the cultures native to Central America, most notably in the mural paintings at Tepantitla, a complex of sacred buildings within the great pre-Columbian city of Teotihuácan, which flourished from the first to the eighth century AD, north of what is today Mexico City. According to Furst, these paintings have been dated to the fifth or sixth centuries AD, when Teotihuácan was one of the largest cities on Earth, with up to 200,000 inhabitants. The main image shows a deity, a highly stylised Mother Goddess (once thought to be the rain god, Tlaloc), from which flows a stream of water, to fertilise the land, but above her is a great vine-like plant with funnel-shaped flowers. Seeds fall from the deity's hands, seeds of *Rivea corymbosa* (see figure 9). *Rivea* has also been found inscribed on the bodies of another Central American god, Xochipilli, the Aztec god of flowers, together with the hallucinogenic mushroom *Psilocybe aztecorum* and *Nicotiana tabacum*, one of two principal sacred tobacco species.[18]

Fig. 9 Mural from Teotihuácan, Mexico, dated to *c.* AD 500, depicting a Mother Goddess and her priestly attendants with a highly stylised Morning Glory plant, *Rivea corymbosa*, the sacred hallucinogenic *Ololiuhqui* of the Aztecs.

Among the Aztecs, the hallucinogenic experience was called *temixoch*, translated as 'flowery dream', and the sacred mushroom was called *teonanácatl*, *teo* meaning god and *nácatl* meaning food or flesh.

The so-called 'magic mushrooms' of Mexico and Guatemala are perhaps the most-studied hallucinogens of the New World. They were used (and in several cases are still used) in the Mayan area, and there is evidence that they have been employed for 3,000 years, notably, according to archaeological discoveries, between ~1000 BC and AD 900. One of the more interesting references, as we shall subsequently see, is that found in a sixteenth-century Spanish account of mushroom use, which refers to them as *xibalbaj okox*, *xibalba* being the Mayan word for 'underworld' and *okox* being mushroom. The Mayan underworld (again as we shall see in more detail later) was deemed to have nine levels but the Mayan phrase meant, in effect, that this particular magic mushroom gave people images of hell or the dead.[19] Many of the other terms associated with mushrooms implied that it was an intoxicant or inebriant – it made people 'crazy', or they 'fell into a swoon'.

In the early 1960s details were published of nine 'beautifully sculpted' miniature mushroom stones and nine miniature metates (grinding stones) dating back 2,200 years. They were found in a richly furnished tomb outside Guatemala City. In Mayan cosmology there are nine lords of Xibalba, the underworld, as described in the *Popol Vuh*, the sacred book of the Quiché-Maya, and so there is a strong presumption that these statues were sacred in some sense: either they were there to accompany the dead person on his or her journey to the underworld; or they were themselves divinities offering companionship and protection. Effigies of mushrooms have been collected all over Central America (El Salvador and Honduras as well as Mexico and Guatemala). According to Furst, some 200 different species have been identified. Many of them, especially those that date to between 1000 and 100 BC, incorporate a human face or figure, or a mythic or real animal, where again the jaguar but also the toad are most represented (the toad, of course, undergoes a major transformation during its lifetime).[20]

Mushroom effigies in fired clay have been found in both Central and South America, some of which appear to resemble *Amanita muscaria*, the fly-agaric mushroom of Siberia rather than *Psilocybe*, the predominant New World psychoactive species. The ceramic art of the Moche of Peru (~400 BC–AD 500) also shows anthropomorphic mushroom effigies, as do gold pendants from Colombia and Panama – again it seems as though a deity is represented.[21]

Several of the early Spanish explorers in Central America mentioned the sacred mushrooms used by the New World inhabitants, and more than one cult, we know, survived into the twentieth century. From this we also know that different species were revered for their different properties. In this way, it also became clear that shamans used mushrooms primarily for divining the cause of illness, though elsewhere they were used for their visionary properties.

Furst says the mosaic of magic mushrooms is now fairly complete, in that we know which ones were most often used and what their properties are. '*Psilocybe Mexicana*, a small, tawny inhabitant of wet pasture lands ... is probably the most important species utilised hallucinogenically in Mexico, but the strongest psychedelic effects seem to belong to *Stropharia cubensis*.'[22] But eight other species (in 57 varieties) were in widespread use, different shamans having their own favourites.

The question as to whether Central American traditions are a carry-over from Siberian practices, or a separate development, cannot be answered directly. The spread of *Amanita muscaria* is native not only to Eurasia but to British Columbia, Washington, Colorado, Oregon and the Sierra Madre of Mexico. In historic times the urine of shamans was considered to posses great magical and therapeutic powers by the tribes of the north-west coast and by the Eskimos. This recalls the shamanistic tradition of reindeer hunters in Siberia, discussed in chapter 3. At the same time, many of the edible mushrooms of central America sprout at times of the year when other foods – such as maize – are not yet ready. Their psychoactive properties could have been discovered accidentally.

The earliest hallucinogenic cactus depicted in ancient American art is a tall member of the *Cereus* family, *Trichocereus pachanoi*. Used by the folk healers of coastal Peru, it is known to contain mescalin and the cactus has been identified on the funerary effigy pottery and painted textiles of Chavín, the oldest in a long succession of Andean civilisations, dating to ~1000 BC. The hallucinogenic cactus is also identified in the ceremonial art of the later Moche and Nazca cultures, giving this sacred psychedelic plant a cultural 'pedigree' of at least 3,000 years.[23]

But the most important hallucinogenic member of the cactus family is a small spineless North American native of the Chihuahuan desert, slap in the middle of the 'Thermal trumpet', as mentioned in chapter 5. This is *Lophophora williamsii*, more widely known as 'peyote'. Peyote was held in great esteem right across Mesoamerica, where it is represented on ceramics from 100 BC on, at the very centre of shamanic rituals over a wide area (even to the north, as far as the Canadian Plains). And its use continued, so much so that it has now been legally incorporated into the Native American Church, which has hundreds of thousands of adherents.[24]

Peyote was described by Richard Shultes as a 'veritable factory of alkaloids', with more than 30 types of their amine derivatives being isolated. The best known by far is mescaline, which is the principal vision-inducing ingredient. Its effects include not only brilliantly coloured images but also 'shimmering auras that appear to surround objects in the natural world ... auditory, gustatory, olfactory, and tactile sensations, together with feelings of weightlessness, macroscopia [seeing things from far away, or from high above], and alteration of space and time perception'. Its continued use can probably be put down to the fact that many of its practitioners came from very isolated groups, such as the Huichols and their cousins the Coras who, despite being discovered by the Spanish, managed through their isolation to resist European influence for many centuries. Even today, therefore, the Huichols regard peyote as sacred, divine.[25]

It should also be said that peyote is more than a psychedelic substance. It is a remarkable stimulant, an effective antidote against fatigue – very useful in a mountain environment. It also has antibiotic properties against a wide slew of bacteria.[26]

Culturally, peyote is notable for its identification with the flesh of the deer. In Huichol mythology, at the time the Great Shaman led the ancestral gods on the first peyote quest,[27] all the ills of the tribe stemmed from the fact that they had forgone the great hunt for the Divine Deer, peyote. Even today every Huichol is familiar with the story, which is recounted to the young when the first ears of maize have ripened in the fields. Among present-day Huichols, says Peter Furst, about half the adult men are regarded as shamans – mostly the heads of families – and although some are more prestigious than others, with more influence among their people, all of these shamans are regarded as having more intense experiences under peyote, travelling to other levels of the cosmos, including the underworld.[28]

The Huichols also make use of another hallucinogen, to which they pray and give the name *Kieri*. This grows in remote, rocky places and consists of white, funnel-shaped flowers and spiny seed pods. This is *Datura inoxia*, though there is some evidence that *Kieri* is also sometimes *Solandra guerrerensis*, equally narcotic if not quite as potentially dangerous. *Datura* was first encountered by the Spanish among the Aztecs, where it was given to those about to be sacrificed so that they would feel less pain. But *Datura* features in a prominent myth of the Huichol in which the shaman, *Kieri*, battles with an adversary, *Kauyumarie*, who shoots several arrows into him. Instead of dying, however, *Kieri* is allowed by his protector, the sun deity, to transform into a flowering plant. Furst says that this myth is a folk legend of a change in ideology, away from *Datura* to the more benign peyote, and may represent an historic account of a battle, perhaps, between rival groups of shamans who used different hallucinogens.[29]

Certainly, *Datura*, a genus that grows in many parts of the world, not just the Americas, is a powerful narcotic: deadly nightshade, henbane and mandrake are all related more or less closely. In the New World, among many peoples, it was added to other hallucinogenic preparations to augment the effect but by itself, beyond its analgesic properties, it can cause unconsciousness and even death. Native Americans were well aware of this and its lethal possibilities no doubt had something to do with the Huichol legend about the fight between *Datura* and peyote.[30]

Properly managed, however, it was undoubtedly useful. Among the Zuñi, where the Rain Priest Fraternity had a special relationship with *Datura*, it allowed curers to perform simple operations while the patient was unconscious – resetting fractured limbs, making incisions to remove pus, opening abscesses.[31]

It is also known that the amount of scopolamine – one of the active ingredients in *Datura* – can vary in strength from 30 per cent to 60 per cent and this may account for some of the negative stories about temporary derangement or permanent insanity that surround its use – people were simply given too much. Properly controlled, and used by experienced shamans, like other hallucinogens it enabled the initiated to travel to other realms of the cosmos, to communicate with the ancestors, to trace and capture the souls of people who were sick, and generally to commune with deities.

Other evidence for the sheer strength of *Datura* comes from the Luiseño Indians in California, who are related to the Uto-Aztecan peoples of the south, and where a puberty rite for young men, or boys, used *Datura*. In one part of the rite, the initiates were given the substance as a drink, the only time in their lives when these individuals would consume this liquid. The *Datura* was normally kept hidden in secret hiding places and was only drunk from a freshly painted mortar where the liquid was prepared, and which was used for no other purpose. Enough *Datura* was taken for each boy (and some girls) to become unconscious. Even in these ceremonies, so carefully calibrated, some initiates died but those who didn't reported visions under the influence, visions in which they encountered an animal, from whom they learned their own personal song, teaching them wisdom and which they would remember always. A few weeks later, the initiates had to pass through a trench, climbing over an animal/human effigy where they mustn't slip or, it was held, they would die early. The trench was said specifically to represent the Milky Way.

Another California tribe, the Cahuilla, regarded *Datura* as 'the great shaman', so that only other shamans could communicate with it during ceremonies. This was achieved by an esoteric 'oceanic language', which only shamans understood, and was spoken by the supernatural beings that lived on the ocean floor. (Underwater volcanoes?) Among the Cahuilla, too, the shaman (called a *puul*) could transcend ordinary reality and fly to other worlds, transform himself into other animals (especially the puma and the eagle), and bring back lost souls.[32] The Cahuilla shamans used *Datura* as a medicinal paste, though here again not everyone shared the same enthusiasm for a substance that was clearly unreliable.

Sacred Snuffs and Toxic Toads

Hallucinogenic snuffs were also widely used, though with a somewhat different distribution: they appear to have been confined to South and Central America, and to the Caribbean. Holly (*Ilex*), acacia and mimosa species may all have been used at one time or another (not forgetting tobacco, see the next chapter) but the *Anadenanthera* and *Virola* families seem to have accounted for most forms of snuff, infusions and even hallucinogenic enemas.[33]

The active alkaloids in both *Anadenanthera* and *Virola* are the tryptamines. These require a monoamine oxidase inhibitor to become active in man, and Indians solved this problem by mixing different hallucinogens together. This had the added effect of making intoxication with snuff extremely rapid. For example, among the Waika (in the Roraima province of Brazil), one man blows the prepared snuff through a long bamboo pipe directly into the nostrils of another man, who almost immediately begins to react. The effect lasts only a short time but during that interval the man will experience instant communication with animals, plants, deceased relatives and other supernaturals. There is some evidence that trained shamans can control their reactions better than others.[34]

The animals most frequently seen under the influence (and this seems to be a characteristic of snuff, that animals are seen) are the bird-feline-reptilian complex, the specific animals being those who, like shamans, can move between the different realms of the cosmos. Apart from the jaguar – always very popular, as the alter ego of the shaman – diving birds are common, appreciated for their ability to reach the underworld, and eagles and condors, which appear to reach the sun.

Snuff pipes and other equipment have been found from ancient Bolivia, Costa Rica, Argentina, Chile, Peru, and Brazil dating back to 1600 BC. In Mexico there was an ancient snuffing complex that dated back at least to the second millennium BC but seems, for some reason, to have died out by AD 1000. Snuffing activity – effigies of people in 'trances' with snuff pipes inserted in their noses – are known from tombs in Colma, western Mexico, dated to 100 BC and from Xochipala, Guerrero, southern Mexico, from 1300–1500 BC. There are also some famous Olmec jade artefacts, called 'spoons' by some archaeologists, that could have been snuff equipment. They too have bird-jaguar motifs and date to 1200–900 BC.[35]

Allied to the hallucinogens are a number of toxic substances associated with frogs and toads, which feature strongly in Native American mythology and legend. These creatures are interesting to us for at least three reasons. In the first place, apart from butterflies, and as was mentioned above, frogs and toads undergo the most dramatic transformation in morphology, from tadpole to adult form and this is perhaps what attracted interest to them in the first place – frogs and toads show shamanistic behaviour entirely naturally. A second reason is that many species of frog and toad secrete substances that are toxic, and in a few cases hallucinogenic. And third, though this is tenuous, there are similarities between New World frog and toad mythologies and those in China, Japan and along the arctic rim of Eurasia, as far as Finland, where the word for toad, *sampo*, is also the word for mushroom.

There is also in North and South America a widespread myth complex that connects the toad to the Earth as the manifestation of an Earth Mother Goddess, who is at once the destroyer and giver of life. On occasions, the toad represents the Earth, and from her body grew the first food – maize in Mexico, bitter manioc in Amazonia.[36] The Earth Mother Goddess is also the benefactress of the first people or 'culture heroes', instructing them in the skills of hunting and the magic arts, and it is her dismemberment which precipitates the origins of agriculture. The Aztecs also had this Mother Goddess/monster theme.[37]

Michael Coe, professor and curator emeritus at Yale's Peabody Museum, found a large quantity of the remains of *Bufo marinus* at the Olmec ceremonial site of San Lorenzo in Veracruz, Mexico, which he published in 1971, and, in view of this toad's high poison content and sacred character, the practice may have paralleled that of the Maya in Guatemala, who are known to have added toad poison to their ritual drink to give it added potency. Elsewhere in Mexico, small, toad-shaped effigy bowls have been found, suggesting that the habit went wider.[38]

There was also the practice of *tapirage*, which occurs in South America. In this procedure, the feathers are plucked from a living parrot and a small quantity of the extremely venomous toad poison, *Dendrobates tinctorius*, is rubbed into the wound which is then sealed with wax. When the new feathers grow back another transformation has occurred – they are of a very different colour to the originals, yellow or red replacing green, for example. This practice extends from Gran Chaco (on the Argentina/Bolivian border) to Brazil to Venezuela. It is not of course hallucinogenic but it *is* magical and it *is* a transformation. Shamans in Guyana use toads and venomous frogs in ritual curing and the Amahuaca Indians of the Peruvian Montaña rub venomous frog poison over their bodies before setting off for the hunt. Their skins have been deliberately burned beforehand so that the poison enters their bodies and can cause vomiting, diarrhoea and hallucinations. In some cases the poison used is the same as that employed in their poison blow-guns: using such a substance as an aid in hunting is a classic shamanistic manoeuvre.[39]

Which brings us back to deer. Deer were important food animals right across the New World. 'But,' says Peter Furst, 'almost nowhere were they only that.' They were intimately bound up with shamanism and their flesh was often sacred. Deer deities stretched from the extreme north of Canada to the deep south of Amazonia and deer ceremonies, says Furst, were 'near universal'. They were never hunted except in a ritual context – this applied, for example, to both the Maya, the Mazatecs in Oaxaca and among the Gé of Brazil. Moreover, the deer was in many places intimately bound up with the use of, and attitudes to, hallucinogens. The best-known aspect of this was the Huichols' practice of equating peyote with the deer, where it was regarded as the 'mount par excellence' to the upper levels of the universe. North of Mexico, deer were linked with tobacco but in the Andes, among the Moche, for example, deer were associated with *Anadenanthera colubrine*, the principal active ingredient in the divine compound known as *huilca*. In the southern plains, deer were involved with the 'ecstatic-shamanistic' mescal bean, *Sophora secundiflora*, where a 'deer dance' formed one of the major ceremonies. The Zuñi too had a deer-maize-peyote tradition, reflected in their ceremonies, one of which involved amassing a collection of wild flowers, flowers that include *Datura* and other hallucinogens and which, the Zuñis believed, attracted deer who 'go crazy with them', and where, on these grounds, their shamans sought to transform themselves into those flowers so that the deer would be 'attracted within range of their arrows'. The Zuñis have a tradition whereby they are reborn three times, then reborn a fourth and final time as a deer.[40]

Furst argues that, during Pleistocene times, before the glaciers melted, deer would have ranged much further south in Eurasia than they did in later times, so that early peoples would have had a much more intimate relationship with these animals. He shows that among the Reindeer Tungus in Siberia the shamans wear a cap crowned by antler effigies and in many places their attire is indistinguishable from the cave paintings in France. The Scythians, much later, had a practice whereby they put antlers on their horses, in the belief that such devices would help them be transported to the other world. Furst, for one, is convinced that early humankind carried these beliefs with them into the New World. They couldn't know that there were far more hallucinogens waiting for them in the Americas but, given this background, it would have been natural for them to make the most of what they found there.

Hallucinogens and Shamanism

The interaction between hallucinogens and shamanism is expanded by the ethnographic evidence. Michael J. Harner, one of the foremost researchers in this area, agrees that 'the American Indian cultures have often preserved an emphasis on shamanism'.[41]

Fieldwork has now been carried out among several tribes, in a systematic effort to explore what is common across cultures and what is confined to this or that tribe, and why. Among the Cashinahua, for example, a small tribe of about 500 people inhabiting the rainforest of south-eastern Peru, almost any initiated male may drink *ayahuasca*, the intoxicating brew of the *Banisteriopsis* or *Psychotria* vines that we met earlier. Consumption occurs roughly once a week and is never taken alone. Research into these experiences shows that certain themes recur, the most frequent of which are: (1) brightly coloured large snakes; (2) jaguars and ocelots; (3) spirits; (4) large trees, often falling trees; (5) lakes, frequently filled with anacondas and alligators.[42] Very often the experience is frightening, but the men persist in their habit because they believe that the visions they see under the influence can act as a warning about ill health, or hunger, or famine, or death, in the days and weeks ahead. After a drinking session, the men spend several hours discussing what the visions meant.

When illness occurs among the Cashinahua, they first try to deal with it themselves, within the family; if that fails, they consult a herbalist, someone skilled in the properties of the plants of the forest; only if that fails repeatedly, do they consult the shaman. In the first instance he tries to draw out the malign substance interfering with the health of the patient, and only if that fails does he resort to *ayahuasca*, to consult the spirits who will tell him what the real cause of the illness is. Treatment often takes the form of chanting or a sucking motion.

Among the Jivaro of the Ecuadorian Amazon, there are two kinds of shaman, the bewitching shaman and the curing shaman. As part of his researches among this tribe, Harner himself took what the Jivaro call *natemä*, a substance also extracted from the *Banisteriopsis* vine. This tribe regards everyday life as a 'lie', according to Harner, real life being glimpsed only under the influence of *natemä*. Here is his (brief) account of taking *natemä*: 'For several hours after drinking the brew, I found myself, though wide awake, in a world literally beyond my wildest dreams. I met bird-headed people, as well as dragon-like creatures who explained that they were the true gods of this world. I enlisted the services of other spirit-helpers in attempting to fly through the far reaches of the galaxy. Transported into a trance where the supernatural seemed natural, I realised that anthropologists, including myself, had profoundly underestimated the importance of the drug in affecting native ideology.'[43]

Harner found that approximately one out of every four Jivaro men is a shaman, the chief power of which resides in their ability to summon up *tsentsak*, spirit helpers who are only visible to shamans under the influence of *ayahuasca*. These spirit helpers can take various forms – giant butterflies, jaguars or monkeys are common – and the services they perform may help the Jivaro on their headhunting raids, or in causing or curing illness. They possess darts – again invisible to all except the shaman in a drug-induced trance – which can kill or injure individuals. Shamans can also suck *tsentsak* from people's bodies, *tsentsak* put there by hostile shamans. Ceremonies take place in the early hours of darkness but the shaman constantly drinks tobacco juice to keep narcotised and his spirit helpers alert.[44]

The name 'Sharanahua' means 'good people'. They inhabit an area of eastern Peru, along the banks of the Upper Purús River, practising slash-and-burn agriculture and subsisting, mainly, on manioc, plantains, bananas, peanuts and maize. Three out of twenty-five males in Sharanahua society are shamans. They too are admired and feared because they can cause harm as well as offer help. They too can suck harmful substances from ill people, and they too can cause illness or misfortune by throwing *dori*, a magical substance, into the bodies of their victims. One of their major techniques of curing – and this is repeated elsewhere – is by singing. The Sharanahua have hundreds of songs, about everything, but the songs are sung in an esoteric form of the language, difficult for others to understand and 'filled with metaphors'.[45] As in other tribal societies, among the Sharanahua the apprentice shaman must undergo a prolonged period of celibacy, shamans frequently take on the form of animals, ceremonies tend to begin in the early hours of darkness, and particular attention is paid to dreams. Janet Siskind, who carried out the fieldwork, believes that experienced shamans can – to an extent – control and even manipulate their visions under the influence. More interesting still, she thinks that many occasions for shamanistic ritual deal less with physical illness than what she called 'alienation' from tribal society.[46]

Among the Campa, another tribe of eastern Peru, Gerald Weiss found that shamans also used a mix of *Banisteriopsis* (also called *ayahuasca*) and tobacco juice. *Ayahuasca* puts the shaman in touch with the spirit world and, among the Campa, the shaman keeps a supply on hand so he can enter this spirit world frequently. Ceremonies once again begin in the early hours of darkness and occasionally consumption of *ayahuasca* will be a group activity, not just of the shaman. But only the shaman undergoes soul flight and he leads the singing designed to offer up cures for afflictions. In this case, however, the shaman is not possessed *by* the spirits, instead he merely repeats what they say. According to Weiss, among the Campa the shaman is the master of ceremonies, the director of the show, 'but not the only virtuoso'. He argues that this is an important change: the shaman is beginning to take on more of the qualities of a priest.[47]

Overall, Michael Harner found there were five themes common to all cultures where shamans used hallucinogens: (1) the soul is believed to separate from the physical body and make a journey, often with the sensation of flight, and frequently to the Milky Way (we shall see why the Milky Way was so important in chapter 20); (2) there are visions of snakes and jaguars, often fearsome; (3) visions of demons and deities; (4) people and locations are seen at a distance; (5) divination occurs – seeing who committed crimes, for example, or which hostile shaman is bewitching sick or dying persons.

All this was augmented by an intriguing experiment carried out by Claudio Naranjo in Santiago, Chile, and published in 1967, in which he gave *yajé* to young, thoroughly modern Chileans who had no real knowledge of anthropology. He found that these individuals mostly wanted to close their eyes in trance 'since the external world appears as of little interest and [as] distracting from the world of visions and inner happenings'. Under the influence of harmaline, there was invariably 'an inclination to think about personal or metaphysical problems with a feeling of unusual depth, insight and inspiration'. Several subjects became nauseous and experienced some form of malaise, but the sensations of rapid movement, and/or of flying and/or weightlessness, were common, together with the feeling of being separated from one's soul, of being at the bottom of the ocean, the centre of the earth or in the middle of heaven. Visions of giant animals, big cats (including the jaguar) and fearsome reptiles were also reported by seven of the thirty-five participants. Sometimes people felt they had been transformed into these animals, sometimes they were wearing masks of the creatures.[48]

It would appear, then, that the psychedelic effects of *yajé* are not entirely culturally determined but that there is something in the narcotic that exerts a physiological effect on the brain which induces experiences that have determined several of the features of shamanism.

Perhaps the most important thing about *yajé*, *ayahuasca*, *natemä* and related chemicals is their strength. There *are* dangers in consuming these substances but, provided they are culturally managed, and not taken alone, the dangers seem manageable and are far outweighed by what are perceived as their benefits.

In contrast, Harner looked at a number of accounts of witchcraft in Europe, many little-known and published in the fifteenth, sixteenth and seventeenth centuries in Latin. Most of these practices had been stamped out by the Catholic Church, especially the Inquisition, but Harner found there was more to it than that. He observed that, in the main, European witches rubbed their bodies with a hallucinogenic ointment containing such plants as *Atropa belladonna* (deadly nightshade), *Mandragora* (mandrake), and henbane, whose active content was atropine, absorbable through the skin. Paintings of the time show in particular women rubbing poles between their legs and Harner argues that this was a particularly efficient (and erotic) way to absorb the substance. Like harmaline, atropine 'sent' individuals on 'trips' and this is where the tradition of the witch's flying broomstick may have originated.[49]

Harner also makes the observation that deadly nightshade, mandrake and henbane belong to the same family as the potato, tomato and tobacco, a family which also includes *Datura*. He therefore offers the theory that the European witchcraft hallucinogens were actually, like *Datura*, too powerful to control and this is why witchcraft never became more organised than it did, why it didn't catch on more than it did, and was one of the associated reasons for why the Catholic Church was so opposed to it.

Here then we have an overlap of sorts with Andrew Sherratt's theory, that in the Old World more moderate psychoactive substances took over from stronger varieties as society became more technologically complex. In the more vegetal New World, where communities kept themselves to themselves, because they were far more dependent on plants with a limited distribution, where there were in any case no domesticated animals who could be used for their energy, or where pastoralism developed, societies remained less technologically complex, more isolated, and hallucinogens of a fairly powerful level of activity were more widely available and perceived as more useful. The whole experience of psychoactive chemicals was different in the New World and the sheer availability of *yajê*, *ayahuasca*, *natemä* and other substances kept the phenomenon of shamanism alive and well.

• 13 •

Houses of Smoke, Coca and Chocolate

The Divine Plant of the Incas

Strictly speaking coca is not an hallucinogenic – it is a stimulant. But, according to W. Golden Mortimer, it is quite unlike any other plant, and its lack of hallucinogenic properties did not affect its importance to the Incas, who regarded it as a 'divine plant ... nature's best gift to man'.[1]

Coca exists in several forms, all native to northern South America (and, in a few cases, to Madagascar). The genus *Erythroxylum* contains 250 species, of which at least thirteen are cocaine-positive, cocaine being the active alkaloid which mainly gives coca the properties that made it so attractive to ancient South Americans (though it is not the only active ingredient – coca also contains phosphorus, calcium, iron and several vitamins). The shrub that is generally used is *Erythroxylum coca* variety *coca*, though *Erythroxylum coca* variety *ipadu* has also been harvested in the Amazon Basin for thousands of years. A third species, *Erythroxylum novogranatense*, variety *truxillense*, was regarded as the best by the Incas, who called it 'Royal Coca', though the active ingredients in this plant are fewer and harder to extract.[2]

Coca grows faster at sea levels but its cocaine content increases at higher altitudes and so it is usually cultivated at between 1,500 and 6,000 feet. Even so, the amount of cocaine in the leaves is very low, so that its effects are by and large beneficial, unlike the commercially concentrated cocaine of our own day (cocaine itself was not isolated until 1859). The most common form of ingestion of coca is by chewing, by forming a small ball or *bola* of leaves, mixed with lime (called *llipta*), to aid the release of the alkaloids, and is held between the gum and the cheek (and sometimes mixed with tobacco). It soon induces a form of numbness in the mouth but coca's properties are legendary in South America, where it has been described as a cure for snow blindness, headaches, constipation, neurasthenia, asthma, to stimulate uterine contractions, to cure open wounds and is even said to act as an aphrodisiac and induce sleep (not at the same time, presumably).

Its chief effect, however, is undoubtedly the contrary, as a stimulant – it suppresses hunger, thirst and, most important, fatigue. Ernest Shackleton took Forced March cocaine tablets to Antarctica in 1909 for the energy boost they gave.[3] Even today, in South America, journeys are still regulated in terms of *cocadas*, the number of *bolas* needed to complete a journey, the effects of a *bola* usually lasting around 45 minutes, or three kilometres on level ground, two when climbing. Coca is also credited with keeping chewers warmer in the cold mountains.

The physiological effects of coca have been variously attributed to its ability to restrict blood-flow to the skin, increasing body temperature, to keeping waste products out of the blood system, to dilating air passages in the lungs and nose.

There is archaeological evidence which dates coca use back to 2500 BC, when it is shown in paintings and figurines which have their cheeks bulging on one side. Coca is found in all manner of ancient graves, as part of the preparation of corpses for the (journey to the) afterlife and the very word, coca, may be derived from Aymara, the language of the pre-Inca Tiwanaku tribes, in which '*khoka*' means plant or tree, the *original* plant or tree.[4]

The Spaniards found that the South American Indians would not work without coca and the Spanish king, after much thought, allowed them access to their favoured stimulant, at the same time instructing the church to forbid its religious use. According to Dominic Steatfeild, coca's earliest use was probably shamanistic, producing a different state of mind for those who chewed it, aiding communication with the ancestors or spirits.[5] At several sites, mummified bodies have been found accompanied by a coca vessel and shell containers in which there was powdered lime, much the same as the *llipta* alkali still used today to help release the alkaloids.

The Incas believed that coca had the powers of rejuvenation and that nothing could prosper without it. Coca was offered in sacrifices, burned so that its smoke would reach the gods. Almost certainly it was also used to help anaesthetise the sacrificial victims ahead of their ordeal. Stars in the sky were named after coca and it was offered to the earth goddess, Pachamama, to ensure plentiful harvest and the good fortune of the army in war. It was offered to the dead and divination with it was also common: the shaman would chew coca and spit on his hand with the first two fingers extended. How the spittle arranged itself and how it fell to the ground, affected the diagnosis of ailments. The Inca rulers were adepts of coca, always accompanied by their pouches, or *chuspa*. Two of the Inca leaders named their wives after coca.

Coca played an important role in initiation rites, helping young men (mainly men) endure the ordeals they had to face. Successful initiates were rewarded with slings for their (poisoned) arrows and a *chuspa* filled with coca leaves.

Coca use was generally confined to the elite, which in this case included the army (another of its properties was that it was felt to give men courage) and the *chasquis*, the imperial runners, for the obvious reason that it helped them perform their duties more efficiently (not just military duties but bringing fresh fish to the court, for example). And it was also available to the *yaravecs*, or memory men, whose job it was to act as human archives and record events on the *quipus*, the elaborate llama-wool devices of knotted strings.[6]

Over time it seems that coca use became more and more widespread. Being such a valued commodity, which grew naturally in remote parts of the Inca empire, it would have been very difficult to stop almost anyone who wanted to from growing his or her own coca plant.

The Smoking Gods

When they first reached the New World, the Spanish priests classified tobacco alongside magic mushrooms, peyote and the morning glories as traditional intoxicants. Judging by its very wide spread of use, from Canada to the Amazon and Bolivia, tobacco may in fact have been one of the oldest, perhaps *the* oldest, psychoactive substance in the Americas. It was consumed in a variety of ways – by chewing, snuffing, even by enema, as well as by smoking – but it was always used ritually, never recreationally, as it is these days.

One of the reasons for this was that the most widely used variety among tobacco's forty-five different species was *Nicotiana rustica*, which is hardier than other forms, growing where other forms cannot grow, and in which, moreover, the nicotine strength is *four times* that in modern cigarettes. This meant that it was far more likely to produce hallucinogenic effects and that it was much more addictive, either than what we would regard as 'regular' tobacco, or than other hallucinogens, very few of which are in fact addictive. *N. rustica* was therefore used mainly for its metaphysical and therapeutic properties, among such peoples as the Aztecs, the early Brazilians and those in the eastern woodlands of North America.

Tobacco was considered by many Native American peoples to be a gift to humankind from the gods, who no longer had it themselves, but craved it, and this is why it was smoked – the smoke rose to the heavens to assuage the appetite of the deities. Shamans consumed more tobacco than anyone else – in some cases prodigious amounts – and the craving this caused in them was regarded as evidence for the craving of the gods. Nowhere did it have a secular use before the arrival of Europeans, even though its distribution (near universal) was much wider than almost any other plant.[7]

Both pipes and enemas have been found from very ancient times. Early Quechua dictionaries mention *huilca* syringes, early Inca accounts refer to enemas, as do those of the Hualca Indians in Veracruz; and they are shown in Moche artworks. Enema rituals were carried out among the Maya, where jaguar deities have been found painted on enema syringes. Men gave enemas to themselves or had women do it for them, according to the art discovered at sites dating to AD 600–800. The enemas themselves appear to have been made of the femur of a small deer, with a bulb of deer bladder rather than rubber. Snuffing tobacco remained common in South America, especially Peru.[8]

Equally careful attention was given to pipes, which were equally – if not more – sacred. Tobacco was smoked by the shamans who lavished great care on their equipment. Pipes would be carved in human or animal forms, no one was allowed to laugh while the pipe was being manufactured, nor could anything be allowed to be broken. Songs were sung during construction, the completed pipes were given names and fired in a pit that was specially dug. In many societies even today the shaman must smoke incessantly, to ensure the gods have enough 'food' and the shamans must never exhale, but 'eat' the smoke 'until it suffuses their entire system'. In this way the shaman becomes 'lightened' by the tobacco and able, via ecstatic trance, to ascend to the heavens to commune with the supreme spirit in the House of Tobacco Smoke.[9]

Tobacco belongs to the *Solanacae* family of nightshades, potato and eggplant. Throughout the New World there were six hundred terms for tobacco, including three in the Maya tongues, *zig*, *kutz* and *mai*. Smoking aromatic herbs (but not tobacco) was also known in the East, in particular ancient India. When the Spaniards first arrived in the Americas they immediately noticed that people smoked wherever they went and that the Aztecs 'composed themselves to sleep with the smoke of tobacco'.[10]

The Indians of Mexico primarily smoked cigars or cigarettes, as we would call them, though they also used pipes, tobacco snuff and even chewed it. These practices may have been adopted from the incense used by shamans. In fact, it may be that the earliest use of tobacco was in the chewing form, used by shamans who mixed it with lime, which releases the alkaloids, making it more potent.[11] Besides being a more rapid way of producing hallucinogenic effects, and of bypassing the stomach and any digestive ailments that might result, tobacco enemas were regarded as very useful medicinally. They deadened the flesh and helped overcome fatigue.

In the Madrid Codex (a 112-page Mesoamerican folding book, probably produced by a single scribe around the time of the Conquest and now in the Museo de América in Madrid), three deities are shown smoking cigars, tobacco playing an important part in religious ceremonies. For the Maya its attractions were that it was aromatic, beautiful in flower and was consumed by fire, 'the great cleanser', disappearing into the great void, 'the abode of gods and departed spirits'. To begin with it was probably used exclusively by shamans but even when it was smoked generally it always retained a pervading holiness, used to seal treaties and bind agreements, and in rites of human sacrifice. The Aztecs believed that the body of the goddess Cihuacoahuatl was composed of tobacco. It was also used as an offering to the war god, Huitzilopochtli, and used in ceremonies declaring war. Tobacco gourds and pouches were symbols of divinity, and gourds have been discovered completely covered in gold leaf. They were worn by priests.[12]

The Huichol Indians of Mexico regard tobacco as a prized possession of 'Grandfather Fire' and carried balls of tobacco tied to their quivers, the balls being burned at the end of successful expeditions. In both North and South America and in the Caribbean, tobacco was believed to augment the powers of the body and it was used as part of the induction into shamanhood. The Menominees of north-east Wisconsin buried their dead with tobacco in order to placate their gods. Francis Robicsek reports that tobacco smoke was used in shamanistic divination by the Venezuelan tribes, the Florida Indians, the Guajiros of Colombia, the Cumanos of the Orinoco, the Arawaks, the Caribs and several other tribes. The Mayans believed that their gods loved tobacco and that comets were the still-burning cigars the deities had thrown away. Several tribes sprinkled tobacco powder on the chest and face of the very ill to protect them from evil spirits, especially those of the underworld. Elsewhere shamans would spit tobacco juice on to sufferers, the mixture being prepared in a vessel decorated, as often as not, with jaguar motifs.[13]

The Aztecs believed tobacco was useful against snake bites, for head colds and abscesses, and their priests carried an ointment of tobacco and Morning Glory seeds to be used against venomous insects.[14] The Cherokees had a kind of tobacco butter that was believed to heal wounds and cure bad humours. Among the Mayans tobacco was recommended as a cure for toothache, chills, lung, kidney and eye diseases and to prevent miscarriage.[15]

In Guatemala the shamans would become intoxicated with tobacco in order to consult the supernaturals and to divine future events. Tobacco's role in shamanistic trance in both North and South America has been documented by Johannes Wilbert of UCLA's Latin American Institute. He argues that many tribes used different species of tobacco for different purposes, the varieties used in ritual being more powerful than those used more generally. He also notes that nicotine is not the only active substance in tobacco – some tobaccos contain harmine, harmaline and tetrahydroharmine, all hallucinogens. Some Indians, such as the Tenetehara in Brazil, smoke enormous cigars and in sufficient quantities to send them into trances.

Robicsek also reports the combination of self-torture and tobacco consumption among the Zapotecs and Mixtecs, when worshipping the god Coquebila. 'In honour of the god they held a fasting period of forty to eighty days, during which time they consumed only a quantity of tobacco and offered blood from their tongues and ears.'[16]

In North America some sixteen plants, other than tobacco, were smoked for pleasure or ceremonial reasons, but they were largely abandoned when tobacco became more plentiful. Some (bearberry, manzanita) may have been hallucinogens and jimsonweed (*Datura strammonium*) certainly was, and was often mixed with tobacco. According to the Dominican friar, Fray Diego Durán's *Book of the Gods and Rites* (1574–76), tobacco was used with *Rivea corymbosa* (Morning Glory) in the manufacture of a magical ointment for Aztec priests which, when it was painted on them, caused them to lose all fear and to slay men in sacrifice 'with the greatest daring'. In lowland South America, tobacco was mixed with *ayahuasca* (see chapter 12), among the Mayans it was steeped with a whole toad (again, see chapter 12), and in North America with peyote.[17]

There is also, among the Mayans, a class of small god, usually anthropomorphic spirits with upturned muzzles, open mouths and prominent incisors, who fall into various categories, one of which is called the flare gods. These figures are associated with the glyph for 'fire' and sometimes with jade celts that the Mayans believed were thrown to earth by the gods through lightning. Sometimes also linked to jaguars, these gods are occasionally depicted as the storm god, which has given rise to the theory that tobacco was somehow mixed up with the maintenance of fire, the storm god being worshipped out of fear because it was in storms that fires went out. Figurines are shown with cigars in their hair because, without the tradition of pockets, and being required to keep fire burning at all times, the hair was one place where cigars could be held.

Smoke also features in the iconography of Tezcatlipoca (Lord of the Smoking Mirror) of central Mexico, a god whose cult was still very much alive at the time of the Conquest. This god, the antagonist of Quetzalcoatl, the Feathered Serpent god, about which there will be a lot more to say, was the patron of sorcerers, princes and warriors. He was also the God of the Sunset and of Sacrificial Pain. His symbol was the obsidian mirror into which the Aztec shamans gazed until they fell into trances and in which they saw the will of the gods. These mirrors are shown on inscriptions emitting coils of smoke. The Olmec had concave iron ore mirrors which John Carlson, the Texan archaeo-astronomer, believes were used as 'burning mirrors' to generate fire.[18]

In his survey of the smoking gods, Robicsek presents many inscriptions where the figures are probably (but not definitely) smoking, some with jaguar cloaks, jaguar-hide belts, while many vases are decorated, for example, with jaguar heads, figures that are half-deer, half-jaguar, and with skeletons smoking, accompanied by dancing (but still ferocious-looking) jaguars.[19] On one vase the jaguar is itself breathing fire, another smoking figure has a jaguar head, and a third wears a jaguar kilt and quetzal feathers (the quetzal is the 'resplendent' golden-green and red-bellied bird native to the forest in central America). On another vase the figure smoking is accompanied by a jaguar-dog, frogs and alligators.[20]

Not enough is known about the origins of tobacco use but it may be that its ability to burn slowly was the all-important factor, the first use of cigars being to conserve fire. The psychoactive properties – discerned simultaneously or subsequently – would have confirmed it as a sacred substance twice over.

Sacred Groves of the Jaguar Tree

'For many pre-Columbian cultures of the Americas, cacao seeds and the comestibles produced from them were literally part of their religion and played a central role in their spiritual beliefs and social and economic systems.'[21] The northern limit of cacao was central Mexico but its distribution took in Guatemala, Belize, El Salvador and Honduras.

The word cacao is a Spanish adaptation of the Nahua *kakawa-tl*. Chocolate is based on another Nahua term, *chocolatl*, though a very late evolution. *Theobroma cacao*, to give chocolate the name Linnaeus chose, is what is called a small 'understory' tree that produces pods containing 25–40 seeds each. Cacao trees are cauliflorous, meaning the flowers are produced directly on the trunk and large branches. It is generally accepted that the plant originated in the Upper Amazon Basin and spread north from there, either naturally or by human agency.

Most scholars believe that only the pulp was used in antiquity. The pulp can be removed from the seeds and made either into a fruit drink or fermented to produce an alcoholic beverage. Cacao was precious because, unlike maize, it could not grow anywhere – it prefers deep, fertile alluvial soils, shaded and humid zones, with heavy rainfall. Because it was so sensitive and precious, wars were fought to control the territory where it grew.[22]

Theobroma is found in archaeological remains from about 1000 BC onwards. To possess cacao was invariably a sign of wealth, power and political leadership – vessels of frothy cacao are frequently shown at the foot of the king's throne on Mayan vases and cacao glyphs occur on elite tombs.[23] As we shall see, cacao seeds were used as a form of money.

Iconographically, cacao was shown as a sacred tree (the 'World Tree', referred to earlier, in the section on shamans, and explored in more detail later), was linked to blood, to political power, to ancestors, to maize and the underworld – the latter being so because the plant prefers shade (maize, preferring open fields, was associated with light). The deceased were provided with cacao for the journey to the next world.

Cacao and maize are, in fact, an important ritual pair in Mesoamerican cosmology. Both formed sacred beverages which, combined with water, served the gods. In Guatemala, Honduras and Mexico, cacao was poured into cenotes, caves, springs and ponds – the underworld is a unifying feature here. Chocolate/cacao was linked with rain, rebirth and ancestors, was associated more with women than with men but also with blood and sacrifice and there are many images of a cacao pod being sacrificed as though it were a human heart.[24]

Because it was precious there were restrictions on its use. People other than the ruling elite, warriors and certain merchants, were forbidden from drinking cacao and it was regarded as bad luck if a common person consumed it. Even among those entitled to use it, it was never drunk unthinkingly. A range of additives was used to flavour the drink – vanilla, chili peppers, honey, aromatic flowers. And it was sometimes mixed with blood and offered in rituals. Kings and queens were buried with it, cacao was used to pay tribute and, according to ethnobotanists Nathaniel Bletter and Douglas Daly, there are 76 substances, in one form of chocolate or another, which act as stimulants, anti-oxidants, diuretics, analgesics, platelet inhibitors, anti-hypertensives, anti-mutagens, neurotransmitters, vasodilators, anti-inflammatory agents, or antiseptic substances.[25]

Various alcohols and vinegars can be made from chocolate and in certain parts of South America, chocolate was mixed with tobacco and/or other hallucinogens. As the pods dry, their surface comes to resemble the pattern on a jaguar's pelt and this has helped give *T. Bicolour* (wild cacao, white cacao) the name 'Jaguar Tree', reinforced by the fact that jaguars are identified with the night, caves and the underworld, the same complex of associations as chocolate. (In the Maya world the word for jaguar and priest are derived from the same root.)

Cacao-associated vessels go back to 1000 BC but its use appears to have spread mostly between 600 BC and AD 400. The word *cacao* is of Mesoamerican origin but did not reach South America in pre-Columbian times.[26]

Chocolate beer (*chichi*) was enjoyed from very early on; in fact this may have been the original use. According to some anthropologists it was employed in non-hierarchical societies to create social debt, 'binding people in asymmetrical social relations'. And, possibly, it was the development of beer that caused the appearance of spouted bottles with flaring necks (a process that perhaps parallels the development of pottery to manipulate maize beer, and also liquids in Europe, which so fascinated Andrew Sherratt – see above, chapters 8 and 10). Condiments may have been added at the time of serving and pots, specially prepared, were smashed after use, adding to the drama of the drinking event, especially for guests, in a potlatch-type ceremony.[27]

Cacao played a part in religion – symbolising fertility and sustenance, sacrifice and regeneration, embodiment and transformation – and was pan-Mesoamerican in scope. The maize god is sometimes shown with cacao pods studded on his skin. The corn cycle was the central metaphor of life and death for the Maya, but humans, maize and cacao all featured, cacao perhaps representing the underworld.[28] The World Tree – that ancient shamanistic feature – is sometimes the cacao tree, cacao being the most privileged fruit grown from the maize god's body, showing its position in cosmology as second only to maize (maize was also added to cacao drinks).[29]

The maize god died each year, when part of his spirit left his body and rose to the heavens. His body was believed to be buried in 'Sustenance Mountain' where it gave birth (out of sight) to the fruits of the earth – cacao first and then other plants. This meant that even one sip of a cacao drink was a sacramental act.

Wealth was measured in cacao pods (as shown in inscriptions) and, in later times it was used as money. Sophie and Michael Coe give the value of various products in terms of cacao beans: a good turkey hen = 100 full cacao beans; a large axolotl (a salamander, a delicacy) = 4; a hare = 100; an avocado, newly picked = 3; a slave = 100; the services of a prostitute = 8–10.[30] They also could be used to pay for a service done, and even to buy one's way out of forced labour.

There were, of course, many other words for cacao, as there were for maize but there seems to have been an intimate association between *iximte* (maize) and *kakaw*, signifying that they were worshipped together.

The drink was prepared by pouring from a great height, so as to produce a foam on the top, which was regarded as the greatest delicacy, the aroma being as important as the taste. The best cacao-drinking vessels were very fine and made of volcanic ash temper; this volcanic link also had religious significance.

Many vessels with caffeine have been found in tombs in Copan. No other substances, other than cacao, are known to possess caffeine. In this area cacao and stone beads were exchanged for salt and cloth. In the Late Classic period (AD 600–800) there was a dramatic increase in the visibility of cacao – for example, it was depicted on stone censers placed in front of temples. Elsewhere cacao pod adornments on pots accompany jaguar gods and diving gods. Many of the censers were intentionally broken as part of a cancellation ritual in honour of deceased rulers. Cacao pods, the jaguar and the underworld are clearly linked.[31]

'Tree crops represent a long-term investment in the land that may tie kin groups to a specific area in ways that maize produced through slash-and-burn agriculture does not.'[32] To begin with only elites were allowed to own cacao orchards though this changed later. The Itzá Maya, an ethnic group based in the north of the Yucatan peninsula, were at the centre of a trade network based on cacao, which they exchanged for feathers, jaguar pelts, slaves and tortoise-shell spoons used to sip the froth of the cacao. They also controlled vanilla and achiote, both flavourings. El Salvador was famous for its production of cacao.[33]

Soconusco (part of the Mexican state of Chiapas, on the Pacific coast) was ideally suited to the production of cacao, maybe from 900 BC on. By AD 1200 they had access to long-distance trade goods, presumably as a result of well-developed production of cacao (the Aztecs conquered the area partly to have more reliable sources). Cacao was part of the tribute it paid, and in the sixteenth century about 1.5 million cacao trees were under cultivation. Only maize is more ubiquitous in ceremonial offerings.[34]

There was, then, a substantial difference between the important plants of the American tropics as compared with Eurasia. In the latter, the grasses (cereals) spread rapidly, aided by the east-west configuration of the landmass, and by the development of the plough and the rest of the traction/secondary products complex. In the New World tropics most of the plants were rooted to one environment, certainly to begin with and for some time after, whether that environment was the rainforest or the mountains, and they did not travel as much or as well as the Eurasian cereals. The lack of pack animals (save for the Andes) also inhibited movement generally.

As we have seen, the presence of domesticated large mammals in the Old World helped to promote a divorce by early peoples there from intensely psychoactive substances. Dispersed pastoral societies, spread more thinly and not grouped into villages or towns, were by necessity more individualistic, coming together intermittently for the advantages that occasional communal solidarity could supply (marriage, defence, attack) and in these circumstances the milder and shared euphoriant properties of alcohol were more suitable than the far more intense and private properties of hallucinogens.

In the New World tropical civilisations – more vegetal, less animal-oriented – it so happened that psychoactive substances were, literally speaking, thicker on the ground than in Eurasia. These psychoactive properties of New World plants were also far more varied than in the Old World, ranging from the intense (and sometimes dangerous) *Datura*, to coca, tobacco and cacao, whose differing effects together created the idea that there was/is a separate realm (even more than one realm) to which access was achieved – uniquely *but reliably* – by ingestion of these sacred substances. Few, if any, of them were taken recreationally, underlining the fact that early people's association with psychoactive plant life in the Americas was much more formal and *intense* than in the Old World.

In Eurasia, plants were not without religious significance but it was the concept of *fertility* itself that dominated worship, naturally enough because of the weakening monsoon. In the New World, on the other hand, the plants themselves were gods, often gods to be feared, conferring on the people who consumed them altered states of consciousness, states that were also to be feared at times, as well as being instructive. In some ways, this meant that New World peoples had, or felt they had, a much more vivid and certain experience of divinity – and other realms of divinity – than did peoples in the Old World. Life in the New World was, therefore, and to this point, both more static and more intensely 'other worldly/religious' than in Eurasia.

Whereas in Eurasia alcohol helped shape civilisation, in the Americas hallucinogens had much more impact. In the New World the existence of a supernatural world was altogether more *convincing*.

• 14 •

Wild: the Jaguar, the Bison, the Salmon

In the South and Central American tropical rainforest, some native Indians believe that certain trees attract particular forms of game. This has never been established scientifically but what is clear is that the dominant animal in these areas, throughout prehistory and until recently, was a carnivorous mammal, the jaguar. In North America, by contrast, the dominant animal was a herbivorous mammal, the bison. Though these animals were very different, as we shall see, they had in common the fact that both were wild, not domesticated. This was an important difference between the Old World and the New.[1]

The jaguar is the third largest cat on earth and the largest in the New World.[2] It is also the dominant predator in the tropical forest, occupying the top slot in the food chain: the forest belongs to the jaguar more than it belongs to man. Notably solitary, it usually stalks its prey alone and at night. 'The victim is brought to the ground by the awesome feet, claws and mighty forearms, and is dispatched by a bite from the extremely well developed canines on the neck or throat.'[3] Moreover, the jaguar is equally at home swimming in water or climbing trees. This means that in some ways the jaguar and man in the jungle are very alike – they tend to hunt the same animals. In fact, for many tribes in the Amazon region men and jaguars are so alike that shamans are thought to regularly transform themselves *into* jaguars, or they are regarded as originally *being* jaguars. In several small statues in the Dumbarton Oaks collection of Olmec art in Washington DC, humans are depicted that appear to be in the process of turning into jaguars – either their hands are claws or their faces are beginning to take on the snarling jaws of an aggressive feline (see figure 11, page 304).

Although jaguars and other felines (such as the puma and the ocelot) are not found everywhere, they do range – in art and in religious contexts – from the north of Argentina to the Gulf of Mexico coast to the south-western tip of California (Panther Cave in Texas dates to between 6000 and 2000 BC). In many of these areas the jaguar is the most prevalent symbol of ritual.[4] They are generally shown with snarling mouths and prominent fangs or canine teeth. Sometimes they are shown attacking humans, occasionally copulating with them. Jaguars are associated with water, rain (its roar is regarded as a form of thunder, indicating the anger of the gods), with the jungle, with darkness and with caves – they are the lords of the underworld. For many tribes, the jaguar is the 'Master of the Animals' and controls the non-human world. In a small number of cases of South or Central American art, jaguars are shown licking hallucinogenic vines. All of this confirms an intimate relationship between early man and the jaguar, one based on fear, respect, and with one of the shaman's main responsibilities being to tame this ferocious creature, a practice that recalls Jacques Cauvin's point about the bull in the Old World representing the untameable forces of nature.

Gerardo Reichel-Dolmatoff, of the Instituto Colombiano de Antropología, in Bogotá, has carried out the most thorough studies of jaguar symbolism in South and Central America and he has demonstrated how, at San Augustín, on the headwaters of the Magdalena River, there is to be found the greatest number of large stone statues in any prehistoric context in the western hemisphere.[5] Mainly located on hilltops and mountain slopes, these statues were both ceremonial and funerary and the great majority are feline. Moreover, apart from one or two, which show jaguars in a more or less naturalistic crouching position, the others – the majority – show a monstrous being, half-man, half-jaguar, with human bodies and what Reichel-Dolmatoff calls 'bestial' mouths with fangs.[6] A second group of statues shows a jaguar in the act of overpowering a smaller figure which represents a human being. In some cases the jaguar is copulating with a woman (figure 10).

Local ethnographic information, also collected by Reichel-Dolmatoff among the Chibka-speaking Páez Indians, who conserve many traits of the ancient belief systems, credit the origins of the Páez with the rape of a young woman by a jaguar, resulting in the 'thunder-child'. Thunder is a central element in Páez life, associated with the jaguar spirit, the concept of fertility and with shamanism (the shaman receives the supernatural call to office from thunder). Reichel-Dolmatoff has found many parallels of this set of myths in South and Central America: the Olmec had a similar legend, of a new race coming into being after the rape of a woman by a jaguar; the Caribs of the Orinoco Plain traced their descent from mythical jaguars; as do the Kogi of the Sierra Nevada and several groups of Tukano in the north-west Amazon. In contrast, such groups as the Arawak, Chocó and

Fig. 10 Monument 4 from Chalcatzingo, showing two stylised felines attacking two human figures, each of whom has a typically deformed Olmec head (see also chapter 17).

Makú, who claim to be descended from other animals, or from caves or rocks, live in terror of those peoples who are said to be of jaguar origin. In particular, they fear that their women will be abducted by men from the jaguar-peoples, perhaps a folk-memory of a real event in the past, and a reminder of the importance of exogamy.[7]

This is reinforced by ideas that the jaguar is never the progenitor of all of the human race, only parts of it, and by beliefs among the Sierra Nevada Indians, who were divided into jaguar and puma clans whose male members had to marry women of the deer or peccary clans.

These clans were held to be intrinsically 'female' because these animals comprised 'the natural foods of jaguars'. Crossing an area traditionally occupied by jaguar-people can cause disease in others not indigenous to the locality. The jaguar has to attack in order to survive and these qualities are regarded as characteristically male. 'The Indians also point out that the jaguar is a great hunter and that this activity implies a strong erotic element, the act of hunting being equated with a form of courting the game animals.'[8] This reverberates with earlier shamanistic ideas, discussed in chapter 3, about hunting as a form of 'seduction'.

The complex of beliefs that surround the jaguar – caves, thunder, rain, fertility, sexual aggression – also constitutes, according to Reichel-Dolmatoff, 'the principal sphere of action of most shamanistic practices'. Jaguars are helpers of the shamans, who can turn into jaguars at will, sometimes to achieve beneficent ends (such as curing disease), sometimes to threaten or kill enemies or rivals. Eventually, after death, the shaman turns permanently into a jaguar. In some tribes the word for jaguar and for shaman are the same, derived from a term for cohabitation.[9]

There is a further level of belief – the jaguar gets his bright colour from the sun in the east, that colour being the symbol of creation and growth. He is associated with rock crystal, particles of which are regarded as thunderbolts and are found where lightning has struck. Jaguar motifs form a large part of the imagery these tribes experience when under the influence of hallucinogens, and shamans in trance may wear crowns of jaguar claws. The drugs themselves are also equated to jaguar sperm or jaguar seed. Narcotic snuffs are kept in tubes made of (sacred) jaguar bones. In trance, the shaman 'talks to the jaguar'.[10]

The headwaters of the three greatest rivers in South America – the Magdalena, the Orinoco and the Rio Negro, a major tributary of the Amazon – all rise within about 160 miles of each other and it is there that the symbolism of the jaguar is at its purest. Nonetheless, it is also incorporated into the more developed religions of the Mochica, Olmec, Chavín, Inca, Maya and Aztec, to name just some of the New World civilisations who came after.

Whereas Amazonian societies had their shamans, the Olmec (1400 BC–400 BC), who had agriculture and an hereditary elite (considered in chapter seventeen), possessed a hierarchy of priests and an organised pantheon of deities, in which, even so, the jaguar played a leading role. For example, in Monument 3, as it is known, in San Lorenzo, near present-day Veracruz, there is a sculpture of a huge jaguar monster copulating with a supine human female.[11] Monument 4 at Chalcatzingo, in the central Mexican state of Morelos, also shows two felines attacking humans who are falling over. Monument 52 at San Lorenzo, discovered in 1968, shows a half-human, half-feline figure with a snarling 'were-jaguar' mouth, its paws resting on its knees. Some scholars regard this as a rain god, or maybe a god of the all-important drainage system – either way, water is involved.

And it may have been among the Olmec that a new deity began to emerge, which was to become an amalgam of the jaguar, the serpent and the eagle. Plumed jaguars are shown on Relief II at Chalcatzingo.[12] This was to become, among the much later Toltecs and Aztecs, Quetzalcoatl, the Plumed or Feathered Serpent god, though Nicholas Saunders thinks the jaguar was involved at the beginning. Certainly, an Olmec ceramic figurine shows a half-human, half-skeletal figure very reminiscent of what early shamanistic experiences are often said to consist of. It is also among Olmec sculptures that we find images, referred to earlier, of humans turning into jaguars.

Are we therefore seeing two things among the Olmec: the association of 'jaguarness' with superiority, which meant that, in a newly evolving hierarchical society, the elite appropriated jaguar qualities; and a transformation from shamanistic society to a post-shamanistic world, with a more complex god, including a more complex relationship between deities and humans, but one where shamanistic elements remained?

In the (slightly later) Chavín civilisation (900 BC–200 BC), there are some impressive stone sculptures of feline heads, half-human and half-feline monsters, the latter becoming a marker of Chavín influence in the Andean world. In fact, the feline deity, so impressively depicted in the art of Chavín de Huántar, a pre-Inca culture dated to 900 BC, came to be seen as the unifying force in the Andean context. At one time Chavín was a famous religious centre, Nicholas Saunders tells us, 'on a par with Jerusalem or Rome'. It was well suited to farming without the need for extensive (and expensive) irrigation and a key to the agricultural success of the area were the 'vertical archipelagos' (mentioned earlier) in which the irrigated valley floor forms the lowest layer, the potato fields occupy the upper slopes and the grazing pastures are even higher. In the view of several archaeologists, against this background

Chavín de Huántar was simply a ceremonial centre, 'a mecca built to serve the feline cult.' This has now been discounted: Chavín was more than a ceremonial centre. But it remained a mecca and one, moreover, where feline symbolism was paramount. It was the location where the idea of an amalgam of felines and eagles seems to have evolved, and a feline staff-bearing god. So here too we may have an incipient post-shamanic scenario emerging, with a priestly caste making an early appearance.[13]

Among the many textiles and pottery of Paracas/Nazca art (200 BC–AD 600), which civilisation grew out of Chavín, feline symbolism remains prevalent, as do feline masks and feline-bird sculptures. On one cloth painting, possibly a ceremonial mantle, an elaborately dressed central figure wears a feline mask and has a curling tail with a severed human head at its tip. The figure has cats' paws for both its feet and hands, in one of which another severed head is held, while in the second hand is an obsidian-tipped knife, indicating ritual headhunting. 'The taking of human heads to "capture" the powerful essence of the victim's soul is a well-documented practice in South America. In recent times the Jivaro tribe of Ecuador [this was written in 1989] used to cut off an adversary's head, peel off the skin and cure it until it became a shrunken head – full of protective supernatural power.'[14]

The feline is ubiquitous in Mochica art (AD 100–AD 800), sometimes as the ocelot or puma, but most often as the jaguar. On pots a snarling jaguar is frequently shown standing behind a human with its paws resting on the human's shoulder. Elizabeth Benson says such pots embody two main themes – those depicting prisoners of war and those associated with rituals surrounding the important coca leaf.[15] 'A link between the narcotic coca leaf and the feline is clearly hinted at by a beautiful golden coca-bag in the shape of a jaguar.'[16] Dried leaves were kept inside, to be used by the shaman or priests in their rituals.

The Mochica were very militaristic and brought large numbers of captured prisoners back to their capitals. The pottery shows long lines of victims tied together by ropes around their necks, and distinctive hairstyles marking them out as prisoners of war. On one beautifully painted pot, a large feline is shown seated in front of a small man, who is typically blind in one eye and has his hands tied behind his back – clearly a captive. The feline is shown as he is about to tear at the man's throat, from which blood appears to be trickling.[17]

The dominant feline in the Andes was the puma (though the jaguar may have been regarded as the father or master of all cats). Puma symbolism was associated with royalty and with militarism – the drums of the troops were played by men dressed in puma skins and there are those archaeologists who believe that the entire Inca state was conceived in the form of a giant puma that bestrode the Andes. War leaders wore puma skins too and dedicated their victories to the animal. Pumas were kept for ritual purposes and great efforts made to tame them. In the Inca province of Pumallacta, pumas were worshipped and, according to some sources, sacrifices of human hearts were made to the feline spirit. Even today puma masks are worn in local fêtes.[18]

Among the classic Maya (AD 300–900), the jaguar was a recurring motif, often portrayed as a deity, the god of the seventh day, *Akbal*, which means 'night'. In Mayan art a rolled-up jaguar skin symbolises the starry night sky and the animal is in general associated with night, dark places, the underworld, danger. But it was also associated with rulership – jaguar pelts were draped over royal thrones, and the thrones themselves could be in the form of jaguars. Nobles wore jaguar skins, warriors and nobles wore jaguar helmets, and carried weapons made of jaguar bones. Luxury gloves were made of jaguar paws.[19]

Mainly jaguar symbolism among the Maya occurred in contexts of warfare, fertility and sacrifice. Knives used in sacrifice and formed of three obsidian blades are intended, according to some sources, to represent the marks made by jaguar claws; and the heart sacrifice that lay at the centre of Maya ritual reached its climax when the victims' bodies were eaten by jaguars.[20] At the site of Chichén Itzá, in northern Yucatán, there are four jaguar thrones, the most famous of which is the 'Red Jaguar' throne, discovered inside the great temple of the Feathered Serpent. This throne is in the form of a life-sized jaguar, brilliant red, with 73 jade discs imitating the rosette pattern on the skin of the real animal. There was, too, a widespread practice of using jaguar names as identifying titles – Shield-Jaguar, Bird-Jaguar, Bat-Jaguar, Knotted-Eye-Jaguar and many, many more.

In the central plaza of Chichén Itzá, there is a platform of the jaguars and eagles, with superbly carved jaguars eating human hearts. Warriors are shown dressed in jaguar skins, and flanked by human skulls. The famous Feathered Serpent god may also conceal jaguar features. According to Joyce Marcus, an expert on Mayan writing, four classic Mayan cities – Yaxchilán, Palenque, Tikal and Calakmul – organised themselves into a military alliance under the sign of the jaguar and in emulation of its aggressive powers.[21]

This association was even more marked among the Aztecs. (Nicholas Saunders describes the jaguar presence as 'ominous'.) The Aztecs inherited some of their more militaristic practices from the Toltecs, famous today for their reconstructed temple-pyramid of Quetzalcoatl, the Plumed Serpent god, around which, originally, a series of jaguars or pumas are seen marching. Alternating rows show eagles crouching and also eating human hearts. There were two elite Toltec warrior castes, one dedicated to the jaguar, the other to the eagle, a practice that was repeated among the Aztecs.[22]

At their great mountain-top temple of Malinalco, west of Tenochtitlán, which may have been dedicated to the two warriors castes, great crouching monumental jaguars line the main stairway which leads to the inner sanctum, dominated by a stone bench with eagles and jaguars carved into it. At the Great Aztec temple at Tenochtitlán itself, one side has a subterranean temple dedicated to eagle warriors (shown marching to the sun) and the expectation is that the other side, not yet excavated, will show jaguar warriors in an equivalent arrangement.[23] This possibility is indirectly supported by the fact that, during the excavations at the *Templo Mayor*, a complete skeleton of a jaguar was found with, between its fangs, a sacred green stone. In the Aztec idea of the cyclical nature of time, with the sun and eclipses separating the cycles, the eclipses are sometimes described as blackened skies with 'flesh-tearing jaguars' descending from the clouds.[24] As Nicholas Saunders puts it, '[This] affords us a brief glimpse into the complex beliefs of the Aztecs, in which jaguars, jade, rain and fertility were all intimately associated. The practice of auto-sacrifice, during which the people would pierce their own skin to collect blood as a ritual offering to their gods, was often accomplished with "spines" or lancets made of jade.'[25] There were regular ceremonies, some called *Tlacaxipehualiztli*, or 'the skinning of men', in which inebriated captives were forced to defend themselves with a feathered stick against jaguar-knights armed with a wooden sword inset with obsidian blades. The obsidian blades, again, were a metaphor for jaguar teeth. Afterwards, when blood had been shed, the captives were sacrificed, with their hearts cut out.

Finally, jaguar symbolism was intimately associated with the identity of the Aztec supreme deity, Tezcatlipoca, Lord of the Smoking Mirror. Among Amazonian societies the spirit world is closely associated with mirrors and to the Aztecs, Tezcatlipoca was able to use his mirror to see into the souls of men, 'in much the same way as the shamans of the Amazon'. In another manifestation, Tezcatlipoca could become Tepeyollotli – the jaguar who lived at the heart of the mountain. Volcanoes, which are usually mountains, are where obsidian is mined so there is a link between jaguars, volcanoes, sacrifice and deities. The pyramid at La Venta which, for a variety of reasons has never been properly excavated, seems to be a model of a volcano, with its shape and the rest of the complex being in the stylised form of a jaguar face.[26]

As Saunders concludes: 'Thus there would appear to be a conceptual link between the shaman as jaguar in tribal societies and Tezcatlipoca as jaguar in the Aztec pantheon. Cultures and civilisations may vary in their size and sophistication, but many of the underlying ideas, and the patterns of thought and association, remain.' Even for the cultures of the imperial Aztecs, the essence of 'jaguarness' was ever-present.[27]

We can go further. We have seen that, overall, there are two overlapping complexes linked to the jaguar. There is the jaguar-volcano-sacrifice-deity complex; and there is the jaguar-night-underworld-water-rain-thunder-war-aggresion-sex-coca-hallucinogen-shaman-superiority complex. Can these be put together in the following fashion to explain the very militaristic, sacrifice-oriented civilisations that developed in Mesoamerica?

In a temperate society practising agriculture, as evolved in the Old World, the annual cycle of sowing, growing, ripening and reaping was practised, based on rhythms that were observed if not fully understood and against the (deep) background of the weakening monsoon. One consequence of this was that worship of fertility was – eventually and for the most part – rewarded. The worshippers observed the sun in the sky and though they knew that movements were reliable, while they didn't understand the reasons for them they could never be certain they would continue. From time to time eclipses occurred and though they never lasted for very long, who was to say that one day they might? The rains were less predictable. They had their own – related – rhythms but again, from time to time they failed, making the need for worship more urgent. But again, and despite episodic variations, eventually the rain gods responded. Essentially, worship in temperate agricultural societies *worked*.

In tropical forest zones, however, where many plants grew throughout the year, fertility was less of an issue. It was not a non-issue but it was less important. What was more important were the threats in the environment – hurricanes, earthquakes, the volcanoes, both above ground and below the surface of the sea, generating super-waves, the increasing frequency of El Niño, generating ferocious thunderstorms with high destructive winds, and the wild animals, in particular the ferocious jaguar. Importantly, these operated on no rhythm, no rhythm that early man could observe anyway, and this is something that applied also with the jaguar. According to Robert Wrangham, 'to judge from records of attacks by jaguars, modern hunter-gatherers are safer in camp at night than they are on the hunt by day' and that must have been even more true in the distant past.[28]

In other words, the aggressive and dangerous jaguar was like hostile and dangerous weather: unpredictable. That so many depictions of the jaguar in South and Central American records show his fangs, or his sexual or other attacks on humans, or draw attention to – or seek to emulate – his sharp claws, and other warlike qualities, underscore that he was what he is: a predatory carnivore, the king of the jungle, of the night. In such an environment, therefore, instead of praying in the hope of making something happen (rain, green shoots, growth), a large element of worship in the New World was to make something *not* happen: to stave off fierce winds, earthquakes, volcanic eruptions, super-waves, and attacks by jaguars. Whereas the main form of Old World fertility religions was *supplication*, in the New World it was *propitiation*.

If one accepts this analysis, one can begin to see how sacrificial rituals in the New World became – as they did become – increasingly bloodthirsty. In Africa today, it is well known that if a lion, say, kills and eats a human, the rest of the tribe must go out and find the man-eater, and kill it, for if they do not the animal will return to where it was rewarded and try to kill again. In the rainforest, however, it would have been much more difficult to know which jaguar had made the kill. No doubt, on many occasions, the tribal members did go out in search of a man-eating jaguar, but an effective alternative, against a deep background of the increased incidence of environmental and other 'attacks', may have been to offer a captive, real flesh and blood but not one of the tribe, either in a sacred ceremony, or by leaving the body to be found by the jaguar, well away from the village. A third alternative was to have a shaman deal with the jaguar.

But here is the main point: *Worship would not have worked.* Or, it would have worked fitfully, at certain times but not at others, with no order to when it did and didn't succeed. El Niños, earthquakes, volcanoes and jaguar attacks occur on no rhythm, are entirely unpredictable. In such an environment, it would have seemed to early man that the gods were not just angry but in addition *dissatisfied* with the offerings being made, which were not sufficient. So they would have offered more, in an increasingly desperate attempt to placate the gods' anger.

This is speculation, of course, and by the time of the Aztecs sacrifice was designed to feed the appetite of the sun. How did this notion of the 'appetite' of the sun arise? Could it have been that, following volcanic eruptions, eruptions of mountains perhaps located beyond the horizon, but whose ash plumes rose high in the sky, the sun was blocked from view, in effect it failed to rise, it failed to appear for a number of days? Alternatively, or additionally, if the sun failed to appear, following an eruption in which many people were killed, this could well have fuelled the idea that the sun 'needed' human blood to raise itself from the horizon in the morning.

We have seen that the Aztec elite warriors were Eagle and Jaguar Societies, and we know that the colour of the jaguar pelt was early on associated with the rising sun. It is certainly possible, therefore, that, as shamanistic practices subsided, jaguar motifs were transferred and transformed into other ideas.

The essential point to keep in mind is that the dominant gods (not all) in the Old World were *responsive* to worship, because of natural rhythms existing in temperate societies but which, at the time, were not understood. In the tropical New World, on the other hand, the gods were unpredictable and seemingly unappeasable. This was a profound difference.

The Science of the Stampede

After ~9000 BC, Palaeo-Indian groups throughout most of North America diversified away from big-game hunting.[29] Some bands adapted to increasingly arid conditions in the west and placed their major emphasis on plant foods. The sparse Palaeo-Indian population of the Eastern Woodlands (east of the Mississippi), on the other hand, pursued deer and other forest game, also relying on seasonal vegetable foods. Most notable of all perhaps, on the Great Plains of North America, there was a switch from hunting several megafauna to hunting just one – the bison. Only on the Great Plains did big-game hunting remain a viable life-way, a life-way that was to survive well into historical times.

The rise of the bison had everything to do with the way the landscape of North America had changed since the late Ice Age. By 9000 BC, there was a vast tract – a sea – of arid grassland that stretched from the southern edges of Alaska to the banks of the Rio Grande, separating what is now the United States from Mexico. This, the 'Great Bison belt', lay mainly in the rain shadow that existed to the east of the Rocky Mountains. Most of the year the plains suffered under a dry air mass. Rain in spring and summer favoured a special type of short grass which concealed most of its biomass underground, retaining moisture in its roots. It was the bison's ability to get at these roots that ensured the species' survival.[30]

Primeval forms of bison reached very impressive dimensions, some with huge horns spreading six feet or more from tip to tip, the remains of which have been found all the way from Alberta to Texas, and from California to Florida. After the ice melted, bison became more plentiful, but also got physically smaller.

Plains Indians had a distinctive lifestyle. For the most part, they moved around continuously in small bands exploiting relatively few locations in a regular cycle. These movements depended to an extent on the distribution of certain forms of fine-grained rock, out of which they fashioned the large tools needed to kill big, mobile animals like bison.[31]

The changeover from Clovis to Folsom tools is associated with the change from really big game to bison. Folsom tools are smaller than

Clovis ones and are heavily 'fluted' – to remind ourselves, they have a concave groove running down most of their length on both faces, and two 'ears' sticking out at the end away from the point. Though no one really knows, the speculation is that the 'ears' were used to haft the tool to a spear and the fluting, though it may have strengthened the tool, as well as reducing its weight and help letting the animal bleed, hastening its death, may also by this time have had a religious function. This speculation stems from the fact that an analysis of early Plains Indian debris shows that many fluted points were broken in the process of manufacture. Would any workman have persevered with so many failures without good reason?

A lot has been inferred from the patterns of human and animal remains that have been unearthed. Such remains as have been found suggest that, to begin with, just after 9000 BC, Palaeo-Indians were less sophisticated in their choice of prey, whereas later they concentrated on killing cow and calf herds. There is also evidence that their stone tools became more standardised later on, suggesting that they had evolved a successful common method of slaughter. Further, it also seems that the Palaeo-Indians regularly killed more beasts than could be eaten at once. Could this mean that they attacked the animals in winter time, in freezing conditions, and then lived alongside the (preserved) animals, eating their way through the cold months? Another finding is that the pattern of remains suggests that human bands were smaller in the south-west than in the north. Does this mean that bison habits were less predictable in the south-west and did this hinder the development of settled, more complex societies?

The best-known fact about bison hunting is the communal drives. Were these regular occurrences, or infrequent ceremonies with a ritual function as well as providing food? Sites with evidence of communal drives – Bonfire Shelter, Lindenmeier, Olsen-Chubbock, all in Colorado, where the animals were driven over cliffs, into sand dunes, into dead-end canyons – have been found, in which herds of up to 250–270 beasts were stampeded. These 'orgies of killing', as they have been called, date to between 10000 and 3000 BC. Not all of these took place in winter, so – again speculating – perhaps different bands of people came together at set times of the year for ritualised hunts. As well as being a very practical solution on the food front, these gatherings would have helped in the exogamous selection of mates, and may also have involved a religious element. Dennis Stanford, director of the Palaeo-Indian programme at the Smithsonian Institution in Washington DC, identified traces of a 'stout wooden pole' near one bison corral, with an antler flute and a miniature projectile point nearby. He argues that the wooden upright may have served as a 'shaman pole' with the associated artefacts part of the ritual paraphernalia.[32] A religious dimension is also suggested by the discoveries at a relatively new site, Certasin in Oklahoma, where an arroyo filled with bison bones from a mass stampede was found. In the centre of this was discovered a skull with a forehead adorned with a red mark in the shape of a lightning bolt.

Judging from eighteenth- and nineteenth-century eyewitness accounts, bison are not particularly fearful of humans provided they are left alone. But they quickly become skittish. 'They can be gently herded a mile or so without trouble, but after that they begin to break and run, when it is almost impossible to stop them.'[33] The eyewitness accounts report that the hunters would slowly ease the animals in the required direction, taking several days about it and sometimes using bison hides as a disguise. They could shout at the animals to move them about, provided they didn't do it too often. Then, when there was nowhere for the bison to go but to their death – over a cliff, into a dead-end gully, into a sand dune – the hunters would capitalise on the propensity of bison to panic. Amid the stampede, and after the animals had fallen over a cliff, or been stuck in soft sand, each hunter would pick out a single beast, and thrust his spear through his target's rib cage into the heart.[34] George Frison, a palaeo-archaeologist from the University of Wisconsin, took a projectile point that had been found in Hell Gap, north of Guernsey, also in Wisconsin, and bound it to a slotted pine shaft with sinew sealed with pine pitch. By either thrusting or throwing the eleven-foot spear he found that he could puncture the hide of a domesticated ox, and that a hard thrust would sometimes penetrate to the heart.

In some ways, killing was the easy part. Once the stampede was over, the hunters faced many hours of arduous labour, sweating in teams, cutting up several animals at once. At first, they rolled the bison on to its belly, cut the hide down the back, and pealed it along the flanks to form a carpet for the flesh. Each animal produced about 470 lbs of meat, 45 lbs of fat and 35 lbs of edible internal organs. One stampede could sustain more than a hundred people for more than a month.[35]

These details, it is true, are based in part on eyewitness accounts from the eighteenth and nineteenth centuries. But it is unlikely that there was much change since the 9000–6000 BC period. The fact that bison hunting endured for so long must mean that it was a successful adaptation to local conditions. And in many areas of North America, as we have seen, agriculture didn't take hold until five hundred or so years before the Europeans appeared. This all shows that there is nothing inevitable about the advent of agriculture. Many archaeologists and palaeontologists regard the invention of agriculture as man's greatest idea, his most innovative invention. Yet, by definition, whole swathes of the New World never had any need for it. The coincidence of a warm, dryish climate, the right type of grass, and an undomesticable but abundant large mammal, conspired to produce in North America the largest area on earth where agriculture never developed. This had enormous consequences for the development of civilisation – indeed, it has consequences for the very concept of civilisation itself.

Salmon and Cemeteries

Around 6000 BC another change, profound but slow, took place. After thousands of years of rising and falling, as the inflow of glacier meltwaters was counteracted by the raising of the land as it was freed from the weight of ice bearing down on it, sea levels at last stabilised. One result was that vast shoals of salmon began to swarm up the great rivers of the north-west Pacific coast to reproduce and die. We know this because, at this date, salmon bones begin to overwhelm all others in the middens, and slate points (for spearing fish) replace microliths. There were rich pickings: the regular arrival of the salmon, in such colossal numbers, was like a harvest, no less abundant and therefore hardly less difficult to bring in than corn.

People learned how to fillet the fish and lay them out on racks to be dried in the sun and wind; or they were suspended from ceilings and preserved in the smoke from the communal hearths. In addition to the salmon, north-west coast Indians hunted seals and otters as well as deer and bear and they collected berries, acorns and other nuts.[36] The sheer abundance of wild foods, and the reliable regularity of the salmon meant that the Native Americans of 'Cascadia' (derived from the

Cascade Mountain range which extends from British Columbia to northern California) were among the first on the continent to occupy permanent villages. We know that they were sufficiently successful and sedentary to afford specialist craftsmen and there is evidence that they traded (obsidian). Their populations increased and leaders emerged but the societies achieved this level of complexity only after 500 BC.[37]

Other than a few post holes, however, the Cascadia villages show no sign of houses.[38] Presumably, this is because the houses were seasonal and therefore too flimsy to leave any traces. Theirs was a semi-sedentary way of life.

Over the following few thousand years, the salmon catches became bigger, as new technology was invented to harvest them in ever greater numbers. Very likely, some clans tried to claim ownership of the best fishing rivers: we infer this because there is evidence of regular fighting between clans – remains of bodies with wounds of violence. (And also with arthritis, a risk from marine diets.)[39]

The Namu site on the central British Columbia coast, near Bella Bella, opposite Hunter Island, is dated to 10000 to 5000 BC with microblades appearing at about 7000 BC, the same time as numerous sea-mammal bones. By 5000 BC, however, their diets had changed to salmon and herring, 'to the point where semi-sedentary winter occupation of the site was the likely norm'. Shellfish came in at about 3000 BC and this salmon and shellfish diet persisted for 2,000 years.

Fish diets, naturally enough, were not as dominant inland as they were on the coast, but along most rivers fish bones are represented more and more after about 5800 BC. Pit-house settlements, semi-sedentary, are usually associated with salmon remains.

The remains of houses and camps do become better documented at about this date. The Koster site in the Illinois River Valley, which famously has fourteen stratified occupation levels, dating from about 8500 BC until AD 1200, shows a seasonal camp that covered about 0.75 acres, with temporary dwellings at horizon eleven, dating to 6000 BC. An extended family group of about 25 people returned to the same location repeatedly, the stone of their tools suggesting they also occupied sites in West Virginia and Missouri, thousands of miles away. Oval graves were found, containing four adults and three infants, buried with three dogs in nearby shallow pits.[40]

Horizon Eight (5600–5000 BC) had evidence of four occupations, with houses measuring 20–35 by 12–15 feet and covering 1.75 acres. The long walls of the houses were made up of wooden posts up to ten inches in diameter, which were laid out in trenches about eight to ten feet apart. Branches and clay filled the gaps in between. Analysis of fish scales (which grow throughout the year) suggests the inhabitants occupied the site from late spring through summer. They harvested only those plants that were easier to cull and this shift, to a narrower range of foods, is mirrored at other Midwestern sites.[41]

An unusual configuration at Windover, near Titusville in Florida, provides a different perspective on the 6000–5000 BC period. Windover was an Early Archaic burial area, where artefacts were deposited in a pond, the dead being immersed within forty-eight hours of death, in peat and water whose neutral chemistry ensured near-perfect preservation. Seven different textile weaves used for clothing were found in the peat, the thread coming from the Sabal Palm and/or Saw Palmetto. Garments – bags, matting and ponchos – were produced, using weaves with as many as ten strands per centimetre – very fine. They also used bottle gourds (*Lagenaria siceraria*), perhaps the earliest example of this type of container to be discovered in North America.[42]

Around this time, too, as well as the remains of the first houses, we see a change from 'free wandering' to 'centrally based wandering'. In other words, instead of wandering from one makeshift campsite to another, clans now started returning to the same base location year after year. This change, at about 4500 BC, went with clans ranging up and down their own segments of river valleys. After this time, however, rivers themselves changed: enough years had passed since the Ice Ages for backwaters and oxbows to form. As a result, aquatic resources became very rich in some areas, so much so that it paid to stay put in one place for months at a time. We know this because middens now began to grow in size.[43]

There are two other clues that sedentary settlement was growing – more substantial dwellings, and cemeteries. Koster has provided the best evidence of relatively permanent houses but elsewhere specially prepared clay floors are seen (Upper Tombigbee River Valley in Mississippi). These floors, we know, were built before 4300 BC.

Formal cemeteries now occur, particularly in the central Mississippi drainage. These cemeteries are usually large, shallow pits with as many as forty people laid out with no particular spacing. The grave goods do not appear to be associated with specific individuals. 'Everything points to egalitarian societies in which individual status was unimportant, where leadership and authority were reflected more in age and experience or skill as a hunter, than in material possessions.' Little differentiated male from female, or social rank, though eight people – mostly men – were buried in clay-capped grave pits, suggesting they were especially important. Another grave contained a bundle of eagle talons, the remains of a bear's paw bones and other objects. A shaman?[44]

The sudden appearance of cemeteries and burial mounds in fertile river valleys almost certainly marks a dramatic change in prehistoric Indian life. Once people mark their lineages and do so by interment, they are laying a claim to the land and the vital food resources it generates. They are saying, in effect, that the land was once owned by their ancestors, that it belongs to them, and that, in order to keep it, they must remain *on* it.[45]

In the central Mississippi drainage, sedentism had emerged as early as 4000 BC. Moreover, with the identification of fixed territories, fixed boundaries may have emerged for the first time, sparking at least sporadic tribal warfare and it is this which may have produced a characteristic form of territorial marking – burial mounds, cemeteries at the top of ridges which could be seen from miles away. This practice continued for 5,000 years, until maize agriculture was introduced throughout the Mid-west, and it remained important in some locations up until historic times.[46]

At about this time too, the proportion of newborn and young animals found among the remains in excavations doubles from about a quarter to a half of the total. This may reflect choice on the part of more sophisticated animal breeders and/or the increase in infant mortality that occurs when animals are corralled in crowded conditions and infectious diseases are more likely (a dilemma that is still with us today).

We see here, then, a picture that is substantially different from that in Eurasia. In the Americas we see the development of sedentism, just as we did in western Asia, but this time it takes the form, mainly, of temporary villages or hamlets, base-camps which are used (and, later on, regularly returned to) for part of the year, culminating in the identification of territory, not by houses as such, permanent houses, but by cemeteries.

Agriculture is invented in the New World not much later than (and maybe simultaneously with) agriculture in the Old World, and with much the same kind of plant (which 'waits for the harvester'), but it does not appear to have been at all the momentous – epoch-changing – event that it was in the Old World, certainly not everywhere. It did not lead, anywhere near as rapidly, to the invention of pottery, metallurgy, and the evolution of villages into cities. This did happen in some locations, which are considered in more detail later, but elsewhere part-year sedentism, herding and hunter-gathering continued for centuries, even millennia, even until historical times.

Almost certainly, this had to do with the abundance of game and other food supplies, and the relatively sparse population, in North America at least. Given this state of affairs, there simply was no *need* for agriculture: part-time sedentism, herding and hunter-gathering were quite efficient enough as enduring ways of life.

But this did have enormous consequences. Hunter-gatherers, constantly on the move, have no need of permanent, still less monumental architecture. They have no need of ceramic pots, which are far too heavy, and too brittle, to carry around from place to place. On top of that, the gods of hunter-gatherers will be different: by definition they inhabit many different landscapes and depend on animals, as much as on plants. The weather and rivers are less important to them, or important in different ways. Their gods therefore differ.

Material goods are no less important to hunter-gatherers and herders but, since they are always on the move, they need lighter, less rigid, more flexible goods, easy to roll up and carry away. So instead of ceramics, they will turn to bags, pouches, nets, ropes, blankets and in turn that means that weaving, sewing and knotting are more important to them than clay working or metallurgy. Of course, it so happens that bags, pouches, sandals (the earliest carbon-dated to 7500 BC), tailored clothing, blankets, nets (if not their sinkers), baskets and textiles are more perishable. And of course, not only the objects are lost; so too are the decorations that adorned them, and any symbolic meanings they may have had.

What we have here, then, in the New World is a clear but *gradual* change in human development: early man responded to the natural world around him, adapting, and with no real *need* for agriculture. As far as his relationships were concerned, he had to cope with wild creatures. They were less malleable but also very different from one another: goats and sheep, or horses and cattle, are more alike than jaguars and bison and salmon. So societies in the Americas differed far more from one another than societies in the Old World, which lived off cereals and domesticated mammals.

America was then, as now, a land of abundance. This has more implications for the Old World than the New. It returns us to the old question, never satisfactorily answered: *why* agriculture developed when it did in the Old World?

• Part Three •

Why Human Nature Evolved Differently in the Old World and the New

• 15 •

Eridu and Aspero: the First Cities Seven and a Half Thousand Miles Apart

In chapter 7 we followed Jacques Cauvin's account of the developments in the worship of the Great Goddess and the Bull in the Natufian and Khiamian cultures of the Middle East, at around 10000–9500 BC, before domestication proper had emerged. Cauvin made it clear that this development was a response to new conditions which had emerged in that part of the world – the Great Goddess was an old, traditional object of worship, but the Bull, as a representative of the male principle, and of sheer power, was also a prominent member of new types of animal that warmer conditions had brought early peoples into contact with. And, as we shall see, these images remained in force – in one form or another – for thousands of years.

In the New World, the earliest evidence we have of religious activity is somewhat later – 7000–5000 BC – but here too the influence of environment was all important, and these images lasted for almost as long. The thought processes were, however, very different.

This early evidence of religious activity in the New World occurs among the Chinchorros of Chile. It would appear that these people had a sedentary maritime existence very early on along the Pacific coast and for specific reasons: there were several unique environmental features to be found in this arid region – a stable landscape with mild seasonal changes, pleasant year-round temperatures, ready access to sources of fresh water and plenty of marine food (they were particularly dependent on sea lions). If they had arrived from the north, from coastal Peru, they would already have had a maritime technology of one kind or another; if they had arrived from inland, from the Andean highlands, then at that latitude they may well have come from Lake

Titicaca and, again, would have had some knowledge of fishing techniques.

The Chinchorros used throwing sticks, harpoons, weights for fish-hooks, stone knives and basketry. Trace mineral analysis of their remains show that they subsisted on fish, sea mammals and coastal birds, and their skulls show evidence of external auditory exostoses, a pathology related to diving in cold waters. They had a sedentary way of life from a very early time, settlements ranging from a few huts at the site known as Acha 2, to about 180 huts at Caleto Abtao at Antofagasta. But it is their practice of mummification that particularly attracts our attention: the Chinchorro practised mummification two thousand years before the more well-known tradition in Egypt. Many coastal sites of South America – from Ecuador to Peru, and dated as early as 8000 BC – show maritime adaptation but nowhere else is there evidence of artificial mummification. This practice began in the valley of the River Camarones and extended to Arica, at the very northern tip of Chile, on the border with Peru, both being part of the Atacama desert which is, according to NASA, the driest desert on Earth. It is these arid conditions that were all-important in a religious context because it is in such circumstances – as in Egypt thousands of years later – that mummification occurs naturally. What a shock these early peoples must have had, to discover that some ancestors, apparently, did not die and putrefy, as most people did, but continued to exist, in a state that appeared to be somewhere between life and death.

Having observed natural mummification, the Chinchorros developed their own artificial mummification procedures, which were exceedingly elaborate, involving a detailed knowledge of anatomy, dissection and desiccation (even facial details and genitalia were intact thousands of years later), according to Bernardo T. Arriaza, an anthropologist at the University of Nevada, Las Vegas, and at the University of Tarapacá, Arica, in Chile. They also showed other cultural innovations, such as post-mortem trepanation, bows and arrows and intentional skull deformation (generally regarded as a device to enhance group identity). The average skull capacity was 1400 cc, the same as is normal today.[1]

There were three forms of artificial mummification, as developed over time. Black mummies were disassembled and reassembled with the soft tissues removed, without leaving any visible signs, and with wooden sticks replacing the longer limbs. A few were left unclothed. In a sense they were as much like statues as cadavers, some showing signs of repainting, suggesting that they were kept on show for a considerable time, before they were buried in groups. Red mummies came later and were painted with clay. Mud mummies, the third type, seem to have been covered in mud, rather than painted, and 'glued' to the floor of the grave pit, using the mud that, in effect, attached the figure seamlessly to the floor, making it part of the ground beneath, suggesting they were buried straight away and never moved. Arriaza speculates that this represents a change in 'theology', that mummies were now seen as 'belonging' to the land, perhaps because the community was under threat from outside.

But arguably the most interesting thing about these mummies, other than their very existence, so early, is that artificial mummification appears to have started *with the children.* Some 26 per cent of the mummies excavated were less than one year of age, a proportion matched by the high incidence of foetuses and newborns found in Chinchorro cemeteries. This is striking because, generally and cross-culturally, children receive less mortuary attention than adults, especially those who never lived, such as the stillborn. The Chinchorro did not put much emphasis on grave goods – they left a little food and a few fishing tools with most adult remains – but they did give more emphasis to those who never achieved their potential. Arriaza thinks the Chinchorro, being maritime people, adopted the fisherman's principle, of throwing back small fish into the sea, to give them another chance at life, to grow bigger. This is what mummification of children was designed to achieve.

And there is compelling evidence that child mortality among the Chinchorro was abnormally high, most likely due to arsenias. For example, the River Camarones, where mummification began, has on average arsenic levels of 1000 μg/L, a hundred times in excess of the 10 μg/L, which is the standard deemed acceptable by the World Health Organisation. It is also well known that chronic exposure to high arsenic levels produces spontaneous abortions, stillbirths and preterm births. Even today, women in Antofagasta (where the arsenic levels are 30–40 μg/L) have significantly lower birth rates than their counterparts in southern and central Chile (1 μg/L). Arsenias also produces keratosis (scaly growths on the skin), and carcinomas of the liver and bladder. Arriaza reports that a modern study of women in Bangladesh, where the drinking water had 100 µg/L of arsenic, showed that birth abnormalities were three times as high as women with no exposure. On this basis, Chinchorro women would have had birth abnormalities thirty times higher than elsewhere.[2]

Mummification, therefore, took on a special significance, perhaps helping to assuage communal grief. As Arriaza puts it, 'The Chinchorro mummies became "living" entities that used the same space and resources as did the living.'[3] The dead, in effect, became an 'extension' of the living, the mummies maybe even becoming part of an ideology that negated death: through mummification (a form of) immortality was obtained, in which the body and the spirit survived. And, as the mummies provided a resting place for the soul, they were considered living entities. Of the mummies found, they divide nearly 50–50 into natural and artificial entities, perhaps showing that the Chinchorro saw mummification as a natural process and they believed they could 'confer' immortality on the chosen few.

In addition to the dangers of arsenic poisoning, in the river valleys earthquakes and tidal waves were and are frequent in the area, creating terror. We know from historical records that the region was devastated by earthquakes in 1604, 1868, 1877 and 1987. In 1868 and 1877 large tidal waves nearly destroyed the town of Arica, when the waves were of such a size that entire ships were lifted into in the middle of the city. Archaeologists have established that the ancient Peruvian site of Huaca Prieta was struck by a tsunami in 700 BC. Some Chinchorro cemeteries are located along the slopes of hills, at higher elevations away from the ocean.[4] Even today, strong inshore currents make fishing less than straightforward.

A final aspect of these burial practices is that hallucinogenic kits have been found in Chinchorro graves. 'This suggests,' says Arriaza, 'that hallucinogenic drugs were consumed by the Chinchorro or by a religious leader in order to communicate with the other world during the ceremony of presenting the finished mummy to the ancestors and for the reintegration of the mourners.'[5] This practice continued much later, as late as AD ~500–1000 along the Atacama desert, where highly decorated snuffing tablets and tubes are found in graves. Mummified bodies found in Arica with abundant hair provided a unique opportunity to test which hallucinogenic plants were used. Analysis by gas chromatography and mass spectrometry demonstrated the presence of harmine, with the *Banisteriopsis* vine (*ayahuasca*, as introduced in chapter twelve) the probable source. Of the 32 mummies examined, three males tested positive for harmine, the alkaloid that aids the catalysis and synergic effects of powerful hallucinogenic drugs. Juan P. Ogalde, at the University of Tarapacá and the leader of the team that carried out this study, suggested that the harmine was related to medicinal practices and not exclusively ingested by shamans. One positive result was found in the hair of a mummified child, supporting the idea either that the drug was being used medicinally or was imbibed through its mother's milk.[6]

The greatest number of highly decorated snuffing implements occurred during the Tiwanaku civilisation, which flourished inland from the Atacama desert around Lake Titicaca in Bolivia, between AD 500 and 900, suggesting that hallucinogenic use increased rather than the reverse. Ogalde also observes that the *Banisteriopsis* vine is an Amazonian plant – it does not grow in the Atacama coastal region. Thus there must have been an extensive plant trade network in antiquity. This underlines how important trance was to the Chinchorro (and hints, perhaps, that they had brought this practice with them when they migrated into the area from the rainforest).

With three out of thirty-two mummies testing positive for hallucinogens, this is a proportion entirely consistent with what we know about the number of shamans in primitive societies. And another link between mummies and shamanism is tantalising, if speculative. We know that a common vision, or dream, by shamans, is of being dismembered, or reduced to being a skeleton, before reassembly. The overlap is there, with the way they disassembled and reassembled the mummies, though we can probably never know how much the ancient Chinchorro made of this.

What we can say is that, as in the Old World, in South America there was an interaction between local conditions on the ground – desert, brought about by the world's overall weather system, which made mummification a natural process, which in turn shaped the Chinchorros' conceptions of life and death, and in turn dovetailed with shamanistic practices, acquired on the great migration south, into Peru and Chile. The fact that the mummies were kept above ground, on show, to 'interact' with the remaining community, suggests to Arriaza at least that mortuary rituals in the Atacama desert could have gone on for weeks or even years. This was not just ancestor worship but a device to have the ancestors present in the community to be guides and go-betweens with the supernatural world, to protect the community in the face of the unpredictable and often hostile environment. It could be that dead shamans were mummified to guarantee their immortality and preserve their supernatural powers.

Although the Chinchorro eventually disappeared (about 1500 BC), their mortuary practices, and views about life, death and the afterlife, survived in evolved form in later civilisations of South America, as we shall see. Even though mummification developed in Egypt some 2000 years after it did in the Atacama desert, its religious significance was very different, and nowhere near as enduring.

The Chinchorro are also important to our story because their early sedentism, on the desert coast, goes against the post-World War Two orthodoxy that agriculture is the crucial transformation in the development of civilisation. The Chinchorro stayed where they were, apparently, for two reasons: because they had stumbled across a reliable source of food (protein) in fish; and because naturally occurring mummification, allied to a high rate of birth defects, provided a convenient religious adaptation to their existential predicament. This recalls the idea, explored in chapter 3, that shamans may have originally been selected because they showed evidence of *meryak* or *menerik*, arctic hysteria, and the potato pottery, discussed in chapter eleven, where individuals with facial mutilations were somehow regarded as special or sacred. Early peoples, lacking modern medical knowledge, may have had very different attitudes to illness – mental or physical – and to physical deformity.

The Maritime Hypothesis

Although the Chinchorro disappeared from the historical record ~1500 BC, as was referred to above, several elements of their culture lived on in South America, underlining just how different was the path to civilisation there.

Traditionally, the world's five 'pristine' civilisations developed as follows: in Mesopotamia roughly 3500 BC (at the time of the secondary products revolution), Egypt ~3000 BC, India ~2600 BC, China ~1900 BC and Mexico ~1200 BC. In all these cases the development of complex society followed a similar pattern, food surpluses sparked a growth in population and in population density, together with the development of specialisations not having to do directly with food production (see chapter 5 for Paul Wheatley's ideas about early urbanisation and religious ceremonial centres). A central element in the creation of this surplus, which fostered the emergence of such non-food-producing specialisms as scribes, artists, craftspeople and maybe early forms of scientist (astronomers or mathematicians, for example), and which help to define civilisation, was irrigation. Irrigation extended the amount of fertile land but itself demanded organisation. It, too, therefore, was an autocatalytic process. A second defining aspect of civilisation was pottery. Complex, sophisticated societies needed to store food, for two reasons: one, to guard against any future failure of the harvest; and two, to trade, to exchange for items the society lacked. The invention of ceramic containers, which were sent abroad, necessitated identification and, as mentioned earlier, the tokens used to do this eventually led to writing.

This view of early man's development in the Old World has come under sceptical scrutiny recently, as we have seen. But it had appeared fairly stable until the early 1970s, the similarity of the Mexican civilisations to the Old World complexes appearing to underline the general picture. But then things began to change, as new evidence emerged to show that the earliest *New World* urban complexes – the very first civilisations – did not develop as a result of surpluses produced by grain, or even, necessarily, by plants. The Chinchorro, who, as we have just seen, settled along the Chilean coast as early as ~6000 BC, were a predominantly maritime people. Arguably they did not have a 'civilisation' – there was no monumental architecture, for example. But that shortcoming does not apply to the inhabitants of ancient Peru somewhat later and further north.

In the early 1970s Michael Moseley, an archaeologist then at Harvard but now at the University of Florida, identified an important 32-acre site at Aspero on the Peruvian coast at the mouth of the Supe River, north of the capital, Lima. What was unusual about this site was, first, that it boasted six platform mounds – in other words monumental architecture; second, there was no sign of either cereal agriculture or pottery; and third, of course, it was on the same stretch of coast and with much the same geographical configuration as the Chinchorro. And, like them, Aspero was a fishing community: studies of middens, and the bone chemistry of such human remains as have been recovered, show that some 90 per cent of Aspero's food was obtained from the sea. There *was* a crop that was grown in Aspero, but it was not grain or any kind of food: it was cotton. Cotton was grown, slightly inland, in order to manufacture fishing nets.

Moseley reckoned that it would have been possible in settlements like Aspero to create food surpluses once the inhabitants learned to cultivate cotton so as to weave it into netting. In turn this would have led to labour specialisation, some people occupying their time fishing, others looking after the cotton plants, and still others weaving the nets out of the fibres. He reasoned also that small amounts of fruits and vegetables would have been grown, but that the chief aim of farming would have been the growing of cotton, the raw material of nets, by means of which the supply of seafood could be maintained. Probably, said Moseley, there would have been a level of society – priest-technicians – who acted as leaders/organisers so as to build the pyramids to the gods that were the main feature of Aspero.

In 1975 Moseley published *The Maritime Foundations of Andean Civilisation*, in which he boldly concluded that, 'The archaeological axiom that only agriculture could support the rise of complex societies is not a universal truth.' In Peru, he maintained, fishermen had evolved many of the elements of civilisation. Full-blown city-states may have emerged only much later, after 1800 BC, with the arrival of grain and pots to store it in, but 'the crucible of complexity lay on the pre-ceramic coast'.[7]

Moseley's theory (the 'maritime hypothesis') was considered heretical to begin with. However, after radiocarbon dating confirmed that Aspero had flourished as early as 3055 BC, his theory began to catch on and by the end of the 1980s it had become the orthodoxy. It was further refined in 1996 by Ruth Shady, a veteran Peruvian archaeologist. She had spent two years excavating other sites in the area, in particular a site known as Caral, twelve miles inland, up the Supe Valley. In the course of her work, she became convinced that Caral and Aspero were part of the same culture and, moreover, that Caral was in fact a true city-state, roughly contemporary with Aspero 'but far larger and more advanced'. 'That', she says, 'is when I realised that I had stumbled across a problem that would change the way we perceive history in my country.' She still held to the basic premise of the maritime hypothesis but, as she discussed it with Moseley, they drew up some elaborations. The most important was that Caral may have started as a colony of Aspero, an agricultural 'satellite', whose purpose was to provide cotton for nets; but in fact Caral occupied the easiest location in all Peru where an irrigation system could be laid out. And, as Caral grew, its yield of cotton grew with it but so too did tropical fruits, beans, chilies, gourds and wood. In this way, enough of a surplus was generated for the Caral population to be able to trade such produce for fish, molluscs and salt from Aspero and other coastal towns, and to support more extensive trade with the neighbouring valleys and the mountains inland. Shady found that Caral received several exotic imports – shells from Ecuador, dyes from the Andean highlands, hallucinogenic snuff from the Amazon basin, across the Andes.[8]

This trade was so lucrative, Shady observed, that Caral quickly outgrew Aspero in both size and sophistication, with as many as 3,000 people occupying the city. There were at least enough inhabitants to remodel the pyramids at regular intervals. This was done in a particular manner: the old structures were filled with reed bags full of rocks (called 'shicra bags'), together with sacrificial objects (clay figurines, even human sacrifices), the old structures then covered with a new skin of carved stone and coloured plaster. Remains at the top of multilevel platforms showed that priests had sacrificed burnt offerings, with each layer of society looking on from below. Further excavation exposed workshops that fashioned jewellery comprised of shells and semi-precious stones. Flutes made of condor and pelican bones were also found, located in an amphitheatre lit by sacred fires. The upper levels of society used stools made of blue-whale vertebrae.[9]

On this account, then, Caral was important for two reasons. In the first place, it was the oldest civilisation in the New World and, at 3050 BC, though not quite as old as Sumer, it was on a par with Egypt and older than the civilisations in India or China. Second, it had arisen by a different route to Old World civilisations: instead of cereal-based-irrigation agriculture producing a complex society, fishing/cotton (and maybe irrigation) had produced much the same. Building on the difference between *milpa* and *conuco* agriculture (chapter 6), this made the early years of mankind in the Americas very different from in Eurasia.

But then, in 2001, and even more in 2004, two American archaeologists produced a major revision. In particular, a few days before Christmas in 2004, the husband-and-wife team of Jonathan Haas and Winifred Creamer announced the discovery of some *twenty* separate major residential centres, stretching over 700 square miles in three valleys about a hundred miles north of Lima. Ninety-five new radiocarbon dates, they said, confirmed the great age of these sites – between 3000 and 1800 BC. They named the area, which included Caral, Norte Chico, and they described it as a whole culture, where people were producing pyramids and sunken plazas, and they insisted it was a civilisation that survived for more than a thousand years. It was certainly something to put alongside contemporaneous Pharaonic Egypt.[10]

There was, however, a controversial twist to these announcements. Haas and Creamer's new dates for Norte Chico led them to counter Moseley's and Shady's argument that Caral (or Norte Chico) civilisation arose along the coast. The inland sites, they said, were contemporaneous with Aspero and the other coastal settlements and, furthermore, it was irrigation, inland, that seems to have been the basis for the development of surplus, and the auto-catalyst for an administrative class. At the same time, they did concede that complex society developed in the Andes *before* ceramics were invented, and that while plants grown with the aid of irrigation, including cotton, squash, chili, beans and avocados were everywhere, there was almost no evidence of preserved corn (maize) or other grain. They agreed with Moseley and Shady that there were numerous remains of fish and shellfish bones recovered from even inland sites, although they pointed out these were less prevalent than on the coast. And they concluded: 'This early culture appears to have developed not only without pottery, arts and crafts but also without staple grain-based food, which is usually the first large-scale agricultural product of complex societies ... The ancient Peruvians took a different path to civilisation.'[11]

How important is this difference? Well, it may be very important in the sense that early man's relationship with his environment determined his ideological – his political and religious – beliefs, and those in turn influenced his future development. For example, fishing is governed by lunar and tide cycles, whereas farm work is determined by solar and rainfall cycles. Long canal systems need large corporate work-forces to maintain them and to work the reclaimed land. Small canals, as in the Supe Valley, can be maintained by mere families or clans. In fishing, small craft, manned by small crews, can net more than sufficient food, 'harvesting' throughout the year. At the time of the Spanish Conquest indigenous fishing and farming were separate professions along most of the Pacific coast, in both North and South America. The populations, moreover, were self-segregated and people married within their respective vocations, and often spoke different dialects. The maritime people tended shoreline gardens to cultivate tatora reeds for their watercraft. They did not grow staples or pay taxes or tribute as farmers did. Instead, there was an essentially symbiotic relationship between the two communities, who used barter to exchange marine protein for cultivated carbohydrates. Presumably, a similar arrangement existed in earlier times.[12]

On the face of it, therefore, it does appear that there *were* important, basic differences in the trajectory by which Caral reached civilisation as compared with the Old World pristine civilisations (and, as we shall see presently) Mexico. Moreover, with the identification, and dating, of Aspero, Caral and other Norte Chico sites, we are at least in a situation where we can now compare some specific *contemporaneous* early sites in both the Old World and the New, and to enquire in detail what differences and what similarities they showed. In each location, the practices/institutions to be discussed were in place by the end of the fourth millennium BC. Given that early humans left Africa as early as 125,000 years ago, to people the world, a gap of 300–500 years in the emergence of civilisation in vastly difference parts of the globe, may be seen as neither here nor there. That in itself is a parallel which is remarkable and shouldn't be overlooked.

Religion Becomes More Important?

In Mesopotamia there were some parallels in climate change with South America and some differences. In fact, the climate change went further than Mesopotamia: in the south-east Aegean, in the east Mediterranean (off Israel), in the Gulf of Aqaba, in the Red Sea and across the Saudi and Yemeni Peninsula, records show that the early Holocene, between 10000 and 6000 BP, was relatively more humid – between 10 and 32 per cent wetter than now, according to some – with a 'pluvial maximum' at 10000–7800 BP, and a weakening monsoon over Arabia from 7000–6500 BP on, producing drier conditions everywhere after that date. And, as happened off the South American coast at much the same time, there was a stabilisation of sea levels at about ~6300–6000 BP. From 15000–5500 BP, but mainly around 8000 BP, the Persian Gulf was progressively inundated from the Strait of Hormuz to Basra, the sea calculated to have encroached at an average of 100+ metres a year, over an area of land roughly the size of Great Britain, but then slowing down and all but stopping. This 'transgression' of the sea, as it is called, would have had a profound effect on ideas about settlement and may have given rise to notions of a 'flood' that were later incorporated into the bible. Areas of the gulf *were* settled – olive-pressing areas and shell middens have been identified offshore, and a 'peak' of around 40–60 archaeological sites – settlements – along the coast appear 'virtually overnight' at ~8000–7500 BP, according to Jeffrey Rose of the University of Birmingham. In the early-to-mid Holocene, Mesopotamian settlements show the fossils of many sea creatures and are not linear in layout, suggesting that there were plentiful marine resources, possibly a marine trade network and, as yet, no need of irrigation.

Until that point, agriculture had flourished in Mesopotamia for thousands of years. Irrigation had not been needed in the north of the country, because of plentiful rain, sufficient to create streams in the desert, and it only began in the south, along the Tigris and Euphrates, when climatic conditions turned drier. 'The climatic variations documented for the middle of the fourth millennium [brought on by the weakening monsoon, according to Clift and Plumb] seem, within a space of two to three hundred years, to have stemmed the floods that regularly covered large tracts of land and to have drained such large areas that in a relatively short period of time, large parts of Babylonia became attractive for new permanent settlements.'[13]

Excavations show that, associated with this climate variation (which, according to Gunnar Heinsohn, remember, may have involved environmental catastrophe), there was a sudden change in settlement pattern, from very scattered and fairly small individual settlements to dense settlements of a much larger kind never seen before. These geographical conditions appear to have favoured the development of *communal* irrigation systems – systems that were not elaborate, not at that stage, but which nonetheless brought about a marked improvements in the yield of barley (which now evolved from the two-row to the six-row mutant), and at the same time taught people the advantages of cooperation.[14]

We must always be wary of assuming that, because events occur in close association, either temporally or location-wise, one causes the other. But it does seem that it was the particular climatic conditions of Mesopotamia – where irrigation could markedly improve crop yields and where there was enough water available (but in the wrong place) – that allowed this development fairly easily and obviously. The crucial point was that though the land was now habitable, there was still so much water available that nearly every arable plot had easy and direct access to it. 'This fact ... must have produced a "paradise", with multiple, high-yield harvests each year.' An added factor was that the southern alluvial plains of Mesopotamia were lacking in other commodities, such as timber, stone, mineral and metals. The food surplus of this 'paradise' could be traded for these commodities, making for a dense network of contacts, and provided conditions for the development of specialist workers in the cities themselves.[15]

This may have been a factor leading to the diverse populations that were such a feature of early city life, going beyond simple kin groups. And it was an exciting advance: for the first time people could become involved in activities not directly linked with food production. Yet such a development would have raised anxiety levels: citizens had to rely on others, not their kin, for essentials. This underlying anxiety may well explain the vast, unprecedented schemes and projects which were perhaps designed to foster a community spirit – monumental, labour-intensive architectural undertakings. For these same reasons, religion may well have become *more* important in cities than in previous configurations.[16] Is this support for Heinsohn's claim that civilisation was born in catastrophe?

The first city is generally held to have been Eridu, a site just over a hundred miles inland from the Persian Gulf and now called Aby

Shahrein. Its actual location was unique, in that it occupied a transitional zone between sea and land. It was near an alluvial plain and close to marshes, which meant that it could easily benefit from three ecological systems – the alluvium, the desert and the marshes, and so profit from three different modes of subsistence: farming, nomadic pastoralism, and fishing. But there was also a religious reason for Eridu (recall Paul Wheatley's *Pivot of the Four Quarters*, chapter five). The city was located on a small hill ringed by a depression, in which subterranean water collected. This surrounding area was never less than a swamp and in the rainy season formed a sizeable lake. It was thus a configuration that conformed neatly to (or was responsible for) Mesopotamian ideas of the cosmos, which pictured the earth as a disc surrounded by a huge body of water. In mirroring this configuration, Eridu became a sacred spot.[17]

The name, 'Eridu', means 'mighty place' and in Sumerian mythology was home of the Abzu temple of the god Enki, the water god. In Sumerian *Ab* means 'water' and *zu* means 'far' and the temple was known as the 'House of subterranean waters', from which the gifts of civilisation arose. Extensive deposits of fish bones were found there, showing perhaps an Abzu cult that endured for millennia, drawing its strength from an underground aquifer, the subterranean waters being the place where the gods lived before humans were created. Petr Charvát, the Czech prehistorian, says that Eridu – where the original chapel, existing over eighteen occupation levels, embracing more than a thousand years, and dated to 4900 BC – was believed to contain the source of all wisdom and that it was the seat of the god of knowledge. He says the 'first intelligible universal religion seems to have been born' in Eridu, in which worship involved the use of a triad of colours in the local pottery. Earthly existence was affirmed by the use of red, death by the use of black, and eternal life (and purity) through white.

In the Norte Chico region there also seems to have been a significant climatic event which preceded the emergence of urban culture, but it was a different event – the stabilisation of sea levels. This would have had at least three effects that we can think of. It would have stabilised the types of food in the sea, enabling fishing techniques to evolve and consolidate. It would have stabilised the coastlines, making settlement more likely, and more permanent, in turn allowing living habits to emerge and mature. And it would have stabilised the gradients of the rivers flowing from the Andes to the Pacific, allowing their courses to settle, which in turn would make living alongside these rivers easier, including the building of irrigation channels. Delta regions would also have stabilised.

In each case, then, urbanisation didn't 'just happen' – it was triggered by a specific climatic event. These events – the weakening monsoon in the Old World, and the increasing frequency of ENSO events in the New World – were described in chapter five.

Another parallel between the two locations arises from the fact that Caral also appears to have been regarded as a sacred site. Ruth Shady believes it was sacred – again, perhaps, because it was an easy location to irrigate, the inhabitants of Norte Chico seeing it as a favoured spot – i.e., favoured by the gods. In fact, what appears to be the oldest god known in the Americas was found in Norte Chico in 2003; it was a clear image on a fragment of a gourd bowl, carbon-dated to 2250 BC and shows a cartoon-like figure, with fanged teeth, holding a staff.[18]

At the same time, what the ideas of Heinsohn, Clift and Plumb have to recommend them is that marked environmental events, catastrophes, would have been sufficiently traumatic to have provoked sudden, large-scale change, which is what we see in the archaeological record associated with the origin of cities. It would also underline Wheatley's point, that the earliest cities were *religious* entities.

A further difference between Eridu and Aspero/Caral was that in order to farm the area of Mesopotamia between the slow-running Tigris and Euphrates, fairly long communally built canals needed to be built, whereas in Norte Chico, with its fast-running short rivers, which cut deeply into the terrain, and are interspersed by arid desert, only short canals were possible or needed, canals moreover that could be dug and maintained by families, rather than wider communities.

It is difficult to know just how important this difference was. By the middle of the third millennium, Uruk was the centre of a 'hinterland', an essentially rural area under its influence, which extended roughly 12–15 kilometres around it. Next to this was an area some 2–3 kilometres wide which showed no influence, and then began the hinterland of the next city, in this case Umma. There were at least twenty cities of this kind in Mesopotamia. Was this arrangement, especially this proximity, the explanation for warfare? At one stage in the evolution of cities, it was assumed by archaeologists that their original purpose was for defence. But, as we have already seen, this argument can no longer be supported. First, even in the Middle East, where city walls were sometimes vast and very elaborate, the walls came *after* the initial settlement. At Uruk, for example, the city had been largely formed around 3200 BC, but the walls were not built until three hundred years later, at roughly 2900 BC.* This may mean that cities developed in tandem with communal irrigation, which in turn produced a rapid rise in population and it was *this*, after several generations, that caused fighting to break out, as competition for arable land intensified.[19]

This perhaps puts into context the fact that both Eridu and Caral were sacred sites – this was the original impulse for city life, some form of communal worship, as Paul Wheatley says, perhaps provoked by catastrophe. In Norte Chico, moreover, and in marked contrast to Mesopotamia, not only are there no walls around any of the cities, nor is there any sign of armed conflict. No weapons have been found, no mutilated bodies, no burned houses. Is this further evidence for the pre-eminence of fish as Norte Chico's food resource? If they were not dependent on arable land, in the way that the Mesopotamian city-states were, then there would be no need to fight over it. Furthermore, the ocean cannot be owned as land can be owned, and the fish off the Peruvian coast were so plentiful, so abundant, that, again, fighting was unnecessary. Added to that, the physical layout of Norte Chico, with its relatively short, steep valleys, more than a day's walk from each other, and interspersed by arid, inhospitable desert, meant that many of the city-states were not contiguous. Here too there would have been less direct contact and therefore less confrontation. Even today, easily watered land occupies barely 2 per cent of the arable coastal terrain in the area. One other difference between Mesopotamia and Norte Chico that has been insufficiently explored is the nature of the surplus each traded. Grain seeds last, not indefinitely but for a considerable period of time. Fish, on the other hand, are perishable unless techniques to preserve them can be found.

* On the other hand, '*Uru*' means a walled area.

Excavations have shown that, in Mesopotamia, at least, these early urban areas were usually divided into three. There was an inner city with its own walls, inside which were found the temples of the city's gods, plus the palace of the ruler/administrator/religious leader, and a number of private houses. The suburbs consisted of much smaller houses, communal gardens and cattle pens, providing day-to-day produce and support for the citizens. Finally, there was a commercial centre. Though called the 'harbour', this area was where overland commerce was handled and where foreign as well as native merchants lived. The very names of cities are believed in many cases to have referred to their visual appearance.[20]

Caral has a central, core zone, with monumental architecture. Around this central zone, even annexed to the mounds, are the elite residential quarters, of mason and mortar construction, with plastered and painted walls. Further out are what Michael Moseley calls the 'lower class barrios', where the houses are made of cane. Stone-built warehouses have been found in Norte Chico, though not at Caral, and plastered stone workshops producing jewellery. Caral also has what appears to be a sunken astronomical observatory and an amphitheatre where musicians played – flutes have been found, made from pelican bones. The fact that Norte Chico had no writing means we cannot be sure it had the same level of complexity as Mesopotamia, but its monumental architecture is every bit as impressive. The importance of monumental architecture lies partly in the evidence it gives for organised religion but also because it presupposes an organised work-force larger than several nuclear families – complex society.[21]

In these first cities, much life revolved around the temple, though they may not have been used as temples all the time. We can say this because they are often found filled with 'ordinary rubbish', unremarkable pottery, for example, and they do not appear always to have been kept clean, as if they were open to the general public at least some of the time. Many fish bones were found among the debris. People associated with the cult were the most prominent members of society. At Eridu and Uruk the existence of temple platforms shows that there was already sufficient communal organisation to construct such buildings – after the megaliths these are the next great examples of monumental architecture. As time went by, these platforms were raised ever higher, eventually becoming stepped or terraced towers crowned by shrines. These are known as ziggurats, a word based on the Assyrian, and probably on an earlier Akkadian term, *zigguaratu*, meaning summit or mountain top. This increasingly elaborate structure had to be maintained, which required a highly organised cult. The name ziggurat also recalls the cultural importance, in Mesopotamia at least, of mountains.

The temples were so important – and so large – that they played a central role in the economic life of the early cities. Records from the temple of Baba (or Bau), a goddess of Lagash, show that shortly before 2400 BC the temple estates were more than a square mile in extent. The land was used for every kind of agricultural use and supported as many as 1,200 people in the service of the temple. There were specialist bakers, brewers, wool workers, spinners and weavers, as well as slaves and an administrative staff. The tenant farmers were not slaves exactly; instead, their relation to the temple seems to have been an early form of feudalism. In addition to the new specialisations already mentioned, we may include the barber, the jeweller or metalworker, the costumier and cloth merchant, the laundryman, brick makers, the ornamental gardener, the ferryman, the 'sellers of songs' and the specialist that interests the archaeologist most – the scribe. Outside the temple area were the residential quarters and an extensive cemetery, where the elite were buried, with never more than two adults to a grave, perhaps indicating monogamy.[22]

In layout, form, dimensions and variety, the monumental architecture of Sumer and Norte Chico was remarkably similar. Their large buildings, temples, appeared suddenly (supporting the argument that environmental catastrophe provoked this change), and both took pyramidal form though this may be the least interesting thing to say about them: without metal or concrete/cement, the pyramid – wide at the base, increasingly narrow at the top – is the only form possible for a large structure. In both Sumer and Norte Chico several pyramids were grouped together at sites – usually with a large pyramid surrounded by six or so smaller ones. In both places, it seems that the larger pyramids were dedicated to the more important gods, the smaller ones to lesser deities. It is possible that subordinate deities had to be placated before access to higher-ranked gods could be obtained.

In Aspero there are six truncated pyramids among seventeen mounds. The largest, called Huaca de los Idolos, measures 40 metres by 30 metres by 10.7 metres high and was topped with summit rooms and courts. The raised platforms have modelled and painted clay friezes. The mounds show evidence of cobble and basalt block masonry and adobe construction. The mounds were composed of successive phases of stone-walled rooms, built by progressive infilling of earlier rooms. The outer platform walls are of large, angular basaltic rocks set in adobe mortar with a smooth outer surface coated with plasters and occasionally painted.

In Caral the largest and most elaborate pyramid is some 140 × 150 metres (450 × 500 feet) and about 20 metres (65 feet) high. The major pyramid has a basal length of 153 metres, a width of 109 metres and rises to a height of 28 metres. El Paraiso, situated two kilometres inland from the mouth of the Chillón River, and adjacent to flood plain cultivation land, was the largest of the Pre-ceramic Period monuments and, at three times larger than any of its contemporaries, was once the largest expression of organisation and labour investment in all of South America.[23] This site comprised more than a dozen mounds spread over a 60-hectare area with a nuclear group of seven mounds, the nuclear group forming a rough U-shape with the largest mounds on the arms, framing a sunken plaza. This layout became common in South American monument formation and its religious significance is considered later.* The two large mounds were used for habitation and 'no artefacts anomalous to domestic use are known'.[24] It is uniformly oriented 25° east of north, perpendicular to the solstice sunrise in 1500 BC. El Paraiso's one-metre-thick stone walls were plastered with clay. Stone was quarried from nearby hills and roughly trimmed. Constructed entirely of monumental masonry, the mound exceeds 100,000 tons in gross weight.

Later structures in the area (i.e., in the third and second millennium BC) became even more imposing. Huaca la Florida, at 250 metres wide, 50–60 metres deep by 30 metres tall, was the largest monument of its time, with low arms extending at the wings for 500 metres, forming a large plaza that could have held 100,000 people. Stone walls at the site were covered with clay plaster and painted yellow and red.[25] Several other indications of change over time are apparent.

* Studies of the 'entoptic' images found painted on the walls of some caves in South Africa included many 'U' shapes.[26]

For example, pre-ceramic sites such as Aspero, Rio Seco, Huaca Prieta and Salinas de Chao are located near the Pacific Ocean, whereas in the initial ceramic period platforms were constructed away from the shoreline, reflecting an evolution from maritime subsistence to irrigation agriculture. Around 1800 BC, the rise in population seen prior to the shift to irrigation agriculture could have been supported by the rich fisheries.

At Caral the dominant cultigens were Guayaba (shrubs, trees), cotton, guava fruit, gourd, and frijol (beans). Potential staples included two specimens of maize, one each of sweet potato/yam and achira (ginger/banana etc.). If this mix is representative of what was farmed, then, says Moseley, it is difficult to understand how horticulture alone could have sustained a large population and, moreover, the scarce remains of staples at Caral raises the possibility that the population could not live there year round. Staple domesticates are important precisely because they can be harvested in such abundance that they can be stored to feed people until the next harvest, which doesn't apply to fruit or beans. So, almost certainly, protein at Caral was obtained by exchanging cotton and cultivated produce for marine products. Fishermen essentially fed the coastal farmers.

This is all confirmed, says Moseley, by the evidence from the Chinchorro. The artificial mummies there, as we have seen, included large numbers of children who underwent privileged mortuary treatment, indicative of early class formation. With so little evidence of farming, and with the chemical analysis of bone indicating that the people obtained 89 per cent of their diet from the sea, the maritime foundation of civilisation in Aspero/Caral can no longer be disputed.

The similarities and differences between Eridu and Aspero/Caral are further underlined by the ways in which each came to its end. Eridu suffered a gradual reversal, lasting from the twenty-first century BC until the last chapter was played out in the sixth century BC. Power passed to the north, to Warka and Uruk, though Eridu was rebuilt later, probably as a temple complex, in honour of its distinguished history. But sand dunes encroached, the salt waters rose, and the city was slowly choked, its legendary fertility a thing of the past.

In contrast, and revealingly, recent evidence shows that, just as the advent of El Niño had helped to create Aspero and Caral (as Clift and Plumb say), in that the erection of temples, to cope with the increased frequency of ENSO events had occurred around 3800 BC, so it was the ever-more-frequent El Niño events that called time on Norte Chico. Again according to Michael Moseley, supported by Ruth Shady and geologists from the University of Maine's Climate Change Institute, a sequence of earthquakes, torrential rains and flood, and the march of sand across once-fertile fields, set in motion a process that brought the civilisation to an end within a few generations.[27]

The Supe people (as the inhabitants of the Supe Valley, part of Norte Chico, are sometimes called) thrived for about 2,000 years but then, around 3,600 years ago, an enormous earthquake, estimated to be at least 8 on the Richter scale, devastated the entire area with a multitude of effects. The earthquake collapsed the walls of the main pyramid complex, as it became part of an enormous landslide of rocks, mud and construction materials. Widespread flooding followed but this was just the beginning. The mountains surrounding the valley were destabilised by the earthquake (and by subsequent quakes) and massive amounts of rock and other debris were sent crashing down. Subsequent El Niños brought huge rains and these rains washed the debris into the ocean where a strong current, parallel to the coast, re-deposited the debris and sand in the form of a large ridge, known today as the Medio Mundo, which sealed off the coastal bays, hitherto rich in marine life, rapidly turning them into sand-filled dead areas. Strong, ever-present, offshore winds then blew this sand back onshore, clogging the irrigation systems and covering the soil in the fields. In the space of a few generations, according to Moseley and his colleagues, what had been a productive (if arid) region, was reduced to an all-but-uninhabitable wasteland.[28]

The trajectories of Eridu and Aspero/Caral are instructive, and for two reasons. They underline the point made earlier, about life in the Old World – religious life and economic life – relating mainly to the slow weakening of the monsoon, and the centrality of fertility in early man's imaginative life, while in the New World the main concern was the unpredictability and violence of natural processes. And second, they highlight how different the natural world was in the Americas, the absence of domesticated animals meaning not just that the New World lacked vital sources of food, but that it also lacked vital sources of energy and was bound to be more restricted in the life-ways that could be pursued. We shall see again and again in later chapters how important domesticated mammals – and in particular pastoral nomadism – were in the history of the Old World.

• 16 •

The Steppes, War and 'a New Anthropological Type'

To call the Bronze Age (roughly speaking 3500–1250 BC) the most interesting and most important epoch in the history of the Old World risks overstating the case. Fundamental transformations undoubtedly occurred at other times. Yet, when we look at the innovations that took place during those centuries, at the changes in weather and climate, at the transformations in material goods, above all at the changes in ideology, there is no denying just how extraordinary a time it was. We may begin by listing those changes/innovations and, as we do, ask ourselves what the long-term consequences were.

In the previous chapter we covered the rise of the first cities, and cities were to be a major feature of the Bronze Age right across Eurasia, both their construction and their demise. For example, in 1946 the American scholar Samuel Noah Kramer began to publish his translations of Sumerian clay tablets and in doing so he identified no fewer than twenty-seven 'historical firsts' discovered or achieved or recorded in the cities of the early Iraqis. Among them were the first schools, the first historian, the first pharmacopoeia, the first clocks, the first arch, the first legal code, the first library, the first farmer's almanac, and the first bicameral congress. The Sumerians were the first to use gardens to provide shade, they recorded the first proverbs and fables, they had the first epic literature and the first love songs. The reason for this remarkable burst of creativity is not hard to find: cities were (and remain) far more competitive, experimental environments than anything that had gone before, and some time in the late fourth millennium BC, people came together to live in large cities. The transition transformed human experience, for the new conditions required men and women to cooperate in ways they never had before. The city is the cradle of culture, the birthplace of nearly all our most cherished ideas.

Tools for Living Together

It was this close contiguity, this new face-to-face style of cohabitation, that explained the proliferation of new ideas, particularly in the basic tools for living together – writing, law, bureaucracy, specialised occupations, education, weights and measures. A measure of the rapidity of the change at this time (and an indication of its importance, necessity and popularity) can be had from the survey reported by Hans Nissen which shows that at the end of the fourth millennium in Mesopotamia rural settlements outnumbered urban ones by the ratio of 4:1. Six hundred years later – i.e., the middle of the third millennium – that ratio had reversed completely and was now 9:1 in favour of the larger urban sites.[1]

The achievements of these cities and city-states were astonishing and endured for some twenty-six centuries. It was in Babylonia that music, medicine and mathematics were developed, where the first libraries were created, the first maps drawn, where chemistry, botany and zoology were conceived. At least, we assume that is so. Babylon is the home of so many 'firsts' because it is also the place where writing ('the invention of inventions') was conceived and therefore we *know* about Babylon in a way that we do not know history before then.[2]

Writing in the Old World began with the practice of clay tokens used to seal and identify containers of goods (and their owners), goods that were traded between the cities of Sumer and elsewhere. In the late 1960s Denise Schmandt-Besserat, a French-American professor of Middle Eastern Studies, at the University of Texas, Austin, noticed that thousands of 'rather mundane clay objects' had been found throughout the Ancient Near East. Regarded as insignificant by most archaeologists, Schmandt-Besserat thought otherwise, that they might have formed an ancient system that had been overlooked. She therefore visited various collections of these tokens and, in the course of her study, she found that they were sometimes geometrical in form – spheres, tetrahedrons, cylinders – while others were in the shape of animals, tools or vessels. She came to realise that they were the first clay objects to have been hardened by fire. Whatever they were, a lot of effort had gone into their manufacture.[3]

Eventually, she came across an account of a hollow tablet found at Nuzi, a site in northern Iraq and dated to the second millennium BC. The cuneiform inscription said: 'Counters representing small cattle: 21 ewes that lamb, 6 female lambs, 8 full-grown male sheep ...' and so on. When the tablet had been opened, inside were found forty-nine counters, exactly the number of cattle in the written list. For Schmandt-Besserat, this was 'like a Rosetta stone'; she had uncovered a primitive accounting system and one which led to the creation of writing.[4]

The first tokens dated to 8000–4300 BC were fairly plain and not very varied. They were found in such sites as Tepe Asiab in Iran (*c.* 7900–7700 BC), where the people still lived mainly by hunting and gathering. Beginning around 4400 BC, more complex tokens appeared, mainly in connection with temple activity. The different types represented different objects: for example, cones appear to have represented grain, an ovoid stood for a jar of oil, while cylinders stood for domestic animals. The tokens caught on because they removed the need to remember certain things, and they removed the need for a spoken language, so for that reason could be used between people who spoke different tongues. They came into use because of a change in social and economic structure. As trade increased between villages, the headman would have needed to keep a record of who had produced what.[5]

The complex tokens appear to have been introduced into Susa, the main city of Elam (southern Iran), and Uruk, and seem to have been a result of the need to account for goods produced in the city's workshops (most were found in public rather than private buildings). The tokens also provided a new and more accurate way to assess and record taxes. They were kept together in one of two ways. They were either strung together or, more importantly from our point of view, enclosed in clay envelopes. It was on the outside of these envelopes that marks were made, to record what was inside and who was involved. And this seems to be how cuneiform script came about. Of course, the new system quickly made the tokens themselves redundant, with the result that the impressions in the clay had replaced the old system by about 3500–3100 BC, the very beginning of the Bronze Age. The envelopes became tablets and the way was open for the development of full-blown cuneiform.[6]

To begin with, there was no grammar. Words – nouns mainly, but a few verbs – could be placed next to one another in a random fashion. One reason for this was that at Uruk the writing, or proto-writing, was not read, as we would understand reading. It was an artificial memory system that could be understood by people who spoke different languages. Writing and reading as we know it appears to have been developed at Shuruppak in southern Mesopotamia and the language was Sumerian. No one knows who the Sumerians were, or where they originated, and it is possible that their writing was carried out in an 'official' language, like Sanskrit and Latin many thousands of years later, its use confined only to the learned. This next stage in the development of writing occurred when one sound, corresponding to a known object, was generalised to conform to that sound in other words or contexts. An English example might be a drawing of a striped insect to mean a 'bee'. Then it would be adapted, to be used in such words as 'be-lieve'. This happened, for example, with the Sumerian word for water, 'a', the sign for which was two parallel wavy lines (≈). The context made it clear whether 'a' meant water or the sound.

Although the first texts which contain grammatical elements come from Shuruppak, word order was still highly variable. The breakthrough to writing in the actual order of speech seems to have occurred first when Eannatum was king of Lagash (*c.* 2500 BC). It was only now that writing was able to convert all aspects of language to written form. The acquisition of such literacy was arduous and was aided by encyclopaedic and other lists. People – in the bible and elsewhere – were described as 'knowing the words' for things, such as birds or fishes, which meant they could, to that extent, read. Some lists were king lists, and these produced another advance when texts began to go beyond mere lists, to offer comment and evaluation on rulers, their conflicts, the laws they introduced: history was for the first time being written down. The list about the date-palm, for instance, includes hundreds of entries, not just the many parts of the palm, from bark to crown, but words for types of decay and the uses to which the wood could be put. In other words, this is how the first forms of knowledge were arranged and recorded.[7]

Lists made possible new kinds of intellectual activity. Among other things, they encouraged comparison and criticism. The items in a list were removed from the context that gave them meaning in the oral world and in that sense became abstractions. They could be separated and sorted in ways never conceived before, giving rise to questions never asked in an oral culture. For example, the astronomical lists made clear the intricate patterns of the celestial bodies, marking the beginning of mathematical astronomy and astrology.

Accordingly, in both Mesopotamia and Egypt literacy was held in high esteem. Shulgi, a Sumerian king around 2100 BC, boasted that

> As a youth, I studied the scribal art in the Tablet-House, from the tablets of Sumer and Akkad;
> No one of noble birth could write a tablet as I could.

Scribes were trained in Ur since at least the second quarter of the third millennium. When they signed documents, they often added the names and positions of their fathers, which confirms that they were usually the sons of city governors, temple administrators, army officers, or priests: literacy was confined to scribes and administrators.[8]

Two schools, perhaps the first in the world, were founded by king Shulgi at Nippur and at Ur in the last century of the third millennium BC, but he referred to them without any elaboration, so they may have been established well before this. The Babylonian term for school or scribal academy was *edubba*, literally 'Tablet-House'. There were specialist masters for language, mathematics ('scribe of counting') and surveying ('scribe of the field') but day-to-day teaching was conducted by someone called, literally, 'Big Brother', who were probably senior pupils.[9]

The scribal tradition spread far beyond Mesopotamia and as it did so it expanded. The Egyptians were the first to write with reed brushes on pieces of old pottery; next they introduced slabs of sycamore which were coated with gypsum plaster, which could be rubbed off to allow re-use. Papyrus was the most expensive writing material of all and was available only to the most accomplished, and therefore least wasteful, scribes.[10]

Not all writing had to do with business. The early, more literary texts of Sumer, naturally enough perhaps, include the first religious literature, hymns in particular. The first libraries were installed in

Mesopotamia though to begin with they were more like archives than libraries proper. They contained records of the practical, day-to-day activities of the Mesopotamian city-states. We have to remember that in most cases the libraries served the purposes of the priests and in Mesopotamian cities, where the temple cult owned huge estates, practical archives – recording transactions, contracts and deliveries – were as much part of the cult as were ritual texts for the sacred services. But the propagandistic needs of the cult and the emerging royal elite – hymns, inscriptions – were the elements that provoked a more modern form of literacy. Texts such as the Epic of Gilgamesh, or the Epic of Creation, may therefore have been used in ritual. But these works, which involved some form of mental activity beyond flat records of transactions, appear first in the texts at Nippur in the middle of the third millennium. The next advance occurred at Ebla, Ur and Nippur. Each of these later libraries boasted a new, more scholarly entity: catalogues of the holdings, in which works of the imagination, and/or religious works, were listed separately. The ordering of the list was still pretty haphazard, however, for alphabetisation was not introduced for more than 1,500 years.[11]

Libraries undoubtedly existed in ancient Egypt, but because they wrote on papyrus (the 'bullrushes' in which the infant Moses was supposed to have been sequestered), little has survived. In describing the building complex of Ramses II (1279–1213 BC), the Greek historian Diodorus says that it included a sacred library which bore the inscription, 'Clinic for the Soul'.[12]

One of the duties of the king in Mesopotamia was the administration of justice and, with the invention of writing, the first law codes were written down, often carved in stone and displayed publicly where everyone could see what the law *was* or have it read to them. The oldest known codes are Sumerian, dated to the second millennium BC and from them we can see that these laws take one of two forms, apodictic and casuistic. Apodictic laws are absolute prohibitions, such as 'Thou shalt not kill'. Casuistic laws are of the type: 'If a man delivers to his neighbour money or goods to keep, and it is stolen out of the man's house, then, if the thief is found, he shall pay double.' The prologue makes it plain that Hammurabi's well-preserved law code, created ~1790 BC, was indeed intended to be exhibited in public. They are not what we would understand as statutes: they are royal decisions, a range of typical examples rather than a formal statement of principles. Hammurabi meant the code to apply across all of Babylonia, replacing earlier local laws that differed from area to area.[13]

A final early effect of writing was that it helped remove authority from the shamanistic type of religious leader: ritual, sacred and supernatural experiences could be written down now, meaning that tradition, rather than personal charisma, began to count for something.

Although archaeologists now order the 'ages' of man into the Stone, Copper, Bronze and Iron Ages, in that order, the first use of a metallic substance was almost certainly iron, around 300,000 years ago, when ochre found favour as decoration. Haematite in particular was popular, possibly because of its colour – red, the colour of blood and life. If the colour, lustre and even the weight of metals made their impact on early humans, it was as raw rocks, or in the beds of rivers and streams that they first encountered them. From this, they would have discovered that some rocks, such as flints and cherts, became easier to work with on heating and that others, like native copper, were easier to hammer into serviceable tools. Gradually, therefore, as time passed, the advantages of metals over stone, wood and bone would have become apparent. However, when we think of metallurgy in antiquity we mainly mean one thing – smelting, the apparently magical transformation by which solid rock, when treated in a certain way through heat, can be transformed into a molten metal. One can easily imagine the awesome impact this would have had on early humans.[14]

Copper ores are found all over the Fertile Crescent region but invariably in hilly and mountainous regions. Archaeologists are inclined therefore to think this is where metallurgy began, rather than in river valleys. The area favoured nowadays is a region 'whose inhabitants, in addition to possessing ore and fuel, had adopted some form of settled life and were enjoying a chalcolithic culture'. This area, between the Elburz mountains and the Caspian Sea, is the front-runner for the origin of metallurgy, though the Hindu Kush and other areas have their adherents too. 'That the discovery was fortuitously made can hardly be doubted, for it is inconceivable that men, simply by taking thought, would have realised the relationship existing between malachite – a rich-blue, friable stone – and the red, malleable substance, which we call copper.' Because such a link was regarded then as magical, the early copper-smiths were believed to have super-human powers.[15]

At one stage it was believed that 'the camp-fire was the original smelting furnace'. No more. Quite simply, the hearths at around 4000 BC were not hot enough. It is not only the temperature that acts against campfires. Not being enclosed, the atmosphere would not have been conducive to 'reducing' (separation). On the other hand, well before the discovery of smelting, much higher temperatures would have been obtained in some pottery kilns. The atmosphere in these baking chambers would have been of a strongly reducing character and modern experiments have confirmed that a spongy copper could be smelted in this way. The accident may have happened when ancient potters used malachite to colour pottery – 'and then got the shock of their lives, when the colour delivered was very different from that anticipated'.[16]

We know that by 4000 BC knowledge of the process had spread to a number of regions in western Asia and that, by 3800 BC, copper smelting was being practised 'comparatively widely' in the ancient world. 'By the early years of the third millennium BC, the people of Sumer had created the first important civilisation known to us in which metals played a conspicuous role.' (The oldest known stock of metal tools dates from 2900 BC). From these dates onward copper was the dominant metal in western Asia and North Africa until after 2000 BC.[17]

Insofar as early metallurgy was concerned, after the discovery of smelting two advances were crucial. These were the discovery first of bronze and second of iron (iron will be considered later). There are two mysteries surrounding the advent of the Bronze Age, certainly so far as the Middle East is concerned, where it occurred first. One mystery lies in the fact that tin, the alloy with copper that makes it much harder, as bronze, is relatively rare in nature. How did this particular alloy, therefore, come to be made for the first time? And second, why, despite this, were advances so rapid, with the result that, between about 3000 BC and 2600 BC, all the important advances in metallurgical history, save for the hardening of steel, were introduced?[18]

In one sense, we should call the early Bronze Age the alloy age. This is because for many years, either side of 2000 BC, and despite what was said above, objects that might be called bronze had a very varied chemical make-up. Alloyed with copper, and ranging from less than 1 per cent to 15 per cent, there could be found tin, lead, iron and arsenic, suggesting that although early people had some idea of what made copper harder, more malleable and gave its tools and weapons a better edge, they weren't entirely comfortable with the precise details of the process. The exact composition of bronze also varied from area to area – between Cyprus, Sumer and Crete, for example. The all-important change-over from copper to real bronze occurred in the first quarter of the second millennium BC. 'Tin differs from copper – and the precious metals – in that it is never found in nature in a pure state. Instead, it is always in chemical combination. It must therefore have been smelted, though (and this is another mystery) hardly any metallic tin has ever been found in excavations by archaeologists. (In fact, only one piece of pure tin older than 1500 BC has ever been found.)'[19]

Though the exact origins of bronze are obscure, its attractions over copper were real enough, once its method of production could be stabilised, and its increasing popularity brought about considerable changes in the economy of the ancient world. Whereas copper was found in a fairly large number of localities, this was not the case with bronze for, as was said above, in neither Asia nor Europe is tin ore widely distributed. This limitation meant that the places where tin was mined grew considerably in importance and, since they were situated almost entirely in Europe, that continent had advantages denied to Asia and Africa. The fact that bronze was much more fluid than copper made it far more suitable for casting while its widespread use in weapons and tools reflects the fact that, provided the tin content could be kept at 9–10 per cent, hammered bronze is usually a good 70 per cent stronger than hammered copper.[20] The edges of bronze tools were at least twice as hard as copper, very important because that meant the edges of daggers became as important as their points, encouraging the development of swords.[21]

Metallurgy was quite sophisticated from early on. Welding, nails and rivets were early inventions, in use from 3000 BC. Gold-plating began as early as the third millennium, soon followed by the lost-wax technique, for making bronze sculptures. In terms of ideas, three uses to which metals were put seem to have been most profound. These were the dagger, as was mentioned above, the mirror, and coins. Mirrors were particularly popular among the Chinese, and the Romans excelled at making them, finding that an alloy of 23–28 per cent tin, 5–7 per cent lead, and the rest copper, served best. Reflections were later considered to be linked to man's soul. As early as the third millennium BC, the inhabitants of Mesopotamia began using ingots of precious metals in exchange for goods. The ingots, of gold or silver and of uniform weight, were called minas or shekels or talents. Money was to have a profound effect on humankind's thinking, but not until later.[22]

The Bronze Age reached its peak around 1400 BC. It was a time when iron was scarce and valuable. Tutankhamun reigned for only a very few years as a pharaoh in Egypt, and died about 1350 BC, but his tomb, famously discovered and excavated by Lord Carnarvon and Howard Carter in 1922, contained – besides vast quantities of gold, jewels and fabulous ornaments – a dagger, headrest and bracelet all made of iron. There were also some very small models of tools, barely an inch long, also made of iron. In all cases this was smelted iron, not meteoric.[23]

The final and all-important fact about the development of bronze metallurgy is that it coincided with the domestication of the horse in the steppe countries of Europe, and with the wheel. Warfare was therefore suddenly transformed – in fact, it changed more rapidly in the late Bronze Age than at any other time until gunpowder was used in anger in China in the tenth century AD.

The Rise and Fall of the Chariot

The wheel, another momentous innovation of the Bronze Age, has already been introduced, in chapter 8. The first vehicles – sledges – were used by early hunter-fisher societies in near-Arctic northern Europe by 7000 BC, presumably pulled by dogs. 'Vehicle' signs occur in the pictographic script of Uruk in the late fourth millennium BC, at the very beginning of the Bronze Age, and actual remains of an axle-and-wheel unit were found at a similar date at a site in Zurich in Switzerland. These vehicles had solid wheels, made either from one or three pieces of wood. From archaeological remains at sites before 2000 BC, the use of these so-called disc wheels stretched from Denmark to Persia, with the greatest density in the area immediately north of the Black Sea. So this may indicate where the wheel was first introduced. Oxen and donkeys appear to have been used at first.[24]

These (four-wheeled) wagons were very slow – 3.2 kph, on one estimate. The (two-wheeled) chariot, however, was a good deal faster –

12–14 kph when trotting, 17–20 kph when galloping. In the cuneiform texts, Sumerian refers to the 'equid of the desert', meaning an ass or donkey, and to the 'equid of the mountains', meaning horse. Three words were used for wheeled vehicles, *mar-gid-da*, for four-wheeled wagons, *gigir*, for two-wheeled vehicles, and *narkabtu* which, as time went by, came to mean chariot. With *narkabtu*, says the British archaeologist Stuart Piggott, 'We come to the beginning of one of the great chapters of ancient history: the development of the light two-wheeled chariot drawn by paired horses as a piece of technology and as an institution within the social order as an emblem of power and prestige.' After the first solid wheels were invented, the spoked wheel was conceived. This had to be built under tension, with shaped wood, but its lightness made much greater speeds possible. Chariot warfare flourished between 1700 and 1200 BC – i.e., the end of the Bronze Age and into the Iron Age.[25]

Robert Drews, emeritus professor of classics at Vanderbilt University in Nashville, Tennessee, describes the chariot as 'a technological triumph of the early second millennium'. Made of light hardwood, with a 'leather-mesh' platform where the driver stood, it weighed barely thirty kilograms.[26] Recent excavations, he says, have enabled us to establish that the chariot became militarily significant in the seventeenth century BC. Before this time they had been used mainly for display and what appears to have made the difference is its coupling with the composite bow, invented on the central Eurasian steppes and made of aged wood, sinew and horn, which together store more energy than wood alone and therefore had a range three times that of the simple (or self) bow. Several score of chariots, says Drews, each manned by an expert driver and a bowman, could overcome a conventional army of infantrymen.[27]

By the middle of the fifteenth century BC, at such battles as Megiddo, kings could deploy as many as a thousand chariots. At Kadesh, in the fourteenth century BC, the Hittite king is said to have deployed 3,500 chariots, though most palace polities would have had armies of several hundred. Chariots were expensive to build and maintain and many armies, Drews concludes, were far smaller in reality than on paper. Composite bows could take years to make (the wood had to age properly) and body armour, or corslets, made from as many as 500 copper scales, were also expensive and took time to manufacture. But while the chariot was the military machine par excellence, it was precious and had its own dedicated bureaucracy – the 'scribes of the stable' and the 'scribes of the chariotry'. 'Everywhere the charioteers have names, while infantrymen are merely numbered.'[28]

Chariots are not plentiful in the archaeological record but in their heyday they were used as mobile platforms for archers, the arrows fired while the chariot was at full speed, a practice that extended from the Near East to India and to China at this time.[29] Records from Nuzi (northern Iraq, half-way between Babylon and Nineveh) show that charioteers were issued with helmets, corslets, a whip, a sword, a bow and quiver with 30–40 arrows. In his Karnak annals, Thutmose (1479–1425 BC), the sixth pharaoh of the eighteenth dynasty in ancient Egypt, specifies that he captured 924 chariots and 502 bows from the enemy. In other accounts of Minoan or Mycenaean battles, records and excavations show that the bow was far more important from 1600 to 1200 BC than it had been before or would be again. In all the kingdoms of the late Bronze Age the composite bow, and the chariot, were the principal offensive weapons. In the Rig Veda and the Mahabharata the chariot dominates the battlefield.[30] The smallest tactical unit appears to have been ten chariots – when they are ordered in the records they are always in units of ten.

The chariot developed as a response to the expanding states, which themselves were an evolution owing to the increased surpluses produced by agriculture. As states expanded in size, they became more likely to encroach on the territory of others, and at the same time needed to exert control – militarily – over their own more widespread peoples. Now, therefore, with the expanded territories, and greater numbers, wheeled traffic became superior to infantry. Carts could transport more logistic material to distant parts of the state, or empire, but it was only when the horse took over from the ox and donkey, about 1800 BC, and the spoked wheel was developed, that the chariot became the prime weapon in Bronze Age warfare. Only now could the much lighter vehicle be driven faster than a man could run and only now was it so light that it could be easily floated across rivers, which otherwise formed natural barriers. The composite bow had traditionally been too costly and took too long – five years-plus – to manufacture, but allied with the chariot the investment was now worth the effort. With a range of 175 metres as opposed to 60–90 metres for a simple bow, chariot armies could bombard infantry (who couldn't afford composite bows in such numbers) from outside their range, and escape comfortably.

As we shall see, this predominance didn't last beyond about 1200 BC but before then the advent of the chariot, and chariot warfare, had brought with it another profound change in the way early peoples thought of themselves.

The Decline of the Great Goddess

For what we see in the course of the Bronze Age is a key change in ideology. As was noted before, albeit in a footnote, religious iconography, especially of ancient religions, can be extremely complex, with many overlapping and contradictory meanings. Nonetheless, scholars are agreed that what we see in the Bronze Age – partly as a result of the rise of warfare, as epitomised by the increased role of the chariot – is a gradual change from worship of the Great Goddess to the worship of male gods. It would be wrong to put this down entirely to the chariot, but the rise of warfare certainly had a role, and we shall see shortly what other factors came into play.

But it remains true overall that, in what are sometimes called the great Palace States of Mycenae (Greece), the Minoan Palaces of Crete, in Egypt, Sumer and the Indus Valley civilisation, from 3500 BC into the beginning of the second millennium BC, worship was mainly devoted to one version or another of the Great Goddess. In Crete the goddesses take the form of the Great Snake Goddess, often shown bare-breasted (and with large breasts), sometimes in a trance-like state, the Goddess of the Double Axe, also bare-breasted sometimes, and sometimes with the double axe in the form of the curving horns of the bull, or associated with the tree of life. On other occasions, the goddess is shown emerging from the earth, clasping corn or poppy seeds, an image familiar in mainland Greece as well, in the form of Persephone, daughter of Demeter, the Corn Goddess, who lives for the winter months in the underworld.[31]

The Bee Goddess, the Goddess of the Sacred Knot, the Goddess of the Animals, the Bird Goddess, the Goddess with the Poppy Crown, the Goddess with the Crown of Doves and Bull's Horns, an image of Two Goddesses and a Child – all these show both old and new ideas, but in all cases the deity worshipped is female. In some cases, at some times, the goddesses were milk goddesses: one sip from the nipple of Juno could confer divinity and immortality. Traces of shamanism are discernible in the images of the World Tree, linking the underworld to the sky, the underworld itself being a shamanistic trace also. A further glimpse is perhaps afforded by the labyrinth drawn on the ground-floor corridor of the Palace at Knossos – an image that lasted across a thousand years. The labyrinth, which has a lady or goddess associated with it, is believed to describe the path of a ritual dance, used as a way to communicate with the goddess, very possibly as a means of inducing trance.[32]

In the Minotaur, Crete also had a profusion of bull images and Anne Baring and Jules Cashford make the point that, in ancient Crete (as elsewhere), the king was ritually sacrificed to ensure that the fertility of human, animal and plant life did not diminish with his (the king's) failing powers, but that, at some stage, the bull was substituted for the king and sacrificed in his place.[33] As we shall see in due course, this was an important substitution.

The goddess culture was a marked feature of the early Bronze Age, say Baring and Cashford, and they trace goddess figures to Anatolia, the Indus Valley, to Egypt and Mesopotamia: Inanna, or Ishtar in Sumeria, Nammu, goddess of the primeval ocean, Ki-Ninhursag, the mother of the gods and humanity, the mother of wild animals and of the herd animals, cow, sheep and goat. (In Sumerian the word for sheepfold, vulva and womb are the same.) The milk of Ninhursag's sacred herd, kept in the temple precincts and fields, nourished the people.[34] Inanna was the Queen of the Earth, of grain and the vine, the date palm, cedar and olive. Inanna and Ishtar were goddesses of sexual love and inspired the practice of temple prostitution. Inanna and Ishtar had consorts in other areas of Mesopotamia called Tammuz and Dumuzi (presumably the same word originally). They were sometimes depicted as 'Shepherds of the People', titles that came originally from their relationship to the Mother Goddess as the Holy Shepherdess, though they could also be accompanied by a golden ram.[35] Inanna was often shown naked.

In Egypt, Isis was queen of heaven, earth and the underworld, often shown as cow-headed or with a crown of cow's horns, and sometimes with her legs parted, giving birth on the back of a pig.[36] Nephthys was shown as a serpent goddess and Sekhmet as lion-headed. Nut was the goddess of the sky and the mother of the sun, moon and stars.

In Babylon, Tiamat was the original mother goddess but the best-known aspect of her is the story as laid down in the *Enūma eliš*, the Babylonian Creation myth (named after its opening words) and discovered in 1849 by Austin Henry Layard in the ruined library of Ashurbanipal in Nineveh (present-day Mosul), Iraq. The version Layard discovered dates to the seventh century BC but the story was first composed between the eighteenth and sixteenth centuries BC, when it was known all over the ancient world. The central element in the story, certainly from our point of view, is the conquest of the primordial Mother Goddess by the sky, wind and sun *god*, Marduk, who takes the tablets of the law for himself.[37]

Her body was distended and her mouth was wide open.
He released the arrow, it tore her belly,
It cut through her insides, splitting the heart.
Having thus subdued her, he extinguished her life.
He cast down her carcass to stand upon it.

The lord trod on the legs of Tiamat,
With his unsparing mace he crushed her skull.[38]

The myth may refer to a real battle, when Babylon overcame Sumer, but it also refers to a much more widespread process that took place during the second millennium BC and culminating at the time the Bronze Age gave way to the Iron Age, after about 1200 BC. This was the almost universal replacement of the great goddess(es) by male gods. This was much more than a change in gender, as we shall see in due course and there were good reasons for it.

One we have already encountered, at least in an initial sense. The domestication of the horse, the invention of the wheel and then the chariot, which transformed warfare, against a background of developing city-states and empires in the Old World, and which may or may not have developed against a sky that was much more turbulent and disturbing than it is now, created a world where war was much more prevalent – more terrible and more far-reaching – than ever before. In such a context, male values became more important than female values, society becoming more 'heroic', with the heroes inevitably being male.

In addition to all this, there was one other factor, one other way of life which developed roughly speaking between 3000 and 1000 BC and never existed, could not have existed, in the New World. It too played an important role in the demise of the Great Goddess. This was the spread of nomadic pastoralism.

The Rise of the Pastoral Nomad

As we saw in chapters 7 and 8, beyond the loess belt in north-western Europe, traditional farming – as developed in the Middle East – was more difficult on the more marginal soils. The development of the plough, pulled by oxen, helped enormously but an alternative adaptation was the evolution of pastoralism – farmers who cultivated some fields but also had flocks of herd animals who could graze on land that was not suitable for cereals (in the hills, for example). The initial effect this had on ideology was shown in the development of megaliths, the symbolic purpose of which was also explored in chapter 9. Elsewhere, and especially to the east, there was a more intensive development of this trend.

One of the more remarkable features of Eurasia – though this is not always appreciated as much as it might be, particularly among the inhabitants of the predominantly maritime nations of western Europe – is the central steppe zone. This area is huge, larger than the continental United States, stretching from the *puszta* (plains) of Hungary for 10,000 kilometres west to east as far as Mongolia and Manchuria, from the Danube to the Great Wall of China. From north to south it extends for up to 600 kilometres, between 58° and 47° north. The principal characteristic, and the cause of the landscape, says Elena Kuzmina, of Moscow State University, is the 'continentality' of the climate and the deficiency of moisture (fewer than 500mm of annual precipitation). This dictates that the prevailing vegetation is comprised of narrow-leaved drought-resistant grasses with well-developed root systems which, when they decay, form soils rich in humus and attract ungulates and rodents (remember this is the very area where many mammals originally evolved – see above, chapter 6).[39] This steppe encounters no real obstacles across its vast dimensions, though it is bordered on the north by dense Siberian forests and in the south, as Haim Ofek pointed out, by very high mountain ranges, from the Caucasus (between the

Black and the Caspian Seas) to the Altai (in central Asia), and including the Hindu Kush (Afghanistan/Pakistan) and the Pamirs (stretching from Afghanistan to China). At its southern edges there is also a number of deserts of which the most terrible are the Taklamakan and the Gobi.[40]

In some ways the steppes are the equivalent of the prairies in North America and the comparison is instructive because the steppes, though dwarfing even the vast reaches of the prairies, also show how much the type of animal – wild or domesticated – can alter the engagement that people have with otherwise similar landscapes.

In the North American prairies, as we saw in chapter fourteen, the bison was the dominant animal, a sizeable skittish mammal too wild to domesticate, but a valuable source of meat and whose hunting was a form of ritual where, besides providing food, small tribes of hunter-gatherers could come together a few times a year for exchange purposes. This system was stable and endured for millennia, until the arrival of the European. The Eurasian prairies, the steppes, would develop very differently.

They were in their configuration a vast east-west highway, a second one, in effect running parallel to that other east-west corridor, the southern coastline of Asia, as introduced in chapter 2, and shown in maps 1 and 4, along which people, animals, manufactured objects and ideas travelled, having a profound effect on the lineaments of Eurasian history: for example, on the formation of the Silk Road, and the rise of civilisation in Iran, India and China (the nomads of the steppe introduced wheeled transport – in particular the chariot, the horse, and metallurgy – to China).[41]

In the words of A.M. Petrov, the Great Silk Road 'is by no means just a road . . . It is a huge, fluid historical and cultural space over which, in ancient times and the Middle Ages, the trans-migration between different peoples from the extreme ends of Asia to the Western countries was realized.'[42] Fernand Braudel, the French historian of '*la long durée*', saw the presence of pastoral nomads as a disruptive force, 'often interrupting periods of slow historical processes, allowing for rapid change and oscillation . . . In [an] epoch that seems to epitomize slowness, these people epitomize great rapidity and unexpectedness.'[43] Gérard Chaliand agreed – he called the steppes Eurasia's great 'zone of turbulence'. We shall, in due course, see how right they all were.

Pastoralism emerges in arid zones where, says A.M. Khazanov, of the Institute of Ethnography of the Russian Academy of Sciences, in Moscow, it is superior to other forms of agriculture. Initially, on the fringes of the steppes, overlapping with areas of settled agriculturalists, sedentary pastoralism emerged, with herdsmen husbandry (much as in north-west Europe). As the fourth millennium gave way to the third, this type of pastoralism was established between the Volga and the Urals and took precedence over other forms of agriculture. Khazanov says it is difficult to imagine the emergence of pastoralism without domestication of the horse and, since the steppes are the natural homeland of the horse, it is no surprise to find that horse bones comprise 80 per cent of the osteological remains of fourth- and third-millennium BC sites.[44]

The crucial change, to full-scale pastoral nomadism, took place, says Khazanov, only in the last half of the second millennium. There are signs that some pastoral nomads started to develop towards urbanisation but abandoned the attempt around 1800 BC. This is when herds grew large, which necessitated great migrations during the year, when dairying developed (because cereal cultivation had been given up), and when excellent horsemanship became the hallmark of what were now pastoral nomads. They also appear to have domesticated the Bactrian camel. Carts carrying their tents might be pulled by up to twenty-two oxen.[45]

Nomad comes from the Latin *nomas*, which means 'wandering shepherd'.[46] And this final change occurred because, during the second millennium, the monsoon was weakening, the climate was drying, provoking nomadism right across the area of the steppe, as far as China, where it was observed of the nomads that 'some of them do not use cereals for food'.[47] Instead, the Chinese used to say, 'They follow water and grass.'

Wandering could be extensive. Nomads in north Eurasia, as an example, spent 4–5 months in winter pasture, two months in summer pasture, and 5–6 months travelling between the two.[48] Estimates vary as to how many animals a family (of five members) needed in order to survive as pastoral nomads, from eight mares, one stallion, one bull and ten cows (20 large mammals in all) to 30–50 horses, 100 sheep, 20–50 goats, 15–25 cows (225 in total) to 800 or even 1,500 sheep. In Inner Mongolia studies show that one person can look after 500 sheep if he has a horse.[49] The steppes were vast but with herds of such size, and given that one sheep needs one hectare, the nomads' requirements were equally vast and therein lay the problem.

Nomadic herds are normally divided into large stock (cows and horses, or camels) and small stock (sheep and goats). In general, small stock was preferred closer to the Middle East, by the peoples who would be called Semitic, large stock was preferred in the deep steppes, where the 'Aryans' or Indo-Europeans were established, though there were fewer cattle and horses in Tibet, because it was so high. The steppe was and is for the most part temperate and that was important in determining migration routes. These were for the most part linear and meridional (i.e., north in the summer, south in the winter, the nutrition value of fodder being two-and-a-half times as poor in winter). Routes changed little over centuries and millennia.[50]

The three traditional products of pastoral nomadism were (and still are) milk, meat and blood and the emergence of this economic package was important in extending food-production to remote and arid areas where other forms of agriculture couldn't work. Strabo recorded that 'even in the Crimea, one of the most fertile areas in the Eurasian steppes, the harvest only yielded thirty-fold, while in Mesopotamia it yielded three-hundred-fold'.[51] Even today extensive pastoralism yields more than agriculture does in many areas.

Of all the domesticated animals, and after the horse, the sheep was perhaps best adapted to the plains because it can graze the ground more closely than other cattle, meaning sheep can get more food from poor areas (though goats are more useful for clearing the tougher vegetation of forests). Sheep are more prolific than goats, producing an average of seven lambs and they can live on drier matter than deer and may even eat soil. By and large they are easy to handle – they flock in herds naturally (lambs play together), they run with goats, and castrated rams give little trouble.[52] In excavations, their remains change morphologically most at around 3000 BC when there is a reduction in bone size and horn size. There was a general change in Eurasia from the use of flax for textiles in Neolithic times to wool in the Bronze Age, and we shall note the growing industrial importance of wool later in the book.

Excavation shows that herd sizes could vary enormously, from 2,000 to 27,000 and that the animals were kept first for their milk and blood and only after that for meat, which was regarded as a prestigious foodstuff, to be consumed only on ritual occasions. That said, when animals were killed, pastoralists tried to do it in the autumn, so as to have fewer mouths to feed during the winter months.[53]

Sheep were not widely worshipped – they figure as tribute in Assyria and there were sheep figurines in India and elsewhere. In general the shepherd's life was regarded as solitary.[54] M.L. Ryder tells us that sheep have changed the landscape more than any other animal, partly because pastoral nomads spread them over huge areas but also because they carried the seeds of plants on their wool and so helped them to spread.

Goats were usually kept with sheep, having been domesticated (in the Anatolian Zagros mountains) at about the same time. Though very similar to sheep, they have three advantages that, in times of difficulty, can be important: their milk has slightly more protein and fat in it than do other milks; goats eat shrubs and small trees, rather than just grassland, as is true of sheep and cows; and their life expectancy (16–18 years) is greater than that of sheep (12–16 years). Goat wool is nowhere near as desirable as sheep wool but the advantages mentioned above (in addition to goat skins, and goat gut, both very useful) may occasionally have been decisive. Since sheep and goats run together quite contentedly it always made sense to have mixed flocks or herds.

No less than the sheep, the domestication of the bull and the cow has had a profound effect on the development of civilisation. In chapter 8 the importance of ploughing and milking was explored but even before that bulls and cows may have been domesticated for religious reasons. Domesticated cattle throughout the world are descended from a single wild species, *Bos primigenius*, the recently extinct wild ox or 'aurochs'. These prehistoric mammals stood two metres high and were three metres long and at their peak they were spread over the entire temperate region of Eurasia, though they never reached Ireland, Scandinavia or the Americas. The world then was wetter (and therefore greener) than today and they inhabited river valleys and marshy forests, becoming complete herbivores, not just grazers, feeding on grass and herbs, tree foliage and even bark.

Our knowledge of the first human contact with aurochs comes from the great cave paintings, as we have seen, and one theory is that, around 7000 BC, when they were first domesticated, they were chosen because the shape of their horns resembled the moon at certain stages of its regular cycle and so they were looked upon as a symbol of fertility and regeneration.[55] They may have been domesticated after the offspring were sacrificed, at which time more docile specimens would have been unconsciously selected. Cattle are naturally gregarious and once a small herd had been created, others would naturally have joined in.

It is possible that cattle were domesticated in three areas – the Near East, the Indus Valley and the south-eastern Sahara – but everywhere they quickly became valuable sources of wealth and a stylised head of a cow was one of the first signs to be used in the Near Eastern tokens that, as we have seen, led to writing. In Hammurabi's Code (see above, page 276) out of 282 laws twenty-nine of the decipherable entries concern crimes against oxen. Later, in Rome, a man was tried – and exiled – for killing an ox before the end of its working life. From the third millennium BC, zebu cattle (with a hump on their back) evolved to tolerate higher temperatures, more humid environments and were less susceptible to insect-borne infections than taurine cattle. For these reasons, they began to be traded far and wide. The Shang in China famously used the shoulder bones of cattle to foretell the future, cracking the bones with hot rods and interpreting the resulting fissures.

Bulls were everywhere admired for their strength and fertility, again as we have seen repeatedly, and were worshipped throughout the Near East, Egypt and India. Many kings had bull titles and in India there is a preponderance of bulls on seals and statuettes. Over time, every major culture adopted the Taurus Constellation in the heavens to represent their specific myths. The title of 'Bull' was translated into – or adapted to – Greek (Tauros), Latin (Taurus), Sanskrit (Vrishaba), Persia (Gav), and Arabic (Thaur).[56] The fullest and best account of bull worship comes from the Rig Veda, the collection of hymns introduced in chapters two and nine, and purportedly written down by conquering Aryans, nomadic pastoralists from inner Asia, who brought their fertility bull-gods into India in the second millennium BC. We are running a little bit ahead of ourselves at this point for, as we shall see, there is considerable controversy as to whether the Aryans ever existed in India and, if so, who exactly they were. But bull and cow worship was (and is) extensive there, as it was in many other civilisations around the Mediterranean, such as in Crete, Greece and ancient Rome, where the taurobolium was a temple dedicated to the sacrifice of bulls.

The cow was much less potent than the bull but still worshipped widely, again most notably in India, which has a special day in November, called *Gosthastami*, set aside for the cow and where the most powerful sacred substance known to Hindus is *Pancha-gavya*, a solution made from the five products of the live cow: milk, ghee, curds, urine and dung, said to have purification properties that can avert evil and bestow blessings on marriage. All Vedic rituals use the products of cattle, Shiva is a bull god, and Vishnu, the preserver of the world, is depicted also as the Lord of the Herdsmen and *Govinda*, 'one who brings satisfaction to the cows'.[57]

Nor should we forget that *Bos* exists in a third 'version', in the form of the ox, a castrated male, who retained his strength even as he lost his wildness, making him eminently suitable for ploughing and pulling carts. To begin with, oxen pulled ploughs by ropes hitched to their most obvious feature, their horns, but the yoke – attached to the animal's even stronger shoulders – was introduced about 3000 BC. Spacious thoroughfares in Mohenjo-Daro suggest that ox carts were used there, as is supported by the discoveries of crude terracotta models of bullock wagons.[58]

Whether or not the Bronze Age sky was as disturbed as the catastrophists say, what does seem to be agreed is that important climatic change occurred in the second half of the second millennium. And it would appear that this change was the final stimulus for pastoralists to abandon agriculture and become fully nomadic. More than that, just as pastoral nomadism was a response to a drier climate, so the great palace states of Eurasia formed as a different response to the same climate variation. This coincidence would come to matter.

In some of the art of the early second millennium people are shown sitting on horses but without saddles and bridles and therefore, says Khazanov, they could not be called real horsemen, who could not have used horses for war or pasturing. Herds at that stage were probably small and manoeuvred with the help of dogs, never straying too far. The appearance of the first real horsemen, he says, occurs towards the middle of the second millennium BC, a time which saw the emergence of the so-called 'steppe-bronze cultures', when pastoralists occupied zones quite deep into the steppe at sites which could be up to 90 kilometres from rivers. But at that time the steppe environment was more humid than it is now, life was more mobile, with families following their herds on foot or in carts drawn by oxen or horses, and perhaps with some people riding.

The best known of these cultures is the Sintashta-Petrovka culture, perhaps more widely known as the Andronovo culture, located to the north-east of the Caspian and Aral Seas, flourishing at ~2300–1000 BC. The pig was completely absent from the Andronovo areas (meaning they were not fully settled) but four important innovations are attributed to them: the smelting of copper and tin, to make bronze; fortified settlements (for a time), to protect their mines; the light war chariot; and three different breeds of horse, one suitable for pulling carts, one suitable for riding and one for pulling chariots. They were typified by their aristocratic warrior class, horse sacrifice, and burials of the warriors *with* their horses. They were the prototype of the nomadic 'barbarians' that were to become such a feature of Eurasian life from the Iron Age on.[59]

The 'Zone of Turbulence' – The Essential Foundation of World Politics

But, no less important, the economic system of nomadic pastoralism is, ultimately, limited. As Khazanov puts it, it is not exactly an evolutionary blind alley but 'the ecological foundations of pastoral nomadism are such that they leave little scope for the development of a complex economy ... pastoral nomadism is doomed to stagnation because its economy is extensive and allows no permanent solution to the problem of balance at the expense of intensification of production. For example, the number of livestock per head among the Hsiung-nu, who in ancient times occupied the territory which now constitutes Mongolia, corresponds almost exactly with the number found in a study carried out in 1918.'[60]

Put more simply, the elements of nomadic pastoralism are self-limiting. When you are travelling round, fodder and hay cannot be stored. One person on horseback can look after a hundred horses – yes, but to feed them you have to keep moving. This in itself brings further problems: in the course of 20–25 years the productivity of pastures can fall by up to a factor of four and their regeneration takes half a century.[61] New territory was needed, therefore, every generation. Travelling constantly in antiquity meant high infant mortality, which further limited population. Milk was the chief source of protein (for if they killed animals regularly, for meat, they destroyed their wealth base) and keeping mammals lactating prevented them from conceiving until lactation was over. Moving around means you need light goods to take with you, and you can't carry much anyway; this, plus the fact that surplus is hard to build up in a nomadic society, meant that craftspeople did not emerge in substantial numbers, so handicrafts or luxury goods were not highly developed, meaning in turn there was less to exchange.[62]

It was this self-limiting quality of pastoral nomadism that created what Gérard Chaliand, one-time professor in the École Supérieur de Guerre in France, called the 'zone of turbulence', an all-important element in Old World history for as much as two thousand years, beginning at the end of the Bronze Age. This zone 'threatened settled peoples from China to Russia to Hungary, including Iran, India, the Byzantine empire and even Egypt . . . Over the Eurasian landmass, the opposition between nomads emerging from central Asia, and settled societies, was, for two millennia, the essential foundation of world politics. They were people without a common language, most of them without writing, of diverse ethnic origins although belonging to a number of major branches (Turks, Mongols, Manchus) . . . but united by a common strategic culture, a culture of the steppe, based on the mobility of the mounted archer using harassment and indirect manoeuvre before delivering the blow. It was a strategic culture that developed the capacity to concentrate far from bases and overcome problems of logistics infinitely more easily than did settled peoples.'[63]

Chaliand was right and Ernest Gellner agreed. 'Minding flocks against depredations by wild animals or, above all, by other shepherds . . . constitutes a permanent training in violence.'[64] As we shall have ample opportunity to observe, the instability and self-limitations inherent in the economy of pastoral nomads was to have important consequences for the political, social and ideological development of the Old World, in ways that were not always predictable. And the first time was at the end of the second millennium BC, when the Bronze Age was giving way to the Iron Age.

In normal times, or quiet times, the nomads traded their horses – the hardy, thickset horses of the steppe – and wool, for grain, tea or silk. Their centre of gravity was between northern Mongolia and Lake Baikal but when they needed to they travelled vast distances. As a result, they became superb horsemen, children being taught to ride from a very early age, and equally accomplished bowmen because their chief fall-back tactic, when times turned harsh, was the raid, in which they harassed settled people without risking frontal attack. Temperatures on the various regions of the steppe could vary from +35 degrees to -40, making the nomads adaptable and hardy. After the enemy had been weakened by these harassments, the nomads would attack more frontally, on horseback, with swords and lances – both invented by them – as were, in time, the horse bit, reins and stirrup.[65] They attacked according to the decimal system – in hundreds, thousands, or tens of thousands.

Although the migration routes of nomads, as they emerged over 2,000 years between 3000 BC and 1000 BC, were fairly settled, the vast distances they travelled, and their requirement of so much pasture to sustain their large herds, inevitably provoked conflict when times were harsh, as they were – increasingly – towards the end of the second millennium BC. Elena Kuzmina tells us that the steppe had been fully exploited by the twelfth century BC and there was no more territory for pastoral peoples to expand into.[66] These conflicts occurred between the different groups of nomads themselves, but as Chaliand says the more important conflicts were with the settled peoples living on the edges of the steppes, all the way from the Black Sea and the northern fringes of Mesopotamia across to China. These conflicts were sharpened by the increasing size of settled societies, aided by their use of the chariot to keep their further-flung territories in line.

Nicola di Cosmo agrees that there was a great expansion at this time, right across central Eurasia, of mounted, warlike nomads.[67] 'The emergence of this new anthropological type,' he writes, 'is attested to by the iconography of tenth-century BC Iran and ninth-century BC Assyria and is confirmed by Assyrian and Greek sources of the ninth and eighth centuries BC, who assign these groups names such as Cimmerians, Scythians and Sakas.'[68] The first riders shown in monumental art were those at the battle of Kadesh, in Tell el Amarna and Saqqara, dating to the fourteenth century BC; armed riders were known in Assyria in the twelfth century BC (divided into those wielding bows and those with lances), and mounted warriors formed part of the Israelites' army in the first Book of Kings.[69]

Robert Drews, in his account of *The End of the Bronze Age*, puts the attacks of the barbarians at the beginning of the twelfth century BC. His barbarians are not always nomads but whether they are or not they did share with them a violent disposition born of climatic change which drove them into conflict with the palace states of the eastern Mediterranean region and in both cases they displayed advances in military technology and tactics, the chief achievement of which was the *overcoming* of the hitherto all-conquering chariot. This was achieved either by the development of the cavalry, or new weapons which transformed the infantry. So far as the cavalry were concerned, people had finally learned how to ride as we understand the term, and thanks mainly to a horseback life on the steppe, aided by the invention of the horse bridle and harness (and later the stirrup), giving people much more control over the horse. Traditionally, firing a bow on horseback had been far more difficult than firing from the platform of a chariot but as horsemen (like the Parthians) became more and more adept, this difficulty was overcome.

At the same time, the invention of the sword, spear, two-bladed javelins and arrows, and the lance – which first appeared in the Andronovo culture in the seventeenth century BC – improved the efficiency of infantries and these, combined with the rise of the metal shield, meant that foot soldiers were no longer as vulnerable as they had been in combat with chariots.

And so, beginning in the twelfth century BC, the end of the Bronze Age and the beginning of the Iron Age, there was what would now be called 'a perfect storm', when developments in weapons, in riding, in climate, in political organisation all came together to provoke a sudden shift across huge areas of the Old World. The most important element in this shift, at least to begin with, was the sudden change in the character of warfare, resulting in the sacking of – according to Drews, though he is not alone in this – some 300 cities of antiquity in a pattern that was repeated from Greece to the Indus Valley.[70] Echoes of this age of destruction are found in both the Mahabharata and the *Iliad*, even in the Old Testament. One Sumerian scribe wrote of the nomads as 'a host whose onslaught was like a hurricane, a people who had never known a city'.[71]

A New Attitude to Death

These 'tribes' are known to historians as Semites and Aryans (and later as 'barbarians') and they brought with them, we are told, their herds of sheep and goats and, the second most important aspect, after the changed character of warfare, their gods. Ideologically, the nomads were very different from settled peoples. Despite their ethnic differences, they shared a similar magical-religious foundation based on a belief in a supreme deity, Tengri, the sky god, plus a host of minor gods, with numerous rituals and prohibitions, and with widespread shamanism, the shaman's task being to question the gods, interpret the signs of the gods' response, and cure the sick.[72] It is this transformation that is represented in the defeat of the Mother Goddess by Marduk, mentioned earlier.

This change in ideology, which would turn out to be momentous (it has been described as 'a crucial pivot in the evolution of human consciousness'), had several interlinked elements. Male gods had begun to grow in importance from the middle of the Bronze Age (2200–1550 BC): in city life, as opposed to village life, and in expanding states, with more far-flung territories, where wheeled transport was needed, and the chariot was the most effective fighting machine, the male role became ever more prominent, the concept of the 'hero' – men with greater powers than others – was born, and the character of the deities followed suit. Furthermore, this development made people more individualistic and was built on by the warrior ethos of the pastoral nomads who, not being settled in any one place, and not being agriculturalists, worshipped the unpredictable forces that mattered to them: storms, wind, lightning (which frightened the cattle even more than it frightened the nomads themselves), the sun and fire (the latter harder to preserve in a nomadic way of life). As was described in chapter 10, they used hallucinogens to visit other regions of the cosmos.

The nomads also lived an essentially *animal* life, in close proximity to the horse and cattle and subsisting on milk, meat and blood (the Scythians, we are told, even 'coaxed' milk from their mares). This meant they had a much less close relationship with vegetation, with the cycles of plant growth. Furthermore, their relationship with horses and cattle was predominantly one of control, even domination, all of which encouraged among them the separation of nature from human life – this was in effect the desacralisation of nature. With life on the steppes and the desert so harsh, constantly on the move, often fighting, the storm and wind gods took on these characteristics as well.

This was underlined by the changing attitudes to death at that time. Because of the conflicts, the constant raiding, violent death became much more common, life was experienced as untrustworthy, and death at the hands of humans became for the first time as likely as death from famine or other natural catastrophes. As a result, death was now looked upon as final, the absolute end of life.[73] The old idea, of death and rebirth, based initially on the regeneration of the moon after its 'death', disappeared in many locations. Just as warfare was widespread on earth, so the sky gods took on a more terrifying aspect in which killing and fighting were common and death the end.

'The moral order of the god culture,' write Anne Baring and Jules Cashford, 'derived from the Aryan and Semitic tribes, was based on a paradigm of opposition and conquest: a view of life, and particularly of nature, as something "other" to be conquered. The manifest world was seen as intrinsically separate from the unmanifest world, which was now placed outside or beyond nature in the realm of the transcendent gods.'[74]

Joseph Campbell (1904–1987), the American mythologist, agreed. In his book *The Masks of God: Occidental Mythology*, he wrote: 'It is now perfectly clear that before the violent entry of the late Bronze Age and early Iron Age nomadic Aryan cattle-herders from the north and Semitic sheep-and-goat-herders from the south into the old cult sites of the ancient world, there had prevailed in that world an essentially organic, vegetal, non-heroic view of the nature and necessities of life that was completely repugnant to those lion hearts for whom not the patient toil of earth but the battle spear and its plunder were the source of both wealth and joy.'[75]

Flowery prose maybe, but these developments would lead in time to the most profound spiritual innovation the world had (and has) ever seen, one that never occurred in the New World, and which is explored in chapter 18.

But not even this completes the account of the transformations that took place in the remarkable epoch we call the Bronze Age. For whereas the chariot was expensive, so that only the most accomplished individuals were able to take part in chariot warfare, and where hundreds were in use most of the time, with cavalry – and especially with infantry – tens of thousands of ordinary men were involved. Under this new system, warrior values were perpetuated but now far more men, in fact almost any able-bodied man who was willing, could be a warrior. While the advent of pastoral nomadic warriors transformed ideology, the advent of large infantries would, in time, provoke its own transformation, this time in politics.

• 17 •

The Day of the Jaguar

At around the time the Bronze Age was giving way to the Iron Age in the Old World, when momentous changes were taking place in both material conditions and in ideology, two civilisations were emerging in tropical America. Although we now know that Aspero/Caral, at Chico Norte, emerged as a proto-urban entity not so long after similar urban structures appeared in Mesopotamia, the urban phenomena did not spread through the New World to anywhere near the same extent as it did through Eurasia. This had something to do with the fact, discussed in chapter eleven, that maize, the New World's most important grain, needed a great deal more manipulation to convert it into a cereal that could produce surpluses, but also because its earliest purpose was to make beer, for use in religious ritual. The trajectory was, therefore, very different. Another reason was that, given the absence of domesticated mammals, and only stone tools, there was a limit in the New World to the amount of forest that could be cleared, and marginal land could not benefit from ploughing.

We have at present no evidence that there were any urban structures built in the New World between the decline of Norte Chico and the rise of the two civilisations that are the subject of this chapter. As Brian Fagan has put it, a vast chasm separates the thousands of farming villages scattered throughout Mesoamerica in 2000 BC and the sophisticated civilisations that arose 'with dramatic suddenness' just a thousand or so years later.[1] This timing has fascinated and perplexed archaeologists because they have long wondered whether there was, in Central America at least, one ancestral culture that gave rise to later states or whether these later states developed independently.

One candidate for a 'mother culture' was identified in 1925 on the island of La Venta, located in a coastal swamp served by the Tonala River, which discharges on to the northern shore of lowland Veracruz, in the Gulf of Mexico. It was there that the Danish archaeologist Frans Blom and his ethnographer colleague, Oliver La Farge, first observed an earthen mound more than eighty feet high, where, they discovered, there were massive throne-like monuments carved in stone with human and feline figures. At first it was assumed that this site was a variant of the Mayan culture, several sites of which had already been discovered to the east of the narrow Mesoamerican isthmus. But then the German scholar, Hermann Bayer, identified them as 'Olmec' because, he said, they came from Olman, the ancient Aztec 'Rubber country' more than 300 kilometres away.[2] With this change in perspective, it eventually became clear that there was a distinctive 'Olmec' style in regard to the statues and jade artefacts that could be found in and around La Venta. They were not Mayan and, again to begin with, archaeologists assumed they must be a later variant.

It was not until Matthew Stirling, of the Smithsonian Institution in Washington DC, excavated at Tres Zapotes, further west, on the very fringes of Mayan territory, in 1938, that the antiquity of what would become the Olmec civilisation was fully understood. Among the monuments he discovered was one, now known as Stela C, which had a jaguar face on one side and a date on the other side, which placed it at 31 BC, newsworthy because that was long before the rise of Mayan culture. As Fagan tells the story, Stirling's dating caused a great controversy in the archaeological profession, for it suggested that the Olmec, instead of being a later manifestation of the Mayan, were in fact much earlier. Undaunted, Stirling transferred his attention to La Venta, where he made a whole raft of further discoveries. These included four massive stone heads, each weighing several tons, yet more thrones adorned with carved figures that were half-feline and half-human, and a spacious ceremonial centre made up of pyramids, temples and wide plazas that had been lined with still more stone sculptures.

These discoveries were made in the years before radiocarbon dating had been invented (in 1947) and for a time the antiquity of the Olmec continued to be a matter of dispute. But then, in 1955, charcoal samples at La Venta were dated to between 1110 and 600 BC.[3] For a time after that, the Olmec culture was regarded, by some archaeologists at least, as the 'mother culture' of Mesoamerica, far older than any of the other civilisations that had been uncovered there. Not many archaeologists share that view any longer but that does not detract from the fact that the Olmec culture *was* the first urban civilisation to develop in Central America.

Water, Rain, Tears

Today, hundreds of Olmec sites are known, but only a handful have been properly excavated, the best known being San Lorenzo, La Venta, Chalcatzingo, El Manatí, Laguna de los Cerros and Tres Zapotes. San Lorenzo and La Venta, the first sites to be explored, are a little less than forty miles apart (a hard day's travelling on foot). The first gravel platforms were constructed at San Lorenzo in roughly 1500 BC and, judging by the stone materials found there – basalt, greenstone and obsidian – the inhabitants of San Lorenzo, who were quite numerous for a few centuries, traded far and wide. Around 1250 BC, they developed a highly distinctive form of pottery, made of kaolin (a clay found in tropical regions, often coloured white or orange and used to make fine ceramics), plus a no less distinctive monumental sculpture, two traditions that emerged 'fully grown', so to speak, with no apparent antecedents in Mesoamerica.[4]

By 1150 BC San Lorenzo was flourishing: its pottery and stone carvings were found all over its hinterland, as far away as Chalcatzingo, 250 kilometres to the north. At that time the city itself had its own imposing ceremonial centre, based around a large platform rising more than 160 feet above the surrounding river basin. In addition there were a number of small pyramids and what the excavators believe was a ball court, perhaps the first to be built in Mexico. The ball court is an institution entirely unknown outside the New World – it is discussed in chapter 21.

Several mysteries remain. One is why, as the excavator of San Lorenzo, Michael Coe, from Yale University, suspects, the San Lorenzo ceremonial complex was built in the form of a huge bird, flying east. Another is why the ceremonial centre itself was never completed, not before the city went into rapid decline, beginning about 900 BC and completed 200 years later. And third, most mysterious of all, perhaps, is that the ceremonial centre contained eight colossal stone heads, each much taller than a man, weighing many tons and using basalt stone mined nearly fifty miles away. Some archaeologists believe that these heads – which famously have flat noses, leading some to think they are of Negroid peoples who reached Mesoamerica very early, and wear helmet-type headgear – are portraits of select individuals, elite rulers of a society where social stratification first emerged. We shall see in just a moment what this implies.

La Venta was slightly later than San Lorenzo: it did not become a major centre until around 1000 BC. It too had colossal heads (weighing 11–24 tons) overlooking a number of mounds and plazas, but in addition it had elaborate tombs for burying important personages, accompanied by extravagant offerings. Here the portrait carvings have been badly mutilated, which may mean that it was eventually sacked by rivals (or it was done deliberately, to prevent the 'power' of these objects being abused). One of the colossal heads at La Venta has been paired iconographically with another at San Lorenzo. The La Venta head holds a cord or rope that binds what looks like the San Lorenzo head, which could mean either that these two dignitaries were related in some way, political or religious, or that the San Lorenzo figure was captured by the La Venta personage.

Recent research has found that the emergence of elites in Olmec society was based on agriculture, which enabled them elite to provide food for their subjects. But the exact form of agriculture was quite complex – more sophisticated than at first appeared. There were in fact two types of agriculture practised in Olmec territory. The first was used in the low uplands, near the main settlements, which produced both maize and manioc crops twice a year, in both the wet and dry seasons. But the Olmec also employed a type of fertile garden located on the natural levees of the rivers, which were flooded and refertilised by the summer rains. These provided high yields year after year.

The two things that stand out at these Olmec sites are the emergence of elite figures and the associated jaguar themes. Five throne-like stone blocks were found at La Venta, each depicting a seated figure, probably a ruler, half-hidden in a niche in the stone, each holding ropes that link him to other seated figures that could be relatives or captives. These 'block thrones' are also decorated with stylised jaguar figures which, the archaeologists say, may have symbolised the supposed jaguar origins of the Olmec, a similar theme to that found in South America. As we shall see, the later rulers of the Maya also sat on jaguar thrones.

The jaguar theme was also widespread among the objects found in the La Venta pyramid, part of the ceremonial complex which, some of the excavators believe, is itself in the form of a stylised jaguar head. The offerings found in the pyramid include jade jaguar masks, a mosaic of serpentine blocks also believed to represent a jaguar face, and other figurines in jade and serpentine – bald, with slanting eyes and drooping mouths, which some think show humans in the process of turning *into* jaguars (see figure 11).

Fig. 11 Were-Jaguar figurine of serpentine, with snarling feline face, possibly representing a masked dancer or shaman transforming into a jaguar.

The distribution of Olmec-style artefacts, thought to have been fairly restricted at one stage, is now known to have extended as far as the Pacific coast, with human figurines and jaguar sculptures discovered at a score of sites stretching from Tehuantepec in Mexico to El Salvador.[5] Chalcatzingo, in Morelos, seems to have been a distribution way-station for various substances that did not occur in Olmec territory but were much valued. It may well have been through Chalcatzingo that the developing Olmec elite reinforced their position, importing luxury objects in obsidian or greenstone. In return, Chalcatzingo showed a strong Olmec influence, not only in the artefacts produced there but in a number of enormous bas-reliefs carved into rock faces.[6] The most striking of these shows a member of the elite seated on a throne inside a cave that appears to be in the form of a stylised jaguar's mouth. Rain clouds emanate from the cave mouth, perhaps an allusion to the breath of the jaguar-god which brings rain. Jaguar figures are found everywhere in Chalcatzingo, not just in the open ceremonial centres but also in the caves in the nearby mountains where the greenstone was mined.

As Fagan also points out, Olmec rulers were the first in Mesoamerica to record their dominance in enduring form and as the colossal statues, carvings, pottery and intricate figurines make clear, the dominant ideology among the Olmec was based on the relationship between humans and jaguars (reliefs of adults carrying jaguar babies, stone axes in the form of half-humans/half-jaguars, rock carvings showing felines attacking humans, see figure 10 page 228). The jaguar was feared and admired in equal measure, feared because of his great strength and cunning, for the fact that he controlled the rainforest at night as much as in the day. He was admired for his (supposed) sexual prowess (few can have seen the jaguar 'in action', so to speak), and his mastery of all environments – water, land and the 'upper world' of the trees. (Jaguars have been observed to attract fish by tapping the surface of a river with their tails.) They were particularly associated with rain and fertility, their roar regarded as a kind of thunder, announcing rain.

As mentioned earlier, several Olmec figurines appear to show humans turning into jaguars with snarling mouths and this has been taken to confirm that Olmec rulers were shaman-kings, elite individuals who, under hallucinogenic trance, could transform themselves into jaguars, from which humans were originally formed. Excavations have shown that species of *Physalis* were grown in some Olmec regions. Being part of the Solanaceae (nightshade) family, this plant can have hallucinogenic properties. These shaman-kings controlled rain and floods by communication with supernatural jaguars. Other objects include whistles, which may have been employed ritually or, as ocarinas, used musically. They were generally carved into animal forms, birds or monkeys, for example.[7]

Recent studies have confirmed the shamanistic nature of Olmec ideology, and that their rulers served shamanic as well as political functions, being seen as representatives of the World Tree, the evidence being found on cave murals at Oxtotitlán and La Venta's Altar 4, and incised on stylised celts or ceremonial axes.[8] Cleared areas have been found in the rainforests of El Manatí which are believed to have been sacred spaces where the houses of shamans were located.

At the same time, new ideological elements were introduced in Olmec settings which were to endure – albeit in much modified forms – for two thousand years and more.

Excess water – swamps, lakes and frequent inundations – was a problem for the Olmecs, rather than water shortage, as it was (and is) in so many other parts of the world. This led them to construct great drainage systems and to devise various cults to ensure a *balanced* rainfall, devoid of inundation that would destroy the harvest. The ceremonies of these cults often took the form of placing axes at the highest point of the house, with the cutting edge pointing at the sky.[9] It was believed that the Lord of the Storm, the Master of Lightning, possessed an axe with which he felled trees in the forest. Obsidian was called 'lightning stone' and where it was found, there it was believed that lightning had struck. (Shamans were known as 'men of hail' in Veracruz.) Ceremonial axes and other objects, such as masks, were thrown into cenotes, natural fresh-water cisterns, in an attempt to control the rains. Rain gods were to prove important to later Mesoamerican civilisations.

Associated with the water cults were further cults of hills and pigmented stone, such as haematite, and when the springs were located in what were seen as sacred mountains (as one of the sources of water), such sites could become cult centres, not lived in but used as places of pilgrimage and worship. There is evidence from some of the objects thrown into springs and cenotes that the worshippers had followed a special diet in their efforts to maintain greater ties with nature.[10] The bones of neonates were also thrown into springs, after being dismembered. To Ponciano Ortiz and María del Carmen Rodríguez, Mexican anthropologists, this suggests ritual cannibalism, further supporting the idea of cult members following a special diet.[11] More than one Spanish chronicle from Conquest times described child sacrifice as a common practice in Mesoamerica related to water and fertility – infants' tears being regarded as akin to rain. Child sacrifice has been practised in many areas of the world, so one need not suppose that the Olmec had in any way 'inherited' this tradition from the Chinchorro, referred to above, in chapter fifteen.

There was an overlap in Olmec ideas between rain and maize deities. Maize was not a staple part of the Olmec diet in all areas (such as La Joya). As we noted in chapter eleven, it took a long time to fully domesticate teosinte and its earliest uses were for beer rather than to make corn. But, beginning around 900 BC, maize did begin to take hold in some areas of the Olmec heartland and as a result, societies, some of which may already have been sedentary, began to build up agricultural surpluses. As a consequence of that, wealth grew, social stratification appeared, art proliferated (presumably with the development of specialist craftspeople) and maize symbolism was produced.

To the Olmecs, the human body served as a model of the cosmos and so the maize god was shown associated with the tree of life, the cardinal directions (related to the axis of the sun's movement), and to other precious materials, such as jade or jadeite and quetzal feathers. The deity was often shown with a human head, sometimes a birds' head, sometimes the head of other animals.

The association with quetzal feathers is particularly interesting. The quetzal is the most spectacular bird of Mesoamerica and its long green tail feathers were especially prized. (The word, *quetzalli*, in Nahuatl, means 'large brilliant tail feather'.) The colour green linked feathers, jade and maize as a symbol of fertility.

And it was in this context that is seen the first glimpse of an image, a concept, a sacred idea that was to become very powerful in Mesoamerica. This is the appearance of the plumed serpent, an image that may have two origins, not necessarily contradictory. We have seen that the Olmec faced problems with excess water. In such an environment, snakes would have been more in evidence than elsewhere. At the same time, the rattlesnake tail in particular is seen as similar to the ear of corn. Both these factors would help explain the conjunction but it also appears true that serpents may have been present far more in Olmec iconography than previously thought. In a paper published in 2000, David C. Grove, from the University of Illinois at Urbana-Champaign, reinterpreted a number of images – cave carvings, paintings, figurines – which had previously been understood to depict jaguars as in fact showing serpents. (This is not as unlikely as it sounds:

Mesoamerican art is – or can be – highly stylised.) There are still plenty of Olmec images that are undoubtedly jaguars but if Grove is right, the jaguar becomes less important at this time and the serpent begins to establish more of a presence in the iconography and its association with maize and quetzal feathers is also begun.[12] In the plumed serpent, therefore, we see a maize stalk (the fertile plant of the surface of the Earth), a serpent (identified with the underworld and life-giving water), and the bird (identified with the heavens, in the sky). We have, in effect, an updated World Tree.

Allied to this are a number of images known to archaeologists as the 'Earth monster'. This is usually shown as a huge mouth, square or square-cruciform. Here, the Earth is represented as both a huge devourer and creator of life, consuming the sun every day, as it does seeds, and then producing a new sun every morning, as it produces corn after the seeds have 'disappeared'. To people living in an evidently flat world, and unaware of the notion of orbits, the sun's apparent trajectory would have been miraculous and the appearance of plants after their seeds had been 'devoured' by the Earth an analogous miracle.[13]

One other change in Olmec iconography has also been observed and dated to 1150–850 BC. Richard G. Lesure, at the University of California, in Los Angeles, finds an important evolution in the pottery of that time, when we see a transformation from zoomorphic imagery – mainly animal heads – to more abstract designs: lines, curved lines, wavy lines, flame-type images. Lesure observes that the animal effigies depicted on the earlier bowls were the creatures that villagers would come into contact with every day. 'Adults killed them, skinned them, ate them. Children poked at them with a stick.' After ~1000 BC, on the other hand, he offers the suggestion that the more abstract designs referred to supernatural entities, imagined as 'fantastic creatures bearing traits drawn from various animals. The subjects of zoomorphic representations changed from the creatures of everyday existence to those of special, numinous experiences; there was a shift from the ordinary to the extraordinary.'[14]

He says it is important not 'to go overboard' with this observation, and that it doesn't occur everywhere in Olmec territory, but he does suggest that the new designs are evidence of a shift in symbolism to a higher-order cosmological concept, one possibility being that people found it increasingly important to signal their membership in large-scale groups or social categories. In support of this he says zoomorphic images were invariably placed on bowls to be used by single individuals, whereas the later more abstract motifs were more likely to be placed on family-sized bowls.[15]

He further concludes that the abstract motifs may have 'belonged' to elite groups, who understood, or claimed to understand, the 'language' of the motifs and this ability unified that elite and helped keep it separate from, and above, the common people, in an 'ideology of inequality'. In fact, he says, there may have been an ideological struggle between the more zoomorphic understanding and the more abstract and supernatural ideas.

What links all these otherwise disparate ideas is *food*: the specialist diet associated with sacred cenotes; the arrival of domesticated maize; the Earth-monster mouth; the changing iconography of food bowls. It may be, then, that what we are witnessing here is a gradual and perhaps fragmented Olmec ideological adjustment to the development of maize agriculture. It was slow but its arrival resulted in more wealth for the community and more *disparities* in wealth. This is reinforced by the carvings of massive heads, likenesses of actual individuals who stood out, and figurines which show signs of ceremonial skull deformation, a common form of status marker.

Old shamanistic ideas continued but new concepts appeared, notably the development of art, monumental art, the idea of more complex gods which, in their fantastic being, embodied the attributes of more than one animal or plant, and a new approach to time, for there is evidence about now of a rudimentary ritual calendar.[16] This would make sense if maize agriculture was beginning.

David Grove says there is no real evidence for warfare between the different Olmec centres and concludes that the political situation 'may have been one primarily of cooperation rather than competition'.[17] He also says that the Olmec people 'do not seem to have conceived of, or to have worshipped, a group of formal deities. Instead, the ruler, through his access to natural forces, mediated with those forces and controlled them. The supernatural creatures depicted in a variety of Olmec art forms represent aspects of those forces (rain, water, etc.) and not specific "Olmec deities".'[18]

This is an important comparison (and contrast) with what was happening in the Old World, where the pastoral nomads worshipped natural forces (wind, rain, lightning), with shamanistic elements, but gave their main god a name (Tengri) and a gender (male), and came into (aggressive) contact with the essentially female Great Goddess of fertility. At that stage, if Grove is right, the Olmec ideology of inequality, and their incipient agricultural deity, comprised a transitional world, also shamanistic and also at least partly the worship of natural forces. But these parallels and similarities are misleading. As we shall see, they led in very different directions.

Kaypacha

Chavín de Huántar is located high in the Peruvian Andes. Its culture flourished between 900 BC and 200 BC and so it broadly overlapped the time period of the Olmec, if a little later. As we saw in an earlier chapter, the Andes comprise a harsh environment but are not without their attractions, in that many of the insects and diseases that thrive in the tropical rainforest are absent there, defence is easier to arrange, and a number of animals and edible plants have adapted to the environment.

In South America before the Chavín phenomenon, as it is called, several urban cultures emerged. Those at Aspero and Caral in Norte Chico (~3000 BC) have already been described. Others occurred at Salinas de Chao, with its eighty-foot pyramid dated to 2000 BC, El Paraiso in the Chillon Valley near Lima, with its U-shaped platforms and plaza, dated to 1800 BC, and Huaca Florida, a few miles inland of El Paraiso, and Sechin Alto in the Lower Casma Valley, each with enormous pyramid mounds and plazas and dating to 1700–1650 BC. No one really knows what system of beliefs operated in these enormous ritual structures but Donald Lathrap argues that these sacred sites operated as symbolic vertical axes between the living and the spiritual worlds. The U-shape was then a focusing arrangement, he says, designed to absorb the sacred energy of the opposing forces of left and right. But no one really knows.

Apart from these U-shaped structures, the early cultures of the Andes shared other features which would find full expression at Chavín. Partly this had to do with their similar location, in mountainous areas where between only 2 per cent and 4 per cent of the land was suitable for agriculture. This meant, among other things, that irrigation – though often critical – only ever consisted of small canals that could be managed, probably, by single extended families. No extensive bureaucracies were needed, as happened in several areas of the Old World. Allied to this, maize, although it is present, is nowhere dominant in subsistence systems, often being used, as was referred to above, in chapter eleven, for beer in ritual contexts, and not as the basis for bread.

In religious iconography, the jaguar and other felines (pumas, for instance) are ubiquitous, often shown in snarling or in other ferocious forms, and there is considerable evidence of ritual decapitation, of the dedicatory burials of (unflawed) children, three-to-five years old (buried below buildings), and hints – more than hints – of cannibalism. Feline figures have been found, for example, at Pacopampa, Punkuri, Cupisnique and Huaca de los Reyes, and 'trophy heads' at Cupisnique, Waira-jirca and Kotosh. At Punkuri a painted clay relief shows a severed head 'oozing blood'.[19] At Pacopampa, the temple plazas – the central realm (between the sky and the underworld) – is represented as *Kaypacha*, a term for 'the world inhabited by human beings and jaguars'. Besides the jaguar, iconographic motifs also stress the eagle and the serpent, or snake. At Pacopampa, for example, jaguars, serpentine figures and avian forms are found together.[20] Here, therefore, we again have animals representing the three 'realms' of heaven, Earth and underworld, though without the maize element that we see in connection with the Olmec. This is what you would expect since so little of Chavín land was suitable for agriculture.

Another great achievement of these early cultures was the domestication and exploitation of cotton. We saw in an earlier chapter how, in Norte Chico, cotton was domesticated in order to manufacture fishing nets but as the centuries passed weaving and dyeing skills developed in the Andes to a higher level than anywhere else in the world, Old or New. Cotton preserved for more than 2,000 years in the Andes has been shown to contain at least 109 hues in seven natural colour categories, and to be so fine the Spanish took it for silk when they first encountered it. Different strains of cotton were developed, which flourished at different altitudes, and the weaving techniques, which were basically symmetrical and angular, were copied as decoration in other materials such as pottery. Anthropomorphic figures, snakes, birds of prey and snarling feline motifs were much in evidence, in bright colours, reflecting the links between the human world and the spiritual world. In these early societies, the community – or *ayllu* – came before the individual and, to judge by the spread of motifs on cotton and pottery, the same beliefs were widespread in South America and very long-lived.

Textile styles took a lot of trouble to appear animated. They were, so far as we know, displayed horizontally, rather than vertically and rarely – if ever – used as garments. Were they danced on, as part of a trance-inducing ritual?[21] Chavín textile knots may well have given rise to the quipu of the Incas (see chapter twenty-three).[22]

And then, in 1991, Chavín de Huántar itself was discovered.[23] This was the work of the Peruvian archaeologist, Julio Tello, in the remote Pukcha River Basin in the foothills of the Andes. (Chavín, incidentally, could come from the Carib word for feline, *chavi*, though the Quechua term *chawpin* means 'in the centre').[24] The main building was a stone-built temple pyramid with carved stelae, monoliths and many articles of pottery, decorated with a wide variety of *forest* animals: felines, lizards, cayman, birds of prey plus a number of part-animal/part-human figures, all executed in the same distinctive style. No less remarkable than the discovery itself was the fact that the designs were similar to those found miles away on the north coast of Peru, and elsewhere in the Andes, on the arid Paracas Peninsula, and as far south as Lake Titicaca.

In all cases, the animals in Chavín art were *tropical* animals. Why should that be? Tello's answer was that Chavín was the 'mother culture' of Andean civilisation. There are ethnographic accounts of shamans making long forays to acquire supernatural wisdom, and even today highland and coastal shamans and healers among the Jivaro or Achuara travel enormous distances to tropical forests which are viewed as a source of medicinal plants and sacred knowledge. Later archaeologists no longer accept that account uncritically but the 'Chavín phenomenon' remains at the centre of the South American trajectory towards civilisation. Daniel Morales Chocano, of San Marcos University in Lima, Peru, has excavated at Pacopampa (north Peru, near Cajamarca) and finds that, between 5,000 and 2,500 years ago (3000–500 BC), the South American forests were much less extensive than today, the climate was drier, provoking the manufacture of ceramic bottles that would not have been needed in a wetter climate. In these circumstances, people would have been pushed outwards from the Amazon area in all directions.[25]

Chavín lies in a small valley that is 10,000 feet above sea level, between the arid Pacific coast and the tropical rainforest of the great Amazonian Basin. The location was chosen, presumably, because the local farmers had easy access to several very different ecological zones, each within easy walking distance.[26] In the valley bottoms there were irrigated maize fields, potato gardens higher up, on the slopes, and grazing llamas and alpacas on the grasslands at high altitudes.

The terrain around Chavín is well suited to high-altitude farming and camelid herding (llamas chiefly, but also vicuña and guanaco). Domesticated camelids only became common outside their natural territory during the Chavín horizon. This shift is probably due to the adoption of llamas as pack animals and their maintenance in small herds for long-distance caravans. (Llamas can carry 20–60 kilograms for 15–20 kilometres a day and, because they follow a lead animal, one driver can control up to thirty of them.)[27]

To begin with, at ~850–450 BC, Chavín was no more than a small village with a shrine, a crossing point for various trade routes between the Andes and the coast. Marine shells were found scattered right across the site, as were pieces of pottery from far and wide, together with a few bones of puma and jaguar. Richard Burger, one of the main excavators at Chavín, argues that these bones were not indigenous to the area but that they were imported for use in ritual apparatus.

After about 450 BC, however, a major change took place at Chavín. For about sixty or so years, the entire population crowded into the temple area, along the river banks. No one knows exactly why this was but it may have been because of a change in ideology, when Chavín became a centre of pilgrimage. This conclusion is drawn because many exotic objects were found at the site, objects drawn from many different locations in the Andes, some of which at least were offerings, together with some everyday objects of use that were brought by travellers and left there.

After this, the town rapidly quadrupled in area until it occupied more than 100 acres, where as many as 3,000 people may have lived. It was now known as a religious ritual centre – a shrine in 'Western' terms – on a par, says Richard Burger, with Jerusalem or Rome of later ages and faiths. Many exotic artefacts were made there and just as many imported, the most notable being pink *Spondylus* shells brought in from the Ecuadorian coast 500 miles away. Some of these shells were decorated with feline motifs.[28]

The temple at Chavín takes the form of a U-shaped structure. Most of the buildings are about forty feet high, open to the east, where the sun rises, but also to the forest, which most of the animals depicted in Chavín art inhabit. The main temple appears to have been rebuilt several times (fifteen phases, in five stages) until it consisted of a maze of interconnected courts and passages and galleries and small rooms, ventilated by vertical shafts.[29] These galleries, as they are known, had no natural lighting, so must have been lit by resin or grease torches. They are connected by short, narrow staircases and they were built, says Burger, with the intention to create a sense of confusion or disorientation, 'in which the individual is severed from the outside world'.[30] The innermost area contains a mysterious white granite monolith contained within a cruciform chamber almost in the middle of the structure. This monolith is long and slender, about fifteen feet high, is located vertically in its slot, and for that reason is known as the 'Lanzón' (for 'lance-like'). One theory is that it was erected before the rest of the building, which was constructed around it.

The Lanzón is carved into an anthropomorphic figure, the chief feature of which is a snarling feline-style mouth, with great fangs. Its left arm is held by its side but the right arm is raised, displaying claw-like nails. Its elaborate headdress is decorated with yet more snarling felines in profile. 'A girdle of small feline heads surrounds the waist.'[31]

The overall shape of the figure – long and slender, essentially vertical – and its position, built into the floor and ceiling, suggests to modern archaeologists at least that it symbolised the deity's role as a conduit between the underworld, the Earth itself, and the heavens. In fact, Tullo found another cruciform gallery – this one much smaller – immediately above the main one and the two chambers were so close (just one stone block separated them) that he raises the possibility that this arrangement served as an oracle, divinations being made in the smaller chamber and evoking a response from the Lanzón.

In fact, it seems likely that there were two principal gods at Chavín. The Lanzón was one, known also by its different name, the 'Smiling God', with its feline head, clawed hands and feet, while the second was the so-called 'Staff God', a figure found carved on a separate stela discovered inside the temple (a staff god was mentioned in chapter 15, regarded as the oldest image of its kind in the Americas – see page 263). This shows a man with a downturned, snarling mouth, a serpent headdress, and holding two staffs, each of which is also adorned with feline heads and jaguar mouths. Richard Burger believes that the Chavín religion was concerned with at least two things. One was an attempt to reconcile the dichotomy between the humid jungle and the high, dry, cold mountain landscape, and the other was the essentially shamanistic idea of transformation, between the human form and animal forms, especially that of the jaguar.

This is further supported by a number of carvings found on Chavín reliefs in the temple complex, one of which shows what Brian Fagan calls 'a jaguar-being resplendent in jaguar and serpent regalia'.[32] In Chavín art, raptor birds, harpy eagles in particular, are depicted as 'were-jaguars' or the 'jaguar of the skies' roaming in search of human heads.[33] Moreover, this figure is shown holding what is recognisably the San Pedro cactus, a well-known hallucinogen and a substance still used today by tribal shamans when exploring the spirit world. As was noted earlier, San Pedro cactus contains mescaline, which helps send the shaman on mind-altering voyages. It grows naturally at Chavín, just a few hundred yards from the temple.

The belief that Chavín priests could transform themselves into jaguars in order to 'contact and affect the behaviour of supernatural forces' was widespread, with a lot of evidence to support it. Hallucinogenic snuffs and beverages catalysed these changes and their consumption was an integral part of Chavín ritual. This role of psychotropic substances is well attested by its widespread representation in Chavín art forms and, as Richard Burger puts it, 'these depictions can be interpreted as providing the mythical antecedent and divine charter for the use of these substances and, ultimately, for the religious authority of temple priests.'[34] Small snuff holders in gold, fashioned into animal forms, have been found at Chavín, as well as effigy vessels in the form of coca chewers, underlining the use of coca leaves during Chavín times.[35]

Furthermore, Burger says, the role of hallucinogenic snuffs in shamanic transformations is clearly expressed in the tenoned heads fitted into the walls of the Old Temple. These represent different stages 'in the drug-induced metamorphosis' of the religious leaders 'into their jaguar or crested-eagle alter egos'. Sculptures show faces with almond-shaped eyes, bulbous noses and closed mouths; they have an unusual hair arrangement – a sort of knot at the top – and wrinkled faces, 'as if they were experiencing the onset of nausea'. This is often the experience of entering trance.

Fig. 12 Head of a mythical priest almost fully transformed into a feline state; the strands of mucus running down from the nostrils signal the involvement of hallucinogenic snuff in this process. Compare with figure 3, page 117.

A second group of tenoned heads, Burger says, 'portray strongly contorted anthropomorphic faces', gaping round eyes, with mucus dripping from their nostrils, 'either slightly or in long flowing streams' – their features and hairstyles suggest that the same group of individuals is being shown (see figure 12 above). As he wryly adds, 'The depiction of nasal discharges in prominent public contexts is alien to Western religious traditions.' Yes, it is, but its significance was known to the early Spanish chroniclers of the New World, more than one of whom observed the Muisca of Colombia putting (hallucinogenic snuff) powders into their noses, which made the mucus run 'until it hangs down to the mouth, which they observe in the mirror, and when it runs straight down it is a good sign'.[36] The flow of mucus caused by the irritation of the nasal membrane by psychotropic substances 'is the most conspicuous external index of an altered mental state'.[37] Gary Urton, Dumbarton Oaks professor of pre-Columbian studies in the Department of Anthropology at Harvard University, says the nose was regarded as a major orifice of the body on Chavín iconography.[38]

A third group of tenoned heads combines human features, such as eye and ear shapes, together with large fangs and other non-human features, while in a fourth group the face is totally transformed into a feline, a raptorial bird, or a hybrid of the two. In some of the sculptures, these phases are linked. Thus in one the face has one bulging eye and one almond-shaped eye; another shows the head of a jaguar with mucus hanging from its nose.

The prominent role of hallucinogenic snuff in this process is further supported by the discovery of small stone mortars at Chavín, and at a number of other sites, such as Matibamba. These mortars are too small, and their depressions too smooth, for them to have been used for grinding grain, on top of which they are carved into the form of jaguars or raptorial birds. Bone trays, spatulas, spoons and tubes have also been found in similar contexts, items often represented in Chavín art, all of which supports an hallucinogenic snuff ritual complex.

A final element in all this is the stone frieze that is such a feature of the Circular Plaza at Chavín. This shows, among other things, an anthropomorphic figure holding the stalk of the San Pedro cactus, a figure that has prominent fangs and feline paws.[39] In the same row, says Burger, there is a pair of carved anthropomorphic figures with almond-shaped eyes and bulbous noses, 'reminiscent of the first group of tenoned heads', and with jaguar tails hanging from their headdresses 'likewise suggestive of shamanistic transformation'. Below these figures are still more representations of jaguars with the taloned feet of the eagle.[40] Burger suggests that these two rows or layers are related, in that the 'flying jaguars' in the lower register may represent the anthropomorphic figures in the upper register in their 'fully transformed' state. In other words, these carvings show their priests in the act of taking snuff and/or consuming the San Pedro cactus mescaline beverage and transforming themselves into supernatural beings.

The San Pedro cactus is less radical in its psychotropic effects than the tryptamine derivatives, which are the active ingredients in epena (a resin from the bark of the Virola tree) and vilca (from the *Anadenanthera colubrine* tree). Burger notes that the modern Jivaro people (on the eastern slopes of the Andes, in Ecuador), use several hallucinogens but that the strongest are consumed only by religious specialists. He also observes that, unlike the San Pedro cactus, which grows locally around Chavín, the plants most commonly used to produce snuff come from the tropical rainforest. The seeds of vilca, for instance, store well and can be widely traded. They may well have been seen as luxury items.

Here too, then, the Chavín reliefs show shamans in the process of turning into jaguars – an ancient theme, as we have seen – under the influence of mind-altering drugs. Burger reports that similar rituals exist to this day in the high, remote Andes.*

A Land of Natural Catastrophe

Chavín appears to have been a major cultural attempt to bring together the very different environments of South America – jungle, coastal region, high altiplano – into one all-embracing ideology. In so doing, it sparked a number of technical innovations, such as painted textiles and wall hangings, new forms of pottery and gold-working, which recorded these ideological developments.

Burger makes the point that 'it would be a mistake to think of the physical environment as a passive, unchanging backdrop'. Peru, he reminds us, is a 'land of natural catastrophe', it was and still is characterised by considerable tectonic activity (as outlined in chapter five), which triggers earthquakes that periodically 'wreak havoc in Peru'.[41] In recent times (recent geologically), earthquakes killed 75,000 in 1970, and in 1746 an eighteen-metre (fifty-nine feet) tidal wave totally destroyed Callao, the main port of Peru. The interplay of tectonic activity and El Niño events (again, see chapter five) makes the area one that is disturbed by what Michael Moseley calls 'radical environmental alteration cycles'.

Burger adds to all this with his observation that Andean religious architecture is designed to act as a focus of 'potentially dangerous supernatural forces'. By the same reasoning, Luis Lumbreras, of the University of San Marcos at Lima, says that the profusion of air ducts and water conduits in the main temple exceeds the practical needs of the community and he proposes that the sound of water draining through the Old Temple would have been distorted and amplified by the ducts and galleries and the resulting sound would have been projected on to the plazas and terraces below. This amplification of sound was later confirmed by experiment and the idea is further supported by

* We should not overlook the fact that there was also a proliferation of secondary gods at Chavín, perhaps reflecting the greater heterogeneity of Chavín society, brought about by the many pilgrimages, facilitated by the newly available llama caravans.

the existence of canals (at Cumbermayo, for instance) built for cult use and not for agriculture.[42]

This was not the only theatrical effect. Chavín lies near the confluence of the Mosna and Wacheqsa Rivers and excavations suggest that the course of the Mosna may well have been changed artificially. Other discoveries provide evidence for elaborate processions at Chavín, with people – depicted with wings and fanged teeth – carrying shell trumpets. Experiments show that the shell trumpets would have produced 'an immense volume of noise' and a cyclical 'attention-commanding beat'.[43] Furthermore, the Lanzón was lit from above, by a ventilation shaft, giving it a mysterious floating aura and the structure of the temple included awe-inspiring large stones, weighing up to fifteen metric tons, brought in from up to 20 kilometres away. Finally, the galleries and internal ('hanging') staircases were built in such a way that priestly figures could emerge, disappear and re-emerge at different points on the temple façade in a careful ritual designed to provoke awe in the watching populace.[44]

What we may be seeing here, says John W. Rick, professor of anthropology at Stanford University, is an attempt, by shaman/priests, to appropriate sacred power from the forces of nature to themselves, by the deliberate (theatrical) manipulation of religious concepts to solidify their authority.

What is most remarkable, perhaps, is the overlap with the (fairly distant) Olmec in terms of jaguar transformations. This need not suggest direct contact between the two broadly contemporaneous cultures – there is nothing really to suggest such a link. Rather, the ubiquity of the jaguar, over a wide geographical area, the universal appreciation of its ferocity, strength and fertility, alloyed with the ubiquity of hallucinogens, which encouraged and facilitated the experience of shamanistic trance and transformation, combined to produce similar ideological practices. These practices went back far in time but, as urban society arrived, the traditional ideas needed modifying.

Another parallel with the Olmec is that a new religious ideology was coming into play at Chavín, albeit a very different kind. This has become known as the 'Chavín horizon', to signify the fact that Chavín ideology spread far and wide, rather as a 'regional cult', its art styles and religious ideas centring around the feline stretching over a wide area and, as Burger says, 'profoundly influencing the diverse peoples who embraced this ideology'. Most archaeologists think that the wide spread of these beliefs was peaceful, more in the manner of the recent regional cults of Africa, embracing different ethnic groups, or the different cults of ancient Greece, all of whom revered the same oracle, rather than the more aggressive proselytising of Western-style missionaries or forced conversions. These widespread rituals fostered openness and universality rather than local political issues favoured by local cults. They encouraged the free movement of at least religious functionaries, pilgrims and merchants trading in religious paraphernalia, across ethnic and political boundaries.

One model for this may have been the Pachacamac regional cult that was in existence just before the Spanish conquest some twenty-five miles south-east of Lima, centred on the mouth of the Lurín River and which, some archaeologists believe, may well have been related to the Chavín cult. The main aspect of the Lurin cult was an oracle situated on the summit of a large adobe platform. Access to the oracle was restricted to religious specialists but was believed to be capable of intervention with the elements, protection against disease, and could predict favourable times for planting and harvesting. 'Divine disapproval was felt when earthquakes occurred and crops failed.' Communities who wanted to establish subsidiary shrines were free to petition the central religious authorities but had to display their willingness to embrace the cult by providing labour to support central cult activities, plus tribute of corn, llamas, dried fish and gold. Eventually, a network of branch shrines was built up as far afield as Ecuador, the subsidiary shrines understood as the wives, children and other family members of the main oracle. The distribution of Chavín art about South America suggests that its cult operated in much the same way as that of Pachacamac.[45]

There was some ideological flexibility the further one went from the centre of the cult. At Karwa, for instance, 530 kilometres from Chavín, the textiles and other decorations were produced in Chavín style, but the feline figures were represented mainly as subsidiary motifs, the cayman and the staff god being represented as the principal deities. Moreover, the staff god was depicted as a female figure, with breasts and a *vagina dentate*.[46] Or, perhaps, on the Pachacamac model, she was seen as the wife, sister or daughter of the staff god. She was also sometimes shown with cotton bolls emerging from her headdress, and she may therefore have been identified as the patron and/or donor of cotton. Karwa textiles also showed a series of concentric circles, together with spirals and 'S'-shaped motifs, which some have suggested are stylised representations of the San Pedro cactus seen in cross-section.

Overall, Chavín-style art is seen as far afield as Ica to Lambayeque to Huánuco to Pacopampa, distances of hundreds of miles. One big difference between Chavín art and Olmec art, which Burger draws attention to, is that Chavín images had no political context or references. 'Unlike Olmec art, the Chavín style does not portray historical personages, scenes of conquest and submission, or the explicit confirmation of royal authority by supernaturals. The unworldly concerns of Chavín iconography are likewise in sharp contrast with the significant sociopolitical content present on . . . later Peruvian art.'[47]

To the north of Chavín, in places such as the Zaña Valley, on the north coast of Peru, similar Chavín motifs are seen – the staff god, felines, birds and anthropomorphic figures in various stages of transformation. Several mortars have been found, to be used for grinding hallucinogenic snuffs, again with the mortars themselves in the form of jaguars, birds and serpents, together with spoons, spatulas and other devices for inhaling snuff.

One other aspect of the Chavín cult deserves mention, which is that a major aspect of its influence stemmed from a number of technological advances that were made at this time. These included the forging and annealing of gold and silver, techniques with no known antecedents in ancient South America, soldering, sweat-welding and repoussé decoration, alloying, embossing and champlevé, resulting in three-dimensional metal objects of great beauty (some people think Chavín art the high point of all pre-Columbian art).[48] Polychrome textiles also showed a marked advance. As Burger observes, 'The ability of a cult to convey or evoke religious awe through artistic or technological devices would have helped to validate its sacred propositions and the authority of its representatives.'[49] The successful extension of the cult would have increased the demand for artisans who would have further enhanced the prestige of its elite. This was all in marked contrast to subsistence techniques, which remained at a rudimentary level.[50]

All of these developments were associated with a rise in social stratification. Very rich burials now begin to occur albeit in a very small number of cases. Their grave goods consist of many gold objects, including gold crowns, necklaces, pectorals, earspools, sheets, beads, pins, finger rings, pendants and small spoons, useful for taking hallucinogenic snuff. Another common device was sets of gold tweezers, with which men removed unsightly facial hair. One burial contained 7,000 stone and shell beads, which may have formed part of a garment which crumbled and disappeared. At some sites the elite individuals were buried on the temple summit in the middle of the religious architecture, suggesting this was as much a religious as a political elite. At other sites, however, religious specialists were buried in distinctive ways – one with a quartz crystal in his mouth, for example, another with a deer bone rattle inserted into his right leg, or with spatulas for inhaling hallucinogenic snuff. None of these accoutrements was especially valuable, suggesting that shamans, though respected and set apart, were not necessarily part of the ruling elite.

And so, although there are many differences between Olmec society and the contemporaneous Chavín, between Olmec art and Chavín decorative motifs and techniques, we do see two phenomena that they have in common and are of central interest to the theme of this book. In the first place, shamanistic devices and techniques continue strong, making undiminished use of various hallucinogenic substances and with continued prominent reference to the powerful – the feared, the admired – jaguar. At the same time, we begin to see in both societies a move beyond the shamanic and a fairly simple relationship with the jaguar. Raptorial birds and serpents – snakes – come more to the fore, as do human supernatural figures in the form of the Staff God in Chavín society. Why, we may ask, should this be? Is it because, with the emergence of urbanisation, and (some) agriculture, the jaguar, while it remained important, ceded paramount importance alongside other animals – birds and snakes – that were more familiar in the newer urban cultures? With the emergence of elites, was the idea of a staff god, a human figure, a way for the elite to bolster its own position at the top of the tree?

In both societies we appear to glimpse the beginnings of what would become the New World's most enduring deity, Quetzalcoatl, the Feathered Serpent god, in which some archaeologists see what is essentially a maize stalk comprised of an amalgam of a bird (the eagle), the serpent or snake, and the feline or jaguar. Such a god would embrace the three traditional areas of the pre-Columbian cosmogony – the sky, the Earth and the watery underworld. We shall follow the development of this phenomenon in the chapters which follow and try to understand what it meant for subsequent pre-Columbian civilisations in the New World, and how it affected their intellectual and social trajectory. It was a transformation no less profound than those under way at the same time in the Eurasia, but it was very different.

• 18 •

The Origins of Monotheism and the End of Sacrifice in the Old World

In 1949, the German philosopher Karl Jaspers published his influential book, *Vom Ursprung und Ziel der Geschichte* ('On the Origin and Goal of History'). In this book Jaspers argued that the years between 900 BC and 200 BC were a pivotal period in human affairs, a sort of Golden Age in spiritual terms when, in four distinct regions, 'the great world traditions that have continued to nourish humanity' came into being: Confucianism and Daoism in China; Buddhism, Jainism and Upanishadic Hinduism in India; prophetic Judaism in Israel; and philosophical rationalism in Greece. This was, he said, the period of the Buddha, Socrates, Confucius and Jeremiah, the mystics of the Upanishads, of Mencius, Isaiah and Euripedes.[1] He called this period the Axial Age.

In some ways, Jaspers over-egged his argument. We now know, for instance, that Zarathustra, one of his 'axial sages', did not live in the sixth century BC, as Jaspers thought he did, but much earlier. However, as this and the next chapter aim to show, Jaspers also somewhat *under*-estimated the changes that took place during this crucial period: it was a greater transformation than even he thought. Many innovations – innovations in *ideas* and *knowledge* rather than in technology – occurred during those years and, for the thesis of this book, this is when a big gap opened up between the two hemispheres. A series of interlinked changes – religious, military, political, economic, scientific and philosophical – transformed humankind's understanding of itself in the Old World in a way that had no parallel in the Americas.

In the Old World, as we saw in a previous chapter, momentous changes took place as the Bronze Age gave way to the Iron Age. As

Robert Drews pointed out, within a period of forty to fifty years at the end of the thirteenth and the beginning of the twelfth centuries, almost every significant city or palace between Greece and the Indus was destroyed, many never to be occupied again. In turn, this brought about a difference in social structure: 'The earlier period was dominated by its kings and its professional elites, whereas in the Iron Age the common man counted for something.'[2]

Drews' basic thesis, as was outlined in chapter 16, was that a radical innovation in warfare suddenly gave to the 'barbarians' a military advantage over the long-established and civilised kingdoms of the palace states. Summarised briefly, his argument – it will be recalled – was that, from the seventeenth century to the thirteenth century BC, the great kingdoms had depended on elite chariot corps, with 'skirmishers' alongside, infantry who essentially tidied up after the charioteers – with their bows and arrows – had done most damage. But, Drews argues, the deployment of this formation eventually became vulnerable to a new kind of cavalry and infantry.[3] With the invention of riding paraphernalia – horse bits, reins, proper saddles and stirrups – cavalries became more manoeuvrable than chariots, and therefore more efficient and powerful, and the infantries, with the long sword and the javelin, also acquired more force. These fighting techniques, combined with guerrilla tactics, developed by pastoral nomads, had never been tried out en masse on the plains and against the great civilisations, but the innovations paid off. As a result, power shifted from the Great Kingdoms to what Drews calls 'motley collections' of infantry warriors from barbarous, mountainous or otherwise less desirable agricultural lands.

The details of these battles need not detain us. Our concern is with two widespread intellectual changes that were brought about as a result of this set of conflicts. One, the subject of this chapter, came in the religious/ideological realm. The other, considered in chapter nineteen, produced major innovations in the social/political/economic/intellectual realm that between them pushed the Old World further and further away from the trajectory of the New World.

The New Spirituality

What happened in the Axial Age was that a number of truly remarkable individuals arose – more or less simultaneously – across a wide swathe of the Old World who had, essentially, the same message. In many ways they were intent on creating a new kind of individual and the spur seems to have been the very great violence that had occurred across exactly that swathe of territory in the late Bronze Age/early Iron Age, a wave of fighting that had essentially been initiated by pastoral nomads whose steppe homeland was drying and therefore yielding much less, forcing them to look elsewhere to subsist. Among other things, this would provoke a marked change in the very concept of religion.

Before the Axial Age, as we have seen, ritual sacrifice was central to the religious quest. The divine was experienced through sacred dramas that introduced another level of existence. What Karen Armstrong calls the 'axial sages' changed this. Ritual was still important but *morality* was now installed at the centre of the spiritual life. The only way to experience 'God', 'Nirvana', 'Brahman', or the 'Way' was to live a compassionate life.[4] Moreover, one must commit oneself to the ethical life to begin with. Then, habitual benevolence not metaphysical conviction, would bring about the 'transcendence' one looked for.[5]

The first people among whom this Axial Age spirituality emerged were the pastoral nomads living on the steppes of southern Russia, who referred to themselves as Aryans, a loose network of tribes who shared a common culture. Over time they developed a number of sky and other gods – Varuna, Mithra, Mazda, Indra and Agni. The hallucinogenic plant, *soma*, or *haoma*, introduced in chapter 10, was also worshipped.[6]

Karen Armstrong says that none of these divine beings were what we would call gods: they were not omnipotent and had to submit to the (higher) sacred order that held the universe together. This arrangement essentially reflected the situation in Aryan societies, where people had to make binding agreements about grazing rights, the herding of cattle, marriage and the exchange of goods, a system of fluid contracts serving a culture that was semi-abstract because it was not tied to one place.

Sacrifice was an important aspect of Aryan ritual. Cattle as well as grain or curds were offered, usually while *soma* was being consumed. Only meat that had been ritually and humanely killed could be eaten. The horses, sheep or cattle were killed, the *soma* distilled, and the shaman/priest laid the choicest parts of the victims on to the fire so that Agni could convey them by smoke to the land of the gods. By a process similar to that known as potlatch among North American Indians, the sacrifice enhanced a person's standing in the community. To begin with, the Aryans had no concept of an afterlife, but by the end of the Bronze Age the view gained ground that wealthy people, who had made a lot of sacrifices, would join the gods in paradise after their death.[7] This all produced the germ of a new idea: that what one *did* in this life was important for what happened in the afterlife.

The original Aryan religion had favoured the values of reciprocity, and respect for animals, an approach that was a hangover from hunter-gatherer days that survived well in pastoral nomadic societies where animal/human contact was crucial. But as the drying of the steppes intensified, conflict and violence increased and with it cattle raiding. This new aggression was reflected in the iconography of Indra, the dragon-slayer who rode a chariot in the clouds – ferocious exploits were much admired. But then, amid this chaos, about 1200 BC, Zoroaster, an Iranian prophet, whose name may mean 'Owner of the Golden Camel', claimed that Ahura Mazda, the supreme god, had commissioned him to restore order to the steppes. As a result, there came about the first sighting of what would become the Axial Age. No less important, Zoroaster claimed that Ahura Mazda was no longer 'immanent' in the natural world – the traditional understanding of divine powers, which infused rocks and streams and plants – but had become 'transcendent', beyond the range of normal perception and therefore different *in kind* from any other previous divinity. Associated with this was the idea of the 'great judgment', on which access to the afterlife would depend. There had been no apocalyptic vision like this before anywhere in the ancient world.[8]

Above all, the spirituality advocated by Zoroaster was based on non-violence. The great conflicts of the late Bronze Age, the conflicts that had imported male storm and sky gods from the steppes and had overwhelmed the Great Goddess, as discussed in chapter sixteen, had provoked a reaction, one that would be widespread and profound. It also embodied another revolutionary belief, that everyone, not just the elite, could reach paradise.[9]

Traditionally, one of the best-known aspects of the Aryans, if not *the* best-known aspect, is their move into India, where they are supposed to have accounted for the demise of the Indus Valley civilisation

(Mohenjo-Daro and Harappa), and where they imported the horse, Indo-European languages, and possibly the cow, cattle rustling being regarded by them as a sacred activity. The problem with this interpretation, certainly from the point of view of indigenous Indian scholars, is that there is very little archaeological support for such a migration. The Vedic hymns do describe a cosmos convulsed by terrifying conflict and the Harappan evidence, such as it is, features mother goddess figurines which appear to have been replaced. But the most recent scholarly analysis of the Aryan idea, Edward Bryant's *The Quest for the Origins of Vedic Culture* (2001), concludes that neither the invasion hypothesis, nor the idea that the 'Aryans' were indigenous Indians, is supported by much convincing evidence. That should be born in mind in what follows.[10]

Judging from the Vedic hymns in the early first millennium BC, the Aryans developed the concept of *brahman*, the 'supreme reality'. Brahman was not a traditional deity, more a power that was 'higher, deeper, and more basic than the gods', a raw force that held the universe together. It followed that Brahman could never be defined or described, only 'sensed in the mysterious clash of unanswerable questions that led to a stunning realization of the impotence of speech . . . The visionaries of India were moving beyond concepts and words into a silent appreciation of the ineffable.'[11] An allied element was a desire to eliminate violence from sacrificial rites and this probably had something to do with the fact that, if the Aryan migration to India is accepted, life was becoming more settled, the economy depended more on farming than on raiding, and there was a growing consensus that the destructive cycle had to stop.

One aspect of this was a marked change in the actual practice of sacrifice. Traditionally, sacrifices had been the dramatic climaxes of ritual, involving the bloody decapitation of the animal, which was understood to replay Indra's slaying of Vritra. But in the new climate Indra was less important now and, in the reformed ritual, the animal was suffocated as painlessly as possible. Already, says Karen Armstrong, at this early stage the ritualists were moving towards the idea of *ahimsa* ('harmlessness') that would become a dominant virtue of the Indian Axial Age.[12]

Added to which, any reference to war was removed from the *Agnicayana*, the hymns which had sacralised the easterly migration of the warrior bands and the conquest of new territory. Attention was now directed away from the external world and instead focused on the interior realm, known as the *atman*, or the self.[13] 'Gradually the word atman came to refer to the essential and eternal core of the human person, what made him or her unique.' This was a marked change in the spiritual quest of India.

The Dangers of Desire

Although life in the Ganges region of north India became more settled as time went by, older traditions still had influence. There were, for instance, some men who took the (for us) extraordinary step of leaving settled family life and instead living rough, letting their hair grow wild and begging for food. They were known as 'renouncers' ('*samnyasins*') and are mentioned in the Rig Veda where they are described as wanderers with 'long loose locks' and 'garments of soiled yellow hue' who were able to fly through the air and 'go where gods had gone before', and see things from far away. Clearly, there are shamanistic ideas mixed in here, though they appear to have had no community role, as early shamans had. The renouncers worshipped Rudra, a ferocious deity with long braided hair 'who lived in the mountains and jungles and preyed on cattle and children'. They liked to wear ram skins over their shoulders and practised the 'three breaths', inhaling and exhaling 'in a controlled manner' to induce a change of consciousness. This early form of yoga, also with obvious roots in shamanism, became central to the spirituality of the renouncers and then of India more generally. And here too, we have the advent of a less violent, more interior, form of religion, underlined by the practice of young *brahmacarin*, who left their families in order to live with their teachers to study the Veda. They too wore an animal skin, but in addition had to spend time alone in the forest, expressly forbidden to hunt, to harm animals or ride in a war chariot. The *brahmancarya*, or holy life, could only be lived by committing no act of violence, by not eating any meat, instead 'sitting by the fire' and exercising controlled breathing. These rites, almost all of them recognisably shamanistic, were believed to internalise the 'sacred fire' that the novitiate would henceforth carry around within him. This form of asceticism, it was said, 'heated up' the individual and confirmed that the new 'hero' of Axial Age India was no longer the warrior but a monk dedicated to *ahimsa*.[14]

This new approach culminates in the scriptures known as the Upanishads, also known as the *Vedanta*, or 'end of the Vedas'. These do include a discussion of horse sacrifice, so the link to earlier practices isn't ignored or forgotten, but the main focus of the Upanishads is no longer the performance of ritual but a consideration of what it means internally. The word 'Upanishad' means 'to sit down near to', a term which implies that there is an esoteric form of mystical knowledge that must be imparted by great men to a few spiritually gifted disciples. The internal references of the early Upanishads contain little agricultural imagery but have many references to weaving, pottery and metallurgy – Indian society of that time was in the early stages of urbanisation. They show furthermore that individuals travelled considerable distances to consult the sages, which meant that transport was improving. The Upanishads embody the view that human beings contain a divine spark at their core, moreover a divine spark that was from the same lineage as the immortal *brahman* that gave life and meaning to the entire cosmos.[15] As a result, the goal of this new spirituality was knowledge of something that was essentially unknowable – the *atman* again. The point was that, in this long, slow quest for self-discovery, people became 'calm, composed, cool, patient'; uncovering knowledge of the self became a journey towards an experience of pure bliss, an *ekstasis*. In time, people formed the view that this was the only way to escape the endless cycle and recycle of pain and death.[16]

Allied to this was the system of thought known as *Samkhya* ('discrimination'). This too focused on an inner light, the practice of Samkhya espousing the idea that *dukkha*, 'the entanglement of life', which inevitably brings suffering, could be overcome by yoga, which Karen Armstrong describes as one of India's greatest achievements. 'Yoga means "yoking", originally used to describe the tethering of draught animals to war chariots.' In effect, yoga is regarded as a way to approach the unconscious mind, believed to be responsible for so much of the pain we experience in life. The followers of Samkhya did not, however, believe they were approaching god; instead they were developing the natural internal capacity of the human being. In particular, they believed their new method enabled them to experience a new dimension of their humanity – the experience of 'nothingness', a new way to free themselves from *dukkha*.[17]

This idea of *dukkha*, the entanglement of life, was to prove powerful in India and by the late fifth century BC it had been added to by the doctrine of *karma*, the belief that human beings were caught up in the endless cycle of death and rebirth, the wheel of desire. 'A householder could not beget children without desire, could not succeed at farming or business without wanting to succeed, in a perpetual round of activity that bound him to the inexorable cycle of *samsara*', rebirth in different forms. The essentially unchanging nature of this doctrine produced despair among many people who as a result longed for a way out, either in the form of a Jina, a spiritual conqueror, or a Buddha, an enlightened one.

But this was not the only factor: there were social/economic elements too. This was a time when ever larger permanent communities were beginning to form, and when iron technology, including the plough, was making it possible to clear greater areas of forest and irrigate them. As a result a greater variety of crops was produced and farmers were growing richer. Disparities in wealth were showing themselves, small chiefdoms were being absorbed into larger political units and the warrior class was again becoming more dominant.[18] In addition, the new political entities, the new states, stimulated trade, aided by the fact that coins replaced cattle as the symbol of wealth, promoting the development of a merchant class, even mercantile empires. Both warriors and merchants tended to ignore priests and traditional sacrificial practices, and in any case in the new cities individualism began to replace tribal identity.

In this system of new social divisions anomie spread, particularly in the cities. Vedic religion appeared increasingly out of touch, not least because cattle were becoming scarce and sacrifice seemed increasingly wasteful and cruel.[19] In such a situation – of general anomie – the goal became not to find a metaphysical truth but to obtain peace of mind.

And it was in these circumstances that, towards the end of the fifth century, a *kshatriya* from the republic of Sakka, in the foothills of the Himalayas, shaved his head, changed into the familiar saffron robe of the 'renouncer', and set out on the road to Magadha, the ancient kingdom in the north-east of India. His name was Siddhatta Gotama and he was aged twenty-nine. He was a popular son and his parents wept bitterly when he left home. He recounted later how, that night, he crept into his wife's bedroom for one last look at her and his son.[20] Gotama was just one of many at that time who was concerned about *dukkha* and the cycle of suffering, but he was convinced that the dilemma could be turned around, that man's predicament must have positive counterparts. He therefore set out on his own quest, a journey that was so successful that later generations called him (and call him still) the Buddha, the enlightened or awakened one.

To begin with, Gotama studied with a number of yogins and explored various states of trance. He concluded, however, that neither doctrine nor trance, for him, achieved any real transformation. All he had left, he decided, were his own insights, and this reality he elevated to become one of the central tenets of his spiritual technique: nothing, he said, should be taken second hand, there must be no reliance on authority figures. Like others before him, and elsewhere, Gotama became convinced that *desire* was at the root of our suffering and for him the way forward lay in reversing that predicament. It was important, he insisted, to eliminate desire, craving and greed from our nature. Selfless empathy, compassion, desiring the welfare of all other human beings – this, he said, was the answer to *dukkha*.[21]

There was no place for God or for gods in his system. Gotama did not make a song and dance about rejecting deities – he simply put them out of his mind. To live morally, he said, was to live for others and, among other things, this meant his method was not a religion for the elite, as the Vedic rituals had been, but 'for the many'. Like Socrates, in Greece (see the next chapter), he wanted to examine himself, his life, to discover the truth within himself and in order to do *that*, paradoxically, he became convinced that one must *behave* as though the self did not exist.[22] Buddhism later split into two schools. Theravada Buddhists retired from the world and sought enlightenment in solitude. Mahayana Buddhism was more involved in the world, in democratic fashion, and emphasised the virtue of compassion.[23]

It would be wrong to give the impression that Buddhism was the only spiritual development in India: it wasn't. With the accession of the Gupta dynasty in AD 320, after a 'dark age', Indian religion became theistic, the people discovered, or rediscovered, the Hindu 'extravaganza of brilliantly painted temples, colourful processions, popular pilgrimages and a multitude of exotic deities'. 'Absolute reality' was now identified with the personalised god Rudra/Shiva, who would liberate his followers from the painful cycle of *samsara*. The fact is that many people needed a more emotive religion than Buddhism could offer, something *reciprocal*, in which their central act of *bhakti*, was rewarded by a God who loved and cared for his worshippers. *Bhakti* was a powerful idea that would find an echo in other parts of the world. It meant 'surrender'.[24]

The *Bhagavad-Gita*, the great Hindu text, was one of the last great books of the Axial Age, being composed some time between the fifth century BC and, according to recent scholarship, the first century AD, marking the end of the era of religious transition that we have been considering, a new insight inspired by revulsion from violence. The message of the text, a conversation between Arjuna and Krishna, before the start of a great war, is that only by imitating Krishna can one detach oneself from egocentric desires. It is essential, Krishna tells Arjuna, to empty oneself of desire. 'The whole material world is a battlefield in which mortals struggle for enlightenment with the weapons of detachment, humility, non-violence, honesty and self-restraint.'[25]

Nothing, Emptiness, Stillness

In China at this time, religious ideology went through several transformations, some similar to what was happening in India and elsewhere, some characteristic of China in particular.

The kings of the Shang dynasty, who had ruled in the Yellow River Valley since the sixteenth century BC, believed fervently that they were the direct descendants of God. At Yin, the Shang capital, the ramparts of the city were only 800 yards in perimeter: the residential complex had room only for the king and his vassals and was surrounded by high walls to guard against flooding or attack. This emphasised that the Shang had a passionate preoccupation with rank and hierarchy, with the Shang nobility devoting all their time to religion, warfare and hunting.[26]

They had no interest in agriculture but landscape had meaning, with mountains, rivers and winds all being important gods. Sacrifice was important, with many soldiers being slaughtered at rulers' funerals. This ritual was important because it was believed that the fate of the dynasty depended on the goodwill of deceased kings. Ceremonies were held in which members of the royal family would dress up as deceased relatives, believing themselves to be 'possessed' by the ancestors they were impersonating.

But Di, the supreme deity, and the ancestors, could be unpredictable and if dissatisfied could bring about drought, flooding, and other disasters. As a result, the Chinese were not at all squeamish about sacrificing as many as a hundred victims in a single ceremony. On occasion, a Son of Heaven might have hundreds of retainers slaughtered with him at his death.

Several of these beliefs and practices are not at all dissimilar to the rituals followed in the New World, as we shall see in chapters twenty and twenty-one but, in 1045 BC, the Zhou dynasty took over from the Shang and China began to diverge towards an Axial Age sensibility. During the Zhou reign, iron was introduced to China which helped them become the longest-serving dynasty in Chinese history. The Zhou spread the view that the Shang had been corrupt, so that heaven had been 'filled with pity for the suffering of the people'. The 'gaze' of Di, the supreme deity, had fallen on the Zhou kings, they said, and they had introduced an ethical ideal into a religion that had been hitherto unconcerned about morality. Heaven, the Zhou insisted, in a quote that is reminiscent of similar ones in the Hebrew bible, was not influenced by the slaughter of pigs and oxen but by compassion and justice.[27]

But we must be careful of making Chinese religion too much like religions elsewhere. The early Iron Age was a time of great weakness in China, with Zhou territory under constant attack from pastoral nomads. This meant that the paramount king was not so paramount after all – he could exert little control over the cities ranged across the central plain. In fact, says Karen Armstrong, the only thing that held these cities together was the cult. And this cult, under the Zhou, had a special character. 'The Chinese would never be interested in a god who transcended the natural order. They were less interested in finding something holy "out there" than in making this world fully divine by ensuring that it conformed to Heaven's prototype.' The king had supreme power most of the time, but only up to a point. At all times, he had to conform to the celestial model.[28]

Towards the end of the ninth century, however, further changes did begin to appear. The decline in the importance of the monarchy was perhaps the most unsettling development and may have had something to do with the fact that, as the poets noted, there was in those days one natural disaster after another.[29] Other changes were self-inflicted. Under the Zhou the Chinese had made great advances in cutting down woods and forests, clearing more and more land for cultivation. But this meant there was now less territory for hunting and for breeding sheep and cattle. The Shang and the Zhou had slaughtered hundreds of animals at their lavish sacrifices and gift-giving ceremonies, but all this was taking its toll on the rich wildlife of the countryside. As a result, the first step towards the Axial Age was taken when ritual laws were introduced to limit the amount of indiscriminate killing. This had the knock-on effect of spreading a new spirit of moderation among noble families. As a result, over time all the activities of the elite were gradually transformed into elaborate ceremony: whatever you did, there was a 'correct' way of doing it. 'Everything had a religious value and there was a ritual attached to it.'

The particular importance of ritual in China was that it led to an important *school* of ritual, located in the state of Lu. There the ritualists understood well the importance of moderation and self-surrender. For instance, they greatly revered Yao and Shun, legendary sage kings of remote antiquity who, it was said, governed by charisma alone, but a charisma quite different from that of the warrior class. Yao was regarded as a gentle man, whose rule had brought about the Great Peace (*dai ping*). Yao and Shun together had produced the *Classic of Documents*, a tacit criticism of earlier rule based on force and coercion and in this way the idea of kingship in China began to change, from fearsome practices to more ethical ones.[30]

The seventh century was a turbulent time in the Yellow River region, but the ritual reform initiated by the literati of Lu had far-reaching significance: the *li*, as the ritualistic approach was known, forced warriors more and more to behave like gentlemen. As a result, wars were usually quite short 'and could not be fought for personal gain', becoming in effect courtesy contests. 'A nobleman lost status if he killed too many people, never more than three fugitives and even then [he] was supposed to shoot with his eyes shut.' Gloating in victory was outlawed. At court all nobles had to make sure they helped improve the beauty and elegance of the ceremonies, and as a consequence they must always be perfectly dressed, well-mannered, and at all times humble. 'By the seventh century these ideals do seem to have transformed Zhou China from a society addicted to rough extravagance into one that prized moderation and self control.'[31]

Then it all changed a second time. During the second half of the seventh century there was another wave of attacks from the pastoral nomads of the north. The uncertainty and turbulence caused by these attacks brought about the rise of the Chu people (722–481 BC), in the central and southern areas of China and they introduced a new kind of aggression. They had no time for the old ritualised warfare; instead, their ferocity was such that, in 593 BC, during a lengthy siege, the people of Song were reduced to eating their own children.[32]

In this chaos, several things (by definition) were happening at once. Amid the new round of violence, noble warriors, although certainly motivated by greed and ambition, were also trying to free themselves from the domination of older families. This was the Chinese way of moving painfully towards a more egalitarian polity. In line with this, and despite the ascendancy of the noble warriors, at least for a time, in Cheng (central China) and Lu (the south-west) fiscal reforms were introduced that improved the conditions for peasants. In the second half of the sixth century law codes were published, which all citizens could use to challenge arbitrary authority. This was accompanied by growing contempt for ritual, and a new taste for luxury, putting an unsustainable strain on the economy, with demand outstripping resources. One result of this was that there was a growth in the number of poor aristocrats.[33]

And it was against this background, at about this time, that a young man called Kong Qiu (551–479 BC) was completing his studies and about to take up a minor post in the administration of the state of Lu. His family were minor aristocrats and newcomers to Lu, who had been forced out of their previous home. Kong Qiu was a clever student who had mastered the study of the *li* by the time he was thirty. By the time he reached forty he had become by the standards of the day a learned man, convinced that he grasped the deeper meaning of the rites he was trained in. Moreover, he believed that, properly expressed, *li* could bring China back to the 'Way of Heaven'. Kong Qiu's disciples would later refer to him as Kongfuzi, 'our Master Kong'. In the west he is known as Confucius.[34]

Confucius, Karen Armstrong tells us, was no solitary ascetic but a man of the world, 'who enjoyed a good dinner, fine wine, a joke and good conversation'. Like Socrates, he did not lock himself away but instead developed his insights in conversation with others. (The famous *Analects* of Confucius were put together by his disciples long after his death.)

Like many others in the Axial Age, Confucius felt deeply alienated from his time, and he convinced himself that in China the root cause of the current disorder and the constant wars that threatened, was a neglect of the traditional rites. If he was like Socrates in being a man of the world, he was like Gotama in not being interested in metaphysics – Confucius always discouraged 'theological chatter'. Instead, he took the view that 'people should imitate the reticence of Heaven and keep a reverent silence'. An undue concern with the afterlife was not necessary, he felt: the point was that people must 'learn to be good here below'. His ultimate concern, and that of his disciples, was not heaven but 'the Way', to tread carefully, not towards a place or a person but to 'a condition of transcendent goodness ... The rituals were the road map that would put them on course.'[35]

Confucius thought that people could become much more fully conscious of themselves and their lives, that self-cultivation was the all-important process in life, and *reciprocal* in nature. In order to enlarge oneself, one should try to enlarge others. 'The Way' was nothing less than a 'dedicated ceaseless effort to nourish the holiness of others'. This also implied egalitarianism. Until Confucianism only the aristocracy had performed the *li*. Now, he insisted, the Way was available to anybody.

Confucius's approach was as much psychological, and individualistic, as political. He wanted dignity, nobility, and holiness to be at the centre of life, and was convinced this could be achieved only by daily struggle. 'Enhanced humanity' was the goal, something that could not be achieved by coercion. Instead, living a compassionate, empathic life took you beyond yourself.[36]

China's Axial Age had some way to go yet. Confucius died in 479 BC and at around that time the country entered another disturbing and frightening era, which historians call the period of the warring states (475–221 BC). Life on the central plain was suddenly more turbulent than ever before, when seven states and pastoral nomads from the north seemed to be constantly at loggerheads, using ever more terrible iron weapons. One result of this horror was a further intensified quest for a new religious understanding. Given that hundreds of thousands of peasants were being conscripted into the new infantries, and that more and more land had to be drained to pay for these expensive campaigns (crops were grown on the newly created fields, crops which could be traded), warrior-peasants became a major factor in economic-political life (as they were to be in Greece, discussed in the next chapter). As elsewhere, aristocratic chariot teams were phased out. By the fourth century BC, cavalry had totally replaced chariots and soldiering became more of a lower-class activity, lessons being learned from the nomads of the steppes. The Chinese also adopted the weapons the pastoral nomads had invented: the sword and the crossbow. This too made the period of the warring states especially fearsome. As a result of all this terrible 'unrestraint', with moderation cast aside, with kings and aristocrats fighting each other, aligning themselves with first one former adversary, then another, no one set of warriors could trust any other, and kings turned increasingly to the 'men of worth', scholars who were literate and specialists in protocol and ethics.[37]

And in these circumstances, there emerged one other teacher in particular who turned his back on militancy and favoured a message of non-violence. Mozi, or 'Master Mo', (*c.* 480–390 BC), headed a brotherhood of about 180 men, a school with strict rules and with a rigorously egalitarian ethic, in which its members dressed like peasants or craftsmen. Instead of fighting, however, they saw it as their mission to intervene to *stop* wars. Mozi argued it was possible to persuade people to love instead of hate and his aim was to replace the ferocious egotism of the warriors with a generalised altruism. The central concept was *Ai*, a deliberately cultivated attitude of wishing everybody well: 'Others must be regarded like the self.' The Mohists were different from others in that they were more interested in *doing* good than in *being* good. Not for them the slow process of self-cultivation; they put their practical skills, logic and willpower at the service of society. Not surprisingly perhaps, during the warring states period Mozi was more widely revered than Confucius.[38]

Despite this period of endemic war, by the fourth century BC the economic and political transformation of China was progressing at a remarkable rate. The cities in particular were no longer the political and religious capitals they had started out as, but had by now become centres of trade and industry, home to thousands of citizens. In this busy, untidy, unprecedented world, other new philosophies/ideologies emerged, one of which was attributed to a Master Yang. His approach was 'every man for himself', a modern-sounding attitude which fitted city life, but in turn it provoked what might be called a neo-Confucian riposte, in particular the notion of *ren*, 'meditation to check the passions and empty the mind'.[39] This was one of the background factors in the development of the Chinese form of yoga, based on the idea of *qi*, understood as the basic energy of life, 'the primal spirit'. Zhuangzi (*c.* 370–311 BC) adopted this idea, arguing that life is constant transformation, something we cannot avoid and that 'the more things change, the more they stay the same' – change, death and disillusion being normal. The Way was inexpressible, the Great Knowledge could never be defined but, Zhuangzi said, egotism was the greatest obstacle to enlightenment.[40]

Meng Ke (371–288 BC), known in the West as Mencius, saw a pattern in history. China, he insisted, had changed for the worse, partly because people had tried to govern by force, whereas for him goodness had a 'transformative power'. He admired practical action and thought everyone had four fundamental 'impulses' (*tuan*) that could grow into the four cardinal virtues: benevolence, justice, courtesy and 'the wisdom to distinguish right from wrong'. He too believed in the golden rule – treat others as you would wish to be treated. 'A sage was simply one who had fully realized his humanity and become one with heaven.'[41]

By the third century BC, as we shall see, the ideas of the Axial Age were more or less in place in other areas of the world. In China, however, the period of the warring states was still in full flood and there was a great longing for peace. Indeed, the political situation was so serious that the Chinese were not interested in scientific, metaphysical and logical matters, as the Greeks or the Indians or the Israelites were: such notions to them in their turmoil seemed trivial. In these circumstances, the last great Chinese sage of the Axial Age emerged, Laozi ('the old master'), or Lao-tzu (whose dates are uncertain, some time between the fifth and first centuries being favoured, though there are those who think he was a legendary figure only). But he too advocated 'Emptiness, unity and stillness', the giving up of desire, the loss of ego, the virtue of non-violence.[42]

The Axial Age in China had many parallels with elsewhere, as should now be apparent, but the Chinese also absorbed one other important lesson – that no school could have a monopoly on truth: the *dao*, the way, was ultimately indescribable. Nevertheless, the merits of Confucianism gradually became clear and in 136 BC the court scholar Dong Zhongshu, mindful that there were too many competing schools, recommended that the six classics, taught by Confucians, should become the official teaching. Even so, in China, it is often said, one can be a Confucian by day and a Daoist at night.[43]

From Baal to Yahweh: the Evolution of the Jewish God

Arguably, the Axial Age in and around ancient Israel was the most influential of all. Not immediately but eventually, and so far as the 'Western' world is concerned. Also, more is known about Israel, because much of the trajectory of its ideological development at that time is recorded in the bible which, as is discussed further in the following chapter, was one of the first books to be written in an alphabetic script.

If the bible is to be believed, the Israelites said that they always felt different from other peoples, and though this may be post-hoc reasoning, there was some truth to it in that they were a semi-nomadic pastoral people, herding 'small stock' sheep and goats, unlike many of the people they came into contact with, such as the Egyptians and the Phoenicians and Canaanites, who were settled. In the early days, Abraham, Isaac and Jacob worshipped El, the High God of Canaan but he was gradually replaced, after about the fourteenth century BC, by Baal, pictured as a divine warrior 'who rode on the clouds of heaven in a chariot', who fought with other gods, often in hand-to-hand combat, and brought rain. In other words, he was a (male) god with many of the same characteristics as Indra. In the very early days Yahweh, who would in time become the great God of Israel, and the model for the great monotheisms of the world, was very similar to Baal. The Israelites would later turn against Baal, just as they turned against other gods, but to begin with they found him inspiring, though they also worshipped other deities. (Yahweh, who to begin with was a god of war, would not become their *only* god until the sixth century BC).

As befitted a semi-nomadic pastoral people, the Israelites had no central sanctuary but carried the Ark of Covenant from one shrine to another at a variety of temples, at Shechem, Gilgal, Shiloh, Bethel, Sinai and Hebron.[44] The Ark of the Covenant famously contained the tablets on which the Ten Commandments were written, which had replaced the Golden Calf that Aaron had made out of the gold of the Israelites' earrings and jewellery, when they had waited forty days and forty nights in the wilderness while Moses communed with God. This story may well reflect the change the Israelites made from earlier forms of worship, of the bull or cow, to later ideas.

But Yahweh was, to begin with, a warrior god (and would always remain a 'jealous' god), who had no expertise or knowledge of agriculture or fertility, which further suggests that, in their past, the Israelites were pastoral nomads of the steppe kind. (Or, it would underline the theory discussed in chapter seven, that Genesis records the transition from horticulture to agriculture.) Fierce warrior gods, storm and wind gods, also bring rain. Yahweh had these elements too but was not worshipped exclusively (most Israelites had local gods they worshipped as well, and Yahweh was himself surrounded by lesser deities). There was only a small minority who always regarded Yahweh as their sole deity.[45]

India had its 'renouncers', China its 'men of worth', while in the Near East – more or less coincidentally – there arose the tradition of the prophet. A prophet is not someone who can foretell the future but someone who speaks *for* a deity, who is graced by revelation from on high. It was a widespread tradition that flourished from Canaan in the west to the Euphrates in the east but more of it was made by the Israelites, possibly because their fortunes were decidedly mixed.

In Palestine, prophets often formed part of the royal court where one of their functions was to criticise the monarch, ensuring he conformed to the pure aims of what was to begin with a 'Yahweh-alone' cult.

The first prophet of consequence, Elijah, came to prominence at a time when Israel was faced by drought. Elijah saw this as a defining moment and insisted that the Israelites must choose between Yahweh and Baal. He devised a famous divine contest, in which two bulls were placed on two altars, one dedicated to Yahweh and the other to Baal. Yahweh won. When called upon, he sent fire from heaven, which consumed both the bull and the altar, followed by much-needed rain which fell in torrents.[46]

But Elijah's Yahweh was a new kind of deity: he was hidden, 'no longer manifest in the violent forces of nature, but in a thin whisper of sound', the still small voice of conscience. The prophets were almost always concerned with social justice and the protection of the weak and this informed the spirituality of the new deity. The catalyst of change, here as elsewhere, was the eruption of violence in the region, which provoked the reaction of building a more ethically based society, and the realisation that the rituals that characterised traditional religions did little in that direction. On top of that, literacy was spreading through the western Semitic world, thanks to the invention of the alphabet, enabling an archive to be created to record ancient stories: the first writings of what would become the Pentateuch were begun.

At that time, politics, social affairs and religion were all mixed in together. Therefore, as the gulf between the rich and poor became sharper, the prophets interpreted this in a religious way. Amos and Hosea, for example, mounted attacks on the government – the king – for not doing more to help the poor, for putting their own interests above those of the people. In such circumstances, they said, Yahweh did not want ritual and sacrifice: he wanted spiritual reform, self-surrender, the abandonment of egotism.[47]

It was not a simple or straightforward process. The evolution of the Jewish God is seen through the coexistence of two sources in Genesis, known as E (for Elohist), that being the name he used for God, and J (for Yahweh or Jehovah). E is regarded as earlier, for at times J appears to be responding to E. J describes God in personal terms, as an individual 'who strolled through the Garden of Eden like a potentate, enjoying the cool evening air'. But in E, God never appeared to human beings, contacting them indirectly either via his 'angels' or as a burning bush. These two styles mark an important transition. It is J who refers to the special relationship between God and the Jews, though there is no mention of the Covenant, suggesting this was a later invention of the sixth century when, during the exile, the Jews became aware of Zoroastrian beliefs in Babylon.[48]

Another new element related to the concept of sacrifice. In the ancient world, firstborn children, or first-appearing crops, were often regarded as the 'property' of a god – and this is why they were 'returned' to him in sacrifice. But in the bible Elohim demands that Abraham sacrifice his son, as a test of faith. This was quite new, partly because it was arbitrary but the test element stands out too. Hitherto, sacrifices had been performed to send energy to the gods, so that they could continue their divine duties (as is true, by and large, in New World religions). But Elohim/Yahweh was an all-powerful God who did not require any input of energy: the test he set Abraham broke new ground. And, at the very last moment, of course, Elohim sent an angel to stop the killing, instructing Abraham to sacrifice a ram instead. This episode may well mark another important transition, when animal sacrifice was substituted for its human equivalent.[49]

This was important also because it opened up a gap between the human and the divine that was much greater than had existed before. Beforehand, the gods of the palace states had been smaller entities: there were more of them, they ruled specific realms, or places, or activities. Now, there was one god, all-powerful, and transcendent. 'Yahweh was no longer the holy one of Israel alone but the ruler of the world.' This idea, that Yahweh could control the gods of other nations, clearly reflected a defiant patriotism that could only have existed in a small nation surrounded by larger more powerful neighbours.[50]

And this was the context out of which, in the seventh century BC, the beginnings of Judaism emerged. Manasseh (687–642) ruled as a loyal vassal of Assyria. He built altars to Baal, transferred an effigy of Asherah into Jerusalem and continued with child sacrifice. The prophets who urged the people to worship Yahweh alone insisted that He would be angered by Manasseh's actions and eventually those actions were overthrown by Hosiah. The house of male prostitutes was sacked, as was the furnace where children had been sacrificed. The rural shrines were closed. A further set of changes was brought about by the Deuteronomists. Possibly assembled by a small coterie of scribes, Deuteronomy was written in the sixth century BC. This book determined that sacrifice could be offered only at one shrine, 'where Yahweh had set his name'. It followed that other temples, outside Jerusalem, where Yahweh had been worshipped for centuries, had to be destroyed. This gave great power to the central cult, which determined that although individuals were permitted to sacrifice animals in their home town, they were not allowed to drink the blood (the life force), which had to be poured away reverently into the ground. The Deuteronomists also established special courts to judge religious cases and the king was stripped of some powers: he was no longer looked upon as a sacred figure. 'His only duty was to read the written Torah; he was subject to Law.'

As Karen Armstrong says, Deuteronomy is in some ways a modern document: it established a secular sphere, an independent judiciary, a constitutional monarch and a centralised state. More important, arguably, God was an abstraction – you could not see Him, nor could you seek His favour by offering sacrifice. He did not live in the temple, as early versions of the Yahweh cult had supposed; the temple now became simply a place of prayer.[51]

All these developments were happening anyway. Then, in 597 BC the Israelites went into exile. Led by King Nebuchadnezzer, the Babylonians besieged Jerusalem, and wreaked even more havoc in 586, when as many as 20,000 Jews may have been taken away. Having lost everything – everything material as well as their self-confidence – it was natural that some of the people of Israel should turn inward. Deprived of everything, the Israelites had to learn to live as a homeless minority. In such terrible circumstances, God had become incomprehensible – and this confined Him and confirmed Him as utterly transcendent, 'beyond human categories'.[52]

Ezekiel now made the most of this desperate plight. Rather than arguing that their God had let the Israelites down, he turned such reasoning on its head and insisted on the opposite: the Israelites had let their God down. Only if they fully repented would God bring them home. And out of this grew the new spirituality. In the P section of the bible (P for 'priestly', who is believed by many historians to have pulled the E and J versions together), all the people must live 'as though they were serving the divine presence in the temple, because God was living in their midst'. This, a new ethical revolution, was based on the experience of displacement. In this way the concept of 'holy' as utterly separate was born. In our contemporary, post-modern terminology, Yahweh was 'other'. An important component of this was respect for the sacred 'otherness' of every creature. 'Nothing could be enslaved or owned, not even land.' Suffering in exile sharpened the appreciation of other people's pain. This was not the same as the Indian idea of *ahimsa* but it was close.

Now the act of sacrifice was changed again, P ruling that the

Israelites could sacrifice and eat only their *domestic* animals, sheep and cattle. P reasoned that these animals were 'clean' or 'pure' because they were part of the community and therefore shared in God's covenant with Israel.[53] 'Unclean' animals that lived in the wild must not be killed.

A final twist to all this was given after the Israelites' return from exile, with the emergence of the Pharisees, a lay group obsessed with ritual purity. The Pharisees were extremely spiritual, believing that all of Israel should aim to be a nation of priests. This too implied that God could be experienced in the humblest home as well as the temple. Moreover, the Pharisees held that sins could be atoned by acts of kindness, not sacrifice, another turn inward. The Golden Rule – do as you would be done by – became the guiding principle. Rabbis taught the scriptures but had no higher status, no greater access to God. In this Rabbinic Judaism, says Karen Armstrong, the Jewish Axial Age came of age. Study became a 'dynamic encounter' with God and revelation could occur at any time a Jew – any Jew – had an encounter with the traditional texts.

Judaism therefore went through various stages of evolution. As we shall see in chapter twenty-two, Christianity itself began as yet another way of being Jewish.[54]

Pagan Monotheism: 'One Uniform Mass'

In defining and describing the Axial Age, Karl Jaspers had concentrated on the aspect he thought was most significant, and newest – the identification of a *new spirituality*, based upon the idea of compassion, non-violence, the Golden Rule, *ahimsa*, essentially the discovery of morality, and the equal access of all – not just the elites – to the new understanding. But, in a sense, that was only half the picture. A second element was the invention, or advent, of the idea of one God, one jealous God who insisted He was totally transcendent.

Throughout history, the Israelites/Jews have become famous for their invention of monotheism, another intellectual development that never occurred in the New World. However, recent scholarship has suggested that such a development was not quite the unique event it has sometimes been made to appear.

The modern concepts, 'polytheism' and 'monotheism', date from the seventeenth century and may well have served to obscure what actually happened.[55] A conference was called at Oxford, in the United Kingdom, in the late 1990s, to consider the ideas and context surrounding the emergence of monotheism, and was published in 1999 as *Pagan Monotheism in Late Antiquity*.[56] This showed that the idea of monotheism was increasingly widespread by late antiquity, for the most part quite independently of Judaism or Christianity, and particularly among the educated in the Greek east. The difference between paganism and Christianity, for instance, say Polyminia Athanassiadi and Michael Frede, is not stark and simplistic. 'The second century BC was a natural watershed: a break with the Hellenistic past and a new beginning. "God being one has many names" permeates Greek religious theory.' The Stoic Cleanthes and Plotinus both adapted traditional worship to a belief in one god only. 'Aelius Aristedes and Celsus both likened heaven to human administration in which many satraps and governors were subject to the Emperor or Great King.'[57]

Nor should we overlook the fact that Christian monotheism (if not Judaism) was articulated in Platonic terms. According to Olympiodorus: 'We ... are aware that the first cause is one, namely God; for there cannot be many first causes.' A close reading of the epigraphic and literary evidence on oracles also shows that the theological interest of the enquirers was dominated by two issues – monotheism and cult (i.e., ritual practices). The priests of Apollo at Claros took the view that the traditional gods of paganism were not God, but his angels. Both Hypsistarian worship and the Chaldean Revelation suggested that there was a 'first principle' and that it was fiery, being identified with the sun. Both Platonists and Aristotelians defined god 'as absolutely immaterial and therefore transcending the world of the senses'. Indeed, this view informed Christian ideas. In fact, say Athanassiadi and Frede, a very substantial portion of late antique pagans was consciously monotheistic, and Christian monotheism is, historically speaking, part of this broader development as much as it was, to begin with, another way of being Jewish.[58]

Society was moving in the monotheistic direction, they say, even as early as Homer, where Zeus is, as they put it, 'a master mind'. The existence of other gods is not denied but they are reduced in status:

this is known as henotheism, a precursor of monotheism. Even earlier, in the Babylonian Creation Epic, *Enūma eliš*, dating to the latter part of the second millennium BC, Marduk comes forward as the saviour of the gods from the oppression of Tiamat and her followers and, as his fee, demands supreme power.[59] Among other pagans of antiquity, Anaximander may be regarded as a monotheist in that he derived everything from a single divine principle. Xenophanes, too, talked of One God, 'not the only god that exists but one that towers above the others'. Parmenides conceived of a goddess who 'steers' everything but he also envisaged lesser deities. Empedocles' system of four divine elements were periodically absorbed under the influence of Love 'into one uniform mass', becoming a single god called Sphairos or Sphere. Herodotus also had the idea of a divine agency which has organised the world 'with intelligent forethought'.[60]

It is by no means easy to distinguish the Christian position from that of Plato, Aristotle, Zeno and their followers. The Stoics made reference to *the* god and Plato's demiurge recognised three principles: God, ideas and matter.[61]

There is, then, say Athanassiadi and Frede, 'a clear sense' in which Platonists, Peripatetics and Stoics 'and thus the vast majority of philosophers in late antiquity' believed in one God. 'They believed in a god who not only enjoys eternal bliss, but in a god who as a god is unique and that he is a first principle which determines and providentially governs reality.'[62] They go so far as to distinguish between 'hard' and 'soft' monotheism. 'Hard' monotheism includes Judaism and Islam, whereas 'soft' monotheism applies to the Greeks (Zeus is supreme but there are lesser deities too, with a divine substance permeating all objects). On this account, Christianity is midway between the two – a 'masterly combination of monism and pluralism'.[63] Perhaps the strongest plank in this argument was the cult of Theos Hypsistos in the second and third centuries AD, Hypsistos being a single, remote deity, worshipped in preference to the anthropomorphic figures of conventional paganism. Theos Hypsistos was worshipped in Thrace from AD 25 on, in and around the Black Sea and Macedonia, and by Jews as early as the third century BC. (The cult was mentioned by Mark in the New Testament.) In an appendix, Athanassiadi and Frede give 293 examples of documentary references to Theos Hypsistos all over the eastern Mediterranean.[64]

The Special Status of the Jews as Pastoralists

We have slightly jumped ahead of ourselves here. Although Athanassiadi and Frede make a good case for saying that monotheism was emerging in general across the Middle East in the first millennium BC, they do little to explain *why* this happened. The emergence of male gods, who defeat or take precedence over female goddesses, as happened in the Bronze Age/Iron Age transition (see above, chapter sixteen), need not by itself have led to monotheism, though perhaps the great military battles of the late Bronze Age did offer a context for conceiving God as a supreme commander. The general move towards compassion, a more egalitarian morality, and the search within for greater self-understanding, as reflected in Buddhism, the Upanishads and Confucianism, may also have spawned the idea of transcendence. But, according to Daniel Hillel, there was a more specific reason so far as the Israelites were concerned.

His view is as follows: 'Because the Hebrews were pastoralists, who inhabited and occupied several ecological realms, they were best placed to perceive the over-arching unity of all creation, a central tenet of monotheism, and to combine the separate deified "forces of nature" into an overall "Force of Nature".' At the same time, he says, a loose assemblage of tribes became a coherent nation and this further underlined the idea of monotheism. A further factor was the failure of both child sacrifice and animal sacrifice to prevent the Jews' misfortune, which eventually produced a change of view among the Jews, with the result that eventually Yahweh told them he didn't need sacrifice.[65]

Pastoral societies, Hillel says, differed from the agricultural societies in that they tended to emphasise and worship 'the brute and procreative prowess of dominant male animals, such as bulls and rams'. As they also depended on rain, they also worshipped a pantheon of animal and rain gods. 'They achieved their overarching unity because they, as pastoralists, experienced more than anyone else the disparate domains of the ancient Near East.' The patriarchs began as herders of flocks of goats and sheep but were forced to migrate by drought and other calamities and, in so doing, encountered other environments, other landscapes. Sheep and goats, Hillel reminds us, must always be within six miles of water – this is the distance they can traverse twice a day, to and from a water source, leading them to cross differing types of countryside. Pastoralists are also more selective than hunter-gatherers in what they kill – males rather than females – and this would have made them more aware of gender and the value of their animals. This may have had something to do with the fact that child sacrifice, so prevalent in the ancient Near East, was eventually replaced by animal sacrifice, when the angel of Yahweh told Abraham to substitute a ram. Because pastoralists limit their population by weaning their children more slowly, children would have been scarcer among them, and correspondingly more valuable. In any case, it was a major step forward towards a more humane view, and an important stage in the evolution of sacrifice.[66]

We can find other aspects of pastoral life in the Old Testament. For example, trade between pastoralists and farmers was well established, but it was always uneasy and the encounter between Cain and Abel may represent a memory of the ancient clash between farmers and pastoralists, which we have already seen as a major force in history.[67] In the same way, as was discussed earlier, the expulsion from Eden is a folk memory of the beginning of agriculture – humans partook of the tree of knowledge, rather than the tree of life. With that transition, humans no longer dwelled idyllically as pastoralists but had become 'the toilsome cultivators of cereals'. It is also known that vendettas feature in pastoral societies – cattle rustling was regarded as a sacred activity, since cattle were the main form of wealth. An 'eye-for-an-eye' is a familiar attitude from the Old Testament and so too is Yahweh's claim that the Israelites shall make no graven image of Him. Pastoralists, always on the move, had much less need of craft or art objects, and a God with no image would have suited them.[68]

Hillel maintains further that the precarious life of pastoral nomads, in which any surpluses could not be stored, compelled them to be constantly on the move, in search of pasture and water, and it may well have been this which induced them to enter Egypt. Here, they found a stable society, with plentiful water. But, to the Egyptians, they may well have seemed the quintessential 'other', and this may have given them the idea that they were special in some way, that they had a covenant with God, and that it was *their* God which was 'other'.[69]

Eventually the Hebrews left Egypt. Hillel calculates that the numbers must have been much smaller than it says in the bible, for the desert could never had supported large tribes. But the important thing is that they left Egypt with their sheep and goats and that, while in the wilderness, they established the calf god. This means that their time in Egypt, which lasted for generations, 'had not destroyed the pastoral tradition'. They had maintained their 'other' status.[70]

The Ten Commandments, which were acquired during their time in the desert, make no provisions for fixed ritual nor make any mention of sacrifice. Hillel believes that by this time the Hebrews wanted to give up pastoralism, and became farmers. Their labile environment made them distrust all others except themselves. But the synthesis of monotheism, he says, took place in the city, which was a symbol for them of national unity, where the new religion set them apart.[71] The importance of the temple was to act as a focus of the nation's religion, where commoners and noblemen and kings alike worshipped Yahweh. El was one of the original names for god, but it was, says Hillel, derived from *ayil*, meaning ram and implying leader or chief.[72]

Changing Beliefs About Animals

Two more profound changes were to overtake people in the Old World, especially the Mediterranean/Near Eastern region, either side of the birth of Jesus Christ. These were the end of sacrifice, blood sacrifice, and the advent of alphabet literacy, giving rise to religions of the book.[73]

The end of public blood sacrifices mattered because it eventually brought about a full-scale reconstitution of religious ritual. There had been no (retainer) sacrifice in Egypt since the First Dynasty (3050–2800 BC), there had been a gradual transition from human to bloody to vegetable offerings in India by the time of the *Satapatha-Brahman* (eighth to sixth centuries BC), and human sacrifice seems to have ended in China in 384 BC.[74] In his well-known study of Near Eastern influences on Greek culture, Walter Burkett describes the growth of 'substitute sacrifice' via a series of Mesopotamian legends in which, for example, a goat or a ram would be dressed up as a human being as a sacrifice to lift a pestilence – the relevant god being successfully deceived.[75]

The Israelites' temple in Jerusalem was destroyed by Titus in AD 70 and it was this which finally brought an end to the practice of sacrifice. (Human sacrifice was outlawed in the Roman empire in 97 BC, but it hung on until roughly AD 400, according to Miranda Aldhouse Green, though it was always rare and the evidence for the later killings is often ambiguous.) With this caesura, the priests were marginalised.[76] Ritual *activity* was also transformed. Blood sacrifice had been at the heart of ancient religions, as we have seen, both among Jews and pagans. But just as the ancient Near Eastern world was reconsidering its position on monotheism, so sacrifice was under review as well. Constantius II was just one who thought sacrifice was folly and focused on its end. There was, besides this, says Guy Stroumsa, a great debate in Hellenic thought about the value and necessity of sacrifice, as the writings of Lucian of Samosata, Theothrastus, and Porphyry show. The latter's *On Abstinence* (from all meat) is the best known generally.[77]

This was all taking place against a background when attitudes to animals generally and domesticated mammals in particular were changing fundamentally. In India, when the sacrifice of animals had been replaced by bloodless offerings, around 1000 BC, there had been a turn to vegetarianism and the doctrine of doing no harm to living creatures had been expressed in Brahmanical lawbooks.[78] There was no turn to vegetarianism further west (not generally anyway, though it did happen here and there); instead, there was a more complex reaction. In Greece, for example, it was observed to begin with that both gods and animals have in common the fact that they were not human. At the same time, animals – domesticated animals anyway – interacted with humans and had animal societies. They shared many characteristics with humans – in the number of limbs, in their love for their offspring, the necessity for carnal union to bring forth new individuals.[79]

The Mediterranean economies depended on animal labour and resources and a hierarchy had emerged based on the perceived affinity that animals had with humans. This involved an important element of reciprocity: agricultural animals were given protection in exchange for their services. At that time they were used mainly for their power, their hides, their wool and their milk – meat was still of minor importance. In sacrifice, they were supposed to be led to the ritual by a slack rope (i.e., not pulled against their will), and a cup of water was thrown over their heads, causing them to nod, and therefore 'assent' to their killing. This would appear to be a dim memory of the hunter-gatherer's reciprocal relationship with the animals he or she kills.

During the first millennium BC, however, attitudes towards animals began to shift. The turning point appears to have come with Aristotle and the Stoics. According to the Stoics, animals are *aloga*, creatures without reason or belief. The Greeks reanalysed animals' psychological capacities, Aristotle concluding that tame animals are superior to wild ones. Since animals had no reason, the Stoics concluded that they were made for the use of humans, a view that was taken over by Jews and Christians and finds expression in the bible.[80]

There was more to it than that, however. Certain animals, it was believed, were capable of acting morally: some, for instance, showed modesty in their sexual life and some did not eat flesh. It followed that if animals were capable of behaving morally, they should be treated with justice. Gryllus observed that 'lions and horses are never slaves to other lions and horses, as man is to man', while Apollonius made no blood offerings and chose not to wear clothes fashioned from animal products. He and others believed that animals were 'ends in themselves' and not, as the Stoics insisted, made solely for the use of humans. Others asked different moral questions, such as: does brutality to animals make men brutal to each other?[81] It was clear that animals had the capacity for suffering, so it was unjust to harm them, especially as the domesticated animals did not harm humans. This caused a big difference to be made in Greece and Rome between tame animals and wild ones.

At the same time, humans began to distance themselves from animals, so as to move closer to the gods. This was one reason why the Romans sometimes looked down on the Egyptians, who worshipped animal gods, a practice they may have adopted because they were frightening or strange or had qualities not found among humans. (The crocodile was worshipped because it had no tongue and in that way was an example of the idea that the divine word does not need a voice. The scarab was worshipped because it rolled its ball of dung similar to the way the sun daily moved across the heavens.)[82] Another reason animals were worshipped may have been because they seemed more mysterious (and therefore wiser) than humans. Moreover, they don't change and so became symbolic of eternity. Ovid asked in *Metamorphoses* what does it mean to be a beast? But even he saw it as a move down to be turned into a beast in metempsychosis. In Rome the arena brutalised the difference between humans and animals.

This all amounted to a change in man's relationship with animals, from reciprocity to domination, which was further reinforced by the introduction, in Mithraism, of the taurobolium. Here the animal was killed over a pit where the priest was placed so that *gallons* of blood showered on him, the idea being that contact with so much blood made the priest even more divine.[83]

Porphyry famously argued, in *On abstinence from animates* (animates being creatures with a soul), that it was not necessary to eat sacrificed animals, that compassion should be shown towards them – the domesticated variety at least – and that material gods want material sacrifices whereas non-material gods want spiritual sacrifices. Animal sacrifice was now defined as the custom of peoples – barbarians – who lacked spiritual insight; in fact, this became a cultural dividing line. To treat animals justly, Porphyry said, improves human nature. (At the same time he defended people's right to kill dangerous wild animals.) Unlike the Stoics or Aristotle, Porphyry conceded to animals both rationality and language, but he insisted that humans were more rational than animals and that the latter were not capable of salvation. The pig, for instance, could 'never look up at the sky'. For him spiritual sacrifice – as practised by the Christians, say – was less barbaric, less degenerate, than animal sacrifice.[84]

Genesis of course gives man 'dominion' over the animals. Indeed, animals in the bible are more like slaves than partners to man (only serpents and asses talk in the bible). Fishes and sheep were both seen as harmless and good. Shepherding was a lowly occupation but sheep were very important economically, the most favoured animals for sacrifice, and shepherds were used in the bible chiefly as metaphors – as teachers who oversee their flocks. Horses were expensive and identified with war and the aristocracy, whereas the sheep and the hen, taking her offspring under her wing, expressed the tenderness that many felt towards the animal world, a tenderness that Christianity would attempt to appropriate.[85]

In Genesis, then, man is made in the image of God, and the boundary between humans and animals was strengthened with, in effect, the Christians 'leaving the animals behind'. With Christianity and the end of sacrifice (see below) animals were excluded from sacred space. This was very different from the New World, of course, where, as we have seen, and as we shall soon see again, the boundaries between shamans and the jaguar, and other animals, remained fluid and ambiguous. In the Old World, St Augustine (354–430) made the irrationality of animals – and their separation from humanity – decisive.[86] Outposts continuing the old ways endured, of course. Among the Slavic peoples, the sacrifice of young animals lingered as late as the twelfth century.

THE PRIVATISATION OF RELIGION

The destruction of the Temple in Jerusalem in AD 70 was a political/military event and, strictly speaking, unrelated to these changing attitudes and beliefs concerning animals. But, as sometimes happens, context and catalyst coalesced neatly. The disappearance of the temple facilitated the spiritualisation of the liturgy by transforming the rites which accompanied sacrificial activity, so that prayers now replaced the daily sacrifices. This further brought about a shift to ritual without priests; worship without sacrifice became more spiritual, ritual could now take place anywhere: the rabbis were teachers but not priests and had no liturgical role. Instead of blood sacrifices, the Jews now made spiritual sacrifices, as reflected in prayer. 'We see here the privatization of religion, a change from a civic religion to the quiet rituals of individual and family worship. The study of the Torah replaces sacrifice.' Furthermore, prayer, fasting and charity became the three pillars after the fall of the Temple. 'Judaism now became a religion of alienation from God ... God is no longer *evoked* but *invoked*.'[87] This too is an interiorisation of faith.

Christianity of course overlapped with Judaism but it was not the same, even then. The appearance of the codex, which many Christians distinguished from the Jewish use of the Torah scroll, allowed the development of *silent* reading, another form of interiorisation which was an essential ingredient in the transformation of culture and in the practice of religion under the Roman Empire. Furthermore, unlike the Jews and many pagans, the Christians had in common neither land nor language nor clothing – all things that hitherto had defined collective identity. Thus Christians conceived a new sort of people unknown until then: 'a people defined by their belief in a single myth, preserved in a sacred writing'.[88] This was further reinforced by the idea that, in order to offer all people salvation through Christ, his revelation had to be *translated*. The first Christians, therefore, in marked contrast to earlier practices, proposed to translate the Scriptures into all possible languages. One effect of this was the emergence of literacy in languages that had previously been only oral, such as Armenian and Gothic.[89]

Books played a rather different role among Christians than among pagans, though Guy Stroumsa says pagans did have the idea of sacred books. But the spread of Christianity brought in other developments. In the east and in the west, the foundation of monastic culture was the uninterrupted reading of the book. On the other hand, in the great cities of the east, Christians succeeded in Christianising Greco-Roman teaching, establishing what was in effect a double culture, embracing two totally different literary traditions, what Stroumsa calls 'the double helix of European culture': the bible and some of the great classical texts of Greco-Latin culture – especially the Stoic and Platonic traditions. 'This ensured that theology became a part of Christianity and a turning point, decisive for the structure of western thinking, the "collusion" between philosophy and theology.'[90] This important development is taken further in chapter twenty-two.

Christianity was of course at root based on sacrifice but on a reinterpreted *idea* of sacrifice. The horror at blood sacrifice expressed by the Christians, coming on top of the destruction of the temple in Jerusalem, was important in changing this ancient practice but we must not overlook the fact that martyrdom had elements of the idea of sacrifice. René Girard in his book, *Violence and the Sacred*, argued early on that Christianity 'put an end once and for all to the sacrificial violence of all religions of antiquity', but added that the *idea* of sacrifice did not die.[91] In effect, he said, the soul became 'the interior temple'.

Another change implicit in all this was that sacrifices had been carried out in public, as expressions of collective identity. The end of sacrifice brought with it the end of civic (public) rituals. Temples had been built on central sites whereas, as Stroumsa points out, 'The new religion had a new geography', which reflected a new idea of the community, which was now much less centralised and more intimate, limited in this way because the new 'text-based faith' had to be heard and discussed. More than hitherto, therefore, beliefs in Christianity (and post-temple Judaism) were intellectual/abstract rather than based on ritual. Whereas pagan religion in Rome involved the observance of ritual, post-temple Judaism and Christianity were above all *internal*.

This implied further, especially with Christianity, that the new religions defined themselves as lying outside political frameworks.[92]

All this amounted to a great rupture in the way people defined their identity, whether personal identity or collective identity. In the Hellenistic world, identity had been conceived essentially in cultural terms, but by the fifth century AD, identity had become a matter understood almost invariably in religious terms.[93] With this, went a profound change in the criteria by which identity was established: the wisdom teacher was replaced by the spiritual master. Whereas the Greek philosophers had offered explanations of the world, the Christian teachers were different from the philosophers, the teachers of wisdom, in that the Christian monk, for example, did not guide his pupil to the point where he could follow his own way (as Socrates, Plato, Gotama and Confucius did) but *accompanied* him on his 'quest for salvation' until he achieved his goal. 'The goal of the spiritual master, then, is to prevent the spiritual disciple thinking for himself... The suppression of will and ignorance and obedience are praised.'[94] Christianity proposed that there was to be no salvation 'except through an intermediary, a master at once human and divine.' Such structures and practices were virtually unknown elsewhere in the ancient world.

The Idea of the 'Other'

Though long, this chapter has in all fairness only scratched the surface of the profound religious changes that occurred during the Axial Age. In summing up, we can say that each Axial civilisation was different, as you would expect, yet went through several similar transformations. Those elements consisted of the following inter-linked features.

- There was first a desire for compassion, justice, ethical advance which put the interests of the poor and weak on a par with, and even ahead of, those of the elites.
- There was an egalitarian spirit abroad, allied to the promotion of moderation, humility, quietness, intimacy and even silence, of putting others before oneself, all no doubt as a reaction to the great violence that raged across Eurasia in the Bronze Age/Iron Age transition.

- The new spirituality was cemented around a new kind of deity, or religious/ethical entity, which (or who) was essentially ineffable, unknowable, hidden, abstract, transcendent and wholly 'other' but which (or who) was equally available to all, as was the 'salvation' this new entity offered for those who passed the test of faith.
- The new transcendent entities did not require blood sacrifice, in societies where domesticated mammals were increasingly valuable, but instead public ritual was downgraded in favour of more intimate forms of worship, typified by prayer and the study of sacred writings.
- Writing itself, in allowing better codified translation, transformed religion too, changing the form of religious hierarchy, pushing it beyond narrow political boundaries and, via the monastic institution, as we shall see, allowed theological and other activities to flourish.

Of all these changes and transformations, the greatest was the idea that God was transcendent, totally 'other', together with the spirit of egalitarianism that this idea allowed to spread. Egalitarianism is essentially seen as a political idea in the contemporary world, and it is an idea that never gained ground in the New World civilisations (though it existed of course in those New World societies at an earlier stage of cultural evolution). But egalitarianism began in the Old World civilisations as a religious idea before it found form in politics (just as urbanisation was a religious idea before it was anything else), though politics and religion overlapped much more in the past than they do now.

Religious transformations were not the only changes in the Axial Age, however. As we shall now see, other changes accompanied the religious ones, a series of steps that was never taken in the New World.

• 19 •

The Invention of Democracy, the Alphabet, Money and the Greek Concept of Nature

The fighting that characterised the end of the Bronze Age in the Old World, while it had profound effects on the development of religion and self-consciousness, even on the very idea of what it means to be human, also stimulated a raft of other changes that were equally transformative and never occurred in the New World. Together they helped push the trajectories of the two hemispheres even further apart.

We have just seen that one of the achievements of pastoral nomads was the evolution of horse-back riding. This had a crucial effect on the transmission of power away from the chariot to the cavalry. In the new cavalry, the horses were ridden in pairs. This was because, before the invention of the stirrup (800–500 BC), it was difficult to fire a bow and arrow and manipulate a horse with one's legs only. Horses were therefore ridden in pairs, with one rider holding the reins of both horses, while the second man loosed off his arrows. This was an advantage in that horses were cheaper than chariots, and if a horse was shot, the two men could escape on the second, much less difficult than transferring from a broken chariot. With the invention of the stirrup, the replacement of the chariot by cavalry was swift.

Even so, in the Iron Age the cavalry was always secondary to the infantry. This came about partly because the limitations of the chariot began to matter (they were not as manoeuvrable as horses, or infantry, and expensive), and with greater urban complexes, with larger overall populations, greater numbers of infantrymen (less expensive and easier to train) were available. For example, one Assyrian army of the period had 1,351 chariots but 50,000 men, thirty-seven men to each chariot: one cannot imagine a chariot, however skilled its rider, being able to account for thirty-seven men. Many of the twelfth-century BC papyri from Egypt refer to great numbers of barbarians, especially Libyans and Meshwesh, who were creating trouble at Thebes.[1] These were almost certainly professional infantrymen. It is known too that Assyria relied on infantry in the early Iron Age, combating tribesmen with at least 20,000 infantry. In Greece, burials and the accounts of Homer both suggest that Dark Age Greeks (1200–800 BC) fought on foot (arrow heads appear almost nowhere among Dark Age grave goods). Recent evidence even suggests that Greek 'knights' rode to battle on horseback, but once there dismounted and fought on foot. And in fact the recent recreations of Greek warfare techniques in the Dark Age, between rival *poleis*, featured massive infantries drawn up in a line, or *phalanx*, of spearmen. 'Dueling nobles are essential for the poet's story, but in reality the *promachoi* [prominent warriors] were much less important than the anonymous multitude in whose front rank they stood.'[2]

Greek infantries of the Dark Age were not impressive by later standards but the crucial point is that an infantry was a community's principal – and usually its only – line of defence. By the end of the eighth century BC, the manufacture of weapons had advanced considerably, and in Greece the *poleis* were increasingly able to equip large infantries in place of much smaller aristocratic squadrons of charioteers. As a result, between 700 and 650 BC, the old-fashioned Homeric-style warriors, who had fought in single combat, were phased out.[3] This was a crucial social as well as military transformation because it meant that warfare was no longer the privilege of nobility. Anyone who could afford up-to-date weapons (*hopla*) could join this prestigious troop, regardless of rank or birth. And so, with the hoplite army, a new equality was born. The hoplite army was now a people's army, drawn from a wider cross-section of the male population than ever before.

This was a major break with the past. 'Hesiod had suggested that it was time to abandon the traditional heroic ideal; the hoplite army effected this severance.' The individual (and invariably aristocratic) warrior, seeking personal glory, was now an anachronism. Instead, the hoplite soldier was essentially one of a team. Hoplite phalanxes were or were not defeated together, en masse. 'Excellence was redefined: it now consisted of patriotism and devotion to the common good. Instead of aggressively seeking his own fame and glory, the hoplite submerged his own needs for the good of the entire phalanx. It promoted an ethic of selflessness and devotion to others.'[4]

This reform changed Greece momentously, if inadvertently; and it laid the foundations of democracy. 'A farmer who fought next to a nobleman in the phalanx could never see the aristocracy in the same way again.' The deference which commoners had for aristocrats dissolved. And it didn't take long for the lower classes to make claims that *their* organisation – the people's assembly – should take a major role in city government. The self-image of the *polis* was radically overhauled by the hoplite reform.[5]

The overhaul had wide-ranging effects. For example, free speech, originally the privilege of the noble hero, was now extended to all members of the phalanx. However, the phalanx used a different language. *Logos* (dialogue speech), direct and practical, was totally different from the allusive poetry of Homer and the Heroic Age. Aristocrats, who were traditionally meant to excel in battle, thought that war gave meaning to life.[6] But *logos* was driven by practical need. Men wanted to know 'What happened?', 'What shall we do?' and it was vital that any soldier felt comfortable in challenging a battle plan that would affect everybody. The *logos* of the hoplites did not replace the *mythos* of the poets – the two coexisted. As more citizens became hoplites, however, *logos* took over as the distinctive *modus operandi* of government. By the seventh century, Sparta, more than Athens, embodied the new ideal. By 650 BC, all male citizens were hoplites, and the *demos*, the people, were sovereign.[7]

The hoplites played a further, albeit indirect, role at the beginning of the sixth century, when the farmers in rural areas of Attica complained of exploitation by the aristocrats and banded together against them. Civil war seemed inevitable and by now the aristocrats no longer had the inbuilt advantage of military superiority which they had traditionally enjoyed. The exploitation of the farmers had deepened at that time owing to the invention, in nearby Lydia, of coins. Their use spread quickly among the Greeks and enabled wealth to grow and more men to acquire land. This land needed defending and, in conjunction with new weapons, played a role in the development of the hoplite phalanx. At the same time, however, the invention of money opened up a much bigger gap between rich and poor.

This gap opened up because land in Attica – desirable though it was in theory – was poor, certainly so far as growing grain was concerned. Therefore, in bad years the poorer farmers had to borrow from their richer neighbours. With the invention of coins, however, instead of borrowing a *sack* of corn in the old way, to be repaid by a sack, the farmer now borrowed the *price* of a sack. But this sack was bought when corn was scarce – and therefore relatively expensive – and was generally repaid in times of plenty, in other words when corn was cheap. This caused debt to grow and in Attica the law allowed for creditors to seize an insolvent debtor and take him and his family into slavery. This 'rich man's law' was bad enough, but the spread of writing, when the laws were set down, under the supervision of Drakon, made it worse, encouraging people to enforce their written rights. 'Draconian law,' it was said, was written in blood.[8]

Dissatisfaction spread, so much so that the Athenians took what for us would be an unthinkable step. They appointed a tyrant to mediate. Originally, when it was first used in the Near East, tyrant was not the pejorative word that it is now. It was an informal title, equivalent to 'boss' or 'chief' and tyrants usually arose after a war, when their most important function was the equitable distribution of the enemy's lands among the victorious troops. In Athens, however, Solon was chosen as tyrant because of his wide experience. A distant descendant of the kings, he had also written poems attacking the rich for their greed. He was given a mandate to reform the constitution.[9]

Solon was a wise man and he was not content to pass a few laws. Rather he thought it more important to make farmers and aristocrats alike aware of their responsibilities, that everyone had a share in the blame for the current state of *dysnomia* ('disorder') as he called it. His real breakthrough, however, was his insistence that the gods did not intervene in human affairs and would not reveal any divine law to help Athenians sort out their problems. At a stroke, therefore, Solon secularised politics, what Karen Armstrong calls an 'axial moment'. In the previous vision of antiquity, justice was part of a cosmic order but Solon would have none of it. For him the city must work in the same way as the hoplite phalanx, in which all warriors acted in concert for the good of the whole. In order to even up the balance between the two main sectors of society – the farmers and the aristocrats – he cancelled the farmers' debts and defined status in a new way: by wealth rather than by birth. Anyone who could produce over 200 bushels of grain, wine or oil each year was eligible for public office.

The hoplites played a still further role – and again indirect – somewhat later, at the beginning of the fifth century, when war with Persia threatened. Persia was a world power then and Athens had unwisely sent help to Miletus, on the western coast of Anatolia (modern Turkey) which had rebelled against Persian rule. Darius (550–486 BC), king of the Achaemenids/Persians, had quashed the rebellion and transferred his attention to the mainland. Faced with a major threat, Themistocles, a general from one of the less prominent Athenian families, was elected magistrate and he persuaded the Areopagus Council (see below) to build a fleet.[10]

The Athenians had no real experience of naval warfare – the hoplites were their pride and joy – so this move was a risk. Nonetheless, they went ahead and built 200 triremes and trained a navy of 40,000 men. This too was controversial, and in two ways. The size of the threat meant now that all able-bodied men were conscripted: aristocrats, farmers, and *thetes*, men of the lower classes, who all sat on the same rowing benches in the triremes. Previously only men who could afford their own equipment had been allowed to join the hoplites; now everyone was part of the military, further widening the democratic ideal. On top of all that the hoplites, used to fighting hand-to-hand and face-to-face, found it dishonourable and demeaning to sit in a trireme, because it meant having their backs to the enemy.

The hoplites must have resented Themistocles' plan, the more so when, in 490 BC, the Persian fleet entered Greek waters, conquered Naxos and landed on the plain of Marathon, twenty-five miles north of Athens. The hoplites set out to meet them and, though they were outnumbered by about two-to-one, they managed through discipline and good leadership to inflict a stunning defeat on the Persians. 'Marathon became the new Troy; its hoplites were revered as a modern race of heroes.'[11]

But Themistocles had cleverly anticipated what would happen next. In 480 BC, Xerxes, the new Persian king, advanced towards Athens with twelve hundred triremes and roughly 100,000 men. In other words, he had six times the number of Greek ships and more than twice the number of men. Even with the aid of other Greek cities – if aid were offered – the Greeks were badly outnumbered.

Themistocles understood these odds and before the Persians arrived he made his move: he evacuated Athens completely, transferring the entire population, including children and slaves, to the island of Salamis. And so, when the Persians reached Athens they found it empty. They enjoyed themselves, looting what they could and setting fire to the Acropolis. Then they moved on to Salamis.

But Salamis was not just another island city. It had one crucial feature, which is why Themistocles chose it. The city could be approached only by a narrow gulf into which not all of the Persian ships could squeeze. In fact, the triremes became gridlocked in the gulf, the great size of the fleet – as Themistocles had anticipated – acting against it, as the ships were jammed together and unable to move. In these circumstances the Athenians picked off the Persian ships, one by one, until by evening the triremes that were still afloat retreated and fled back home.

Salamis was another axial moment and so here we see the hoplites being involved in four important changes that occurred in Greece at that time: the development of democracy; the secularisation of politics; the development of the disciplined exercise of reason and logic in which rational thought is abstracted from emotion; and – to an extent – the experimental approach to life, knowing when to dispense with tradition and use new thought patterns generated by new conditions.[12]

In view of what was to follow, it is worth taking time to remind ourselves what other aspects of democracy the Athenians introduced. From the point of view of the argument of this book, the most important element was the Council of Five Hundred, initiated by Cleisthenes in 508–507 BC, which included not only aristocrats but men of modest means.[13] Then there was the fact that Athenians established the idea, and practice, that democracies require public spaces, not as religious theatres but open to all, where matters of common concern can be defined and lived. 'For many, the *agora* – the main public space in Athens – served as a second home.' Life in Athens was anchored in a polytheistic universe of gods and goddesses though in the early 440s Protagoras of Abdera told Athenians that man was the measure of all things, including the deities, who perhaps did not exist, except in men's minds. John Keane, in his celebrated history of democracy, also makes the point that the Greek system of gods – many individuals behaving (and misbehaving) in all sorts of ways, human and superhuman – themselves inhabited a form of democracy, where negotiations took place, where they were open to persuasion, where their minds could be changed. This is why many Athenians thought of their democracy as a system for establishing and enforcing the will of the deities, 'who in turn authorized the exercise of human powers'. The assembly regarded itself as sovereign but also as divinely mandated.[14]

No conflict was seen between the exercise of democracy and the existence of slavery. Business in Athens was seen as quite separate from (and, status-wise, below) politics: the two were kept physically separate. Slaves were used to round up citizens who should have been in the assembly but weren't. Seating in the Council of Five Hundred was egalitarian, all business was carried on face-to-face, speakers stood on a small platform, the better to be heard, decrees were written down and deposited in the city archive. Heralds and archers were on hand to prevent disagreements leading to violence, because people were expected to trade in what the Athenians called frank speech (*parrhēsia*), which was seen to be a great discharger of friction. The chief executive officers of the administration of the Council of Five Hundred were chosen by lot. Citizens had to serve on juries and everyone was equal before the law. Ballots were in secret.[15]

Above all, perhaps, democracy highlighted the contingency of things, of events and of peoples. 'The originality of democracy lay in its direct challenge to habitual ways of seeing the world, to living life as if everything was inevitable, or "natural".' Open-ended government produced nail-biting cliff-hangers and stimulated a sense of scepticism about power and authority; life was open-ended. Athenian democracy managed to trigger radical questioning of who gets what, and pulled the rug from under the high-and-mighty. Monarchy, tyranny and oligarchy were rejected and could not be defended as 'natural'. Men were not the same and it was recognised that being born well was a fluke. Athenian theatre showed this. Plays depict individual qualities, even in slaves. The ubiquity of perplexity was shown in all characters. Scripts could be rewritten, with endings that were unknown.[16]

Many, of course, were against democracy, including Plato and, formally, it didn't last. In 260 BC the Macedonians captured the city and democracy disappeared for centuries. But the form by which men and women govern themselves was not the only legacy of Athens' democratic age: the contingency and open-endedness revealed by democracy, the role of persuasion, of equality, of secularisation and radical questioning of convention had stimulated other activities that could not be so easily crushed.

And in fact there were two other principal aspects to the legacy of democracy. One was that the Greeks were the first to truly understand that the world may be known, that knowledge can be acquired by systematic observation, without aid from the gods, that there is an order to the world and the universe which goes beyond the myths of our ancestors. And second, that there is a difference between nature – which operates according to invariable laws – and the affairs of men, which have no such order, but where order is imposed or agreed and can take various forms and is mutable. Compared with the idea that the world could be known only through or in relation to God, or even could be known not at all, this was a massive transformation.[17]

From Dancing to Metaphysics

The question of order is interesting. One of the innovations of that time that generated new thought patterns had nothing directly to do with politics or military affairs. It was an intellectual invention all by itself and which, historians tell us, separated the Mediterranean/European world not only from the New World but from the East – China in particular – as well. This was the introduction of the alphabet.

Big claims have been made for the 'alphabet effect' and some of these are no doubt overblown. Still, the alphabet was important – revolutionary – for two reasons, one socio-political, the other religious. What made it so transformative was the ease with which people could learn to use it. In other cultures it was (as we shall see) in the interest of the scribes to keep the rest of the population ignorant of the secret of writing: those who were literate had a great advantage over those who were not, who often looked upon scribes as possessing what were, in effect, divine powers. But the development of the alphabet ended the dominance of the literate elite.[18] In place of a complex syllabary of some 6,000 characters (and a complex grammar that had to account for these characters), an alphabet consisted of between twenty and thirty signs, signs that children and the less intelligent members of a society could – and did – master with ease. Possibly the alphabet can be traced back to Egyptian hieroglyphics, which included a complete set of twenty-four signs for the twenty-four Egyptian consonants (which in turn may have begun life as clicks in click languages).[19] But the Egyptians never took the next step, to a proper alphabet which, as we have seen, was itself a democratising device.

This aspect of the alphabet had further knock-on effects. As it lost its elitist associations, writing was used for more and more different activities: it is another reason why, in places like Greece, philosophy, theatre and history writing flourished: as more people could read, so there was an increasingly bigger market for the production of written material. By the same token, the sheer simplicity of the alphabet allowed people to systematise knowledge and this too benefited a wider range of citizens. Information, knowledge, was easier to store and easier to retrieve.

The alphabet also encouraged abstract thinking. Because its signs were totally removed from the entities they represented (unlike cuneiform, for example), the alphabet encouraged people to see beyond what was particular in nature and to seek out what was the 'essence' or universal.[20] And, as Leonard Shlain has pointed out, divining the laws that unite seemingly disparate events is the essence of theoretical science. This aided the investigation – and understanding – of nature. In other words, the arrival of the alphabet occasioned a subtle but profound change in human thinking.

Nor was this the only change. The introduction of alphabetic literacy had a profound impact on religion, so that the alphabet provides a link between this chapter and the last, which concerned the spiritual changes that overcame mankind in the Old World during the Axial Age. Alphabetic literacy, according to some scholars, encouraged men and women to turn away from the worship of idols and animal totems that represented the images of nature, and begin paying homage to the much more abstract *logos*.[21]

Ernest Gellner, in his celebrated book, *Plough, Sword and Book: The Structure of Human History* (1988), argued that the transcendent is born at this point 'for meaning now lives without speaker or listener'.[22] The concept of the 'other' acquires a genuine independence; concepts that had once been 'danced out', as he puts it, 'and thus tied to a community', now came to be written out in doctrine, available to all and binding, independent of community. Metaphysics was born and made to underwrite culture, 'a new situation altogether'.[23]

There is some dispute over the earliest alphabet. Traditionally the Phoenicians were credited with this achievement, based mainly on the account of Herodotus, who in the fifth century BC wrote that they had introduced a number of accomplishments into Greece, of which the most important, he said, was the art of writing. Later, archaeologists uncovered evidence for an earlier alphabet in Canaan around 1600 BC.* But if that is true, it was an uncommonly long time before its use became widespread and, given its undoubted usefulness, this seems unlikely. Added to that, Phoenicia and Canaan do not seem – on the face of it – to be the type of cultures in which an alphabet might be conceived. The Phoenicians were not an agricultural people – their land rarely stretched more than ten miles inland – and their trading cities were dotted around the Mediterranean, a good distance from each other. They never excelled culturally, save for naval design, and their only literary legacy appears to be the alphabet itself and the word for book, derived from the city of Byblos.

The most vivid account of the Phoenicians comes not from themselves but from the Romans, who famously laid siege to the important Phoenician outpost of Carthage in North Africa. Because their deity, Moloch, could only be appeased by human sacrifice, they threw several hundred children – drawn from the finest families – on to their sacrificial fires. This is apparently confirmed by the great number of funerary urns found at Carthage containing the bones and ashes of children.

Another indirect piece of evidence is that the Phoenicians instituted no religious reforms and yet, as we shall see, the introduction of alphabet literacy had a profound effect on religion. Their gods, instead, were the harsh Storm-Ruler-God and the fierce Warrior-Sexual-God, similar to other cultures throughout the Middle East at that time.[24]

Nor does Canaan suggest itself as a place of origin. Many letters written by Canaanites were discovered at Tel el Amarna, dating to 1450 BC, but they are all in cuneiform. There *are* a few Canaanite inscriptions in alphabetic script but their contents do not suggest a

* Chinese script dates from 1300 BC, possibly earlier.[25]

high level of literacy or advanced thought. In Egypt the alphabet does not appear until much later.

Another theory, no less plausible, is that alphabetic writing first surfaced in the Sinai. This, known as Proto-Siniatic, was first discovered in 1905 at the Serabit al Khadem temple, in the Sinai itself, and is thought to have been left by the Seirites, who worked in the copper mines for the Egyptians. They are known in the bible as Kenites and Midianites and are the people with whom Moses 'sojourned' in the desert when he was exiled from Egypt. The first two letters of the Semitic alphabet are *aleph* and *bet*, the Semitic words for, significantly, 'oxhead' and 'house'. These theories have been challenged by more recent discoveries in Palestine and Ras Shamra in northern Syria.[26]

And the addition of vowels and word breaks was not made first in Phoenicia but in Hebrew and Aramaic and greatly refined and improved by the Greeks (the earliest Greek inscription dates from the eighth century BC). The addition of vowels, to create a phonetic alphabet, allows a one-to-one correspondence between the written and spoken language.[27]

Robert K. Logan, in his book, *The Alphabet Effect*, says that the alphabet encouraged the development of mathematics, codified law, and deductive logic, all of which made possible the development of modern, Western, abstract science. He argues that because phonetic alphabets allow closer parallels between the spoken and written language, this encouraged the development of prose and therefore of narratives, which in turn made possible a new and more accurate account of history. This, he says, would have been especially important to nomadic people like the Hebrews who, moving on all the time, with little sense of place, would have benefited from setting down a written history. Logan further argues that the Ten Commandments in the Old Testament comprise three separate innovations in the life of the Israelites:

- The first use of an alphabet script
- The first adherence to a codified system of law and morality (no mention of law is made in the Torah until we encounter Moses)
- The first acceptance of a complete form of monotheism

As a result, the *word* of God becomes a revelation.

Moreover, alphabetic literacy places a stress on linearity and uniformity, and in so doing, Logan says, encourages the centralisation of social functions.[28] The fact that the alphabet could be – and was – taught to young children encouraged the development of religions of the book, where the beliefs set down *in* the book characterised the religion, rather than some other form of religious identity, such as birthplace or the practice of ritual. This would in time give rise to the notion of religious intolerance and the possibility of conversion, neither of which features had characterised earlier religions.

Apart from this, Logan says, the most striking aspect of the alphabet effect was the proliferation of abstract thought that it brought about, the new abstractions it provoked, seen most notably in the Greek world and the flowering of philosophy, drama and science, for which classical Greece is so famous. The alphabet also allowed knowledge to be systematised as never before. Abstraction and systematisation led directly to the development of logic and to ever more sophisticated analysis. Analysis and logic led to *re*-systematisation, which encouraged more reflective *observation* and *that* led to the discovery (and exploration) of nature. The narrative quality of prose likewise led to linear notions of cause and effect, another central ingredient of incipient science.[29]

Is it possible, some scholars have asked, that it was the Hebrews who invented the alphabet and is that why the mysterious origin of the Ten Commandments is so important to what was to become the first religion of the book – the Old Testament being the first book written in an alphabet? And does the alphabet play a part in the development of monotheism, a more abstract, more internalised form of religion in which – again for the first time – a book plays a most important part?

Competitions in Wisdom

If politics – democracy – is the most famous Greek idea that has come down to us, it is closely followed by science (*scientia* = knowledge, originally), which was to produce a quite radical concept of nature. Quite a lot of scholarship recently has explored early concepts of nature. The most important point, alluded to several times already, is that the existence of domesticated animals, in particular herding mammals, encouraged a relationship of *dominance* between human beings and other forms of life, a relationship especially set down in the bible where humankind is given (by God) 'dominion' over the animals. This is in contrast to hunter-gatherer ideas of nature, where humankind is fully *a part* of nature, which, as Tim Ingold has shown, is perceived as filled by *personages* of which humans are but one kind.[30] In this way, humans became separated and apart from nature. This separation, in turn, helped the Greeks look upon 'nature' as 'out there', and it was this 'out there-ness' which, with the aid of the systematisation encouraged by the alphabet, gave rise to science.

This most profitable area of human activity is generally reckoned to have begun at Ionia, the western fringe of Asia Minor (modern Turkey) and the islands off the coast. According to Erwin Schrödinger, there are three main reasons why science began there. First, the region did not belong to a powerful state, which are usually hostile to free thinking. Second, the Ionians were a seafaring people, interposed between East and West, with strong trading links. Mercantile exchange is always the principal force in the exchange of ideas, which often stem from the solving of practical problems – navigation, means of transport, water supply, handicraft techniques. Third, the area was not 'priest-ridden'; there was not, as in Babylon or Egypt, a hereditary, privileged, priestly caste with a vested interest in the status quo.[31]

In their comparison of early science in ancient Greece and China, Geoffrey Lloyd and Nathan Sivin argue that the Greek philosopher/scientists enjoyed much less patronage than their contemporaries in China, who were employed by the emperor, and often charged with looking after the calendar, which was a state concern (as was also true, to an extent, of the Mesoamerican civilisations). This had the effect of making Chinese scientists much more circumspect in their views, and in embracing new concepts: they had much more to lose than their counterparts in Greece, with the result that they seldom argued as the Greeks argued. Instead, new ideas in China were invariably incorporated into existing theories, producing a 'cascade' of meanings; new notions never had to battle it out with old ones. In Greece on the other hand there was a 'competition in wisdom', just as in sports contests (sport was itself seen as a form of wisdom). Lloyd argues that there are far more first-person-singular statements in Greek science than in Chinese, much more egotism, individuals describe their mistakes more often, confess their uncertainties more and criticise themselves more. Greek plays poked fun at scientists and even this served a useful purpose.[32]

What these Ionians grasped was that the world was something that could be *understood*, if one took the trouble to observe it properly. It was not a playground of the gods who acted arbitrarily on the spur of the moment, moved by grand passions of love, wrath or revenge. The Ionians were astonished by this (it is often said that the Greeks 'discovered' nature) and, as Schrödinger also remarked, 'this was a complete novelty.' The Babylonians and the Egyptians knew a lot about the orbits of the heavenly bodies but regarded them as religious secrets.

Mott T. Greene, in his analysis of Hesiod's treatment of volcanoes in the *Theogony*, has shown how the author achieved early on a sort of half-way stage in natural knowledge. While still referring to volcanoes as gods, his powers of observation, and his descriptions so carefully based on those observations, allow him to be well aware that there were different types of volcano. Hesiod's gods differed in their natural properties.[33]

The very first scientist, in the sixth century BC, was Thales of Miletus, a city on the Ionian coast. However, science is a modern word first used as we use it in the early nineteenth century, and the ancient Greeks would not have recognised it; they knew no boundaries between science and other fields of knowledge, and in fact they asked the questions out of which both science and philosophy emerged. Thales was not the first ancient figure to speculate about the origin and nature of the universe but he was the first 'who expressed his ideas in logical and not mythological terms, who substituted natural causes for mythical ones.[34] As a merchant who had travelled to Egypt, he had picked up enough mathematics and Babylonian astronomy to be able to predict a total eclipse of the sun in the year 585 BC, which duly occurred, on the day we call 29 May. (For Aristotle, writing two centuries later, this was the moment when Greek philosophy began.) But Thales is more often remembered for the basic scientific-philosophical question that he asked: *What is the world made of*? The answer he gave – water – was wrong, but the very act of asking so fundamental a question was itself an innovation. His answer was also new because it implied that the world consists not of many things (as it so obviously does) but, underneath it all, one thing. In other words, the universe is not only rational, and therefore knowable, but also simple. Before Thales, the world was made by the gods, whose purpose could only be known indirectly, through myths, or – if the Israelites were to be believed – not at all. This was an epochal change in thought (though to begin with it affected only a tiny number of people).

Thales' immediate successor was another Ionian, Anaximander. He argued that the ultimate physical reality of the universe cannot be a recognisable physical substance (a concept not so far from the truth, as it turned out much later). Instead of water, he substituted an 'undefined something' with no chemical properties as we would recognise them, though he did identify what he called 'oppositions' – hotness and coldness, wetness and dryness, for example. This could be seen as a step towards the general concept of 'matter'. Anaximander also had a theory of evolution. He rejected the idea that human beings had derived indirectly from the gods and the Titans (the children of Uranus, a family of giants) but thought that all living creatures arose first in the water, 'covered with spiny shells'. Then, as part of the sea dried up, some of these creatures emerged on land, their shells cracked and released new kinds of animal. In this way, Anaximander thought 'that man was originally a fish', that species could *transform* themselves into others. Here too it is difficult to overstate the epochal change in thinking that was taking place – the rejection of gods and myths as ways to explain everything (or anything) and the beginnings of observation as a basis for reason. That man should be descended from other animals, not gods, was as great a break with past thinking as could be imagined.[35]

Anaximander was fascinated by the present order in the world (which implies some dissatisfaction with it) and was fascinated too by how the present order was established. By analogy with embryology he concluded that the current order had developed and had not always been that way. This was a crucial collective achievement of Ionian positivism, that humans can change and in so doing affect their future. Interestingly, Anaximander wrote in prose, not verse, to underline the break with mythical thinking and he formed the view that the universe and the earth had mathematical and geometric qualities. He thought the heavenly bodies were like chariot wheels – arranged in a circular formation and that there was nothing divine about them. Possibly he got the idea of the circularity of the universe from the shape of the *agora*, where men sat in a circle as an egalitarian arrangement where everyone could be heard equally.

For Anaximenes, the third of the Ionians, *aer* was the primary substance, which varied in interesting ways. It was a form of mist whose density varied. 'When most uniform,' he said, 'it is invisible to the eye ... Winds arise when the *aer* is dense, and moves under pressure. When it becomes denser still, clouds are formed, and so it changes into water. Hail occurs when the water descending from the clouds solidifies, and snow when it solidifies in a wetter condition.' There is not much wrong with this reasoning, which was to lead, a hundred years later, to the atomic theory of Demokritos.[36]

Before Demokritos, however, came Pythagoras, another Ionian. He grew up on Samos, an island to the north of Miletus, off the Turkish coast, but emigrated to Kroton, in Greek Italy because, it is said, the pirate king, Polykrates, despite luring poets and artists to Samos, and building impressive walls, headed a dissolute court that Pythagoras, a deeply religious – not to say mystical – man, hated. All his life, Pythagoras was a paradoxical soul. He taught a wide number of superstitions – for example, that you do not poke a fire with a knife (you might hurt the fire, which would seek revenge). But Pythagoras's fame rests on the theorem named after him. This particular theorem (about how to obtain a right angle), we should never forget, was not merely an abstraction: obtaining an absolute upright was essential in building. This interest in mathematics led on to a fascination with music and with numbers. It was Pythagoras who discovered that, by stopping a lyre-string at three-quarters, two-thirds or half its length, the fourth, fifth and octave of a note may be obtained, and that these notes, suitably arranged, 'may move us to tears'. This phenomenon convinced Pythagoras that numbers held the secret of the universe, that number – rather than water or any other substance – was the basic 'element'. This mystical concern with harmony persuaded Pythagoras and his followers that there was a beauty in numbers, but it was a fascination that also led Pythagoras to what we now call numerology, a belief in the mystical meaning of numbers and an elaborate dead-end.[37]

The Pythagoreans also knew that the earth was a sphere and were possibly the first to draw this conclusion, their reasoning based on the outline of the shadow during eclipses of the moon (which they also knew had no light of its own). The varying brightness of Mercury and

Venus persuaded Heraclitus (very close to the later Pythagoreans) that they changed their distance from Earth. These orbits added to the complexity of the heavens and confirmed the planets as 'wanderers' (the original meaning of the word).

This quest for what the universe was made of was continued by the two main 'atomists', Leucippus of Miletus (fl. 440 BC), and Demokritos of Abdera (fl. 410 BC). They argued that the world consisted of 'an infinity' of tiny atoms moving randomly in 'an infinite void'. These atoms, solid corpuscles too small to be seen, exist in all manner of shapes and it is their 'motions, collisions, and transient configurations' that account for the great variety of substances and the different phenomena that we experience. In other words, reality is a lifeless piece of machinery, in which everything that occurs is the outcome of inert, material atoms moving according to their nature. 'No mind and no divinity intrude into this world ... There is no room for purpose or freedom.'[38]

Anaxagoras of Klazomenai was partially convinced by the atomists. There must be some fundamental particle, he thought: 'How can hair come from what is not hair, or flesh from what is not flesh?' But he also felt that none of the familiar forms of matter – hair or flesh, say – was quite pure, that everything was made up of a mixture, which had arisen from the 'primordial chaos'. He reserved a special place for mind, which for him was a substance: mind could not have arisen from something that was not mind. Mind alone was pure, in the sense that it was not mixed with anything. In 468–467 BC, a huge meteorite fell to earth in the Gallipoli peninsula and this seems to have given him new ideas about the heavens. He proposed that the sun was 'another such mass of incandescent stone', 'larger than the Peloponnese' and the same went for the stars, which were so far away that we do not feel their heat. He thought that the moon was made of the same material as the earth 'with plains and rough ground in it'.[39]

The arguments of the atomists were strikingly near the mark, as experiments confirmed more than two thousand years later. (As a theory it was, as Schrödinger put it, the most beautiful of all 'sleeping beauties'.) But, inevitably perhaps, not everyone at the time accepted their ideas. Empedocles of Acragas (fl. 450), a rough contemporary of Leucippus, identified four elements or 'roots' (as he called them) of all material things: fire, air, earth and water (introduced in mythological garb as Zeus, Hera, Aidoneus and Nestis). From these four roots, Empedocles wrote, 'sprang all things that were and are and shall be, trees and men and women, beasts and birds and water-bred fishes, and the long-lived gods too, most mighty in their prerogatives . . . For there are these things alone, and running through one another they assume many a shape.' But he also thought that material ingredients by themselves could not explain motion and change. He therefore introduced two additional, immaterial principles: love and strife, which 'induce the four roots to congregate and separate'.[40]

The Ionian positivists believed that contingency – chance – played an important role in human affairs, that life may have started in the Nile, where they observed that fresh alluvium was laid down each year and, from its thickness, could have been accumulating for 10,000–20,000 years; they noticed fossils and understood them for what they were; they grasped that weather had natural causes rather than divine ones; and even that natural catastrophes were exactly that.

They were not as completely modern as all this makes them sound. Empedocles' thought was characterized (by E.R. Dodds) as a typically shamanistic amalgam of magic and naturalism, and Greek mythology, in its tradition of the separation of the sky and earth, recalls the very earliest myths discussed in chapter two. But in Heraclitus's idea that humans have the capacity to increase their understanding and the general Ionian belief that man must live in conformity with nature, this was a significant change in humankind's mental life.[41]

One other important aspect of these various figures: Hesiod was a merchant's son (so he tells us), Xenophon was an aristocrat, Heraclitus, according to some accounts, was a king who renounced his throne, Cleanthes started life as a boxer. In other words, although many of the Greek philosophers were wealthy men, their world was by no means closed; the hierarchy, as befitted (and reflected) a democracy, was not rigid.

As ever, we do well not to make more of Ionian positivism than is there. Pythagoras had such an immense reputation that he was credited with many things he may not have been responsible for – even his famous theorem, which may have been the work of later followers. And these first 'scientists' have been compared to a 'flotilla' of small boats headed in all directions and united only by a fascination for uncharted waters.

The Origin of Philosophy

The birth of reflection in Ionia, what some modern scholars call Ionian Positivism, or the Ionian Enlightenment, occurred in a dual form: science and philosophy. Thales, Anaximander and Anaximenes can all be regarded as the earliest philosophers as well as the earliest scientists. Both science and philosophy stemmed from the idea that there was a *cosmos* that was logical, part of a natural order that could, given time, be understood. Geoffrey Lloyd and Nathan Sivin say that the Greek philosophers invented the very concept of nature 'to underline their superiority over poets and religious leaders'.[42]

Thales and his immediate followers had sought answers to these questions about natural order by observation, but it was Parmenides, born *c.* 515 BC in Elea (Velia) in southern Italy, then part of *Magna Greciae*, who first invented a recognisably 'philosophical' method, as we would understand that term today. His achievement is difficult to gauge because only about 160 lines of a poem, *On Nature*, have survived. But he was a great sceptic, in particular about the unity of reality and the method of observation as a way to understand it. Instead, he preferred to work things through by means of raw thought, purely mental processes, what he called *noema*. In believing that this was a viable alternative to scientific observation, he established a division in mental life that exists to this day.[43]

Parmenides became known as a sophist. To begin with, this essentially meant a wise man (*sophos*), or lover of wisdom (*philo-soph*), but our modern term, philosopher, conceals the very practical nature of the sophists in ancient Greece. As classicist Michael Grant tells it, sophists were the first form of higher education – in the western world at least – developing into teachers who travelled around giving instruction in return for a fee. Such instruction varied from rhetoric (so that pupils could be articulate in political discussion in the Assembly, a quality much admired in Greece), to mathematics, logic, grammar, politics, and astronomy. Because they travelled around, and had many different pupils, in differing circumstances, the sophists became adept at arguing different points of view, and in time this bred a scepticism about their approach. It wasn't helped by the sophists' continued stress on the difference between *physis*, nature, and *nomos*, the laws of Greece. (It was in their interests to stress this division because the laws of nature were inflexible, whereas the laws of the land could be modified and improved by educated people – i.e., the very students they taught, and received income from.)

The most renowned of the Greek sophists was Protagoras of Abdera in Thrace (*c.* 490/485–after 421/411 BC). His scepticism extended even to the gods. 'I know nothing about the gods, either that they are or they are not, or what are their shapes.' Xenophon was also sceptical: he asked why the gods should have human form. On that basis, horses would worship horse gods. He thought there might just as easily be one god as many. Protagoras is probably best remembered, however, for another statement, that 'the human being is the measure of all things: of things that are, that they are; and things that are not, that they are not.'[44]

This is how philosophy started, in particular how the concept of nature became integral to the great changes brought about in Greece during the Axial Age – democracy, science, secularisation. The Greek Axial Age would be political, scientific and philosophical, not religious.[45] This is why there are three great philosophers whose names everyone knows: Socrates, Plato and Aristotle.

'Moneytheism'

As was mentioned earlier, people had begun using ingots of precious metals in exchange for goods as early as the third millennium BC, in Mesopotamia – when uniform weights of gold and silver, known variously as minas, shekels or talents, came in – were very useful when large warehouses of goods were being traded.[46] But gold remained much too scarce for the average person wanting to sell, say, a basket of wheat, or buy a goatskin of wine. Money proper was invented only once, in Lydia, in what is now Turkey but was then a small state adjacent to Greece, which was the first beneficiary of this revolutionary innovation.

There is no mention of money in Homer, nor do markets feature as places of importance. Markets did exist – in Mesopotamia, China, Egypt and several other parts of the world – but it was only in Lydia, between 640 and 630 BC, that their kings recognised the need for very small and easily transported ingots 'worth no more than a few days' labour or a small part of a farmer's harvest'. The ingots were of a standard weight and were stamped with an emblem – a lion's head – that verified their worth, even to illiterates. In doing so, the Lydian kings, in Jack Weatherford's words, 'exponentially expanded the possibilities of commercial enterprise' and transformed the world, opening up fresh dimensions for new segments of the population.[47]

The first coins were made of electrum, a naturally occurring mixture of gold and silver, which made redundant the need to weigh the gold or silver each time, something that the less well off couldn't do because they could rarely afford weighing scales. The opportunities for cheating were also reduced.

The innovation sparked a trade explosion in Lydia and also saw the introduction of the retail market, where anyone could bring goods they wished to sell. Marketing became so important to the Lydians that Herodotus called them a nation of *kapeloi*, meaning both 'merchant' and something altogether less wholesome. But the advent of money sparked widespread social changes, not least in the status of women who, through their crafts, for example, accumulated their own wealth and were now able to select their own husbands. New services also were introduced as a result of money – the first known brothels were built in ancient Sardis, and gambling was invented.

Money and markets spread quickly across the Mediterranean but it was the Greeks who profited first and most. 'With the spread of coins and the Ionian alphabet, a new civilisation arose in the Greek islands and along the adjacent mainland.' Coins provided a stability to commerce, it was easily stored and transported, and made possible the organisation of society on a much more complex scale than is possible in kinship-based communities, the use of money not requiring the face-to-face interaction and the intense relationships of kin-based societies. Money made possible more social ties but in making them more widespread, faster and more transitory, it weakened traditional patterns.[48]

The impact of coined money was political but not only political. In Solon's reformation of the traditional basis of status, money was instrumental in democratising the political process. More than that though, the vibrant spread of commerce among the Greeks, which this transformation inspired, produced a raft of new temples, civic buildings, academies, stadia and theatres, along with a glorious body of art, philosophy, drama, poetry and science. In parallel, the centre of the classical Greek city moved from the palace or temple to the *agora*, the marketplace. Wealth generated by commerce stimulated the proliferation of leisure time, so that the elites could expand their civic life in such forms as sport, philosophy, the arts and good food. Then there was the fact that the money system brought with it the need for a new kind of mental discipline – people needed to count and use numbers long before they became literate. Counting and calculating produced a tendency towards the rationalisation of human thought that, says Jack Weatherford, 'shows in no traditional culture without the use of money'. Thinking became less personalised, he says, and more abstract. The exactitude in the money culture forced a decidedly logical and rational intellectual discipline not needed before.

A knock-on effect that was no less important was that the marketplaces of the Mediterranean were the focal points for discussing a new kind of religion. The Greek tongue spoken in the *agoras* from Iberia to Palestine was not the classical Greek of Aristotle, nor yet the ancient Greek of Homer. Instead, it was a 'pidginised shop Greek'. And this was the language that the followers of Jesus used to spread their ideas. In cities such as Ephesus, Jerusalem, Damascus, Alexandria and Rome, the early Christians wrote down their stories in this market Greek. Originally lampooned as 'God's poor Greek', these writings became the New Testament.

The combined effects of money and markets, in a process some have termed 'moneytheism', went further. Before the Greek commercial system swept the Near Eastern world, each country had its own gods, each different. The near-universal commercial culture, however, opened the way for the rise of a common religion, available equally to all people. 'Christianity blazed through the cities of the Mediterranean as a totally new and revolutionary concept in religion.' It was a uniquely urban religion with none of the fertility gods or weather gods of the sun, wind, rain, and moon normally associated with farmers. Christianity 'was the first religion that sought to leap over the social and cultural divisions among people and unite them in a single world religion'. In doing this, the early Christians were acting in much the same way as merchants were using money to create a universal economy.

Monotheism, money, the alphabet (and the 'linearity' that went with it), plus the concept of 'nature' as an entity in itself, separate from humanity … none of these phenomena ever emerged in the New World. The trajectory they propelled is continued in chapter twenty-two.

• 20 •

SHAMAN-KINGS, WORLD TREES AND VISION SERPENTS

As we have seen, in chapter 17, in the first millennium BC, in the New World, there were two prominent civilisations: the Olmec in Mesoamerica and the Chavín in South America. Numerically at least, this compared poorly with the civilisations of North Africa, the Mediterranean, Near Eastern, Indian and Chinese civilisations of the Old World, not to mention the several pastoral nomadic cultures. In the next thousand years, however, during the first millennium AD, the New World went some way towards making up this deficit. Between AD 1 and AD 1250, many cultures flourished up and down the Americas.

The Nazca culture was one of several traditions which preceded and in many ways set the stage for the Incas. Several of the cultures we shall be examining, which dominated the coastal areas and the highlands for many centuries, collapsed in sudden decline as the forces of drought, earthquake and/or El Niño devastated political and economic conditions without warning. This is a pattern we shall see time and again.

The Nazca comprised a confederation of minor kingdoms that flourished from the Chincha River in Peru, south to the Acari Valley, near the southern border of what is now Peru, and Chile. Nazca populations were relatively small because the rivers of the area were small, without much run-off, but the people responded to frequent drought by building long (500 metre) tunnels to feed aquifers that brought water to specially built storage tanks. In this way, between AD 1 and 750 they developed an elaborate pottery and textile tradition, which combined cotton and alpaca wool. In 2009 Lidio Valdez, of the University of New Mexico at Albuquerque, reported the 'unprecedented finding' of dozens of decapitated heads buried inside a carefully located structure at Amato, an early intermediate site in the Acari Valley. Several of the heads were associated with cervical bones showing cut marks, wrists and ankles that were tied together, and 'parry fractures' on the skulls, strongly indicating 'outright violence' and conflict, that fitted with walled settlements and buffer zones between them.[1]

Above all, however, the Nazca are famous throughout the world for their 'lines', great designs on the desert floor that, to an extent, remain a mystery to this day. This part of the desert is covered with a relatively fine layer of sand and small pebbles, which the Nazca were able to clear away to create their lines. Some are scrawny, some are as wide as an airport runway, some run for five miles in dead straight lines whatever the terrain, flat or hilly. Some are triangles, some are zigzags, some are spirals, some make no sense on the ground but when seen from a helicopter reveal themselves as birds, as monkeys, as spiders or plants – there is even a whale. How and why would the Nazca wish to create images that can only be seen properly from the heavens? Were they there for the benefit of the gods in the sky?

Many archaeologists have tried to understand the lines and their meaning. One theory was put forward by Paul Kosok, originally a professor at Long Island University in New York state, after he had observed, by chance, the sun setting at the exact end of one of the lines near the village of Palpa. This suggested to him that the lines had an astronomical function, an idea that was carefully followed up by his German colleague, Maria Reiche, who had been in Peru working for the German consul. She spent years measuring the alignment of the lines, often over-nighting in the desert among her beloved formations. She concluded that the layout of the lines reproduced ancient constellations of the stars and that their positions on the ground point to areas of celestial activity above the distant horizon. From her researches, she argued that the original Nazca first built models of the great designs, then used lengths of sisal cord to arrange them on the desert floor (this is how they knew what the large figures looked like from above).[2] Later researchers did not entirely support Reiche's work – few alignments could be linked to the state of the heavens at the time the Nazca flourished. It was not until the 1970s that archaeologists discovered drawings on the ground that occurred *outside* Nazca territory, elsewhere in the Andes, showing animals, humans, or abstract symbols. All these have now been widely studied and have been found to extend over more than 800 miles of terrain, some of which are dead straight for as much as twelve miles. Some radiate from hills, some lead to well-watered areas (and may, therefore, have served as pathways), and many figures were drawn over by later lines, as if they were of only transitory importance. Some are believed to show transformed shamans in the process of acting as intermediaries between the two realms of the Nazca world.[3]

It is fair to say that no one has a convincing answer to the meaning of the lines but some of the latest ideas come from Johan Reinhard, explorer-in-residence at the National Geographic Society, significantly a mountaineer as well as an anthropologist. He has inspected the 'geoglyphs', as they are now called, over all 800 miles of their distribution, and observes that they are associated with lakes, rivers, the ocean – and the mountains. He further makes the point that mountain gods are everywhere in the Andean region, where they were believed to protect humans and their livestock, and figure prominently in rain-making rituals, given that mountains were the source of the rivers from where water came. Many Christian churches in Bolivia, for example, lie at the end of Nazca-type lines, lines at the end of which, even today, the local headman will make offerings to encourage rain.[4] In support of this, recent excavations by Helaine Silverman at Cahuachi – an enormous 370-acres site, dominated by a central pyramid – have uncovered ceremonial centres, mounds, cemeteries and shrines, where several lines point directly at this ritual centre. Cahuachi was not an urban centre, but a place of natural springs, a ritual location that flourished and then disappeared. Human heads were found there, which appear to be trophies, so there may have been other rituals practised at Cahuachi. The lines themselves may have been sacred so that, as pilgrims approached the ritual centre, they did so by special routes, transforming themselves – by dance, elaborate costume and hallucinogens – in a shamanic process that applied to everyone. This is supported by the very latest theory, of Tomasz Gorka of Munich University, who found along the lines anomalies in the Earth's magnetic field caused by changes in soil density. He believes these changes may have been caused by people constantly walking back and forth in prayer rituals. 'This activity was closely connected to the placing of ceramic vessels along the lines, perhaps as offerings.'[5]

It now seems that the Nazca culture collapsed due to a combination of an El Niño event and environmental degradation. The latest studies show that, around AD 500, the pollen of huarango trees (*Psosopis pallida*) gave way to the pollen of maize and cotton. This suggests the beginning of agriculture but the huarango trees performed a vital function in the area: they have roots which reach as deep as sixty metres underground, in search of water, and are therefore very efficient binders of the soil. Where the pollen transition takes place, the scientists found the remains of many huarango tree stumps. It seems therefore that, in the transition to agriculture, the Nazca farmers cut down the very trees that kept the soil in place. Once the trees – and their all-important roots – had gone, the area was especially vulnerable to extreme weather.[6]

Pumas and Potatoes

For many years, research in the Peruvian and Bolivian highlands was hampered by politics – guerrillas occupied large parts of the area and made life very difficult for archaeologists. But recently one breakthrough has followed another and among the more spectacular discoveries has been that of Tiwanaku, or Taypi Kala, 'The Stone of the Centre', once a city of 50,000 people on the southern shores of Lake Titicaca.[7] The site was first occupied around 400 BC but large-scale construction didn't begin until around 100 BC, when it continued for nearly three hundred years. It was contemporary with another South American civilisation on the coast, the Moche, whom we shall meet next. But Tiwanaku outlasted the Moche, not collapsing until around AD 1000–1100.

At its height, around AD 650, Tiwanaku was a place of palaces, plazas and vividly coloured temples boasting many gold-covered bas-reliefs. It was 'an architectural masterpiece, marked by its many gateways and massive masonry buildings'. The gateways, fashioned from single slabs of rock, formed religious entities, in particular the 'Gateway God', possibly a solar deity, which was positioned above the doorway, wearing a headdress with a sunburst motif, with many projecting rays that culminate either in circles or puma heads. The god wears a tunic and skirt, boasts a fine necklace and holds two staffs adorned with condor heads. He is surrounded by three rows of winged figures with human or bird heads.[8]

There is abundant evidence that Tiwanaku religion revolved around human sacrifice. The remains of dismembered bodies have been found throughout the area and ceramics often show warriors with puma masks decapitating their enemies and holding trophy heads. Human heads with their tongues torn out were used to decorate belts.

Aside from the sacred gateways, a large artificial platform, fifty feet high and 650 feet long, dominated the city. At the summit was a sunken court around which the priests lived. When it rained the water was directed through the sunken court, out on to the surrounding terraces and into the temple, sluicing then into a great moat that surrounded the ceremonial precincts. Some archaeologists believe that this set-up was meant to represent a sacred island and it was on this massive terrace (known as the Akapana) that Tiwanaku's elite appeared, dressed in the manner of gods, or as condors or pumas, wearing sacrificial knives hanging from their belts, alongside the trophy heads of their victims. The bodies of at least a dozen sacrificial victims have been excavated near the bottom of the platform.[9]

Other plazas and courts exist at Tiwanaku, some decorated with stone sculptures in the form of human heads or skulls, some with monoliths in human form, some with images of peoples the Tiwanaku conquered, raising the possibility that their gods, as well as their subjects, were held captive there.

Recent excavations have shown that agriculture at Tiwanaku was much more sophisticated than previously thought. They created raised fields, long ridges that covered the flood plains separated by ditches. When these were recreated, they produced dramatic results: for when frost descended on the altiplano, the raised ridges were protected by the water in the nearby ditches which kept the air warmer and helped form a protective mist in the early mornings. Potato harvests using this system were both larger and more reliable.[10] After massive droughts in the sixth century AD, Tiwanaku's rulers invested huge resources in reclaiming altiplano land in this ridged-field system, achieving yields perhaps four times what modern farmers achieve and supporting a population of somewhere between 40,000 and 120,000 people.

The llama and alpaca provided wool and some protein for food and the former were used as animals to carry pottery, textiles, wood carvings and gold objects far from the city. By Andean standards Tiwanaku was both large and long-lived, surviving two or three hundred years after the Moche had disappeared, until – probably – the area was hit by a drought that lasted for decades. Not even the ridged-field system could save them.

Punishment, Prisoners and Precious Metals

Copper and gold ornaments first appear in the archaeological record, in an area stretching from Ecuador to Bolivia, as early as 1500 BC. But it was not until the Moche civilisation appeared, about AD 100, that it achieved its apogee – metalsmiths who could weld and even electroplate gold and copper. They also developed sumptuous textiles and elaborate pottery, making this culture one of the jewels of pre-Columbian America. As Brian Fagan correctly says, the Moche never invented writing but we have a vivid record of their lives.[11]

Moche was never a large kingdom, stretching at the most for 250 miles along a narrow strip from the Lambayeque Valley to the Nepena Valley on the north-west coastline of Peru. Most people lived in the valleys of short (fifty-miles) rivers. The kingdom's success was due to the efficient management of water – irrigation canals and ridges, many canals being fortified. Much effort was spent in keeping the canals free of debris, and agriculture was further aided by the industrialised collection of guano, droppings by seabirds that were so rich that Peru exported tons of it each year until the end of the nineteenth century and the revolution in modern nitrogen fixing. This substance was regarded as so precious that anyone who approached the seabird nesting areas during the breeding season was executed.[12]

The central ceremonial feature of Moche culture was the *huaca*, truncated pyramids, the greatest of which, the Huaca del Sol, the Pyramid of the Sun, rises 135 feet above the plain. The pyramids were made of adobe sun-dried bricks – 143 million went into the Huaca del Sol – each stamped with a symbol indicating which gang (or kin group) had made it and which quarry it had come from. Michael Moseley, who excavated the Pyramid of the Sun, calculates that it took a century or so for the entire construction.

These sacred pyramids served as temples, even symbolic mountains, where the elite lived in palaces and presided over human sacrifices which were the main aim of their military campaigns (battles were designed to capture the enemy, not kill him – he would be sacrificed later). The elite were also buried in the pyramids, with an array of fine grave goods. Tens of thousands of people lived around the pyramid/temples in a loose agglomeration, a different kind of 'city' or 'urban structure' from those in Mesopotamia or Mesoamerica, but still a sizeable population.

The Moche were America's first accomplished metalworkers and their first creators of elaborate pottery. They understood the lost-wax technique, to produce three-dimensional figures in precious metal. They developed a number of techniques to vary the colour of gold, using salt or soda, to produce a vast array of objects. By the same token, the range of their pottery has been compared to classical Greek vases, many of which are as much sculptures as drinking vessels, in the form of houses, blind people, curers treating their patients, anthropomorphic animals, musicians, humans making love with deities, a jaguar attacking a man. 'Prisoner vessels' were a particular genre, showing victims seated, with ropes around their neck, and their hands tied behind their back. The pots were placed next to the sacrificial victims and ceremoniously broken beforehand.[13] The Moche were the first pot-makers in South America to produce ceramics from moulds. The pots are painted with scenes of battles, sometimes showing decapitation, or lines of prisoners passing before a ruler. According to one archaeologist who made a study of 125,000 Moche art objects, every one has a symbolic meaning.

Moche iconography shows many images of coca users, use of the San Pedro (psychoactive) cactus, a strange fruit known as *ulluchu*, not yet identified but possibly hallucinogenic, being associated with coca and feline figures, which in Moche society represented thunder, lightning and rain. There was also a plant, *espingo* (*Nectandra*), which was added to *chicha* beer and made the shaman act, according to one conquistador, 'as if mad'.[14] Papaya (*Carica candicans*) was also widely used, having properties that prevented blood coagulating, so that it could be employed by the priest/shaman after the sacrifice had been completed.

The Moche appear to have had many gods but their supreme deity was a sky god creator, with feline fangs, who lived – significantly enough – in the mountains.[15] A major theme in Moche iconography was punishment inflicted on human individuals. In some cases sacrificial victims were killed by being left exposed on mountain tops, sometimes after being flayed.[16] The idea of punishment was designed specifically to make the victim cry out in pain, so as to be heard by the gods, and to deprive them of the strength to summon up malevolent forces. The Moche believed in a mythic being, or *amaru*, who lived in a lagoon high in the mountains. Every so often, the *amaru* emerged with extraordinary violence and destroyed everything in its path, leaving a record of its wrath on the landscape. 'The *amaru's* appearance announces the disorders that provoke ancestral wrath and the lack of respect for ritual.' This is a fairly obvious reference to a volcano, with the reference to ritual designed to reinforce the role of shaman/priests.

There was also what Elisabeth Benson calls 'god-the-son', a more active form of deity, with a feline mouth and a jaguar or sunrise headdress. He may have been inherited from the Chavín and was much concerned with the doings of humans. But here as elsewhere the dominant sacred image is of the jaguar, who decorated temples, often anthropomorphised, sometimes on top of humans, where it is uncertain whether the animals are attacking the humans or copulating with them.[17] Feline figures were common on gold and shell jewellery.[18]

Later there is a subtle change as a new god appears 'to share power with the fanged deity'. He is a warrior, clad in a warriors' armour and headdress. He may be accompanied by a jaguar but in any case himself has fangs in his mouth and snakes that radiate from his head.

The elite of the Moche were buried with many precious objects (ten golden heads of felines, others showing spiders) together with wives, concubines and other retainers who were deliberately sacrificed alongside them.

The Moche elite were all warriors and in their art there are countless scenes of battles and in particular the sacrifice of prisoners. In this case, archaeologists believe there may have been an association between sacrifice and fertility, in that Moche art is famous for its erotic content, in particular showing men and women making love, humans making love with gods, and men with prominent phalli. In sacrifice where death occurs by strangulation or decapitation, the penis can become erect – Moche priests may therefore have formed a conceptual link between these apparently similar processes.

As with the Chavín peoples, the animals in Moche art are predominantly those of the rainforest – jaguars, pumas, monkeys and ocelots – and they are often combined with plants and mundane objects to represent fantastical images: helmets or weapons with legs, for example, which Brian Fagan says 'almost certainly' represent shamans' visions.

Moche civilisation disappeared suddenly around AD 800. Studies have shown a prolonged drought in the area between AD 562 and 594, and then the region was hit between AD 650 and 700 by a great earthquake that affected many areas in the Andes. Mudslides blocked the canals and disrupted life on the coast, and were followed by a major El Niño event which brought torrential rain and violent winds that swept away entire villages and towns and killed the anchovy harvests. The elite moved north, abandoning the Huaca del Sol but half a century later another El Niño brought yet more mayhem.

Michael Moseley and his colleagues believe that the El Niño which destroyed Moche civilisation was greater even than the one in 1997–98, the effects of which began in Peru and lasted for eight months, killing ~2,100 people and causing $33 billion of damage, and in which the incessant rain and mudslides lasted for weeks on end as far away as Kenya, Poland, California and Madagascar. It is known that the effects of an El Niño event can last for up to eighteen months but Moseley and his team argue that the part of Peru that was hit by the mega-El Niño shortly after AD 600 created after-effects that lasted for much, much longer.[19] Steve Bourget, at the University of Texas at Austin, argues that prisoner sacrifices at the Huaca del Sol were performed in times of crisis associated with torrential El Niño rains, since some of the victims were killed during periods of heavy downpours.[20]

An El Niño event may have caused the collapse of Moche civilisation but a further factor appears to have been a commoners' revolt against intolerable burdens placed on them by the elite, when the residences of the Pampa Grande nobles in the Lambayeque Valley were burned. As we shall see, this is not the only time commoner revolts occurred in pre-Columbian Latin America.

The Mayan Milky Way: The Drama of the Night Sky

The Classic Maya civilisation, as it is called, is, with the Aztecs and Incas, the best known of tropical American cultures. It flourished long before the other two, between 200 BC and AD 900 across as much as

100,000 square miles of the Yucatán Peninsula in what is now Guatemala and south-eastern Mexico. There were in fact about fifty independent (city) states, ruled over by a small elite, and tens of thousands of villages.[21] They traded widely, their kings were regarded as divine and they built grand urban complexes with pyramids and other monumental architecture, including ball courts.

Theirs was a world where, as with the Olmec, the natural world and the supernatural world were closely intermingled. Indeed, the classical period itself may have been initiated after a massive volcanic eruption, one of the greatest in the Holocene, at about AD 200. For them, as for others in Mesoamerica, the cosmos was made up of three levels – the Upper World of the heavens, the Middle World, inhabited by humans, and the mysterious Underworld. Linking these three layers was the World Tree, or *Wacah Chan* (literally 'raised-up sky'), which had its roots in the Underworld, its trunk in the Middle World and its branches and leaves in the Upper World. The souls of the dead could pass via this tree to the other levels.

The actual position of this World Tree was fixed – or personified – by the body of the king, who could bring the tree into existence 'as he stood in a trance on top of a temple pyramid'. This act, in which the king performed the most sacred deed of kingship, was achieved through trances and the shedding of his blood, which opened a doorway or portal into the spiritual world. Portals were an important concept in Maya religious belief, being openings or gateways to the 'Other World' beneath the Earth, known as Xibalba. Caves could be portals but so too were cenotes, a major geographical feature of the northern lowlands of Yucatán formed, some people believe, 65 million years ago when a giant asteroid hit the earth in the Gulf of Mexico. Cenotes can be very deep water cisterns, and were equipped with potentially dangerous ladders (if they broke) being built down into the deepwater levels, and with many votive offerings being cast into the waters. At the Cenote of Sacrifice at Chichén Itźá the remains of many sacrificial victims have been discovered.[22]

Linda Schele and David Freidel, two of the foremost Mayan scholars, also argue that the World Tree was seen in the sky, the Milky Way being regarded as the Wakah-Chan and within which many other images were discernible, mostly relating to Mayan ideas of the original creation of the universe. During the course of the night, the Milky

Way, as the World Tree, turns in the heavens, to form a canoe-shaped entity which was believed to carry the First Individuals to the Place of Creation. In other words, on certain important nights, great cosmic transformations took place in the Upper World.[23] (The Milky Way is both darker in places and lighter in patches in the tropics compared with the northern skies.)

Clouds of incense and smoke were generated in the course of these rituals, the coils of smoke fashioned naturally so as to resemble a serpent high above the pyramid, rising to the heavens. 'This was the Vision Serpent, the Feathered Serpent, perhaps the most powerful symbol of Maya kingship, the tortuous path between the natural and supernatural realms.'[24] The Vision Serpent was generally evoked via a bloodletting ritual, most often by passing a rope through a hole cut in the tongue. This painful procedure elicited a trance state in which the Vision Serpent would appear, the creature being 'the conduit through which the ancestors came into the world and spoke to their descendants'.[25] The Milky Way and Vision Serpents are discussed in more detail in the next chapter.

Mountains, Masks and Memories

Bloodletting appears to have been the most fundamental ritual in Maya life. Everyone offered blood at all occasions, from the birth of a child to marriage and funerals. Usually, it was taken from the tongue or the penis, using either a fish spine or an obsidian blade to make the incision. Most of the time a few drops were sufficient but on important occasions elaborate cleansing rituals were devised in which people (men and women) would draw a 'finger-thick rope' through their tongues or penises to stimulate ready flows of blood which dropped on to sacred papers that were then burnt. (This, too, is explored more fully in the next chapter.)

The Maya believed that gods could take the form of mountains or other features of the sacred landscape, as shamans or as trees. Thus the pyramids sometimes represented sacred mountains, and the images of kings were carved on stelae to create a number of standing stones that, together, comprised a 'forest of kings'. Even today, says Linda Schele, shamans still make models of the sacred landscape out of sticks and corn stalks which they place in caves or at the foot of sacred hills to communicate with supernatural forces.

The Maya famously had a sacred calendar, or rather two sacred calendars. They did not invent this device – it was very ancient (beginning ~600 BC), based on the early Mesoamerican counting system, a system ordered around the number 20, probably derived from the number of fingers and toes on the human body.* From Olmec times onwards, at least, priests devised a *tzolkin*, or sacred year of thirteen months each with 20 days, giving a 260-day cycle. The derivation of this cycle, used nowhere else in the world, still remains obscure but the best 'guesstimate' so far is that it stems from the average gestation time of humans, which is 266 days (conception to birth, 280 from last menstruation to birth), and, roughly speaking, the basic agricultural cycle in Maya territory. Ancient priests may therefore have seen something special or sacred about the 260-day period.[26] If this is true, it would confirm the relatively late discovery of the link between coitus and birth, as discussed in chapter seven.

But they also used a second cycle, the *haab*, which was developed later and consisted of 18 'months' of 20 days each (20 again), and one short month of just five days, making a total of 365 in all. This too may have been a farming calendar but, unlike the calendars of other civilisations across the world, the Maya *tzolkin/haab* cycle did not make use of the phases of the moon, to divide the year into twelve, and as a result gradually went out of alignment with the seasons by a quarter of a day every four years, meaning it was 1,460 years before it was back in step.

Two other features of this system – again quite different from anything elsewhere in the world – deserve mention. One, around the time of the birth of Christ, the two calendars were meshed together, in an interlocking calendar, which produced a great cycle of time of 18,980 days, or fifty-two years. And two, they developed the so-called 'long-count', a calendar which used the 52-year system to calculate backwards in time to the creation of the world which, according to them, took place on 11 August 3114 BC in the calendar we use today. This enabled

* This was a system of dots (=1) and bars (=5) and the zero (the sign for which was somewhat like the drawing of an eye). Peter S. Rudman, in his book, *How Mathematics Happened: The First 5,000 Years* (2007), includes the Mayan system as a form of abacus notation, meaning we can be reasonably confident the Mayans did abacus addition.[27]

Maya kings to locate themselves in the grand sweep of history and establish their genealogy, which added to their legitimacy to rule. Underneath it all, however, the Maya had a cyclical idea of history and believed that, in the fullness of time, events would repeat themselves. (See above, chapters two and five, for the possible origin of the idea of cyclical time.)

Linda Schele and David Freidel have chronicled the rapid birth of kingship in Maya society. Kingship, they say, was essentially of a shamanistic nature, the divine *ahua* opening a portal to the supernatural world via bloodletting. Kings bore responsibility for the weather, for disease, even for death.[28] They observe that this idea of kingship seems to have emerged in Cerros, until about 50 BC a small village on the Gulf of Mexico. At that date, however, the archaeological evidence indicates that, for some reason, the inhabitants deliberately abandoned their homes in the centre of the settlement and replaced them with a T-shaped temple that stood on a pyramid of earth and rock, where shamanistic rituals were carried out. (This was reminiscent of the Olmec architecture.)[29] The façades of the monumental buildings were plastered (an innovation of the Maya) and decorated with political and religious messages that described the role and functions of kingship.

There was a massive staircase from the plaza to the shrine at the top of the pyramid, where there were huge post holes, in which were located large tree trunks to symbolise World Trees and the cardinal directions, in relation to the rising and setting sun. This is where the king-shaman carried out his bloodletting and went into a trance in preparation for his ritual meeting with the gods and the ancestors. The lower levels of the pyramid were decorated with huge masks of snarling jaguars, representing the Jaguar Sun God and the Ancestral Twins of Maya legend (described in chapter twenty-one).

According to the archaeologists, the temple precinct in Cerros grew in magnificence in just a generation, to three times the size of its original construction, with the raised platform on the pyramid reaching a height of 52 feet. Other platforms and ball courts were built nearby and for a time Cerros was an impressive centre. Unfortunately, it didn't last and the city was destroyed, possibly by a rebellion of the people, in a pattern that had occurred at Olmec San Lorenzo and was to recur time and again in Mesoamerica.

The ultimate significance of Cerros is that it forged the hierarchical nature of Mesoamerican society that was to be repeated at the great classic Mayan cities of El Mirador, Tikal and Uaxactún. The hierarchy in Mayan society was marked. There were four ranks, with the elite enjoying a distinctive diet that resulted in them being, on average, ten centimetres taller than the rest of the population.[30] We know more about El Mirador, Tikal and Uaxactún because of the inscriptions that have now been deciphered and because *Popol Vuh*, the Mayan 'Book of Counsel', which gives us the Mayan view of the cosmos, in which the main players are the great shaman-kings.

El Mirador, the first of the great Mayan cities, lies among the swamps and low hills of the Petén (north Guatemala), and flourished between 150 BC and AD 50, covering in all some six square miles. It had a large pyramid at its centre, plus many other ceremonial buildings and plazas, all raised on platform mounds, and was controlled by a highly organised elite, which employed its own artisans, priests, engineers and traders. Some of the earliest examples of Mayan writing appear at El Mirador but it hasn't been deciphered and for that reason (among others) the causes of the rapid decline of the city in the first century AD are not understood.

Tikal and Uaxactún, barely forty miles away, filled the political vacuum. The former, located in a swampy region, had started out as a small village around 600 BC, trading obsidian and quartzite. Gradually large public buildings appeared – platforms and plazas – adorned with plaster masks so typical of the developing Mayan style. Notably, too, the royal tombs of this time contain the bones of individuals who were larger and more robust than the average citizen – again the nobility were taller and ate a better diet than commoners. Royal individuals were buried with the paraphernalia for bloodletting, but the royal headdress continued to be used for generations.

The development of Uaxactún paralleled that of Tikal, with six temples built on a small acropolis, with massive stucco sculptures and masks, depicting a sacred mountain and monster sitting in primordial waters and with the heads of the Vision Serpent on either side. Jaguar masks flank the stairways leading to the central edifice. 'It is here, at Uaxactún, that we see Maya kings memorializing themselves for the first time.'[31]

At the time of El Mirador's demise, Tikal and Uaxactún were more or less equals. Tikal was ruled by a dynasty founded by Yax Moch Xoc between AD 219 and 238. He and his successors depicted themselves in the inscriptions as shamans with an ancient royal ancestry but also with a close relationship to the mysterious jaguar. Here too, as with the Olmec, the jaguar was represented as the master of the Underworld, as the powerful, preternatural symbol of kingship, of bravery in war, and religious authority in general. Sometimes, sacrificial victims are shown cowering at the *ahud*'s feet, the capture of noble prisoners being an important criterion of royal power, and of the king's sacred ability to nourish the gods with noble blood.

The equality of Tikal and Uaxactún produced a rivalry which climaxed during the reign of Great Jaguar Paw, the ninth ruler of Tikal's long-lasting dynasty. Great Jaguar Paw captured Uaxactún on 16 January AD 378 and this appears to have been warfare on an unprecedented scale, undertaken for territorial conquest rather than for captives to provide blood for the gods.[32]

Teotihuácan, in the Valley of Mexico, was also coming on stream at that time and may have influenced events in Yucatán. Certainly, by the late fourth century artefacts from Teotihuácan are common at Tikal as it forged widespread trade links.

New war cults took hold in Mayan life at this time, for Tikal's leaders now timed their wars for specific points in the cycle of Venus, a planet associated with war, and the Tlaloc-Venus costume became a standard part of royal regalia. In this way, Smoking Frog from Tikal became the ruler of Uaxactún, the battle memorialised for generations.

Tikal now entered a period of great prosperity until, in the mid-sixth century, it was overthrown by the rulers of Caracol, who in turn embarked on years of conquest, eventually being overtaken by a newly resurgent Tikal in the late seventh century, when a number of new massive buildings were constructed. Its rulers were depicted as under the protection of the Jaguar God and large numbers of captives of the wars were sacrificed in elaborate ceremonies. The latest evidence shows that Tikal finally declined because it ran out of resources. Traditionally, its temples used wood from the sapodilla tree, very strong but easy to carve. After AD 741, however, sapodilla was replaced in temple construction by logwood, 'a smaller, gnarly tree that is almost impossible to carve'. David Lentz, the palaeoanthropologist who carried out the study, says the temple builders would never have accepted logwood if they had not run out of sapodilla.[33]

Palenque, the most westerly of all Maya cities, and unique in having a distinctive four-storey tower, is also distinguished by having the most complete written history, with the most detailed inscriptions. These show that its recorded history begins with the accession of Bahlum Kuk (Jaguar Quetzal) on 11 March AD 431, a dynasty that was to endure until 799. Another of Palenque's unique distinctions was the discovery, in 1952, of a hidden staircase in the heart of the Temple of Inscriptions, itself the most famous mortuary temple. The stairway descended into the bowels of the pyramid to the tomb of Pacal, its great ruler who had reigned for sixty-seven years, the outside of his sarcophagus containing a magnificent carving showing Pacal's fall down the World Tree into the Underworld. His son built yet more monuments, the inscriptions now regarded as attempts to rewrite history, to glorify still further the dynasty and make it appear older than it really was. (Place was important in the Mayan system. The location of the World Tree, where the sky was raised up, was regarded as the Navel of the World, where all began.) But Palenque also collapsed, in the ninth century, its people returning to farming in villages 'around the ruins of their once-great city'.[34]

Copán, in modern Honduras, was at the southern end of Mayan territory but it too was relatively long-lived. It is also known for the fact that it had more inscriptions and sculptures than any other Mayan city. Its pyramids and plazas, its altars and stelae, cover an area of about thirty acres. The site had been inhabited in some form since about 1400 BC but it wasn't until AD 400 that it became a city, with monumental buildings, including a ball court. The inscriptions (many contained on a fabulous Hieroglyphic Stairway), show that all the kings of Copán claimed descent from a founder, Yax K'uk Mo' (Blue Quetzal Macaw) in about AD 435. Several rulers had long reigns, some lost their heads in wars, and among the artefacts found in royal tombs were sacrificial knives and stingray spines for bloodletting. The city collapsed in about AD 830.

It seems that during the late eighth and early ninth centuries AD, chaos reigned in the Yucatán Peninsula. In the south, the cities collapsed but not in the north, where they lasted until the Spanish Conquest. Why this should be so is unclear. One theory is ecological: environmental degradation, crop failure, drought, malnutrition and starvation. Certainly eighth-century skeletons from Copán show signs of distress. More intriguing are the political theories, that in rigidly hierarchical societies, where the charisma of rulers depended on their elaborate ritual displays, based on shamanistic beliefs, such rulers actually only had a weak control over events (especially 'supernatural' events like volcanoes or earthquakes) and were never able to establish stable political entities which, eventually, collapsed. Moreover, they had no draft animals, no wheeled carts, nor the technology to clear good roads through the forest, nor horses to ride large distances. Under such systems, the size of Mayan political entities was naturally limited and the need for sacrificial victims for their religious beliefs may simply have imposed too many burdens on such societies.

Lightning and the Cloud People

North and west of the Mayan area, in two valleys, two cities, or city-states, emerged that were to have major consequences. These were Monte Albán and Teotihuácan. In the Valley of Oaxaca, where Monte Albán would develop as the centre of the Zapotec civilisation, maize and bean agriculture had been present by about 2000 BC and one site, San José Mogote, boasted some simple buildings as early as 1350 BC. Shell trumpets, figurines of dancers and masks were all found there, and fish spines linked to bloodletting. By 1000 BC, the central precinct extended for 50 acres and by 600–500 BC the first hieroglyphics and calendars appear, together with were-jaguar designs, and the Feathered Serpent, almost certainly 'inherited' from the Olmec.

At about this time, at a strategic point where the three arms of the Oaxaca Valley intersect, a major centre was established. The new settlement, Monte Albán, grew quickly and was soon home to 5,000 people, forming the first real city in Mesoamerica. Two things stand out about the site. One, it is without arable land or water. Two, its pyramids and palaces and plazas are so immense and intricate that they must have formed some sort of symbolic landscape. Here, nearly two miles of stone walls were built, not as fortifications but as an attempt to isolate the lofty citadel from the valley below. Richard Blanton, of Purdue University in Indiana, believes that Monte Albán was a 'neutral' city on an infertile hill set up to act as a political capital, to provide defence at times of trouble for people who, otherwise, did not need to come together. There is, however, no sign that the citadel was ever attacked, so it seems more likely that it was a purely symbolic centre, symbolic of power and dominance, in an area where the local peoples didn't compete, because it was economically worthless.

The city rose in importance and size and reached a population of ~15,000 by 200 BC. By then a powerful elite ruled over the Valley of Oaxaca, surrounding the immense plaza at Monte Albán with stone temples and palaces, notably the Temple of Danzantes, between 500 and 200 BC. This contains a number of stone slabs which show male nudes in strange 'rubbery' poses, as though they were swimming or dancing, with down-turned mouths in the Olmec fashion. No fewer than 140 figures are shown in this way, but the feeling now is that these are not swimming or dancing figures, but corpses, noble enemies slain by the Monte Albán elite: some have scroll motifs on their groins, as if they had let blood from their penises. As Brian Fagan puts it, 'the connotations of ritual bloodletting provide a strong link to the underlying shamanistic beliefs that fueled and nurtured Mesoamerican civilisation.'[35]

The city, which straddled three hills, was elaborately hierarchical, with elite tombs for those of noble birth, fifteen residential areas, each with its own plaza, enormous central precincts and imposing ball courts. There is also the mysterious Mound J, an arrow-shaped structure that points south-west and is honeycombed with vaulted tunnels. There are forty slabs on the lower walls of this building showing sacrificed corpses and the mound points to a bright star at certain times of the year. Is this a war memorial to a war fought on a propitious date?

Monte Albán reached the height of its influence between 200 BC and AD 200, when its population was about 25,000, after which it did not expand but traded with Teotihuácan and remained prosperous until the eighth century when it declined, after the Zapotec leaders lost popular support.

The Zapotec civilisation is often considered alongside the Mixtec, or Mixteca, slightly to the west, on the Pacific coast of Mesoamerica, with its capital at Tilantongo. The Mixteca were the most highly stratified people in Mesoamerica, placing immense importance on birth order, with rulers occasionally marrying their full siblings to ensure high rank for their offspring. Barbro Dahlgren's analysis of pre-Hispanic codices lists eleven royal marriages between a male and his brother's daughter, fourteen with a sister's daughter, one to a half-sister, and four to a full sister. Increasing endogamy generally among the Mixteca eventually caused adjacent valleys to separate out. And their rulers became so differentiated from ordinary people that there was a separate vocabulary for the various parts of the royal body when contrasted with those of ordinary citizens.[36]

Both the Zapotec and the Mixtec, who developed a drought-resistant form of maize, worshipped great natural forces: winds, clouds, lightning, thunder, fire, earthquakes. The pottery of both civilisations featured fire-serpents and were-jaguars. The most feared natural phenomenon appears to have been lightning among the Zapotecs, rain among the Mixtecs. In the Zapotec sierra south of Miahuatlán, communities believed in four types of lightning 'that reside on certain hills oriented to the major world directions'. Their year was divided into four *cocijos*, or 'lightnings'. Kent Flannery, Joyce Marcus and Ronald Spores say that these four world directions, the idea of a rectangular universe with four quarters, each related to a distinctive colour, is so widespread among the Indians of North, Central and South America, as well as vast areas of Asia, 'as to suggest that it may have been part of the cultural baggage of the first immigrants to cross the Bering Strait'.[37]

The Zapotec were essentially animistic, considering many things to be alive that we consider inanimate. Marcus says they were not, strictly speaking, monotheistic but they did recognise a supreme being without beginning or end 'who created everything but was not himself created'. It was this supreme being that had created lightning, sun, earthquakes, clouds and so on. Lightning was the most powerful supernatural to the Zapotecs, the most feared and, say Marcus *et al.*, the oldest. On the summit of a mountain, long before the dawn of the world, lived Cocijojui, the old lightning of fire. At the foot of his throne he had four immense jars, in which he kept (shut up) clouds, wind, hail and rain. Each of these was watched over by a lesser lightning. Under certain circumstances, according to legend, these forces would be released as people petitioned the gods. Clouds were regarded as ancestors, though people had been jaguars at the start of the world. Clouds were the form to which all people would return, and the rainbow was given the name, *Pelaquetza*, quetzal serpent. *Cocijo*, the word for lightning, was incorporated into more than one ruler's name.

Sacrifices were offered (often in remote, rugged mountains) to calm these forces, prisoners of war in particular, or children. Bloodletting was practised, with stingray spines or obsidian blades, the blood being caught on grass or bright feathers, then offered to the gods. The flesh of victims was cooked for eating and their skulls sometimes exchanged as gifts. Their priests had the power to put themselves into ecstatic states, both Zapotec and Mixtec using alcohol and hallucinogens in their rituals. Pulque, made from the sap of the maguey plant, was a favoured alcoholic drink, the main hallucinogens being the mushroom, *Psilocybe* spp., Morning Glory (*Rivera* spp.) and *Datura* spp., or jimsonweed.[38]

Underlying the Mixtec and Zapotec civilisations, according to Richard Blanton, was an intimate link between religion and warfare. As time went by, the elites of these societies took on an ever-more powerful religious role, in which elites manipulated the ideology, in particular the promotion of 'inter-polity conflict'. This warfare increased cohesion in their communities, and provided opportunities for ever-greater tribute payments. But, according to Blanton, the chief point of religion was the promotion of conflict – war – because it legitimised the elite's control over ritual, and maintained their status.

We may ask why this was necessary – constant war was clearly a high-risk strategy. One answer is that war was a threat that the elite *could* manipulate, at least to an extent. They could choose their opponents, and when to fight. In contrast, the threats from nature, the 'supernatural' threats – lightning, torrential rain, hurricane winds, earthquakes and volcanoes – could not be predicted *or* controlled. In such a scenario, the elite's standing, as ritual specialists, as calendar specialists, would from time to time be under threat because their methods, whatever they were, did not work. By 'seeing' threats in other political groupings, the elite could reassert itself and counter any tendencies on the part of the non-elite segments of the population to question their authority. This explains both the propagandistic nature of their art at the time and their constant preoccupation with genealogy, in particular genealogy that linked the elite to the gods.

Even so, it is not quite clear whether their decline was due to environmental degradation, catastrophe, economic crisis or because people began to question the dominant ideology. Also, their decline coincided with the rise of Tula (see chapter 23).[39]

A City of 600 Pyramids

Finally, among the Mesoamerican cities, Teotihuácan. The people in the Valley of Mexico had a long-standing relationship with the Olmec, through trade, and two cities benefited from about 1000 BC on. These were Cuicuilco and Teotihuácan, though the former was destroyed by a nearby volcano, whereas the latter prospered. By the time of Christ its population was 40,000 and by AD 500 it was between 100,000 and 200,000, making it one of the largest cities in the world.

Teotihuácan was immense but, unlike many cities, it did not grow haphazardly. Instead it was laid out as a symbolic landscape of artificial mountains and foothills separated by open spaces. The city contained no fewer than 600 pyramids, 500 workshop areas, 2,000 apartment compounds, a great marketplace and numerous plazas anchored by the north-south axis of the three-mile long Street of the Dead. Even today it is an awe-inspiring achievement.

The city's prosperity and *raison d'être* seems related to its proximity to a major obsidian source, the green-black obsidian glass-like stone being much prized by Olmec, Mayan and other stone workers, for sacrificial knives and mirrors. But Teotihuácan was also on an important trade route and its large-scale irrigation infrastructure, in a swampy area, provided sizeable maize and bean crops. All of which made it a sacred city on top of its other advantages. It may well have been a place of pilgrimage.

This could have arisen because the great Pyramid of the Sun was built over a natural cave in the volcanic lava – caves, as we have seen, being regarded as portals to the Underworld in Mesoamerican belief systems, and caves being centres of shamanistic ritual, making it a logical place to build a temple/pyramid. The great pyramids were built in the late second century AD and by AD 500 Teotihuácan covered more than seven square miles. Its mercantile influence was felt all over Mesoamerica.

It was a painted city, the surfaces of its streets and plazas being covered usually in white or red pigment and it was kept meticulously clean. There were many murals, maize and water being common themes as subject matter, but so too its great gods, Quetzalcoatl and Tlaloc. Quetzalcoatl, the Feathered Serpent, was the primeval deity of

Mexico, a second god being Tlaloc, the god of rain and water but also, in some places, the god of war. Both these gods would be developed later among the Aztecs.[40]*

Jaguars are also a common theme at Teotihuácan. 'They seem on first acquaintance to perform a whole variety of trained animal acts, such as straddling corn-grinding tables, wearing flowers and feathered ruffs, blowing on shell trumpets, swimming among waves, and shaking rattles,' writes the art historian George Kubler. One ceremony centred on a jaguar-serpent-bird icon – this was first seen at Teotihuácan, then later elaborated at Tula of the Toltecs. Most of the jaguar images at Teotihuácan show people wearing jaguar costumes, sometimes a complete pelt, sometimes a headdress. Either way, the jaguar is always associated with bird and serpent images. Whereas this three-pronged association occurs only with humans at Teotihuácan, later on Toltec and Aztec depictions link the jaguar to the eagle. Judging by the iconography and artistic style, the forerunners of the Teotihuácan jaguars are the Olmec jaguars. In the excavation of the Pyramid of the Moon, Japanese archaeologists found bound sacrificial warriors along with traces of wooden cages and the remains of jaguars, wolves and falcons, which had been entombed alive.[41]

This underlines the inextricable mixing of jaguar ideas and serpent ideas in Mesoamerican belief. The jaguar, as we have repeatedly seen, represented power and the fertility of the earth, the serpent that which comes from water. These ideas would find later expression in Quetzalcoatl and Tlaloc (again, see the next chapter).[42]

Between AD 500 and 750, Teotihuácan's fortunes began to decline and, interestingly, the images produced in the city show a great preoccupation with warfare, with armed deities and priests shown bearing shields and spears. In about 750 the city collapsed and the central ceremonial areas were burned. One reason may have been environmental degradation, in that much of the surrounding land was cleared and emptied to make lime and mortar for the extensive building projects, which may have accelerated erosion, bringing about the loss of agricultural land. To this may have been added drought, weakening an already weak state and making it vulnerable to semi-nomadic neighbours.

* Rain was dramatic in the tropics. Uniquely, in that region the initial signs of thunder can be heard around the time leading up to the first overhead passage of the sun.[43]

*

As we have repeatedly seen, as one city-state, or culture, faltered in Meso- and South America, another rose to prominence. This happened after the Moche disappeared on the Peruvian coast, as a new state, Huari, appeared in the Ayacucho region of the highlands. This eventually became home to between 20,000 and 30,000 souls and was a sharply segregated community, divided into kin groups, social classes and different occupations. Each region of the city had its own residential quarters and its own plaza, for ceremonies, the elite being separated from the artisans, such as potters, who had their own area.[44]

The Huari were master engineers, who built long canals linking highland springs with steeply terraced fields enabling them to grow maize in plentiful quantities. The city lay astride a major trade route over the Andes and for this reason, perhaps, their deity was the Gateway God, adapted by them to be a maize deity also, images showing corn ears growing from his headdress. The Huari successes in engineering enabled them to colonise land that others could not make much of until, eventually, their territory stretched for 600 miles across a number of mountain basins, in each of which the elite lived and worked with the local inhabitants, but kept themselves separate in administrative enclosures walled off from everyone else. Although they had good trading links with peoples on the coast, they never colonised that area, preferring to maintain their engineering superiority in the highlands.

Huari flourished from about AD 500 to 900 after which it declined, once more, perhaps, as a result of internal revolts.

The Pueblo Phenomenon

In North America ('Turtle Island' to many indigenous Indians), no civilisation of the kind that flourished in South or Mesoamerica ever existed. One reason, as was discussed in an earlier chapter, was that wildlife was so abundant that earlier life-ways – hunting and gathering – were sufficient and efficient ways of existing for as many as 800 tribes.[45] A second reason may have been that the New World grain, maize, did not reach North America until relatively late: it took time for strains that had emerged in the tropics to adapt themselves to the very different conditions of North America. A third reason was that, in many areas, water was a problem and this further inhibited the development of maize agriculture.

Nonetheless, by the time of the millennium we are considering – AD 1–1000 – in what is now considered to be the south-west and the Midwest, there were small towns, impressive architecture, fine ceramics and evidence of astronomical alignments. Small herbaceous plants, like spearmint and onion, were 'encouraged' with small, ad hoc canals which led from rivers to naturally occurring stands of plants. Recently also, ridged fields have been identified, so as to cope with cold environments, not unlike those found in South America. But by AD 200, maize sustained hundreds of village communities, which existed in the same place for generations, isolated but with pit houses dug into the ground.[46] Because of the extremes of temperature in North America, very hot summers and very cold winters (the 'thermal trumpet', see page 96), villages developed in a distinctive way, being constructed of a series of rooms directly abutting one another, this being a thermally efficient way of coping with such a distinctive climate. These are the structures known to archaeologists as pueblos.

The most striking of these was the Anasazi town of Pueblo Bonito in Chaco Canyon (in New Mexico, near Albuquerque). In fact, Chaco Canyon, with stark, vertical cliffs many hundreds of feet high, was home to thirteen pueblos, each with a *kiva* or ceremonial room, plus another 2,400 archaeological sites nearby of one kind or another. Collectively, these show that the canyon had been inhabited for as much as 8,000 years, but the pueblos themselves did not emerge until between AD 700 and 900 Most of them were built in a semicircular fashion, the main reason appearing to be that in this way each dwelling unit was equidistant from the sunken kiva, the focal point of ceremonial life, in which some people worshipped a plumed serpent. The kivas were sunken so as to represent the primordial underworld 'from which people emerged to populate the Earth'. According to one Hopi origin myth, 'In the beginning there was only Tokpella, Endless Space ... Only Tawa, the Sun Spirit, existed, along with some lesser gods. There were no people, then, merely insect-like creatures who lived in a dark cave deep in the earth.' According to the rest of the legend Tawa led these creatures through two levels of the world and eventually they climbed up a stalk, through a doorway in the sky into the upper world. 'The gods gave them corn and told them to place a small *sipapuni* on the floor of each kiva' (the *sipapuni* being a small hole to represent where the primordial creatures emerged into this world). Thus the kiva represents the layered structure of the cosmos, and was the place where both the business of the community was discussed (when to plant the corn, when to harvest it) and where its elaborate ceremonies were conducted.[47] It was the equivalent of the World Trees further south.

Pueblos and kivas could reflect astronomical alignments. For example, the great kiva at Casa Rinconada in Chaco Canyon has a main doorway that faces celestial north, the point around which the stars seem to rotate. It also had four post holes that housed great tree trunks symbolising the four trees that the first people climbed up to reach the world. At the solstice sunrise the first rays entered the doorway and shone into a niche that marked the northernmost journey of the sun. Here, as elsewhere in North America, Indian astronomers were also shamans, who used hallucinogens to pass from this world into the spirit world, who turned into animals (wolves in North America, rather than the jaguars of further south), and were sometimes depicted in art as flying between the realms. Human sacrifice was not common but not absent either.

Eventually nine major semicircular 'Great Houses' were built in the canyon, all in place by the eleventh century, with many being constructed next to major drainages, to benefit from seasonal floodwaters. The great Pueblo Bonito consisted of 800 rooms surrounding the semicircular plaza. The size of the undertaking can be gauged from the calculation that each room needed four entire pine tree trunks to make the beams which supported the rooms, each pine taken from forty miles away.[48]

One early mystery has been 'solved'. It was estimated that although there were a total of 6,500 rooms in the Chaco pueblos, the soil in the region could support a population of barely 2,000. It seems therefore that Chaco was, for many people, a ceremonial centre only, not unlike Monte Albán, or Chavín de Huántar, a place of pilgrimage where people came to worship but most didn't stay very long. And this puts into context the discovery of a number of tracks converging on Chaco from the outside. These unpaved prehistoric roadways, often dead straight, sometimes lined with stones, sometimes dug inches deep, link Chaco to about thirty other settlements and some are as much as 40–60 miles long, forming a network that stretches for 400 miles. This network is less well known than the Inca road system, or the lines of Nazca, but in its way is every bit as impressive.

Chaco was a trade centre as well, turquoise being the main precious substance in the area, used for ritual objects of many kinds. But the Chaco (Pueblo) Phenomenon, as it is called, extends over more than 25,000 square miles of what is now New Mexico and Colorado and appears to have been linked by an extensive ideological system, now lost, but perhaps related to ideas of sacred landscape.

The system collapsed towards the end of the twelfth century. This probably followed a period of prolonged drought that we know affected the San Juan Basin for more than half a century after AD 1100. But here too there is a lingering suspicion that the Chaco communities overextended themselves, in that at least 20,000 pine trees went into the major pueblos, and this does not include the quantities of timber that would have been used up in heating during the harsh winters. So did woodcutting strip hillsides of their tree cover, leading to soil erosion? As we have seen, it happened elsewhere.

Mesa Verde lies in the south-west region of Colorado, a three-hour drive from Chaco Canyon in north-west New Mexico. Here, on a snowy day in 1888, two cowboys searching for stray cattle came across an extraordinary site – a high rocky overhang with a cave beneath it, the cave filled with more than 200 rooms and 23 circular kivas. Later archaeological investigations revealed that Mesa Verde was begun around AD 600 with a population, to begin with, of 150, living in small pit houses. By the ninth century the community had grown and built large kivas and by the eleventh century it had a population of about 2,500 but with an additional 30,000 living nearby. The Anasazi, the people who inhabited Mesa Verde, were good engineers and built a series of ditches and reservoirs to guide and collect water. The culture flourished, to the point where, between AD 1150 and 1250, some thirty cliff dwellings were clustered in three canyons, with 550 rooms and sixty kivas. They farmed the land above the canyons but in ~1300 the Anasazi abandoned the area. Here, the most likely reason was drought, the same drought, post-1150, that wreaked havoc in Chaco Canyon.[49]

The Mogollon people, who lived in the arid and mountainous parts of Arizona and south-west New Mexico from about AD 200, were longer-lived than most – they did not die out until 1400 or even later. They lived in relatively small pueblos along the banks of rivers and stood out by the standard of their pottery, of which the best known was produced by the Mimbres people (named after a river). Each village had expert women potters who built up fine, thin-walled bowls from coils of fine river clay. The pots were fired then painted with brushes made of the yucca plant and it is the designs on these pots that have become famous, making them the finest ceramics in all of ancient North America. They show everyday life (people fishing), animals and insects, humans in a variety of clothing, flute players, dancers wearing animal masks, in some sort of shamanistic ritual. The Mogollon people were also interesting for the fact that they are known to have imported copper bells from Mexico, raising the possibility of trade links (and therefore the exchange of ideas) between North America and Mesoamerica.[50]

Not all the ancient peoples of North America built pueblos. Many others were mound builders. One of the earliest to be formally excavated was at Snaketown, on the Gila River, near Sacaton in Arizona. Although this is a desert landscape, the river was a fertile oasis attracting deer and water birds in profusion. On top of which, the Hohokam people, who occupied Snaketown from perhaps 200 BC to AD 1450, were expert irrigation engineers, building canals up to three miles long to water their fields. Some archaeologists think the Hohokam people were migrants from Mexico, which would help explain some aspects of their culture, though other archaeologists play down these ideas.

Their houses were not very sophisticated – pit houses and, above ground, structures built around poles, filled in with brush. But the houses were grouped next to central plazas surrounded by low mounds up to fifty feet across and three feet high, topped with clay and on which were built their altars and shrines. Nearby – and this is what suggests a Mesoamerican link – were two ball courts, about 130 feet long and 100 feet wide. However, the Hohokam ball courts are much less elaborate than the Mesoamerican variety (often built of sand rather than stone) and there are no signs of human sacrifice associated with the game. The trade goods at Snaketown are from the Pacific coast and the Gulf of Mexico and a variety of macaw feathers have been found, part of a dance ritual costume and they too could have come from Mexico.

Brian Fagan argues that some 'pervasive' ideas undoubtedly drifted northwards from Mexico, but that the Anasazi, Mogollon and Hohokam cultures of the North American south-west were indigenous and characterised chiefly by harsh conditions, cycles of drought and occasional plentiful rainfall which together prevented the development of elaborate civilisations. Mostly, they had shamanistic belief systems.[51]

Maize in North America

When European settlers first encountered the great earth mounds of Ohio and the Midwest, in the early nineteenth century, most of them refused to believe that Native Americans were capable of such achievements, and the idea of a 'vanished race' took hold. It was a notion that did not outlast the century, as researchers from the Bureau of American Ethnology excavated more than 2,000 mounds between Wisconsin and Florida and concluded they had *all* been built by indigenous tribes and their ancestors.[52]

The best known of these are the Adena, Hopewell and Cahokia cultures. In 1720, the French explorer Le Page du Pratz called on Chief Tattooed-Serpent of the Natchez, a group of about 4,000 people in seven villages built around a large mound. During the course of the Frenchman's visit, the chief died suddenly, so du Pratz was able to witness his funeral. He recorded that Tattooed-Serpent lay in state, dressed in his finest regalia, with his face painted and wearing a 'crown of white feathers mingled with red'. His weapons and his ceremonial pipes were placed by his side. At the main ceremony, in the temple, his two wives and six other sacrificial victims were made to kneel on ritual mats. Their heads were draped with animal skins as they ate tobacco pellets 'to stupefy them'. They were quickly strangled by chosen executioners, after which, the temple was burned to the ground.[53]

Mound-building burials appear to go back to about 2000 BC, with communities burying their dead on ridges often overlooking river valleys. These were usually associated with mythical founders of races and ancestors, ways to claim the land.

One of the main centres of mound-building came from the 'Adena' people in Ohio – of the 2,000 excavated mounds, as many as 500 are Adena mounds. To begin with people were buried singly, and corpses added to the original one. Then, around the time of Christ, simple graves were replaced by larger burial chambers, containing more than one body, and lying under circular houses. Copper bracelets, marine shells and finely carved pipes accompanied the dead. Judging from the quality of the copper goods and the elaborate pipes, these bodies were of important people who were also shamans, the people who helped organise the community.

By the early centuries AD, the Adena complex had been replaced by the Hopewell, named after a town in Ohio, in which the finely dressed dead were buried in mounds, accompanied by exotic materials from all over North America: copper, silver, quartz crystal, which had been passed along narrow trails (travelled on foot) from as far afield as Florida, the Gulf of Mexico, the Rockies and the Great Lakes.

Sometimes the mounds hint at something more than burial chambers. At Newark, Ohio, for instance, the mounds are linked by a network of circles, squares and octagons joined by avenues. As with the Nazca lines in Peru, the builders of these monuments could only have truly appreciated them if they could have seen them from a helicopter. The average mound is thirty feet high and a hundred feet across, with a volume of half a million cubic feet. It has been calculated that each would have required 200,000 man-hours of earth moving. The most impressive mound of all is the Great Serpent Mound, where a snake modelled in earth lies with its sinuous body and coiled tail on top of a low ridge. Its mouth is gaping open, as if about to swallow another large oval mound.[54]

Analysis has shown that hundreds of people were cremated and their ashes buried in the Hopewell mounds. Their leaders and their shamans received different treatment, being buried in log-lined tombs, sometimes in special burial houses, together with ceremonial artefacts that identify their clan, usually named after game animals and/or birds of prey.

Despite their elaborate burial customs and their impressive mounds, the Hopewell people subsisted almost entirely by hunting, and by gathering wild plant foods. About AD 500 their way of life went into decline, with various reasons being suspected: the climate turned colder, driving game animals elsewhere; war, after the introduction of the bow and arrow; the development of maize-and-bean-based agriculture.

This latter was certainly a major factor in the rise of Cahokia. Maize spread into eastern North America during the first millennium AD but it did not become a major staple until about AD 1000, when both maize and beans developed hardier strains and higher yields (see above, chapter eleven). After that, though, it spread rapidly. In fact, the evolution of Cahokia is possibly the only area of the New World where the Old World pattern of agriculture leading to civilisation is at all closely paralleled.[55]

Cahokia was the largest town and ceremonial centre ever built in ancient North America. It is now dwarfed by Eastern St Louis, near where it is located, but between AD 1050 and AD 1250 Cahokia occupied more than five square miles, 'a tangle of thatched houses, earthen mounds and small plazas' built in the heart of the American Bottom, a low-lying flood plain south of the confluence of the Illinois River and the Mississippi. It is here that Monk's Mound was constructed, the largest earthwork structure ever built in prehistoric North America, rising in four terraces to more than 100 feet and covering sixteen acres. This is modest by Mesoamerican standards but it still looks impressive against the flat country that surrounds it. It once boasted a large temple that would have been the focus of communal life. The plaza next to it, and various burial mounds and charnel houses were all surrounded by a high wooden palisade that effectively isolated the 200-acre central precinct from the outside world. To one side stood a circle comprised of 48 wooden poles, 410 feet across, with a central observation point for calibrating the position of sunrise at the equinoxes and solstices.[56]

The chiefs who ruled over these arrangements were buried, at least in one case, on a platform of some 20,000 shell beads, plus 800 arrowheads, copper and mica sheets, plus fifteen polished stone discs. Around him were three other men and women buried nearby, and the bodies of other relatives perhaps sacrificed at the time of his interment. Beyond that were the remains of four decapitated men with their hands cut off and as many as fifty young women aged between thirteen and eighteen lay in a pit nearby, probably strangled to death.

Cahokia was the main settlement in an agglomeration or network of nine other settlements all large enough to have their own chiefs and burial mounds, and beyond them another forty small, palisaded hamlets and farmsteads, all within the American Bottom. 'Her artisans' copper ornaments, masks and fine pots with effigies of humans and animals were known to shamans and chiefs hundreds of miles away.'[57]

In general, Cahokia and its associated satellites are attributed to the Mississippian culture, now held to extend beyond the Mississippi Valley into Alabama, Georgia, and as far south and east as Florida. Its flourishing completes this chapter: AD 900 right up to European contact. Other centres emerged, such as Moundville, with large mounds and palisaded communities overlooking the river. Burials indicate that about 5 per cent of the population comprised the elite, who lived in special areas and who, when they died, had their bones stored in special charnel houses, together with regalia that indicated their elevated status.

Mississippian societies do not appear to have ranked as full-blown states, in the Aztec or Inca sense. In the Midwest all households, even that of the chief, farmed land and fished the rivers. There were few full-time craftspeople, though every village would have found someone more adept than others at working in clay or other materials. But they do seem to have had a tribute-paying hinterland, showing that they did have power over other peoples, so perhaps this was a society beginning the transformation to statehood, which hadn't occurred by the time of European contact. Studies show that tribute was collected from no more than nine miles away. Without domesticated animals, to carry the tribute, or wheeled carts, or horses, on which warriors might have been mounted, the size of the tribute-hinterland was naturally limited.

The tribute was still important, though. Exotic luxury goods would have been incorporated into ritual activities, their sheer exoticness adding to the legitimacy and authority of the religious hierarchy.

There is some slight evidence that the general ideas of the Mississippian cultures derive to some extent from Mexico, though there are as many opponents to this notion as there are adherents. Many of the figurines show long-nosed gods, with weeping eyes and a constant preoccupation with wind, fire, sun and human sacrifice. These ideas are very old and extend over vast stretches of the southern half of North America. Weeping eyes are associated in Mesoamerica with appeals and prayers for rain and where, on occasions, young girls were sacrificed, as mentoned earlier, because their tears were believed to bring rain. Another belief system centred around the fire and the sun: it was the task of Mississippian chiefs – identified with the sun – to rekindle the fire, perhaps during ceremonies honouring the new maize-growing season.

In the end, although the Mississippi centres were located in a temperate zone, the make-up of North America, the converging nature of its mountain system, the 'thermal trumpet' (discussed in chapter five), mean that it suffered exceptionally hard winters with sharp frosts. This naturally confined the number of people the land could support, and true civilisation was simply not able to form there.

In the first millennium AD, therefore, the western hemisphere was home to many developing cultures, and several civilisations. Perhaps the most interesting thing about these civilisations, in the context of this book, was their paramount concern with social stratification, the determination of the elites to hold on to control of ritual, their obsessions with war and the ubiquity of sacrifice.

No one would deny that many Old World societies were very hierarchical. But the birth of the democratic idea, via war, and the end of sacrifice, are notable differences. We shall see later on where these differences were to lead.

• 21 •

Blood Letting, Human Sacrifice, Pain and Potlatch

We now turn to a number of practices and institutions that existed in the New World and not in the Old, and consider their significance. Several of these involve types of violence that seem extraordinary to us today. I stress *types* of violence rather than *levels* because the modern world has of course experienced unprecedented amounts of aggression and brutality. In the ancient New World it is the *form* violence takes that seems so unusual to us, moreover violence that was often directed against the self. First, however, we need to consider the context of this violence, namely the gods that were worshipped in the great Mesoamerican and South American societies and assess if, and how, they differed from deities in the Old World.

The basic Maya cosmos we have met already. It consisted of an Upper or Over World, a Middle World and an Underworld, the latter entered through caves or bodies of standing water such as lakes or cenotes.[1] The Mesoamericans believed that the sea lay under the land and formed lakes as well as the oceans. The Middle World was oriented in four cardinal directions, the principal one being the East, where the sun rose, while the West was associated with black, with night and with death. At the centre stood the World Tree, personified by the ruler, which linked all three worlds.

The sacred Maya scriptures, *Popol Vuh* and *Chilam Balam* – which we know as copies transcribed into the Latin alphabet – and the four surviving indigenous codices, contain literary narratives as well as divination almanacs and, taken together, confirm that there was in Mesoamerica an inseparable link between deities, the calendar and astronomy. Rooted probably in Olmec religion, Maya beliefs had three main elements – the cult of the jaguar, shamanism, and what is known as *nahualism*, the belief that each deity had an animal double through which it could come into direct contact with human beings by means of hallucinogenic rituals.[2] By Maya times, shamans had been supplemented by a priestly caste, in charge of worship and who also guarded the temples and, with the scribes, engaged in divination.

The most important gods were Chac Tlaloc, the rain and storm god who could be very cruel, Hunab-Ku, conceived as an abstract entity, higher than other gods and so detached that representation was impossible, and Itzámná, worshipped as the first god, the creator god (humans were first formed from maize), an important protector god, and the inventor of writing. Unlike other gods, Itzámná's role was always positive – he was never linked with death or war.

Hunab-Ku had elements similar to Old World abstract deities but he was never remotely the only god worshipped in Mesoamerica: indeed there were innumerable other Mesoamerican gods of this period and they fell into one of four categories: worldly phenomena, anthropomorphs, zoomorphs, and animals. Some gods were immutable, some aged, all had particular attributes when shown in art, though those too could change and vary. Boundaries between the different gods were not always sharp.[3]

The worldly phenomena reflected the essentially shamanistic view that mountains, rivers, caves and other inanimate objects (as we would say) were in fact alive with spiritual power. However, they did not behave like gods in the sense of having a personality and being able to interact with humans. Instead, they formed part of a sacred environment, a sacred cosmos, with human beings living out their lives within their sacred aura. Blood, clouds, lightning, maize, smoke and mist were all sacred, all shown in Mayan art by scrolls, the squiggly lines of which appear to have represented their mysterious power. They needed to be interpreted according to context.

Planets, animals, plants and death all appear as gods, mostly in half-human and half-animal or half-plant form, none of which is so very different from early gods in the Old World. The two major differences which distinguish New World religions are, first, that they changed much less radically than in the Old World. The jaguar continued to be worshipped, as did the sun and the moon, as did ideas about water and the Underworld. This may have been because of the second great difference – the greatest difference of all, and possibly the greatest difference there was between the Old World and the New. This second difference was the means by which New World peoples *accessed* their gods. We saw earlier that it was the prerogative of shamans to enter the supernatural world, usually by means of trance, and as often as not induced by hallucinogens. This meant above all that the religious *experience* was much more vivid, much more absorbing, much more *convincing* than the techniques of worship in the Old World (animal sacrifice, ritual, prayer). Social conditions in the New World changed more slowly than in the Old World, for a variety of reasons as we have seen, and will see again, later on, but the sheer *vividness* was surely one reason that beliefs in the Americas changed only slowly and with difficulty.

Which brings us to the subject matter of this chapter, for there *was* change and, when that change is considered, one can see why it took the form that it did.

Among the Mayans, trance still formed the centrepiece of the religious experience, and it remained as vivid as ever. But in addition to hallucinogens, various new means of inducing trance were evolved, means that were more dramatic, that were a better form of theatre, and therefore more suited to the elaborate public ceremonials of city-based cultures with larger populations. These developments in the New World came in the realm of violence and pain.

'Floating in Blood'

The first of these practices was bloodletting, in which individuals deliberately pierced various parts of their own anatomy with knives or stingray spines in such a way as to produce copious amounts of blood, which they would then let fall on to sacred pieces of paper, which were then burned. The most frequent areas of the body which were perforated during bloodletting ceremonies were the ear, the tongue and the penis.

Bloodletting imagery pervades Classic Maya art.[4] Archaeologically, too, there is much evidence for it – for example, stingray spines are often found in tombs, located in the pelvic region of the remains of dead bodies. They were perhaps carried on belts which perished over time.

Bloodletting appears to have been fundamental to Mayan rulership – Linda Schele and Mary Ellen Miller describe it as 'the mortar of ritual life'. Blood was let at the birth of a child, at funerals, at the dedication of buildings, when crops were planted, but most of all it accompanied a ruler's accession to the throne. The pain seems to have been integral to the institution. There are descriptions of people who were accustomed to the ritual not actually appearing to feel any pain, though alongside these are accounts of tobacco or hallucinogenic drugs being used, sometimes taken in enema form, presumably as a way of stupefying people as well as helping them acquire visions. Alongside these accounts are others whereby the *massive* blood loss entailed in these ceremonies itself brought about trance-like states, under the influence of which adherents experienced the burning smoke of the bloodstained sacred papers, ascending to the heavens in twirls, as 'Vision Serpents', sacred creatures out of whose mouths gods and ancestors would appear. The ability to summon up these images was part of the power of rulers just as shamans in earlier times had been able to enter the supernatural world and achieve contact with gods and forebears.

Even when blood was not being let, the Maya wore bloodletting paraphernalia. They wore cloth strips and knotted bows on their arms and legs, through pierced earlobes, in their hair and in their clothing. Sacrificial paper made from the felted bark of the fig tree was used first as cloth and then torn off in strips to be used in bloodletting ceremonies, before being saturated in blood and burned as offerings to the gods.

The Mayans believed that blood was the main ingredient of the Middle World, between the Underworld and the Sky Upper World, and that their gods needed human blood, carried upwards by burning. Smoke and blood were indistinguishable to them, different forms of the same sacred substance. The Mayans believed that the gods had used part of themselves to create the earth and its creatures and they had to repay this largesse by reciprocal offerings of the most important element in their bodies – their own blood.

As Schele and Miller show, some of the most dramatic representations of ritual bloodletting are found in two series of lintels on buildings at Yaxchilán, near what is now the Guatemala-Mexico border, at the southern edge of Petén. These show a series of stages in a ritual possibly having to do with the accession in AD 681 of Shield Jaguar and his principal wife, Lady Xoc. In the first scene of the sequence, Shield

Jaguar is shown wearing the shrunken head of a past sacrificial victim tied to the top of his own head. His wife kneels before him, and is depicted drawing a thorn-studded rope through the open wound in her perforated tongue. She lets the bloodstained rope fall into a basket at her feet which is already full of blood-soaked paper strips.

A later lintel shows 'the consequence and purpose' of the rite. 'The same woman, still kneeling, gazes upward at an apparition, a Tlaloc warrior, emerging from the gaping mouth of a Vision Serpent (figure 13). (Tlaloc was the goggle-eyed god of rain, fertility and water, but also of hail, thunder and lightning, who was feared as much as admired, the patron of the calendar, who demanded child sacrifice.)

In this image, Lady Xoc holds in her left hand a bloodletting bowl; in her right hand, she grips a skull and a serpent symbol. It is notable that the serpent's sinuous body rises upward through a scroll of stylised blood, 'indicating that the vision originates in the blood itself'. During the accession rite, the ruler's wife underwent a bloodletting so she could communicate with a warrior ancestor, perhaps seeking guidance.[5]

According to Schele and Miller recent psychiatric research has shown that endorphins – chemically related to the opiates and produced by the brain in response to massive blood loss – can induce hallucinogenic experiences. The Mayans are known to have used some hallucinogenic drugs but do the practices – and the vision serpents shown on these inscriptions – imply that they had discovered they could experience visions *without* the use of drugs? If so, how might this development have originated?

Bloodletting and pain were not the only devices used by the Maya. Their elaborate ceremonies were carried out against the background of imposing architecture and accompanied by music, dance, and elaborately costumed participants. A larger crowd of these participants, 'wearing bloodletting paper or cloth tied in triple knots', sat watching the accession ceremony on the terraces. According to accounts by Spanish missionaries soon after the Conquest, these participants would have prepared themselves by fasting, abstinence and ritual steam baths so that, at the climactic point in the ceremony, the ruler and his wife would appear from inside the most imposing building and, in full view of the rest of the assembled people, he would lacerate his penis and she her tongue. By means of ropes drawn through their wounds, the blood

Fig. 13 Lintel 24 from Yaxchilán, showing, at left, the ruler Shield Jaguar holding a giant torch over his wife, Lady Xoc, who pulls a thorn-lined rope through her tongue; and at right she witnesses the appearance of the double-headed Vision Serpent, a warrior emerging from the front.

Fig. 14 Stone relief of post-ball game sacrifice scene from the north-eastern wall of the South Ball Court, El Tajín, Veracruz, Mexico, AD 850–110.

was carried to the paper strips. The strips of saturated paper were removed to braziers where they were burned, creating great columns of black smoke. 'The participants, already dazed through deprivation, public hysteria and massive blood loss, were culturally conditioned to expect a hallucinatory experience. The rising clouds of swirling smoke provided the perfect field in which to see the Vision Serpent.'[6]

The final act in the sequence of Yaxchilán inscriptions shows Shield Jaguar, dressed in (cotton) armour and carrying a stabbing knife. His principal wife stands by his side, blood still oozing from her wounded mouth, but she now holds her husband's jaguar helmet and shield. She is helping him to prepare for battle, when he will be expected to take captives, which he will then sacrifice, in the final act of the accession ritual and which will consolidate his royal power.

This sequence of rituals appears to have been fairly ancient, stretching back at least to a stela dated to AD 199. This inscription further suggested that the earlier parts of the ritual occurred fifty-two days before the final sequence. In this early carving, the king – Bac-T'ul – has a vision serpent draped over his shoulder, the serpent with a flint blade forming its tail – identifying it as a sacrificial being – and with sacrificial victims, their bodies cut in half, falling down the (World) Tree that grows out of the king. This is the king making physical contact (shaman-like) with supernatural entities.[7]

The paraphernalia associated with bloodletting was itself regarded as sacred – the stingrays and flint blades used to make incisions were often decorated with the 'Perforator god'. Carvings and clay models depict the king, or soon-to-be-king, incising his penis, with a rope around his neck. This device shows that, for the duration of the ceremony, the figure is a penitent who has adopted the role of captive, the lowest rank of all in Maya society, and the person who will be sacrificed at the culmination of the ritual. Some figurines show individuals screaming in agony while bloodletting takes place, others show no such feelings. Possibly bearing pain stoically was admired though at times vessels were employed which may have contained drugs used to help ease the suffering (again, see below, this chapter). Other figurines show men dancing with large strips of bloodstained paper wound around their penises, still others wear jaguar-pelt skirts.[8] The maize god was also brought into the ceremony, underlining what was said earlier, that the Maya regarded maize as the substance of man's flesh and blood. Elsewhere, figurines which show bloodletting have lines drawn on their faces in the pattern of a skull. This suggests that this priest/shaman anticipated a particular role he would enter during the vision part of his ceremony. (We saw earlier that hallucinations induced by opiates in the rainforest could summon up visions of skeletons; page 220 above). One inscription dating from AD 350–500 is taken from the lid of what is called a cache vessel. These vessels, essentially two deep plates or bowls set lip to lip, were placed under floors during the dedication rites of new buildings, and contained offerings such as decapitated heads, flint blades, stingray spines and thorns. The blood-letting scene shown on this particular lid has liquid-type motifs in the background showing, say Schele and Miller, that 'the entire image floats in blood'.[9]

Torture, Agony, Captive Sacrifice

Records of warfare in Mesoamerica are known from as early as the fourth century AD, and the eighth century appears to have been especially disfigured by conflict. These records often show captives under the feet of their conquerors and this encapsulates an important point, that Mayan warfare (like Moche and Mixtec warfare) was fought in order to *capture* (and not kill, at least not immediately, on the field of battle) the warriors of other polities, those captives being needed in bloodletting and sacred sacrificial ceremonies. Rulers who succeeded in capturing distinguished warriors often identified themselves afterwards as 'Captor of . . .' whoever it was.[10]

The inscriptions also show that the taking of captives did not come easily but occurred only after aggressive hand-to-hand combat. The losers were stripped of their armour and finery and taken back to the city of their captors. Once there they could – if they were particularly distinguished – be kept alive for some time, being forced to take part in bloodletting ceremonies and/or tortured. Eventually, however, there was no escape: they would be sacrificed.

In the inscriptions, captives are often identified with their names written on their thighs. They are invariably depicted at the foot of the inscription, without much clothing, being trodden on or lying on the ground, or having their hair held. Sometimes they are tied up, invariably they are shown in an attitude of humiliation. In contrast the victor is shown resplendent in his cotton armour, often with a jaguar helmet, jaguar boots and leggings, a jaguar cape, and wearing the (shrunken) heads of previous victims.

In earlier times (AD 100–700), costumes of the victors were not limited to jaguar pelts but had a much greater variety of motifs and materials – using feathers, for example. From the eighth century on, however, as well as in later Mesoamerican art, 'warriors in bird suits often fall prey to those in feline costumes'. It even came to the point, say Schele and Miller, that the depiction of warriors in bird costumes symbolised defeat, an idea that was to be reflected, as we shall see, in chapter twenty-three, in the Aztec belief that the current era was begun when Tezcatlipoca, a god conceived sometimes as a jaguar, defeated Quetzalcoatl, a feathered serpent. 'Perhaps during the eighth century, the notions of cosmic cataclysm found to be prevalent among the Aztecs at the time of the Conquest were formed.' Was this linked to the widespread wars of the eighth century? And is there a parallel of sorts here with the Bronze Age/Iron Age transition in the Old World, discussed in chapter sixteen?[11]

During the classic period too, many powerful Maya kings came to believe that the position of the planet Venus could be a guide to victory in battle, and they would invoke its assistance. The dates associated with this invocation suggest that the most propitious time for battle was when Venus made its first appearance as (what we call) the Evening Star, visible after the 'superior conjunction', when Venus passes behind the sun (again, see below).

Just as bloodletting and the accession of kings was linked in Maya ritual, so too was the sacrifice of captives. On one stela from Piedras Negras (in Petén, Guatemala) we see a captive bound to a wooden scaffold with a lattice base. The captive then had his heart torn out and his body was left at the foot of the scaffold, which was covered with cloth and jaguar skins by attendants. The new king stepped over the captive – *his* captive – leaving bloody footprints on a cloth-covered ladder. He reached the throne, at the top of the scaffold, where he received recognition by the public. In other words, captive (bloody) sacrifice sealed the accession ritual.[12]

Other images show that this climax came only after protracted bloodletting rituals, not to mention outright torture. One inscription commemorates a battle that took place this time during an 'inferior conjunction' of Venus, when the planet passed in front of the sun and when, a few days later, Venus reappears as the Morning Star. This image shows king Chaan-Muan displaying nine captives and a decapitated head. It also features a figure either pulling out the finger nails of one captive, or else cutting off the ends of his fingers, so that blood streams down the captive's arms. He has sunken cheeks, which may mean his teeth already have been pulled out – again blood flows from his mouth. Other captives are shown inspecting their hands and howling in agony. Yet another figure, already dead, has cut marks on its flesh indicating, as Schele and Miller put it, the 'cat-and-mouse' torture that preceded death. Still other figurines depict captives that could have been scalped and disembowelled, while one had kindling strapped to his back, suggesting that he was about to be set on fire. Finally, in yet another stela, showing eight captives bound together, the captor/ruler is shown with what appears to be an entire shrunken human body strapped to his chest.*

Schele and Miller conclude by observing how striking it is that 'in the representation of warfare in their art, the Maya addressed no issues of material gain. Instead, they cast warfare and sacrifice in terms of ritual that upheld the cycle of kingship.' More than territory (to which they were not entirely indifferent), they needed captives, for their blood to nourish the gods.[13]

A Sport of Life and Death

The Mayan ball game was a sport that was tough to play, requiring physical hardship, and could be fatal in its consequences. It involved the skilful manipulation of a heavy rubber ball and concluded as often as not with the sacrifice of the defeated. Invented in the second millennium BC, it proliferated throughout Mesoamerica to the point where, at the Conquest, there were more than 1,500 ball courts across the region: Cortés was so impressed that he took a troupe of players to Europe in 1528.[14]

* On the other hand, it might be just a figurine sculpture.

The rules of the game and the length of the ball court varied from location to location, averaging 120 feet by 30 feet, and they were often shaped like a capital 'I', with parallel masonry walls (sometimes vertical, sometimes sloping) enclosing a long narrow playing alley of earth or stone that fitted between two end zones. Usually whitewashed or painted in vivid colours, the courts were also decorated with tenoned stone heads, of jaguars, raptors and serpents. In some sculptures of teams, they have their eyes closed, indicating, say the ethnologists, that the ball game is taking place in the Underworld, and is being played by dead people. Carvings of *tzompantli*, or skull racks, show that the heads of decapitated losers would be displayed near the ball courts to advertise the consequences of defeat.

The games were played for pleasure, for religious reasons, and as pre- and post-war rituals, even for forging alliances and legitimising rulership. According to Michael Whittington, the city of Cantona in the Puebla area of Mexico had 24 ball courts, the largest number anywhere. This site flourished between AD 600–1000.[15]

As with so much else in Mesoamerica, the game appears to have begun with the Olmec. Early on, female ball players with strong secondary sex characteristics (large breasts and painted red nipples) were known, and they had nothing special in the way of status markers – they were just ordinary people. However, females disappear from the record about 1300 BC, when the game seems to have acquired a religious significance, operated by the elite. (Another case – reminiscent of the Mixtec – of the elite appropriating religion for its own ends.)[16]

Teams consisted of from one to four players who had to control the ball without touching it with their hands. Points were scored if the ball bounced more than twice or was hit out of the court. The highlight – and ultimate aim – of the game was to hit the ball through a stone hoop tenoned into the wall, high up (maybe as much as twenty feet up). This hoop, however, was only slightly larger than the ball, so that such scoring was rare and if it occurred usually signalled the end of the game. Balls were mostly made of solid rubber that was cooked and coloured black and were somewhere between 12 and 18 inches in diameter, making them more like a modern medicine ball than a hollow basket ball. They would have weighed about 8 lbs, except in Chichén Itzá (between Valladolid and Merida, in Yucatán), or El Tajín (on the gulf coast, near Veracruz City), where the balls had hollow cores formed by human skulls inside. The bouncing ball made a great deal of noise and this too formed part of the excitement (animated squiggles on certain inscriptions suggest both sound and movement of the balls). Ball courts have been found as far north as the south-west of what is now the United States. They were usually situated at the very heart of the sacred centres of ancient cities, integrated into important temple complexes.[17]

There were several plants in the New World which produced gummy resins, including the prickly pear cactus (*Opuntia* spp.), sometimes used to make gods, or as incense during worship. But rubber, *Castilla elastica*, grew throughout Mesoamerica and in the northern Andes and was in ceremonial use as early as 1600 BC.[18] It had several applications – to haft stone axes to wooden handles (holding them together and helping to absorb shock), as the tips for ceremonial drumsticks, as body armour, as footwear, as waterproof seals for containers, as the wicks for candles (held in jars made of copal), as hollow or solid figurines; and it had many medicinal uses, for cold sores, earache, as suppositories and in treatments for fertility and urinary problems.

Many of its uses were religious and/or magical. For example, melted rubber was spattered over the pages of bark-paper books to simulate (and encourage) drops of rain that were understood to be sent to earth by the Tlaloques (tiny supernatural beings who assisted the rain god).[19] In this context, rubber was mainly associated with water gods. Several rubber objects were dredged up from the Cenote of Sacrifice at Chichén Itźá and dated to somewhere between AD 850 and 1550. Often the rubber-painted items were burned at the climax of ceremonies, producing 'a dense, sweet-smelling black smoke'.[20]

When Europeans arrived in the New World, until that point the only balls they had experience of were made of wood, leather or cloth, and so their elasticity was limited. The Europeans were astonished at the 'jumping and bouncing' qualities (as Fray Diego Durán put it) of Mesoamerican balls.

The *ulquahuitl*, or rubber tree, can grow to a height of sixty-five or eighty feet, likes the company of other trees and prefers a humid climate. The white latex sap was collected by scoring a diagonal fissure in the greyish-brown bark and waiting for the tree to 'bleed'. It was this association with blood that made rubber a sacred substance. Latex is viscous and whitish, though on exposure to air it turns thicker and changes colour, first becoming yellowish, then grey, and can finally turn black. Chemically, it is a polymer composed of carbon and hydrogen. Once collected, the latex is boiled with the juice or roots of certain other plants that grow in much the same region, notably species of the *Ipomoea* 'Morning Glory' families, which we have met before, as hallucinogens (and therefore also sacred to the ancient Mexicans), but which in this case add small quantities of sulphur to the mix, making rubber more pliable, more durable, and even more bouncy. The *Codex Mendoza* records that the province of Tochtepec (now Oaxaca) had to send to Tenochtitlán some 16,000 rubber balls twice-yearly as tribute.[21] The balls were constructed from simple lumps or in strips wrapped around each other, gradually building into a sphere. At other times the strips were rolled up like ribbons.

The religious associations of rubber balls were with movement and vitality. In the Nahuatl language there appears to be a close association between 'rubber' (*olli*) and 'movement' (*ollin*).* At the same time, the Yucatec Maya word, '*k'ik*', means both 'blood' and 'rubber', and in the *Popol Vuh* the ball in the ball game is called *quic*, a Quiché term that also means 'blood'.[22] The round shape of the ball also associated it with heavenly bodies. In several instances depiction of balls in ancient paintings have inscriptions on them, invariably associated with the deities which represent the sun at the moment it descends into the Underworld, where it will die, only to be reborn later. In other representations, the ball may be superimposed with a skull or the decapitated head of a player who lost the game.

Contestants in the game wore special protective devices, because the ball was so heavy – their arms, wrists and knees were covered – and they also wore wooden, and in some cases stone, yokes around their waists, with protuberances attached, to help guide the ball. Sculptures of ball players show they could be maimed during the game (swollen eyes, for example) and this no doubt added to the excitement.

Most of the drama, however, was surely bound up with the gladiatorial nature of the contest in which the main purpose was to try the strength of a captive and test his desire to avoid death. Among the

* The Spanish word for rubber, *hule*, is said to derive from this term.

Aztecs, for example, war captives, already weakened by deprivation, were set successively against one another until only one remained. Losers were invariably sacrificed, their hearts ripped out and offered to the gods. Occasionally, and this takes some getting used to, their decapitated heads were used as the ball. Though gruesome, the inscriptions leave no doubt that this happened.[23]

The Aztec nobility played the game themselves, for pleasure, even fielding their own teams, with heavy gambling. Lower-ranked people apparently would risk their own children as wagers, and in some cases were even willing to be sacrificed themselves if they lost. According to Spanish chroniclers, great players were esteemed and could become the consorts of kings.[24]

But mostly the game had a religious significance. This is shown by the fact that the Maya version of the game also featured in their mythology. It is described in the *Popol Vuh*, in which the ball game was played by the Hero Twins. In this legend the brothers, 1 Hunahpu and 7 Hunahpu, were the best ball players on earth. But their relentless bouncing of the heavy rubber ball when they played the game disturbed and angered the gods who lived in Xibalba, the Maya Underworld, so much so that the malicious deities lured the twins down to a game. Through deceit the gods won, and the twins were sacrificed, one being buried, the head of the other hung on a tree as prominent evidence of the gods' victory. Later, the daughter of an Underworld noble walked by the hanging head and spoke to it, whereupon the head spat into her hand, miraculously impregnating her. Outraged, her father commanded that she be sacrificed, forcing her to escape to the Middle World, and hide in the very house where the Hero Twins had grown up. There, she eventually gave birth to twins herself who, when they were older, found the ball game equipment of the original twins, learned how to play the game and this time deliberately provoked the gods of the Underworld so that they – angered all over again by the noise – invited the second set of twins to another game in the Underworld. This time the twins outwitted the Underworld gods, dismembering them before taking their own place in the Over or Upper world as heavenly bodies, identified as the sun and the moon, or the sun and Venus.[25]

This legend underlines the religious significance of the ball game, how it was conceived as yet another portal to the Underworld, how the captives taking part, though they were decapitated if they lost, were nonetheless part of a ritual which would take them to the heavens. In the Mayan manner the game was also considered a cosmic metaphor in which the ball, in being hit back and forth, replicated the movement of the heavenly bodies. The ruler, who may have started the game, was thereby seen as the agent who set the course of the heavens.

There was also a second form of the game, or a second stage, which was played against stairs, or stepped structures and was distinct from the game conducted on the courts proper. In this form, humans – possibly the losers in the games proper – were trussed up like balls themselves, and rolled down the stairs. This was surely a very painful process, and possibly fatal. Alternatively, it was the final ceremony prior to sacrifice itself.[26]

Sacrifice, say Schele and Miller, was a deliberate sequel to what took place on court, 'an integral part of the ritual ball game cycle'.[27] Glyphs associated with the ball game are often inscribed on staircases and risers. Defeated and decapitated captives are shown on inscriptions with streams of blood pouring from their necks, after their heads have been removed, the streams transformed into serpents which 'water' the land (figure 15).

In some cases the players are shown dressed as warriors, in jaguar suits defeating victims in bird suits.

There are some references to altered states of consciousness in connection with the game. At Chichén Itzá, one inscription shows a victorious player holding the head of a defeated opponent, with the rest of the loser's body kneeling in front of him. Out of the neck of the kneeling player sprout seven serpents, but surrounding the scene is a plant, possibly a species of *Datura*, well known for its psychoactive properties. Water lilies are also shown, in abundance, on objects associated with the ball game. The most obvious association of water lilies is with water and fertility, but Michael Whittington says it is also significant that the rhizomes (horizontal, underground stems, from which roots proliferate) of water lilies are hallucinogenic and this raises the idea of entering a different reality via consciousness-altering substances. Water lilies are also a violent emetic, giving them an association with ritual purity. In the ball court relief at Teotihuácan,

Fig. 15 A decapitated ball game player, whose spurting blood has been turned into serpents.

in the Tepantitla Palace, there is a depiction of Tlaloc with a water lily in his mouth and very possibly a hallucinogenic species of Morning Glory (*Turbina corymbosa*). Whittington observes that a *Datura* is probably represented in the relief on the great ball court at Chichén Itźá.[28]

Furthermore, ball game yokes sometimes show carvings of toads, especially the giant toad of Mexico, *Bufo marinus*, which contains hallucinogenic toxins. Decorated yokes first appear in the Protoclassic period bearing images of toads and felines, all identified with the Underworld.[29]

Possibly we can never recapture the entire meaning of the ball game. Were the representations of psychoactive substances evidence that hallucinogens were used by the players, either during the game, or afterwards, during sacrifice? Or were they placed there as metaphors, metaphorical decoration, reminders that the ball game was itself a portal to an alternative world?

The ball game was most widespread in the late Classic period, when more than half the courts were built. In general open courts evolved into closed ones.[30] The game passed into oblivion in Yucatán after the fall of Chichén Itźá though as we shall see it regained its popularity further north, later, among the Aztecs.

Dark Shamans and the Meaning of Pain

Bloodletting, sacrificial practices (such as hearts being ripped from living bodies), and the ball game, are the most studied aspects of organised violence and pain in the New World. For many years it was believed that such violence was practised only relatively late in the New World chronology, among the Toltecs (AD 800–1000) and the Aztecs (AD 1427–1519), for example. Scholars were not minded to explore practices elsewhere, believing that the early Spanish accounts were exaggerated by European religious zealots. Research and discoveries over the last thirty years, however, have totally transformed this view. We now know that organised violence was practised throughout all areas of Mesoamerica, from very early on, and from Costa Rica to Panama to central Mexico to Monte Albán (Oaxaca, southern Mexico), and right across northern South America – in Colombia, Nazca, Aspero, and among the Chavín, Chimu, Moche, Huari, at Tiahuanaco, throughout the Inca empire and many other places as well. According to John W. Verano, each year brings a significant new discovery.[31] Again, it is not so much the level of violence that fascinates researchers, so much as its organised nature and the specific forms of brutality that existed, and attitudes and practices in regard to the associated pain.

We now know that child sacrifice was widespread, as was mass sacrifice, as was the decapitation of captives and the taking of trophy heads. We know that the forms of killing involved the ripping out of hearts from live captives, evisceration – the removal of entrails – decapitation, strangulation and garrotting, flaying, and being burned alive. We know there was teeth mutilation, that the ears of children in the cradle could be cut and there are several images of *self*-decapitation, though whether these images recall actual events or represent ideal moments in a religious ritual isn't certain (figure 16).

It has also become clear in recent years that, in addition to the ball game, which often resulted in the sacrifice of the losers (and in some cases, in Tajín in Veracruz, for example, of the winners, on the grounds that the gods would be more likely to accept a petition from a victor), there were extremely brutal boxing matches, in which the contestants wore protective uniforms, including headgear (often in the form of

Fig. 16 Stone relief of post-ball game self-sacrifice scene from the north-eastern wall of the South Ball Court, El Tajín, Veracruz, Mexico.

jaguar masks), and fought with stone balls in their fists, or stone knuckle-dusters called *manoplas* which could inflict fatal injury.[32] Images show boxers badly bruised and deformed; sometimes they are shown fighting jaguars who also wield stone balls in their fists. Alcohol was consumed at these events by the participants, who took part, apparently, while they were inebriated (figure 17).

Such was the prevalence of these events and occurrences that archaeologists and anthropologists have concluded that pain, which has no real meaning outside the medical realm in our own day, was laden with symbolic and ideological meaning in ancient America.

Pain, in fact, seems to have been related in an almost philosophical way with certain negative events in life. For example, Steve Bourget has produced evidence for a direct association between (at least four) sacrifices and torrential rain on the north coast of Peru, which may have been El Niño events. (Archaeology indicates that many other coastal sites in South America were damaged or destroyed by water in prehistory.)[33] Among the Pachamac, the events of rain and sacrifice were also clearly related, though it is not specified as to whether sacrifice was designed to bring rain or avert it. Bourget found that there was a greater representation of sea lion remains associated with sacrifice, and with *lomas*, bands of fog vegetation appearing on certain slopes exposed

Fig. 17 Boxers in combat, showing protective clothing and stone balls in their hands.

to the prevailing winds during El Niño events. He argues that the *lomas* announced the impending El Niños and that, during the actual ENSO event itself, when the seas adjacent to the land warmed up, the normal foodstuffs of the sea lions were removed, they became angry and competed more directly with humans for food, causing great rivalry. Humans were therefore compelled to kill them in greater numbers than usual.

In any event, El Niño events changed sacrificial behaviour. The form of child (or youth) sacrifice known as *capac hucha* was often carried out at high altitudes and at least in some cases (though by no means all) the children had zigzags painted on them, as though in dedication to lightning. Boxing contests were also sometimes staged on mountain tops.[34]

Weather, it has also emerged recently, was looked upon by many in Mesoamerica as a form of illness (and therefore implicitly painful) and natural disasters were regarded as diseases writ large. Flooding, for example, was regarded as an earthly form of diarrhoea, and drought as akin to dry skin. Weather shamans understood these links (flooding, for example, may have brought on water-related diseases, quite apart from any more or less obvious analogies), and violence in one form or another (bloodletting, for example, blood being associated with rain) was recommended as treatment, for both the catastrophe and the associated disease.[35]

The phenomenon of the 'dark shaman', or *kanaimà*, has recently been identified, in which the object is to create an atmosphere of terror and control in rural communities, the maleficent shaman transforming himself into a jaguar or other animals 'for the purpose of implementing assault sorcery in order to punish and destroy an enemy'.[36] The dark shamans were believed to influence bad weather – thunder, lightning, wind, floods, droughts, earthquakes – which could cause illness. They also poisoned (and in some remote rural areas of Guyana still poison) their prey in a way designed to provoke anxiety and terror, the *coup de grâce* being an attack on the victim from behind, the victim identified as a maleficent force, in which a sharp forked stick was thrust into the anus of the target, then forcibly withdrawn, bringing with it a section of the rectum and severing it 'so as to produce a painful, wasting death'. The episode was not complete until the dark shaman had visited the grave of the maleficent target and tasted a sample of the decaying corpse, 'by means of which he destroys the victim's soul'. In many societies, jaguar-shamans were regarded as major sources of illness and, in return, if they could be captured they were themselves decapitated. Decapitation itself often played an important part in the defeat of maleficence, though it could also be associated with fertility and cosmogenesis.[37]

Dark shamanism may have been a newish development. If so, it contrasted with a very old shamanistic tradition which was still strong in Mesoamerica at this time – namely that, among the Maya, males sometimes donned female attire for bloodletting ceremonies. Much the same phenomenon occurred in Chile where males also assumed female guise to practise their craft. By the same token, Peruvian tapestries of this period, and Nazca pottery too, both show shamans turning into jaguars, condors and other animals, and (on cups) using hallucinogenic drinks. Shamanistic traditions remained strong.[38]

Still on the subject of pain and violence, there continues to be some doubt as to the exact nature of warfare in Mesoamerica. Was it the 'Western' kind, for conquest, or just to take prisoners who could be sacrificed? In the inscriptions of the Mesoamerican region there are several glyphs for war, which might indicate different kinds of battle.

We know of 'flowery wars', low-intensity operations, to take a few captives or to probe the strength of the enemy – essentially propaganda skirmishes. There were also 'destruction events' and 'shell-star' events, the latter indicating in all probability territorial conquest. Without horses, or chariots, or metal weapons, it may have been simply too difficult for certain groups to hold territory they had taken in warfare, especially when their main aim was captives for sacrifice (though, as we shall see, the Incas and Aztecs managed it). And war, as has been noted, seems to have been more common in the eighth century, possibly having to do with climatic variation.[39]

The idea that pain, violence and sacrificial killing had a different meaning for the inhabitants of the ancient New World, compared with our own day, lies in several pieces of evidence. For instance, there is evidence that families who had sacrificed their children achieved a higher status as a result; there is evidence, noted above, that sometimes the *winners* in ball games were sacrificed; other evidence lies in the fact that in the gambling that was linked to the ball games (in the Aztec world, for example), people would gamble their own lives on the result; we know also that the high priests at Tlaxcala (a state that adjoined the Aztecs) tried to outdo each other in passing a higher number of sticks through their tongues – records show that some succeeded in passing 405 sticks, others in 'only' 200;[40] there is also evidence that in some cultures trophy heads could be used as balls which 'could jump, roll and fly'. Andrea Cucina and Vera Tiesler report that, in the lowland Mayan centres, 'it was relatively easy to buy the sons and daughters of local slaves or orphans ... when these means of acquiring infants were not successful, the natives' deep faith induced them to offer their own sons or nephews.' Suicide is unfortunately common enough in our own day, but the ancient inscriptions showing people beheading themselves, if they are to be believed, are also evidence of a very different attitude to pain. Finally, given what has gone before, what are we to make of the evidence from Adriana Agüero and Annick Daneels, that no one in Veracruz (where they carried out their research) lived more than an hour's walk from a ball court? Watching painful, bloody and deadly contests was extremely popular.[41]*

* As it was of course in ancient Rome.

Many images of contestants or sacrificial victims do not show suffering – at the most they perhaps show people in a state of stupor. But other images do show people in pain, with contorted facial features, pear-shaped tears on their cheeks or cries emanating in scrolls from their mouths (see figure 18).

Fig. 18 A wounded ball player lying outside the playing field, crying (tear) and moaning (scroll).

This has all led Jane Stevenson Day, among others, to conclude that, 'Rather than destroying a community, violence actually bound its members together as co-actors in a bloody ritual performance – bloodshed, pain and death were one side of the coin, and sustenance, regeneration and life the other . . . Repetitive dramatic performances of decapitation may have reduced sensitivity to its horror and engendered a fatalistic acceptance . . . The presence of organised violence and the use of blood sacrifices for the benefit of an entire community are basic to the world view of most cultures in the pre-Columbian Americas.'[42]

The Great Star and the Wasp Star

Throughout Mesoamerica, as we have seen, there developed sophisticated calendrical systems, certain elements of which were probably inherited from the Olmec and Zapotec civilisations. This was discussed in the previous chapter but not mentioned there was the 'long count', the date that was always given at the beginning of an inscription, which usually consisted of five parts: *baktun*, *katun*, *tun*, *uinal* and *kin*. Kin stood for '1', the day (*kin* also means 'sun'), the actual unique date, reached by counting the total number of days that had passed since the date the Maya considered their 'zero year'. Under this system a *uinal* was a month of twenty *kin*, a *tun* was eighteen months, 360 *kin*, a *katun* was twenty *tun* and a *baktun* was twenty *katun* or 144,000 *kin*. Apart from the *kin*, the basic unit may have been the *katun*, 7,200 *kin* or roughly twenty years. This was not far off the length of a generation, close to the limit of individual memory and, perhaps, an average time between natural disasters – earthquakes, El Niños, volcanic eruptions.

On the basis of the *kin-baktun* system, the Mayan 'zero hour' has been calculated, as mentioned earlier, as 2 August, 3114 BC. The choice of this date is still a mystery but it may have had a propaganda value, stressing that the Mayan rulers could identify their ancestry a long way back into the remote past, adding to their authority. This date may also have served to separate what we might call historic time – the time of real people – from mythic time. Mayan mathematics envisaged other units of counting – the *pictun*, for example, was twenty *baktun* – and extended, theoretically, millions of years into the past. Such devices may have helped link the rulers of Classic Maya Mesoamerica with the gods who inhabited mythic time in the remote past and, again, cemented their authority.

Another aspect of the Mayan understanding of time concerned the role and behaviour of the planet Venus. We have already seen, in chapter twenty, that the Maya attached particular importance to the Milky Way, seeing in its broad shape the World Tree which, in the course of the night, rotated, becoming the canoe in which the first individuals were created. The Mesoamericans were fascinated by transformations of this kind, perhaps as a result of the transformations shamans underwent during trance.

The attention the Maya gave to Venus may well have stemmed from its prominence in the night sky and its habit of appearing and disappearing – just as the moon waxed and waned – in a regular cycle. The Mayans knew that Noh Ek, the 'Great Star', or Xuc Ek, the 'Wasp Star', as it was variously called, operated on a synodic revolution (its apparent 'year') of 584 days (the current measurement being 583.92 days), divided into four periods. The first period lasts 240 days, during which time Venus is the Morning Star. The second period corresponds to the superior conjunction, when the star disappears for approximately three months. In the third period it reappears as the Evening Star, for another 240 days. Finally, there is the inferior conjunction, when it again disappears, this time for two weeks. The Mayans knew that five Venus 'years' corresponded to eight solar ones, and they knew that their calculations, which were extremely complex, needed to be corrected every so often.

The significance of Venus, for the Mayans, was that it was associated with war. The glyph for 'war' was usually the sign for Venus plus the name of the city they were planning to subdue. Perhaps the fact that Venus reappeared so consistently, after periods of 'defeat' (when it vanished), accounts for this warlike association. In any event, murals painted at Bonampak (in Chiapas, Mexico) and elsewhere show victorious warriors, dressed in robes studded with symbols of Venus, beheading captives and offering their remains to the gods.[43] Copán (north-western Honduras) had a Venus temple with a slit window, through which the court astronomers could time the appearance of the Evening Star. Bathed in the light of the Morning Star (before the sun rose), Maya kings would carry out sacrifices, in so doing cementing and renewing their power by aligning themselves with the order of the heavens, which their astronomer/astrologists could predict.

The difficult-to-calculate cycle of Venus, its disappearance and reappearance, would have appeared magical to the uninitiated, maybe even more so than the behaviour of the sun and the moon (and which might account for its prominence in the records). The astronomers' – and their rulers' – ability to foretell the disappearance and subsequent reappearance of Venus, would have been an impressive demonstration of shamanic predictive/divinatory power. According to other Mesoamerican traditions, Quetzalcoatl – the Toltec and Aztec god whom the Maya knew as Kukulkan – was transformed into the Morning Star after being expelled from Earth. Nor did it go unnoticed that Venus, in its Morning Star period, rises just before the sun, as if announcing its arrival.

In fact, in Mesoamerica entire cities were oriented astronomically. Tenochtitlán is a good example, where excavations have shown that the *Templo Mayor*, the Great Temple, was situated so that the rising equinox sun shone its light into a notch between the twin summits of the main building. Once the sun had arrived, the town crier would signal the time to begin the first rituals of sacrifice that accompanied the start of each of the eighteen months of the year. Many animal victims, in wooden cages, together with human sacrificial victims, often bound or wearing goggle masks associated with the storm deities, were found in the excavations, laid out in an east-west direction.[44] These were complicated rituals, incompletely understood by archaeologists, but they seem to link fertility, maize symbolism, warfare and sacrifice in a sacred landscape, of which the Mesoamerican city was fully a part. The calendar was central to these rituals.

Writing as Propaganda

Research into the writing systems of the New World now recognises that there were four civilisations in which some form of script developed – the Zapotec, the Mixtec, the Maya and the Aztec. In all cases, these writing systems were essentially hieroglyphic: nowhere in the New World did writing similar or equivalent to cuneiform develop, nor was an alphabet ever introduced in pre-Columbian America. As we shall see, this was because the *purpose* of hieroglyphic writing is, in general, different from that of other forms.

We cannot be certain that Mayan writing – the best understood – developed from earlier forms, Olmec or Zapotec, but it seems likely, if only on the grounds that some of the symbols overlap and that writing emerged nowhere else in the New World.

In general, Mesoamerican scripts are written in vertical columns and are meant to be read from left to right and top to bottom. Symbols are known as glyphs, from the Greek *gliphein*, to engrave or carve. Each glyph has a central element, surrounded by qualifiers. The system proved difficult to decipher partly because epigraphers and linguists were more used to deciphering Indo-European languages, which those of Mesoamerica are not, and partly because it took a while to realise that the script was 'mixed' – that is, some glyphs were logograms (where each symbol expresses a complete word), some are ideograms (where each symbol expresses a thing or an idea), and still other glyphs are syllabic. In addition to this, the Mesoamericans liked to express each word in more than one way.

Maria Longhena, for example, explains how the glyph for 'jaguar' (*balam* in Maya) may be represented either by a drawing of the head of a jaguar, or even the whole silhouette, or by writing the three syllables, *ba-la-ma* (in which, just to confuse matters, the last 'a' was silent).[45] A further possibility was to place the jaguar's head between the glyphs for '*ba*' and '*ma*'. Still other complications arose because some words had quite different meanings in different contexts – *chan*, for instance, could mean 'snake', 'sky', or '4'.

At the present stage of research (this is also a relatively new field) logographic syllables seem to be more common than syllabic-phonetic symbols. Maya word order was also slightly different: 'He captured – Jewelled Skull – Bird Jaguar' would be, in English: 'Bird Jaguar captured Jewelled Skull.' With these provisos, however, the Mesoamerican hieroglyphs, in having a main idea surrounded by various qualifiers, were not at all dissimilar in principle from Egyptian hieroglyphs.

Where the differences arose were in the uses to which writing was put in these areas of the New World, which set them quite apart from cuneiform writing in Mesopotamia which, as we have seen, began for economic/trade-related reasons.

Joyce Marcus, in her recent magisterial survey of four Mesoamerican writing systems, shows that our understanding of the inscriptions has gone through several phases.[46] At one time, when decipherment was in its early stages, researchers were impressed by the large number of calendrical dates (see above), leading some to think the Mesoamericans worshipped time. Then, around 1949, it was realised that many of the personages referred to in the writing systems were real people – rulers, not gods – and so the prospect of a proper Mesoamerican history was opened up. But Marcus's message is: 'not so fast.' Her view is that Mesoamerican societies were not really literate at all, not in the modern Western sense. Instead, her argument is that in the four Mesoamerican cultures she studied, the rulers had a monopoly on truth, they made a distinction between 'noble speech' and 'commoner speech' and literacy helped distinguish the ruling class from commoners.

She says that from the moment rank societies or chiefdoms arose, there was 'intense competition between elites . . . Reinforced by mythologies that provide separate origins for low-status and high-status individuals, such competition led to endemic raiding between villages, and even to conflict among members of the same high-status families . . . Mesoamerican hieroglyphic writing grew out of this competition for prestige and leadership, and almost immediately became one of the vehicles by which competition escalated . . . Mesoamerican writing did not appear until societies with leadership based on hereditary inequality already existed . . . '[47] She says this explains why the earliest texts show slain enemies, why place names do not appear until there were actual states. She says, plainly, that writing in Mesoamerica was devised as a form of propaganda, in a society where myth, history and propaganda were more or less the same thing, much as had been the case in ancient Egypt, but was much less true in Mesopotamia.

She distinguished two types of propaganda – vertical and horizontal. Vertical propaganda was contained on major public monuments, in which the elite sought to convince the commoners of their legitimacy to rule (details of battles with many captive being taken and sacrificed, skull racks showing the price to be paid for not accepting the legitimacy of the elite). Horizontal propaganda, on the other hand, was directed from elite to elite and was much more intimate in nature – codices, for example, and genealogies, designed to impress on others a particular family's claim to succession, many examples of which, Marcus shows, contained falsifications designed to improve the status of the people who commissioned the inscriptions.

In addition there was *agitation* propaganda, designed to destabilise the claims of the current incumbents in a society, and *integration* propaganda, with the opposite intent, designed to increase the stability of the regime.

Overall, Marcus concludes, there were eight themes most often addressed by pre-Hispanic texts across the four cultures with this form of writing: identifying vanquished rivals; identifying the limits of the ruler's political territory; identifying the conquered places paying tribute; the genealogical right of the ruler to rule; the date of their inauguration; their marriages to important spouses; the birth of their heirs; and the various honorific titles they could claim. On top of this, Marcus shows the systematic nature of the falsifications in the texts: one ruler of Palenque claimed to be the descendant of a woman who took office when she was 800 years old, and who had given birth when she was over 700 years old; Aztec documents show that Xolotl ruled for 117 years and Tezozomoc for 180 years.[48] A statistical analysis of 1,661 Mixtec male nobles' names shows that they deviate significantly from the pattern that would be obtained if every one was named after the day of their birth, as nobles were supposed to have been. This suggests that their birth dates were deliberately falsified to coincide with what were felt to be propitious dates.[49]

Mesoamerican societies were not literate, Marcus concludes, 'because literacy was not the goal'. The important distinction in

Mesoamerica was not between myth, history and propaganda but between noble speech and commoner speech. 'Noble speech, like the *ma'at* of the Egyptian pharaoh, was by definition true no matter how improbable. Commoner speech was confused, uninformed, full of falsehoods. Hieroglyphic writing, therefore, was the visible form of noble speech.' Put another way, writing was a skill used to maintain the gulf between ruler and ruled – this was accomplished by a myth of separate origins. It was broadly similar to the Egyptian idea of 'a great culture', a court-created tradition that 'became the instrument of royal rule'. And there is a strong echo here, of course, with the Mixtec elitists' linking of war and religion, as outlined in the previous chapter.[50]

This puts into context the Mesoamerican calendar. Scholars, Marcus says, should acknowledge the *potential* accuracy of dates in the Long Count system 'without claiming that the system's *goal* was accuracy.* After all, accuracy could hardly be the goal of a calendar that had someone giving birth at 754 years of age and of royal ancestors taking office at 815.' She quotes one telling example where the age of a ruler, determined osteologically, differed markedly from what the written record claimed. In still other cases, rulers took office before humans were known to have entered the New World.

Marcus's work is a salutary reminder that New World epigraphic research is still a relatively new field. But her analysis does suggest that the phenomenon of writing in Mesoamerica was somewhat different from what occurred in Mesopotamia and which led eventually to alphabetic script. Both share the term 'writing', which is accurate enough as far as it goes. But the difference it conceals is, arguably, more revealing.

Fighting with Food

Outside Mesoamerica, in what is now the continental United States, there is one other institution unique to the New World that we need to consider: potlatch. We need to be particularly careful what we say

* The Greek 'Antikythera' mechanism, discovered on a wreck dating from AD 87, had an assembly of toothed gears, nineteen rotators of the main wheel coinciding precisely with 235 rotations on another, just as 19 seasonal years fit precisely with 235 lunar synodic months. This would seem to be a mathematical equal to the Mayans.[51]

here because a similar institution has existed at times in Oceania and, moreover, potlatch appears to have been confined to a relatively small area – the north-west Pacific coast – before European contact, which helped it spread quickly into some other adjacent areas.

Potlatch is a Chinook jargon term meaning 'give'. In a potlatch, hosts give gifts of wealth to formally invited guests in amounts that, to many early (European) observers and missionaries, seemed wasteful or even economically harmful. Archaeological evidence shows that the ceremonies may have begun as early as 1500 BC in the Strait of Georgia region, between Vancouver Island and mainland Canada, with the practice originally extending along the north-west coast from the Eyak, near the Copper River in Alaska, to the Chinook, near the Columbia River in Oregon, and among many tribes in the adjacent parts of the Arctic, sub-Arctic and plateau (see frontispiece map in the following reference).[52]

Early on, Franz Boas, the famous early twentieth-century anthropologist, working with the Kwakiutl on Vancouver Island, British Columbia, described potlatch as a way to acquire rank and as an investment which produced interest because potlatchers knew that the recipients of their largesse would, at a later date, have to repay their hospitality with even more largesse. But this seems to have confused two processes. In planning for potlatch, the hosts might borrow gifts, which had to be repaid with interest, but the potlatch itself did not need to be repaid with interest, or even repaid at all. On this basis, others disputed Boas's conclusion, arguing instead that potlatchers gave out of pride, not greed or in an attempt to 'break' their rivals through increasing amounts of giving.[53]

In *The Gift: Forms and Functions of Exchange in Archaic Societies* (1967), Marcel Mauss concluded that potlatch was a stage in the evolution of society, in effect an aberration, 'the monster child of the gift system', in which, as Ruth Benedict put it, the societies were dominated by a single 'megalomaniac' idea, with groups trying to outdo each other in the giving and even the destruction of ever-larger amounts of property. It was, in some ways, a very unusual form of conspicuous consumption.[54]

Most anthropologists now seem to accept that potlatch is related to rank, that it is a way for individuals to express their status and self-esteem, emphasising social and economic hierarchy, and at the same time helping in the distribution (and redistribution) of goods. There is also some evidence that potlatch was related to the decline of warfare, a substitute for physical conflict, a case of 'fighting with property'. The more considered view now (although the practice varies from tribe to tribe) is that potlatch serves to identify publicly the membership of a group and, within that, the status of its leading members. 'Competitive potlatching' does occur but only when two clan members claim the same social rank. The practice has more recently been found to extend outside the Pacific north-west coast area, even to circum-Pacific rim areas, but whether this spread post-dates the Conquest isn't certain.[55]

What links potlatch with the other practices described in this chapter – though perhaps not too much should be made of this – is the concern with status, with rank. With the emergence of chiefdoms, and early states, status in the New World became a major concern. Warfare, or a concern with warfare, was widespread but in a world without the horse, or the wheel, wars were localised and, because all transport was on foot, it was much harder to hold on to territory gained in conflict (when that was the aim, which it wasn't always). Polities therefore remained small and people did not, by and large, exchange ideas about each other's gods. Everyone kept much more to themselves and there were no large-scale conflicts over wide areas, as there were in the Old World, stemming from the wide distribution and changing environmental conditions of the horse-riding pastoral nomads. There were no 'international' languages, like Arabic or Latin, to allow people to move freely across wide areas.

There is one other reason why potlatch deserves its place in this chapter. Its very existence shows that food was abundant on the north-west Pacific coast, thanks to the teeming salmon rivers discussed in chapter 14. This contrasts with the Mesoamerican civilisations, where the differences in height and diet which separated the nobility from the commoners shows that food, and food surpluses, were much more of a problem and a factor in hierarchy. This serves to remind us that 'civilisation' is not always the 'advance' it is made out to be. It also helps us understand the painful religious rituals of the Mesoamericans which were designed, as much as anything, to maintain social divisions and reinforce the privileges of the rulers and nobility. Pain, on this understanding, was a form of propaganda.

Possibly the most important point to be raised in the last two chapters is the link between religion, war and elite status. The lengths that the New World elites went to, in order to maintain their privileged status, even pursuing warfare so as to manipulate threat levels, is extraordinary and distinctive. As introduced in the previous chapter, this seems related to the basic nature of religion in the New World: the widespread threat of 'supernatural' forces, such as earthquakes, volcanic eruptions, hurricanes and so on which were impossible to control, whereas war, to an extent, was. This trajectory is taken further in chapter 23.

• 22 •

Monasteries and Mandarins, Muslims and Mongols

In chapter 18 we explored how monotheism evolved in the Old World and what some of the more important consequences were, producing phenomena which never occurred in pre-Columbian America. In the early development of monotheism, the experiences and ideas of the Israelites – the nomadic pastoralist Hebrews – stood out. However important that development was, historically speaking, many historians now take the view that it was the advent of Christianity – the birth, death and resurrection of Jesus – that was, intellectually speaking, the more important development in the long run. They argue this because, according to them, the Christian concept of God, although it may have begun as a new way of trying to be Jewish, became over time even more radically different from anything else that had gone before. It was, on this account, and according to a raft of figures which includes St Augustine, St Thomas Aquinas, A.N. Whitehead, Henri Pirenne, Ernest Gellner, John Gilchrist, David Lindberg and Rodney Stark, why Europe forged ahead in the millennium and a half leading up to Columbus's discovery of the New World.

On this view, theology – as we understand it in the modern world – can exist only where human beings embrace the view that there is but one God and that He (invariably a He) is a *rational* creature, as understood by the Greeks. On this view, the modern world – the world that discovered the Americas – is made possible only by an environment in which classical Greek rationalism is taken over by a Christian understanding of a rational God.

Rodney Stark has put the case most bluntly in his recent book, *The*

Victory of Reason: How Christianity Led to Freedom, Capitalism and Western Success (2005). He begins by arguing that the gods of polytheism 'cannot sustain theology because they are far too inconsequential'. Indeed, theology cannot exist unless we accept that God is conscious and rational, a supernatural being of unlimited power and scope, 'who cares about human beings and imposes moral codes and responsibilities on them'. Only in this way do such questions arise as why sin is allowed to exist or when does an infant acquire a soul.[1] To underline his case, Stark argues that there are no theologians in the East. Taoism, Confucianism and Buddhism, for example, envisage 'the way' or 'nirvana' but these are unsatisfactory phenomena, he says – one might meditate 'for ever' on such an essence but there is little to reason about, whereas in the West, as David Lindberg put it, God is the epitome of reason. Ernest Gellner agreed, adding that although the development of the 'Transcendental' was a wholly new stage in human thought, it yet required a figure like Jesus, a divine intervention in history, because 'men may dread a high god, but they will not quake before a high concept ... abstractions will not inspire awe'.[2] This is the role of miracles, so much more important in Christianity than other religions. So too is free will.

Only with Christianity, beginning with St Augustine in the fifth century, do we encounter the idea that God has given us free will. This is an enormous innovation because at a stroke it vitiates such ideas as astrology and fatalism. As a direct result, new doctrines are allowed to emerge on the basis *of reason*. As Clement of Alexandria warned in the third century, 'Do not think that we say these things are only to be received by faith, but also that they are to be asserted by reason. For indeed it is not safe to commit these things to bare faith without reason, since assuredly truth cannot be without reason.'[3]

Christianity stands out because, unlike Muhammad or Moses, 'whose texts were accepted as divine transmissions and therefore have encouraged literalism', Jesus wrote nothing, so that the church fathers were required to *reason* about the meanings of his remembered sayings: the New Testament is an *anthology*, not a unified scripture, like the Qu'ran. As a direct result, a 'theology of deduction' began with Paul, who accepted that 'our knowledge is imperfect and our prophesy is imperfect'. This is to be contrasted with the second verse of the Qu'ran, which proclaims itself 'the Scripture whereof there is no doubt'.[4]

What this implies, therefore, is that, from the beginning, Christian theologians subscribed to the view that the application of reason can yield an *increasingly accurate* understanding of God's will. This is a crucial difference between Christianity and other faiths, what Gellner called its 'socially fertile anxiety'.[5] Augustine, early on, noted that although there were 'certain matters pertaining to the doctrine of salvation that we cannot yet grasp ... one day we shall be able to do so', a view which still obtained in the thirteenth century when Gilbert of Tournai wrote, 'Never will we find truth if we content ourselves with what is already known ...' It followed, therefore, on this account, that Christians, much more than those of other faiths, were committed to progress through rational means, a doctrine which reached its apex in the *Summa Theologiae* of St Thomas Aquinas, published in Paris in the late thirteenth century. This attempt to arrive at 'logical proofs' of Christian doctrine assumed also that God was the epitome of reason, one of the important side-effects being that the bible 'is not to be understood always literally'.[6]

Throughout the Middle Ages, therefore, the idea flourished that a rational God – a god who favoured an orderly and unified Nature – gradually revealed himself *as humans gained the capacity to better understand.* Indeed, given that the universe is God's personal creation, it '*necessarily*' had a rational, lawful, orderly stable structure, *awaiting* increased human comprehension. This was the key to everything that happened in Christendom and is why, according to scholars like Whitehead, Nathan Sivin (University of Pennsylvania) and Joseph Needham (University of Cambridge), real science arose only once, in Europe. 'Only in Europe did alchemy develop into chemistry. Only in Europe did astrology develop into astronomy.' Whitehead was just one who said the 'widespread faith in the possibility of science ... [was] derivative from medieval theology ... [that] there is a secret, a secret which can be unveiled ... It must come from the medieval insistence on the rationality of God.' Whitehead insisted that the gods of other religions, especially in Asia, 'are too impersonal or too irrational to have sustained science'. On this account, science arose as the handmaiden of theology.*

* This somewhat overstates an interesting argument: for example, Christians continued to believe in the virginity of Mary, whose popularity peaked in the twelfth century; and there was no shortage of adherents who opposed other achievements of rationality, such as those who opposed Copernicus's findings.

In China, for example and in contrast, the mandarins prided themselves on following 'godless' religions, where the supernatural was understood as an 'essence' or 'principle' governing life – such as the Tao. In marked contrast to Christian theologians, Joseph Needham, in his massive Cambridge-based study of Chinese science, concluded that Chinese intellectuals 'pursued "enlightenment", not explanations'. In China, 'the conception of a divine celestial lawgiver imposing ordinances on non-human Nature never developed.'[7] Mott T. Greene agreed: 'It is a characteristic of enlightened knowledge that it cannot be put into words.'[8]

Likewise with the Greeks where, says the historian A.R. Bridbury, such figures as Parmenides conceived the universe as being in a static state of perfection. Platonic idealism, in particular, acted against the idea of change or innovation, and much the same was true of Islam, where all attempts to formulate natural laws were condemned as blasphemous, because they denied Allah's freedom to act. Moreover, as Caesar Farah put it, Arabic intellectuals treated Greek learning almost as holy scripture, to be believed rather than tested and investigated. This too was in marked contrast to Christian scholars who, by the time they encountered Aristotle and other Greeks, had, as Nathan Sivin and Geoffrey Lloyd have shown, acquired the habit of disputation.[9]

Politically and morally (psychologically), Christianity focused on the individual – the idea that sin was a personal matter promoted individualism and at the same time the idea of moral freedom, that actions have consequences. 'Most important of all was the doctrine of free will. Unlike the Greek and Roman gods, who did not concern themselves with human misbehaviour (other than failures to propitiate them in an appropriate manner), the Christian god is a judge who rewards virtue and punishes sin ... The God that treats all equally is fundamental to the Christian message: all may be saved.' Pope Callistus (died 236) had been a slave.

Another important corollary of the doctrine of free will was that, in theory at least, theologians could propose new doctrines without provoking charges of heresy. Christianity, of course, was not always as tolerant as this makes it sound but, at the same time, 'To say that sages or saints in times past may have had imperfect or limited understanding of religious truth is rejected out of hand by Buddhists, Confucianists, Hindus and Muslims.'[10]

Religious Capitalism

A second – and perhaps still surprising – influence of Christianity was on the development of capitalism, beginning early in the ninth century. This was an achievement of Catholic monks in the great monasteries who made a number of innovations despite having put aside worldly things. Their motivation was a concern to ensure the continued economic security of what were becoming vast monastic estates.

Augustine had been one early source who had taught that wickedness was 'not inherent' in commerce. As with any activity, he said, it was up to the merchant to live within the church's teaching. But by the ninth century, as a result of several technological changes discussed immediately below, monastic estates were no longer operating on subsistence agriculture. Rather, they became increasingly specialised and found themselves able to sell their produce at a profit, which brought with it a cash economy. They found a ready outlet for the reinvested profits with noble families, to whom they became, in effect, a form of bank. And it was this financial creativity of the monasteries, what the American sociologist Randall Collins called 'religious capitalism', which turned them into an institution that provided the backbone of the medieval economy.

It was helped by the fact that, throughout the medieval era, the church was by far the largest landowner in Europe, its liquid assets and annual income far surpassing not only those of the wealthiest king but probably that of all Europe's nobility added together. As well as receiving many gifts of land, as endowments, the orders reclaimed still more poor land, by drainage, resulting in extensive property holdings over a large area. Cluny may have had a thousand priories by the eleventh century but it was far from being the only behemoth: it was by no means uncommon for monasteries to establish fifty or more outposts. As Paul Johnson has pointed out, many Cistercian houses farmed 100,000 acres or more.[11]

Some monasteries specialised in wine, others in fine horses, others in grain, still others in sheep and cattle. On to this early system were grafted three more important developments. One, a meritocratic management system appeared with talented administrators able to make plans for the future. Next, there was the all-important shift from barter to cash – a transformation that occurred mainly in the ninth century.[12] Third, there arose the phenomenon of credit. Cash and interest were easy enough to understand and calculate and the historian Lester K. Little has shown how a monastery like Cluny, for example, was able, in the eleventh and twelfth centuries, to lend huge sums to Burgundian nobles. In turn this gave rise by the thirteenth century to mortgages, a system in which the borrower pledged land as security, with the lender (the monastery) collecting all income on that land during the course of the agreement. In the event of default, the monastery acquired yet more land.

Nor was this all. On top of everything, the monks offered their liturgical services, for which they also charged, further adding to their income, 'raising it to luxury levels', says Lutz Kaelbar. (Henry VII of England, for example, paid for no fewer than 10,000 masses to be said for the state of his soul.) One knock-on effect of this was that the monasteries were now able to hire their own labour for the fields they owned so that, by the thirteenth century, 'many monasteries resembled modern firms' – well administered and quick to respond to changes in the market or in technology. This in itself was a marked evolution in ideology.

Just how important this change was may be seen by comparing the Christian attitude to work that emerged in the monasteries with, say, the Chinese mandarinate, 'who grew their nails to fantastic lengths to emphasise they did no manual work'. The developing Christian attitude was supported by doctrinal advances – Aquinas, for example, declared that profits were 'morally legitimate', and even found time to justify interest rates. In other words, as Stark puts it, the church made its peace with capitalism, mainly because the church's own institutions were doing so well out of it.

It would be truer to say, probably, that the church *re*invented capitalism, in that Rome at its peak had been a successful fully monetarised economy; and although many scholars have stopped using the Dark Ages as a concept, there seems no doubt that standards of living did drop markedly between AD 400 and 900. (A hoard found at Hoxne in Suffolk in Great Britain from the late Roman Empire had 14,000 coins, while the Sutton Hoo treasure a few miles away and dated to two centuries later had 40.)[13]

If the church had moved a long way in terms of economics, it also played an important role in providing a moral basis for democracy, certainly far beyond anything envisaged by classical philosophers. Bernard Lewis, the great Western scholar of Islam, has conceded that the idea of a separation of church and state 'is in a profound sense Christian', something that could not exist in the Muslim world. 'In most other civilisations religion was so much an aspect of the state that rulers often were regarded as divine.' But Jesus himself, in the New Testament, stipulated the separation of church and state: 'Render therefore unto Caesar the things which are Caesar's; and unto God the things that are God's.' Paul further argued that Christians should always obey secular rulers, unless ordered to violate a commandment. Augustine believed that states were necessary to maintain an orderly society, but still lacked overall legitimacy – as he put it, 'What are kingdoms but great robberies?'[14] The divine right of kings was nowhere endorsed by the church and, in fact, by insisting on the secularity of kingship, the church improved its position and yet made it possible to highlight the natural history of worldly power, of what would come to be called *realpolitik*.

There was one final factor concerning the interaction of the church and the European landscape that would prove fundamental. As the economic historian Eric Jones (among others) has pointed out, Europe's physical layout – mountains, many rivers, peninsulas, islands, inland seas – resulted in a continent with a host of small political units: during the fourteenth century there were about a thousand independent statelets in Europe which had three important consequences. In the first place, it made for weak rulers alongside which the centralised papacy was welcomed as a strength (see immediately below). Second, this great diversity and endemic weakness made for intense creative competition. Third, the great number of political entities meant that there were many other places people could go if they couldn't get on where they were. It was a state of affairs that made for at least some very responsive governments. Well-led, smaller statelets played off imperial, papal and Byzantine ambitions, added to which the independent developments in trade produced a wide *dispersion* of power, all of which strengthened the hold of the nobility, clergy and military elements, of the traders, bankers, manufacturers and workers guilds rather than centralised monarchs with top-heavy bureaucracies. Thus arose the twin developments of post-religious capitalism and an early form of democracy, most vividly seen in the Italian republics of Venice, Milan, Florence, Lucca, Pisa and Genoa.[15]

A Supranational Entity

The other side of the coin, in such a disparate Europe, was the role that the Christian Church played in the vital *unification* of the continent. At the time, the name Europe (Latin: *Europa*) was rarely used. It was a classical term, going back to Herodotus, and though Charlemagne called himself *pater europea*, the father of the Europeans, by the eleventh century the more normal term was *Christianitas*, Christendom.

The early aim of the church had been territorial expansion, the second had been monastic reform, with the monasteries – dispersed throughout Christendom – leading the battle for the minds of converts. Celibacy ensured that hereditary priestly castes did not emerge, as happened in India, for example, and certain New World societies. Out of all this arose a third chapter in church history, to replace dispersed localism with central – papal – control. Around AD 1000–1100 Christendom entered a new phase, partly out of the failure of the millennium to provide anything spectacular in a religious, apocalyptic sense, partly as a result of the Crusades which, in identifying a common enemy in Islam, also acted as a unifying force among Christians. All this climaxed in the thirteenth century with popes vying with kings and emperors for supreme control, even to the point of monarchs being excommunicated.

Around and underneath this, however, there developed a certain cast of mind, which is the main interest here. The problems of the vast, dispersed organisation of the continent-wide church, the relations between church and monarch, between church and state – all these raised many doctrinal and legal matters. Because these matters were discussed and debated in the monasteries and the schools that were set up at this time, they became known as scholastic. The British historian R.W.S. Southern was most intimately involved in showing how Catholic scholars, as a 'supranational entity', aided the unification of Europe.

The role of the scholars was immediately obvious in the language they used – Latin. All over Europe, in monasteries and schools, in the developing universities and in bishop's palaces, the papal legates and nuncios, educated men everywhere, exchanged views and messages in the same language. Peter Abelard's enemies perceived his books to be dangerous not only for their content but for their reach: 'They pass from one race to another, and from one kingdom to another ... they cross the oceans, they leap over the Alps ... they spread through the provinces and the kingdoms.' Because of this, papal careers were notoriously international. Frenchmen might be seconded to Spain, Germans to Venice, Italians to Greece and England and then to Croatia and Hungary, as Giles of Verraccio was between 1218 and 1230. In this way there was in Europe between AD 1000 and 1300 a unification of thought, of the rules of debate, in the ways of discussing things and in agreeing what was important, that did not occur anywhere else on Earth. And it was not only in strictly theological matters, but was felt in architecture, in law, and in the liberal arts.[16]

Theology, law and the liberal arts were, according to Southern, the three props on which European order and civilisation were built during the twelfth and thirteenth centuries – 'That is to say, during the period of Europe's most rapid expansion in population, wealth and world-wide aspirations before the nineteenth century.' These three areas of thought each owed its coherence and its power to influence the world to the development of schools of European-wide importance. Both masters and pupils travelled from all regions of Europe to these schools and took home the sciences which they had learned. Coherence was achieved because the men who created the system, besides using one language, Latin, all used the same, ever-growing body of textbooks, and they were all familiar with similar routines of lectures, debates and academic exercises and shared a belief that Christianity was capable of a systematic and authoritative presentation and *could be improved*.[17]

What had been inherited from the ancient world was very largely uncoordinated. The scholars' aim now was to restore 'to fallen mankind, so far as was possible, that perfect system of knowledge which had been in the possession or within the reach of mankind at the moment of Creation'. This body of knowledge, so it was believed, had been lost completely in the centuries between the Fall and the Flood, but had then been slowly restored by divinely inspired Old Testament prophets, as well as by the efforts of a range of philosophers in the Greco-Roman world. These achievements had, however, been corrupted once again and partly lost during the barbarian invasions which had overwhelmed Christendom in the early Middle Ages. Nevertheless, many of the important texts of ancient learning had survived, in particular those of Aristotle, albeit in Arabic translations and glosses. It was understood as the task of the new scholars, from about 1050 onwards, to continue the responsibility of restoring the knowledge that had been lost at the Fall. '[W]hat could legitimately be sought was that degree of knowledge necessary for providing a just view of God, of nature and of human conduct, which would promote the cause of mankind's salvation ... The whole programme, thus conceived, looked forward to a time not far distant, when a two-pronged programme of world-wide return to the essential endowment of the first parents of the human race would have been achieved so far as was possible for fallen mankind.'[18]

The theory of knowledge on which the scholastic system was based – that all knowledge was a *reconquest* of what had been freely available to mankind in its prelapsarian state – brought with it the idea that a body of authoritative doctrine would slowly emerge as the years passed. By 1175 scholars saw themselves not only as transmitters of ancient learning, but as active participants in the development of an integrated, many-sided body of knowledge 'rapidly reaching its peak'. In stabilising and promoting the study of theology and law, the scholars helped create a fairly orderly and forward-looking society. Europe as a whole was the beneficiary of this process.[19]

Still another factor where Christianity affected the creation of Europe was the Crusades. To begin with, the Christians had hoped to recapture the Holy Land and convert the Muslims – by force if necessary – to Christianity. The first Crusade was proclaimed in 1095 but by 1250, when Europe discovered the Mongols, and their vast numbers and reach, not to mention their preternatural skills with the horse, and their great trading network across the steppes and deserts of central Asia, it became apparent that, as Southern put it, 'There were ten, or possibly a hundred, unbelievers for every Christian.'[20] This drove the (West European) Christians back on themselves, to study Arabic thinking and, through their translations, Greek philosophy and science. By then the Moors were in Spain, where the reconquest would play its part in what came later, in the discovery of the Americas.

The Crusades, however successful or unsuccessful they may be regarded in religious/ideological terms, did promote change in trading patterns, in particular the exchange of fine European woollens for Eastern spices and silks. The opportunities opened up by the Crusades also helped advance agriculture, mining and manufacturing in north-western Europe; it was a time of rapid urbanisation, which promoted the innovation of trade fairs. In many areas the growth of the cloth industry and banking went together: demand was high, the product was anything but fragile, could be easily replaced if a ship sank; it was a good risk.[21] The Crusades, ironically and perhaps paradoxically, helped shift the centre of gravity of Europe to the west and north.

Flanders was, in any case, highly suitable land for raising sheep, but weaving underwent a technological breakthrough in the eleventh century that increased the productivity of workers by three and five times, as the horizontal loom was replaced by the vertical one.[22] As a result Flanders' trade, mainly local to that point, began to be exported much more widely, so that by the end of the Middle Ages Bruges, and to a lesser extent Ghent, were filled with foreign traders and bankers – French, Italians from all over, Portuguese, English, Scots, Germans.

Another knock-on effect of the Crusades was to initiate a frenzy of shipbuilding. From 1104, with the building of the Venice Arsenal, states began to take an interest in building ships, which had hitherto been private concerns. Great advances were now made in the design, and size, of ships, both those with sails and those with oars. Some could even carry 1,000 paying passengers or pilgrims. Charts, compasses and astrolabes were introduced, as were convoys, to reduce the risk of losses, but these required new forms of capitalism, notably the *fraterna*, where the risk-money was shared among the brothers in the family business.[23]

As should now be clear, the so-called Dark Ages were anything but. And this too is part of the arguments put forward by such scholars as Michael McCormick, Carlo M. Cipolla, Robert Lopez and others, that the Dark Ages have been often misnamed – and during that time, and directly attributable to Christianity, Christendom's technology and science overtook and surpassed the rest of the world. A close reading of the Middle Ages, they say, shows that there was at the time an extraordinary outburst of innovation: the water mill from the sixth century; the plough from the seventh; the crop rotation system from the eighth; the horseshoe and the neck harness from the ninth. In the same way, says Carlo Cipolla, the use of the mill proliferated to other uses, from beer-making in 861, through tanning in 1138, paper-milling in 1276, to the blast furnace in 1384. In 1086, the Domesday Book recorded 5,624 mills for 3,000 communities in England. There is no reason to believe that England was technologically more advanced than the rest of Europe, though watermills naturally tended to congregate there because there were a lot of rivers in a small area. Hence wool and cloth manufacture became a major feature of England and Flanders. Most of the watermills belonged to monasteries, Robert Lopez observes, which also built an increasing number of dams, the water power they generated being used for sawing lumber and stones, for turning lathes, grinding knives and swords, fulling cloth, drawing wire and making paper. 'The idea of paper had begun outside Europe but its introduction spread rapidly in the new scholastic climate.'[24]

Wind and horse power proliferated, thanks to the invention of iron shoes and nails. The horse collar was invented, and harnesses enabling teams of horses to be used, two abreast rather than in single file. Wheel brakes were introduced, and axels that swivelled, also allowing for more flexible transport. The experience in casting bells was adapted to casting cannons, in 1325. The church promoted fish – not just on Fridays but on feast days, which totalled then about 150 days a year. As a result, artificial lakes and ponds were created, in which Cistercians were especially active. It was monks who found that the bottoms of fish ponds soon became extremely fertile so they would be drained every so often.[25]

Of course there were other factors than the church in accounting for Europe's rise. Carlo M. Cipolla, the Italian economic historian, notes that Europe may have differed from the East in having a larger proportion of the population who were unmarried, which helped avoid the break-up of estates and reduced the number of large families, both factors which helped ameliorate poverty.

It was during this time too that agriculture was transformed by the three-field system. In *The Rise of the Western World*, Douglas North and Robert Thomas argue that in the High Middle Ages – the years between 1000 and 1300 – Europe was transformed 'from a vast wilderness into a well-colonised region'. There was a marked population increase which meant that, in effect, Europe was the first region in the history of the world to be 'full' with people. This was aided by the layout of its main rivers – the Danube, Rhine and the Rhône/Saône – which led deep into the heartland. Together, these factors had a number of consequences, not the least of which was to begin a change from the old feudal structure, and to give more and more people an interest in property, in owning land. It was this wider ownership of land which would, before too long, lead to a rise in specialisation (at first in the growing of crops, then in the services to support such specialisation), then to the rise in trade, the spread of markets, and the development of a money economy, so necessary if surplus wealth were to be created, and which were the circumstances by which capitalism developed beyond the monasteries.[26]

Under the two-field system all arable land had been ploughed but only half of it planted to crops, the other half being left fallow to recuperate its fertility. The three-field system now divided the arable land into three parts. Typically, one field was ploughed and planted to wheat during the autumn, the second ploughed and planted in the spring to oats, barley or legumes, such as peas or beans, with the remainder being ploughed and left fallow. The next year the crops were rotated. This led to a massive 50 per cent rise in yield, at the same time as spreading agricultural labour throughout the year, and reducing the chance of famine through crop failure. Thanks to yet another piece of serendipity, domestic mammals – sheep and cattle in particular – could graze the fallow land and manure it, thereby helping to speed its return to fertility. Using the fallow field for grazing also had a dramatic effect on medieval economies. Manure was highly prized but sheep were the most blessed of all animals, providing milk, butter, cheese, meat, wool above all and even their skins made clothes and parchment. Fleeces in fact were the major industrial raw material in medieval times, woollen cloth industries dominating these early days of capitalism. Looms, carding and fulling machines were all either invented or improved.[27] This period also saw a change from oxen to horses as the beasts of harness, the latter being 50 to 90 per cent more biologically efficient.

These twin developments, of significantly more people having a stake in the land, together with the idea that there was no more to go around, had two psychological effects, say North and Thomas. It helped make people more individualistic: because he or she now had a stake in something, a person's identity was no longer defined only by his or her membership of a congregation, or as the serf of a lord of the manor. And it introduced the idea of efficiency, because now that Europe was 'full' resources could be seen to be limited. Allied to the increased specialisation that was developing, and the burgeoning markets (offering tempting goods from far away), this was a profound social-psychological revolution which, in time, would lead to the Renaissance.[28]

In addition to the theologians, two scholars in particular may be singled out for their contributions to the idea of the West. The first is Robert Grosseteste (*c.* 1186–1253). A graduate of Oxford, who studied theology at Paris, Grosseteste is best known for being chancellor of Oxford University. He was a translator of the classics, a biblical scholar and bishop of Lincoln. But he was also, and possibly most importantly, the inventor of the experimental method, which initiated an interest in exactness, and led in turn to a concern with measurement which produced a profound psychological and social change, which occurred first in the West in the thirteenth–fourteenth centuries. Exactitude was helped by the introduction of eyeglasses which came into use in 1284 and had a marked effect also on productivity. Hitherto many people had been washed up at 40; now they could see and use their experience.

It was at this time too that the clock was invented (the 1270s). Until then, time had been seen as a flow (helped by the clepsydra, or water clock) and clocks were adjusted for the seasons, so that the twelve hours of daylight in summer were longer than the twelve hours of daylight in winter. With the clock people could now coordinate their activities more or less exactly. (The Chinese and Muslims eschewed clocks because they secularised time and the Mandarinate/Immams didn't want that.) Now clock towers began to appear in towns and villages, and workers in the field timed their hours according to the bell that sounded the hour. In this, exactitude and efficiency were combined.[29]

The second scholar who helped to lay the fundamentals of the West was Thomas Aquinas (*c.* 1225–1274). His attempt to reconcile Christianity with Aristotle, and the classics in general, was a hugely creative and mould-breaking achievement. Before Aquinas the world had neither meaning nor pattern except in relation to God. What we call the Thomistic revolution created, at least in principle, the possibility of a natural and secular outlook, by distinguishing, as Colin Morris puts it, 'between the realms of nature and supernature, of nature and grace, of reason and revelation. From [Aquinas] on, objective study of the natural order was possible, as was the idea of the secular state.' Aquinas insisted there is a natural, underlying order of things, which appeared to deny God's power of miraculous intervention. There is, he said, a 'natural law', which reason can grasp.[30]

The recovery of the classics could not help but be influential even though that recovery was made within a context where belief in God was a given. Anselm of Canterbury (*c.* 1033–1109) summed up this changing attitude to the growing power of reason when he said, 'It seems to me a case of negligence if, after becoming firm in our faith, we do not strive to understand what we believe.' At much the same time, a long tussle between religious and political authorities climaxed when the university of Paris won a written charter from the pope in 1215, guaranteeing its independence in the pursuit of knowledge. It was a scholar at Paris, and Aquinas's teacher, Albertus Magnus, who was the first medieval thinker to make a clear distinction between knowledge derived from theology and knowledge derived from science. In asserting the value of secular learning, and the need for empirical observation, Albertus set loose a change in the world, the power of which he couldn't have begun to imagine.

Aquinas accepted the distinction as set out by his teacher, and also agreed with Albertus in believing that Aristotle's philosophy was the greatest achievement of human reason to be produced without the benefit of Christian inspiration. To this he added his own idea that nature, as described in part by Aristotle, was valuable because God gave it existence. This meant that philosophy was no longer a mere handmaiden of theology. 'Human intelligence and freedom received their reality from God himself.' Man could only realise himself by being free to pursue knowledge wherever it led. He should not fear or condemn the search, as so many seemed to, said Aquinas, because God had designed everything, and secular knowledge could only reveal this design more closely – and therefore help man to know God more intimately.[31]

Other contemporaries at Paris, Siger of Brabant, for example, argued that philosophy and faith could not be reconciled, that in fact they contradicted one another and so, if this were the case, 'the realm of reason and science must be in some sense *outside* the sphere of theology' (italics added). An important break had been made.

In his recent account, Stark – building on Whitehead, Lopez, Cipolla, Lindberg and others – insists that the three elements discussed above – Christian ideals and accumulated layers of Christian thought, many small political units and, within them, diverse, well-matched interest groups – 'occurred nowhere else in the world, and created a political and intellectual freedom that were the necessary preconditions for the development of the modern world'.[32]

The Zone of Turbulence Again

This is not the whole picture. There is no shortage of scholars who argue that, by the eleventh, and even more by the twelfth century, many disparate parts of the Old World were already integrated into a system of exchange from which all apparently benefited equally. In region after region across Eurasia, for example, there was by that time an efflorescence of cultural and artistic achievement: Sung celadon ware in China, turquoise-glazed bowls in Persia, furniture inlaid with gold and silver in Mamluk Egypt, cathedral building in west Europe, the great Hindu temple complexes in India. On this version, Europe until this point was the least developed region of the Old World and perhaps had the most to gain from the new links being forged.[33]

Moreover, between AD 1250 and 1350 an international trade economy was developing that stretched all the way from north-western Europe to China. The main goods traded were cloth – silk, wool, linen and cotton – and spices grown mainly in the east, from India onwards, and which were used to flavour the meat diet of the cultures further west, in a world without refrigeration. Trade was facilitated by the fact that, although people spoke many tongues, in fact Arabic covered a wide area, thanks to Islam's religious conquests from the seventh century on, as did Greek and the vernaculars of Latin and Mandarin Chinese (although Arabic didn't travel as far as Islam did, in Indonesia, for example). The currencies were not the same everywhere: silver was valued in Europe, gold in the Middle East, copper in China, but this was not insuperable either.

At this point, if any country had the lead it was, according to Janet Abu-Lughod, for example, China, whose level of metallurgy in the twelfth century would not be equalled in Europe until the sixteenth, nor its paper-making and printing technology, which were several centuries ahead of those in Europe. Still more to the point, she says, the invention of paper money and credit took place first not in the monasteries of Europe but in China, where the introduction of paper bills in the ninth century, credit, the pooling of capital and the distribution of risk began and spread, first to the Arab world, then to the Mediterranean, then to western Europe.[34]

Scholars such as Abu-Lughod argue therefore that Europe did not so much pull ahead by the mechanisms outlined by Stark – rather, they say, the East dropped behind, and it did so for three reasons. The first was the progressive fragmentation of the overland trade routes across the Eurasia steppes. In the first millennium AD, the pastoral nomads of the steppes had continued the pattern of their forebears in the first millennium BC, with repeated forays and attacks on the more settled civilisations and peoples who rimmed the great grasslands. During the previous millennium in fact there had been repeated incursions of this kind. The Huns under Attila (fifth century) had sped overland as far as Germany at the collapse of the Roman empire. Later, the Seljuks, another Turkic tribe, had pressed westward and by the twelfth century controlled virtually all of Iraq, the Fertile Crescent, and Egypt, while yet another group, the Khwarzim Turks (also twelfth century onwards), held Transoxiana. As Abu-Lughod puts it, echoing other scholars, '[This] inhospitable terrain was the place of origin for a long succession of groups that left it to plunder richer lands. From earliest times, nomadic groups poured out of this marginally productive zone, seeking better grazing land, more space, or a chance to appropriate through "primitive" accumulation the surplus generated in the more fertile oases and trading towns.'[35]

From the ninth century on, at least in theory, there were three routes connecting Europe with Asia, all of which passed through the Near East 'land bridge' and which, by the second half of the fourteenth century, lay largely in ruins, as we shall see. The northern route went from Constantinople across the landmass of central Asia; the central route connected the Mediterranean with the Indian Ocean via Baghdad, Basra and the Persian Gulf; and the southerly route linked the Alexandria-Cairo-Red Sea complex with the Arabian Sea and then the Indian Ocean.

But it was not really until the thirteenth-century advance of the Mongols, under the leadership of the self-styled 'world conqueror', Genghis Khan ('the very mighty lord'), that the northern route was consolidated. Despite the fact that, at the time, overland transport was about twenty times more expensive than sea travel, Genghis Khan and his pastoral nomads for a time guaranteed the safety of travellers on the northern overland routes.*

As has already been noted (chapter 16), pastoral nomadism is limited as a lifestyle choice, especially as the relatively infertile steppes dried out. The tolls exacted for the protection of east-west caravans suited the shape of the Mongol empire (see map 2). Despite the expense of overland travel, they soon understood that the security they provided allowed merchants at the least to *calculate* their costs more or less accurately and, moreover, these overland routes were not subject, as sea travel was, to the monsoon in the Indian Ocean and whose strongly seasonal nature could keep ships in harbour at the wrong time of the year for up to six months, adding unconscionably to the length (and therefore the cost) of journeys. The northern route was also much shorter from inland China.

And so, in the thirteenth century the Mongols opened up for a time a route across central Asia that broke this domination of the more southerly routes. The steppes were even now a barren, empty, harsh landscape, in which travellers had to take with them provisions for, say, twenty-five days' travel at a time, the whole journey taking, roughly speaking, 275 days to go from Tana, on the north side of the Black Sea Basin, to Peking in, say, seven long-haul stages. In spite of these problems, however, Muslim and Jewish traders still crossed the vast area of the steppes in great caravans, while the unification of the enormous region under the Mongols actually reduced the number of tribute gatherers along the way, and made for greater safety.[36]

To begin with, western Europe was ignorant of this area and system;

* John Larner says Bahrain to China was 70 days by sea and 274 overland from Tana, on the Sea of Azov, to Cathay.

nor were the Chinese any better informed – cotton, according to the Han Chinese, 'was made from hair combed from certain "water sheep"'. Stories of fantastic peoples, without mouths, or with faces between their shoulders, passed both ways. This was offset, gradually, by papal envoys, seeking converts, and the famous exploits of the Polos, who were given safe conduct through Mongol lands. This so-called Pax Mongolica gradually brought about a flourishing of Mediterranean-Mongol trade, the primary item of which was silk, which now reached Champagne via Genoa. Overall, the unification of much of central Asia under the Mongols put Europe and China in direct contact for the first time in a thousand years but the unintended consequence of this unification, as we shall see, was a pandemic that set back the development of a world system by as much as 150 years.[37]

When Genghis died, in 1227, his territory was divided among several successors and these factions – predictably perhaps – were soon at war with each other. Not even the calm established by Kublai Khan later in the thirteenth century (under whose safe conduct the Polos crossed all of Eastern Asia) could overcome the fighting. Arab Asia survived the Crusades more or less intact and even the capture of Baghdad by the Mongols in 1258, but it fell to Tamerlane around 1400, and the steppe route was again sundered. So the northern steppe route between East and West flourished and faded, flourished and faded, and finally fragmented. This obviously hindered and helped trade alternately, though a quite different consequence was yet to come.

The second factor which sapped the development of the East was the withdrawal of China. Chinese history on the eve of the conquest of the Americas is paradoxical. China had traditionally preferred the interior, Asian route to the west but, when the pastoral nomads threatened (this is why the Great Wall was originally built), they turned to the sea. In the early fifteenth century, a large fleet of 62 ships set out to visit the ports of the Indian Ocean and a second voyage, of 48 ships, was mounted in 1408, visiting Champa, Malacca and Ceylon. Five other missions followed between 1412 and 1430, reaching Borneo, the African coast, the Persian Gulf and, according to some, a few, the Americas in 1421. But then these visits ended, the Ming withdrew their fleet and terminated their relations with foreign powers. The Chinese by this time were far more technologically sophisticated than were the Europeans and her abrupt withdrawal, according to some authorities, had a decisive effect on the fact that the East now dropped behind the West.

There are two explanations given for this sudden and dramatic reversal. One view stresses the significance of the Confucian ideology, which demeaned worldly striving and commercial/industrial gain. The other view stresses that the Chinese elite was divided into a bureaucratic Mandarin class, which controlled the state apparatus but did not engage in (and in fact looked down on) trading and business, while the merchant class – because of this division – had no access to power. Whatever the reason for China's withdrawal, that withdrawal surely mattered.

Wool and the Plague

The third factor was the plague, between 1348 and 1351, and which affected Asia far worse than Europe, 'changing the terms of exchange because of differential demographic losses'.[38] According to William McNeill, by the start of the Christian era there were 'four divergent civilized disease pools' in the Old World – China, India, the Middle East and the Mediterranean – each of which consisted of some 50–60 million people and had reached a relative equilibrium with its environment, including endemic diseases. The relative separation of these areas from one another (and the delay in travel occasioned by the seasonal monsoons) prevented the transfer from one system to the next of 'strange' (exotic) diseases to which the populations had not built up resistance. Between AD 200 and 800, there were outbreaks of measles, smallpox and bubonic plague that afflicted China and Europe in particular because, being at the end of the transport chain, those areas had least experience of these diseases and had built up less immunity. (According to some historians, it was plague in Rome that allowed the Barbarians to attack.)[39] But gradually the Old World adjusted, according to McNeill.

Then, following the great success of the Mongols in stabilising and opening up the steppes to trade and travel, a new route was established for infectious diseases to travel by. Now, despite the expense of overland travel, a communication network capable of travelling one hundred miles a day for days on end *on horseback across the grasslands* created what McNeill called an epidemiological human web that 'in all probability' carried wild rodents of the steppelands to the Volga and the Crimea, bringing with them bubonic plague. 'Not only did the Mongols have little resistance to the disease, but their mounts [horses, camels and pack-asses] offered a safe harbor for the rapid transport of infected fleas to the burrows of underground rodent colonies in their northern grasslands. There the bacillus could survive even the ravages of winter.' It seems that in 1331 the plague travelled from China to the Crimea, where the Black Death broke out in 1346 among the armies of a Mongol prince who laid siege to the trading town of Caffa. This compelled his withdrawal but not before he had entered Caffa itself, from where the plague spread by ship through the Black Sea and Mediterranean.[40]

This was crucial because, as an inspection of the map will confirm, each of the East-West routes had to traverse the relatively narrow land bridge between the northern tips of the Red Sea and the Persian Gulf, and the Black Sea: Iraq, Egypt, Palestine, Syria. The cities of the land bridge were particularly badly affected because, being surrounded by deserts, their inhabitants could not escape. In Cairo, it has been estimated, the death rate in 1348 and 1349 was 10,000 a day. This was despite the fact that the pre-Islamic Arabs had an early idea of contagion because of the spread of mange among camels. They knew how to counter the disease, by separating the animals, and that it was not just divine activity.[41]

Nor was this all. Not only were the traditional routes from East to West being ravaged by plague, so too were the great European rivers of the Danube, Rhine and Rhône, their communities all decimated. The effects of this were two-fold. First, there was pressure on the ports of the *western* Mediterranean to explore alternative routes to the East. Second, the spread and pattern of the plague (worse in Asia and the Middle East) helped account for the rise of northern Europe.

A rival theory still indicts domesticated mammals. On this account, plague was actually anthrax murrain, a cattle disease caused by forest clearing, which reduced the amount of venison available and provoked instead an enormous increase in cattle ranching, in congested conditions. At least ten cattle herds belonging to medieval abbeys or priories in Britain are known to have been contaminated with anthrax murrain in the decade before the Black Death.[42]

The personal consequences of the Black Death were horrific. By the time the plague lifted in Europe three years after it had arrived in 1347, a third of the population – about 30 million people – had succumbed. But the economic and social consequences were more complex and less devastating than they might have been. To begin with, there was a labour shortage and wages rose. The labour shortage was so sharp, however, that, as the price of helping out their 'superiors', many serfs became free tenants. And, as tenants, they had greater motivation, with the result that food production dropped less than food demand. This was deflationary and made living on the land less and less attractive, so that there was a net migration to the cities, even though the plague had been worse there. In these new circumstances, prosperity returned more quickly than it might otherwise have done and here the woollen industry in particular benefited.

'It was woollen cloth that first brought capitalism to north Europe.' Wool was tougher, cheaper and more reliable than flax, which could only be grown then in the regions from the eastern Mediterranean to India, badly affected by plague. Sheep were more adaptable and lived for many years, and were less susceptible to flooding, drought or cold. Flanders and Britain, as we have seen, became known for manufacturing the finest woollens in the world, which generated more income than any other goods made in Europe. Cloth manufacture was favoured there because the low, waterlogged land was unsuitable for growing grain, and good-quality English fleeces were not far away to make up for any shortages. 'This industry accounted for the rise, successively, of Bruges, Ghent, Antwerp and Amsterdam, with Mediterranean wine, spices and silks, to be exchanged for woollens.'[43]

Then there is the fact that the routes between the Mediterranean ports and the Low Countries took ships out into the Atlantic – the open sea – a sailing and navigational experience that was helped enormously by the technological improvements and innovations stimulated by the Crusades and which would prove even more useful in the years ahead.

More tentative is the link between the plague and the Renaissance. The Black Death hit many of the great cities of north Italy hard and this is the charged background for Giovanni Boccaccio's *Decameron* (*c.* 1350–1353) about the flight of the fashionable Florentines into the safer countryside, where they hope to continue their glittering lives. One might have expected the plague to provoke soul-searching rather than Boccaccio's more earthy and earthly stories, but perhaps widespread death, as Norman Cantor says, weakened traditional faith and set off a quest for a more naturalistic understanding of nature.[44]

However that may be, in the millennium and a half since Jesus, Christian ideology had contributed disproportionately to the rise of Europe, to its thinking, its economy, its innovation in all directions and, at the same time, right across Eurasia, in the fourteenth and fifteenth centuries, and not for the first time, horses and sheep, domesticated mammals, proved important – vital, in both good and bad ways – to the defining events that constituted the history of the Old World.

• 23 •

The Feathered Serpent, the Fifth Sun and the Four *Suyus*

When the Spanish first arrived on mainland America, as opposed to the islands of the Caribbean, there were two civilisations that were prominent – even dominant – in the New World. These were the Aztecs in what is now Mexico, and the Incas in Peru. Both were flourishing at the time – each had elaborate capital cities, organised religion with ritual calendars and associated artworks. Both societies were rigidly divided into social classes and each had successful methods of food production. But there was more to the Aztec and the Inca than met the eye.

The Aztecs were reached first, in 1519, thirteen years before the Incas. When the Spaniards crossed the ring of mountains which surrounded Tenochtitlán and descended into the Valley of Mexico, and beheld the astonishing cities which formed the core of the Aztec empire, with its network of shallow lakes surrounded by active volcanoes, they could scarcely believe their eyes. So elaborate were these cities, and of such a size, that some of Cortés' soldiers could not be sure whether what was before them was real or an hallucination. But what the Conquistadores discovered soon enough was that, despite being themselves capable of a very practical and calculating brutality, the Mexica, as the Aztecs were also known, 'presided over a city of pyramids and sacred temples that reeked with the blood of human sacrifice'. This also took some getting used to. 'The dismal drum sounded again,' wrote Bernal Díaz del Castillo, one of Cortés' disaffected aides in his *A True History of the Conquest of New Spain*. '[It was] accompanied by conches, horns, and trumpet-like instruments. It was a terrifying sound, and we saw [the captives] being dragged up to the steps to be sacrificed. When they had hauled them up to a small platform in front of the shrine where they kept their accursed idols we saw them put plumes on the heads of many of them; and then they made them dance with a sort of fan ... Then after they had danced the [priests] laid them on their backs on some narrow stones of sacrifice and, cutting open their chests, drew out their palpitating hearts which they offered to the idols before them. Then they kicked the bodies down the steps, and the Indian butchers who were waiting below cut off their arms and legs and flayed their faces ...'[1]

Later, in 1529, a young Franciscan missionary, Bernardino de Sahagún, stepped off the boat on the Gulf of Mexico, set about learning Nahuatl, the language of the Aztecs, and consulted with prominent elders to compile a history of this strange people. The elders made available to him a number of codices – birch-bark books – that they had hidden from others and, over the next two-to-three decades, Sahagún compiled his famous twelve-volume compendium, *General History of the Things of New Spain*. Given how much of the original indigenous material of the Americas was destroyed, we are fortunate that he did.

One of the many things that Sahagún observed early on was that most of the great Aztec families claimed to be descended from Toltec lineages, though the Toltecs had long disappeared as a political or ethnic or cultural entity. These ancestors were regarded as great warriors, heroes whose ideal society had served as a blueprint for the Mexica state. As Sahagún put it, 'The Tolteca were wise. Their works were all good, all perfect, all marvelous ... in truth they invented all the wonderful, precious, and marvelous things which they made.'[2] By Aztec accounts (mainly oral, it is true) the Toltecs were tall, good-looking, artistic and talented people who invented the calendar and were 'extremely righteous'.

Everyone likes to come from a fine lineage and the Aztecs were no different but modern archaeology has confirmed that this glowing picture of the Toltecs is, to put it mildly, something of an exaggeration. Since the time of the Second World War, excavations have shown that Tula, in Hidalgo Province, fifty miles north of Mexico City, is in fact Tollan, the Toltec capital and that it was, by Mesoamerican standards, nothing to write home about: at its peak the city could boast a population of 60,000 at most, and so was dwarfed by both Teotihuácan and Tenochtitlán.

There had been a small village at Tula since about AD 650, but it didn't achieve any prominence until around 900, its prosperity based on its access to the prized green obsidian deposits once in the control of Teotihuácan. Tollan/Tula also achieved lasting fame as the place where the cult of Quetzalcoatl, the Feathered Serpent god, blossomed, this deity becoming a powerful religious and political force – in fact, probably the most enduring religious-political force in Mesoamerica.* It was here that the 'I'-shaped ball court evolved, skull racks, and elaborate serpent-shaped columns decorating doorways.

Geoffrey Conrad and Arthur Demarest remind us that 'Mesoamerican religion is exasperatingly complex', and so it is. Moreover, as Conrad and Demarest also say, Mesoamerican gods were not really deities in the Western sense: 'Rather they were divine complexes that could unfold into myriad aspects depending on specific temporal and spatial associations.' Finally, and an equally important point, they were 'ever-threatening'.[3]

Quetzalcoatl stands out in Mesoamerican history for several reasons. His name means 'Feathered Serpent' (see figure 19).

Coatl means serpent while the quetzal was – and is – probably the most colourful bird of the rainforest. Some scholars, such as Nicholas J. Saunders, say that this deity was originally an amalgam of three animals – the jaguar, the serpent and the quetzal. As such it was a manifestation of creatures which between them inhabited the three realms of the Mesoamerican cosmos – the Sky Upper World (the quetzal), the Middle World (the jaguar), and the watery Underworld (the serpent). Whether or not it was there to begin with, the jaguar appears to have dropped out of the iconography later on, perhaps because it was worshipped on its own so strongly elsewhere and perhaps because, as we shall see, the green feathers of the quetzal were adapted parsimoniously to stand also for the leaves of the maize plant, as organised agriculture became an ever more important feature of the middle realm of the Mesoamerican cosmos. Quetzalcoatl also acquired an association with Venus, possibly because the disappearance and reappearance of the planet at one point during its complex cycle lasted eight days, more or less the same interval as between the planting of

* 'Probably' because we must never forget how much original indigenous written material was destroyed by Conquistadores, compromising our understanding.

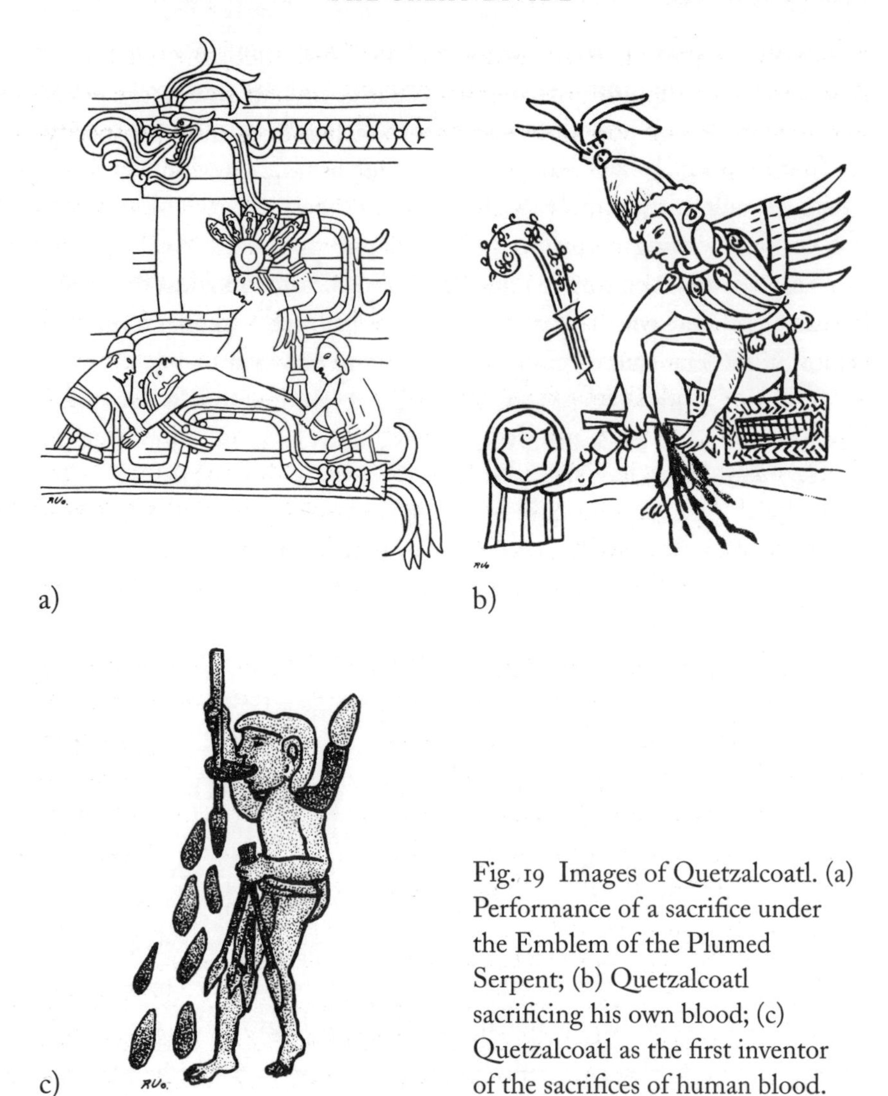

Fig. 19 Images of Quetzalcoatl. (a) Performance of a sacrifice under the Emblem of the Plumed Serpent; (b) Quetzalcoatl sacrificing his own blood; (c) Quetzalcoatl as the first inventor of the sacrifices of human blood.

maize seeds and the first appearance of green shoots.

David Carrasco, professor of the history of religions at Princeton University, has described temples to Quetzalcoatl in at least six Mesoamerican cities – Teotihuácan, Cholollan, Tula itself, Xochicalco, Chichén Itźá, and Tenochtitlán. This underlies that Quetzalcoatl was an *urban* deity and that it sometimes had different names – the Maya, for example, called it Cuculcán and among the Quichés of Guatemala he/it was known as Gucumatz (Gux=green feathers; Cumatz=serpent).

Now that we know maize was domesticated as a foodstuff much later than used to be thought, the stability of Quetzalcoatl among the Maya in classic times (AD 200–900) suggests that their long-term prosperity, and the flourishing of their culture, was due as much to domestication of the maize plant as anything else.[4]

But there is more to say about Quetzalcoatl than this. Possibly, there is a link between the serpent qualities of Quetzalcoatl and the Vision Serpent of the Maya shaman-kings, evoked through trance via blood-letting. But the advent of maize worship was a step – probably tentative at first – beyond shamanism, towards a priesthood. Priests of the Quetzalcoatl cult were not judged by their access to other worlds, which only they could visit and see, but by their ability to ensure a good harvest, which everyone could see for themselves. Catastrophes still occurred from time to time, of course, which could be attributed to dark shamanism. So shamanism continued but it now existed alongside a powerful priesthood.

And this is where Tula comes in. As has been mentioned before, Mesoamerica in the eighth and ninth centuries was disfigured by constant warfare, perhaps brought about by climate change, either catastrophic or more gradual, which provoked straitened economic times across the region. This was the period when, among other things, the classic Maya culture collapsed and when, after AD 750, Tula itself was responsible for the demise of Teotihuácan. It is tantalising to note, therefore, that in the wake of these troubles, at the beginning of the post-classical period (AD 900–1000), something new occurred in Mesoamerica. This was the advent of a cultural hero, what many scholars think was a 'flesh-and-bones' figure, an actual real-life ruler who also adopted the name and identity of Quetzalcoatl. He was known to his contemporaries as Ce Acatl Topiltzin Quetzalcoatl and in Mesoamerican legend he was awarded all sorts of qualities: he was a charismatic lawgiver, a creator of the cosmos, a founder of the ideal kingdom, a bringer of wisdom and civilisation.

This is noteworthy above all because of the timing. Topiltzin Quetzalcoatl, to give him his convenient shorthand name, and to distinguish him from the deity, ruled (if indeed he did rule), beginning around AD ~968 – that is, soon after the terrible eighth and ninth centuries, when certain accounts say he emerged as a forceful leader able to help regenerate civilisation following a series of cataclysms, perhaps brought on by natural disasters wreaking devastation (the wrath of the gods), but provoking in turn widespread fighting and man-made destruction. If this is what happened (and so far what is offered here is speculation, though it fits the facts as we know them), then the next development is doubly fascinating.

We know that at Xochicalco, for example, there is a glyph showing a hand which is gripping a rope attached to a calendrical sign which is being pulled out of one position and into another.[5] This suggests an attempt to resynchronise the calendar, which would have been a major event, brought about perhaps because a natural catastrophe, or a series of them, had caused a seeming change in the natural rhythms that had hitherto been observed. Moreover, the ancient sources of Mesoamerican history describe a set of episodes – widely known to the Aztecs, for instance – in which Topiltzin Quetzalcoatl, while still being a cultural hero of great importance, did fail in his aims in one important respect. During his reign a religious controversy broke out concerning the appropriate victims for sacrifice. Quetzalcoatl had turned against human sacrifice and instead proposed that quail, butterflies, snakes and large grasshoppers be substituted. But the warrior classes, who had their own god, the bloodthirsty god of war – Tezcatlipoca, 'Lord of the Smoking Mirror' – objected. According to legend, Smoking Mirror got Topiltzin so inebriated that he slept with his sister and in disgrace was forced to flee Tula with his followers.

According to one version of the legend, Topiltzin Quetzalcoatl reached the Gulf of Mexico, where he sailed away on a raft of serpents, vowing to return (a vow which was to have fateful consequences). In another version, he decamped to another city, possibly Chichén Itźá where, again according to the indigenous histories, we see 'the forceful appearance . . . of human sacrifice under the patronage of the feathered serpent'.[6] In this version, Quetzalcoatl, having failed in Tula and been forced to flee, has completely reversed his position, and adopted the new aggression which characterised the Toltecs and which the Aztecs were to emulate.

There are two points to make about these events, always acknowledging of course that they are sketchy and distant in the extreme. The first is that, if Topiltzin Quetzalcoatl *was* a historical figure, then this set of events offers a parallel of sorts with what happened in the Old World at the end of the Bronze Age. Was the widespread fighting, and the associated killing in Mesoamerica in the eighth and ninth centuries, so bloody, so awful, that Topiltzin became convinced that enough was enough? Why else would he propose that animals be substituted for humans in the sacrificial rites? His ideas went against centuries of New World traditional practices. However, whereas in the Old World such abolitionist thinking helped give rise eventually to the Axial Age, in Mesoamerica Topiltzin failed: he lost the argument and he lost his position.

Second, if this set of events is anywhere near what happened, then the Toltec civilisation that the Aztecs so admired was born out of catastrophe, possibly both natural and man-made. As we shall see in just a moment, Aztec cosmology envisaged history as a set of repeated catastrophes bringing one age after another to a cataclysmic end. The rupture which provoked the religious controversy in Tula seems to have been a major event, remembered and enshrined in Aztec cosmology. Perhaps they idealised the Toltecs because of the new beginning that they represented after the disasters of the eighth and ninth centuries.

Quetzalcoatl was not the only god of the Aztecs, and not always even the most important. But his durability, his ability to stay the same and change all at the same time, the fact that he incorporated so many aspects of Mesoamerican cosmology and history – agriculture, Venus, actual people, sacrifice and its meaning – meant that his power only increased as time passed and the priesthood arrogated to itself greater privileges. Quetzalcoatl's many qualities meant that priests had to undergo a long and austere training even before they were allowed to indulge in shamanistic rituals. Throughout the classical and post-classical eras, and thanks to Quetzalcoatl (but not only to him), priest-shamans became a class, an important estate within the society. As we saw with the Mixtecs, the elite controlled access to the gods, jealously preserving their privilege.

With Topiltzin Quetzalcoatl gone, the warrior elite now controlled Tula and, via a number of campaigns, they expanded north. They maintained extensive trade networks and irrigated much new land, settling their people in garrisons which extracted tribute from subject peoples. This too was an approach adopted subsequently by the Aztecs.

Even archaeologists who specialise in Mesoamerican societies have termed the Toltecs 'grim', as unashamedly militaristic. For example, everywhere in their ceremonial architecture, 'Fierce Toltec warriors strut, men carrying feather-decorated atlatls (spear throwers) in their right hands, bundles of darts in the other. They wore quilted armour, round shields on their backs, hats topped by quetzal feather plumes. Great stone warriors stand atop a six-stepped pyramid.' In several places, chacmools adorn the temples – these are reclining figures with round bowls sunk into their bellies, receptacles for the hearts ripped out of sacrificial victims captured in war. An enormous, 131-feet 'Serpent Wall' on the north side of the pyramid at Tula adds to the grim picture – it shows serpents consuming humans 'in a bizarre *danse macabre*, their heads reduced to a skull, the flesh partially removed from the limb bones'. Elsewhere on the temple there are carvings of jaguars and coyotes, eagles are shown devouring human hearts, while the effigy of a god, Tlahuizcalpantecuhtli, another deity associated with the planet Venus, is shown emerging 'from the fangs of a crouched jaguar adorned with feathers and equipped with a forked serpent's tongue'. These 'ardently militaristic' inscriptions at Tula are new, says Brian Fagan, their brutality perhaps indicating an especially intense competition for land and natural resources that may have erupted due to frequent droughts or crop failures, bringing about political uncertainty.[7]

This is supported by the fact that Tula appears to have suffered a violent end during the reign of a ruler named Huemac, in the late twelfth century. The pyramid dedicated to Quetzalcoatl was brutally dismantled, the city torched and the population driven into the surrounding countryside. But the Toltec legacy was kept alive among the nomadic Chichimec in particular, a people who struggled to maintain supremacy over the fertile lakeside territory of the Valley of Mexico. The Aztecs themselves were to begin with a small group of nomads who wandered into this valley. Because the best land in the area had already been settled, in about 1325, more than a century after Tula fell, they chose a tiny hamlet in the swamps of Lake Texcoco as their unpromising foundation. Within a hundred and fifty years, they had transformed it.

Rains of Fire, Vigilante Jaguars, Deluges and Hurricanes

Brutal and militaristic as the Toltecs were, the Aztecs were quite possibly more so. Bernal Díaz, who served as a *rodolero* (a 'shield' man) under Cortés and claimed to have been in 119 battles, and who wrote his own, very famous account of the invasion, *The True History of the Conquest of New Spain*, was appalled as much as his colleagues were by the extent of human sacrifice among the Aztec. He recorded their priests had many ways of carrying out sacrifice: by shooting the victims with arrows, by burning or beheading them, by drowning them, by throwing them from a great height on to a bed of stones, by skinning them alive or crushing their heads. But the most common method was to rip out the victim's heart (figure 14, page 418). Their gods, such as Huitzilopochtli, the god of war and the sun god, and Xipe Totec, the planting god, were seemingly insatiable. 'In honour of Xipe Totec young men would don the flayed skin of a sacrificial victim, and wear it until it rotted off. A new, clean youth emerged, a symbolic emergence of a fresh sprout from an old maize husk.'

In the actual sacrifice itself, the priests painted the victims with red and white stripes, then reddened their mouths and glued white feather-down to their heads. Dressed in this way, the victims were lined up at the foot of the pyramid steps, before being led up one by one, symbolic of the rising sun. Four priests held the victim down over the sacrificial stone while a fifth pressed hard on his (or her) neck, causing the chest to stand out. The leading priest thrust his obsidian knife swiftly through the rib cage and tore out the heart while it was still beating. Figures are hard to be sure of. Two Conquistadores say they saw 136,000 skulls on a rack but most scholars believe this to be a gross exaggeration.[8]

Although these are practices we may find abhorrent today, there was – as should now be clear – a traditional coherence to Aztec beliefs. In whatever way the idea of sacrifice was conceived, by the time of the Aztecs the practice reflected the notion that the sun god, in his daily passage across the heavens, had to be sustained by the nourishment of human hearts. (If a volcano had erupted in the remembered past, killing many people and blotting out the sun, which then gradually reappeared, one can see how such beliefs may have arisen.)

At the time the Europeans discovered the Aztecs, its king/emperor, Moctezuma II, ruled an empire of at least five – and perhaps fifteen – million souls, stretching from the Gulf of Mexico to the Pacific and from northern Mexico to Guatemala (see Appendix I for a discussion of pre-Columbian population levels). And although, in the early Conquest, Aztec books had been burned, because the European Christians believed the Aztecs to be pagans and heathens, later on some hidden codices were found and the systematic study of Aztec culture was belatedly begun. The codices were not written in the modern sense; they were, rather, *aides-mémoires*, prompt books for the elite members of Aztec society who had been entrusted with preserving oral traditions. On the basis of this, Bernardino de Sahagún and other friars collected an amalgam of legend and historical fact, 'a jumble of officially sanctioned genealogies and blatant political propaganda that confuse scholars to this day'.[9]

The first we hear of the people who would become the Aztecs is as a small tribe probably living under the rule of the Toltecs on a mythical island in a lake named Aztlan, from which they took their name. They were then 'semi-civilised' farmers who, some time in the twelfth century AD, migrated for reasons unknown southwards into the Valley of Mexico. At least seven clans formed this early group, who wandered for some years in the mountains and woods and encountered other groups in a succession of small wars. From time to time they stopped at places with strong Toltec associations.[10]

According to legend, the Aztecs were transformed from this minority position by the spiritual leadership of their war god, Huitzilopochtli, known also as 'the humming bird of the south'. This god was believed to have appeared to one of the tribe's priest-leaders and instructed him to search for a place where 'a great eagle perched on a cactus'. There, it was said, the Aztecs should build their capital city. The priests found the sacred place and recognised the symbolism of the image: the fruits of the cactus were red and shaped roughly like the human hearts that Huitzilopochtli devoured; and the eagle was the symbol of the sun, Huitzilopochtli himself. 'Less than a century and a half later, the greatest city in the Americas lay on this spot.'[11]

This place, Tenochtitlán, which means 'rocks growing among prickly pears', was a swampy island at the southern end of Lake Texcoco, which at the time filled quite a bit of the Valley of Mexico. But, by digging a network of canals and creating very fertile raised fields above the surface of the lake (known as *chinampas*), and by forging clever military and diplomatic alliances with a succession of neighbours, the Aztecs grew in power and influence until, in 1426–28, they felt sufficiently emboldened, under the leadership of Itzcoatl and his nephews, Moctezuma I and Tlacaelel, to turn on their former masters, the Tepanec. Following this great war of independence, they consolidated their position by forming a Triple Alliance, which comprised the three lake cities of Tenochtitlán, Texcoco and Tlacopan. They were by now a bellicose people, their warriors feared throughout the Valley of Mexico. They made a point of burning all the codices of their defeated rivals, so that their histories were lost, and declared themselves the true heirs of the equally admired and equally feared Toltecs. Their god, Huitzilopochtli, was transformed by propaganda into a ferocious warrior, demanding sacrifices. The warrior class assumed ever-greater privileges and there was a marked growth in fanaticism – perhaps the Aztecs' most fearsome asset.

Theirs was an hierarchical society. The central institution was the *calpulli*, a unit of social membership, a residential unit whose members owned land communally. In the mid-fourteenth century there were fifteen such units in Tenochtitlán. Each ward had its own school and temple and fellow members of each *calpulli* often fought alongside one another in warrior squadrons.[12] As time went by, leadership of the *calpulli* tended to become hereditary. But differences in rank were now stressed and privileges in dress, education and ownership became more differentiated as Aztec society evolved, like the Mixtec and the Toltec before them, into – essentially – a war machine. Their main motive force was an apocalyptic vision 'of the constant struggle between the forces of the universe', in which their role was to avoid the final cataclysm by furnishing the sun with the vital energy to be found only in blood.

This too was a move beyond shamanism. The priesthood was judged on a daily basis by its ability to sustain the sun in the sky, which in turn depended on victory in battle, which brought captives to be sacrificed. Everyone could see the success (or otherwise) of what the priests – who recommended when war was to be waged – achieved. Here too dark shamanism continued at times of catastrophe or cataclysm, but most of the time the priesthood, part of the nobility, was the more powerful caste.[13]

Their military conquests were carried out in the name of the war god, the sun god Huitzilopochtli, epitomised by the ruler and fanatically supported by the nobles and warrior knights. When Moctezuma Ilhuicamina ('the Angry Lord, the Archer of the Skies') took power (1440–1468) he declared that war was to be considered 'the principal occupation of the Aztecs, war designed not only to expand the imperial domains, but to ensure a constant supply of prisoners to satisfy the insatiable Huitzilopochtli'.[14] Under this new policy, hundreds of prisoners (and possibly thousands) were sacrificed each year.

A variety of weak and strong rulers followed Moctezuma Ilhuicamina, during which time parts of Guatemala and El Salvador were added to the Aztec homeland, each conquest intensifying the appetite for human sacrifice. In 1487, the forces of Ahuitzotl (1486–1501) put down a rebellion among the Huaxtecs (who occupied the coast of the Gulf of Mexico), the annexation coinciding with the completion of the Great Temple in Tenochtitlán. The joint celebrations which followed involved lavish gift-giving and the sacrifice of no fewer than 20,000 captives, some bound together by ropes through their noses, who were formed into four lines running down the steps of the temple 'and out along the four causeways of the island city'. Every one of them had his heart cut out and the entire ceremony lasted four days. (The Aztecs frequently had ceremonies lasting twenty days.)[15]

According to witnesses, the warriors responsible for taking the captives usually sponsored a feast for friends and family 'in which the flesh of the victim's limbs was served in carefully prepared stews'. Leaders from all over Mesoamerica were invited to witness this spectacle and plied with gifts, which included, on one account, 33,000 'handfuls' of exotic tropical bird feathers. Some victims seem to have resisted sacrifice but, according to a monk travelling with Cortés, most of them allowed the sacrifice to take place without 'uttering a word'.[16]

The Aztecs were a very religious people and inhabited a world where every activity, however mundane, had its symbolic aspects (the mass-production of religious artefacts underlines this). Their New Year started with a ritual dedication to Tlaloc, the rain god, who had been 'inherited' from the Toltecs. (Even today the rain clouds gather at Mount Tlaloc.) This was followed by worship for several months of Xipe Totec, the spring deity. September to March (as we would say) were dedicated to warfare, hunting and fire – these were the dry season gods. The dual-system calendar, introduced in chapter twenty, was used and each day had its name god with a specific meaning. Mayahuel, for example, was the god of intoxicating drink and therefore the patron of days with the rabbit sign, 'for the drunkard weaved and strutted about in the same erratic way as the rabbit'.[17] Some days were propitious, others definitely not. As already noted, it fell to Aztec priests to decide when the corn should be sown, and when wars should be declared.

Merchants were known as *pochteca*. Foodstuffs were brought into the city's markets (which had a state-controlled entrance) from the raised irrigated fields on the lake, with a whole class of artisans depending on the merchants for their supply of exotic raw materials. Cacao, as we have seen, was a prized luxury, its consumption confined to the nobility. (In moderate amounts, it was said, 'it gladdens one'; on the other hand, excessive consumption 'makes one dizzy'.) Cacao beans also served as a form of money, as did cotton cloth and copper tokens.

The major market, though, was not in Tenochtitlán itself but at its satellite, Tlatelolco. With as many as 20,000 farmers, merchants and visitors arriving via dozens of canals, their small canoes brought goods right to the heart of what was the largest entrepôt, by far, in all of the New World. Bernal Díaz tells us he saw merchants of gold and silver, of precious stones and exotic feathers, of embroidered goods. Male and female slaves were traded, along with 'rope, sandals, otters, jackals, deer, mountain cats and other wild animals'.[18] When the Spanish arrived they marvelled at the orderly nature of the market, from where the noise of collective activity could be heard 'more than a league off'. A court sat each day to settle disputes.

Enterprising *pochteca* ventured further afield, jointly charged with acting as the eyes and ears of the Aztec state. Though they could only aspire to the nobility in times of war (not so infrequent), they could enhance their social position by throwing elaborate feasts, or by buying a slave and sacrificing him to Huitzilopochtli. Alternatively, they could hand out chocolate and hallucinogenic mushrooms in a process not dissimilar to potlatch. 'He who eats many of them sees many things which make him afraid, or make him laugh ...' remembered one of Sahagún's mushroom-takers.[19]

They used the dual-system calendar, from which it followed that every 52 years the cycles coincided and the Aztecs regarded this as an especially risky period, when time itself had to be renewed. To do so, they extinguished all fires, and people were required to destroy all their possessions and stay awake 'in fear' until the priests, secluded on a sacred mountain near Tenochtitlán, rekindled a flame 'in the chest cavity of a sacrificial victim'. This act launched a new cycle of time.[20]

Such a view was also reflected in the Aztec creation legend that Four Suns lasting 2028 years (39 × 52) had preceded the present era of the fifth sun. The actual place where the gods had their beginning wasn't known, but it was believed that they had gathered in darkness, to bring forth the sun, and with the result that 'all the gods died when the sun came into being'.[21] In the first age, or Sun, the giants who lived then were devoured by jaguars after 676 years. The people of the second age were swept away by a great hurricane and changed into monkeys. The people of the third age were destroyed by a mighty rain of fire and transformed into butterflies, dogs and turkeys. In the fourth age the people were swept away by a great flood and turned into frogs. Each sun or era lasted for a multiple of fifty-two years and, says David Carrasco, each age took its name and 'character', in the Aztec system, not from its creation but from its destructive elements. 'Each beginning is sure to result in catastrophe and there appears to be no end in sight to this divine antagonism, these rains of fire, vigilante jaguars, deluges and hurricanes. The forces of nature collapse with violent totality upon the little populations.' After the Fourth Sun, the world was in darkness for 52 years, 'When no sun had shone and no dawn had broken.'[22]

The Aztecs were also convinced that the era in which they themselves lived, the world of the 'Fifth Sun', would one day end in violent destruction. This fearful event could only be averted, or postponed, if the gods were offered sacrificial victims, so that it became the Aztecs' sacred duty to feed the sun daily with *chalchiuhuatl*, 'the precious liquid', human blood. Feeding the sun was in particular the responsibility of the warriors. Aztec poets equated warriors in their cotton armoury as like 'tree blossoms in spring time'.[23] The entire Aztec civilisation – education, art, poetry – was organised around this one sacred duty, to secure enough captives that the gods would receive sufficient nourishment.

The focal point of this worship was the Great Temple in Tenochtitlán, famously excavated in the 1980s. From modest beginnings, the temple was rebuilt six times and grew increasingly grand. Some 6,000 objects were discovered in the temple, in 86 'caches', mostly tribute or spoils of war from far-flung parts of the empire. The design and layout of the Great Temple revealed that the Aztecs believed the earth lay at the centre of the cosmos, surrounded by water, as Tenochtitlán itself was. Above were the heavens, below was the Underworld, while the Middle World, the terrestrial level, had a central point – the Great Temple itself – which was the symbolic pivotal point, the 'portal' where a vertical channel linked the three levels. It was here, via the vertical channel and portal, that the supreme ruler interceded with the gods. At the summit of the pyramid of the Great Temple, there were two shrines, dedicated to the two supreme gods, Huitzilopochtli and Tlaloc. Huitzilopochtli was the god of war, who demanded heart sacrifice, Tlaloc the underworld god of rain, who demanded child sacrifice, and most of the cache offerings came from under his shrine, including a complete jaguar skeleton.[24] This confirms the view of Enrique Florescano, the prominent Mexican historian (and Master of the College of Mexico), that in the most ancient times the important gods of Mesoamerica were those of the netherworld, the powers that 'managed the forces of destruction, decadence and death, as well as the miracles of regeneration'.[25]

Potentially rebellious nobles were kept in line by being forced to attend court daily and by being made to supply warriors and foodstuffs for the frequent campaigns. And they were often married off to the women of the royal 'harem', so that the elite was increasingly interlinked by blood.

Brian Fagan argues that the Aztec state wasn't as monolithic as the Roman one, to which it is sometimes compared, but more a patchwork of alliances unified by a tribute system. 'There was no standing army but there were tax collectors: for example, twenty-six towns had to supply one of the royal palaces with firewood. Other tribute payments included gold dust, tropical bird feathers, jaguar skins, tree gum.' There were two classes of warrior elites, the Jaguar Knights, who wore pelts of the animal, and the Eagle Knights, resplendent in feathers of the rare and predatory Harpy Eagle. Rivalry between the two ensured that both kept up their standards.[26]

Sacred Cruelty

But does all this make the Aztecs sound rather more bloodless than was actually the case? David Carrasco, the professor of the history of religions at Princeton who was introduced earlier, thinks that it does. He goes so far as to say that much of the scholarly work on ritual sacrifice has ignored the Aztec case, that Aztec practices have troubled modern readers for centuries and that although we have been aware of this 'shocking practice (for us)' for almost half a millennium, 'the scholarly community has been remarkably hesitant to explore the evidence and nature of large-scale ritual killing in Aztec Mexico'.[27]

He has no such reservations himself, conceding that religious violence has a curious fascination 'for both the scholarly and the lay public' and in his book, *City of Sacrifice: The Aztec Empire and the Role of Violence in Civilization* (1999), he explores four elements that have been neglected by others: the *increase* in sacrifice between 1440 and 1521, the anxiety and paranoia at the root of Aztec beliefs, the phenomenon of flaying, and the sacrifice of women and children – in other words, he tackles head-on the most extreme aspects of ritual violence in ancient Mexico.

Carrasco puts the increase in sacrifice down to the anxiety the Aztecs felt about their universal order (cosmic life was an unending war) which 'was intensified to the point of cosmic paranoia'.[28] The temple was rebuilt and enlarged as rebellions were put down, and sacrifices carried out on an ever-larger scale to deter and threaten other potentially rebellious peoples. It was now that tribute took on the form, not of feathers, or gold, or maize, say, but of sacrificial victims. Brought to Tenochtitlán, these victims were ritually transformed into gods, and then slain. These sacrifices were not just more numerous, he says, but more widespread, carried out across the city, not just at the *Templo Mayor*. Indeed, the city walls were plastered everywhere with the blood of victims, as another sign of incremental domination. Some festivals lasted for months, in some cases victims being allowed privileges before they were slain, such as a number of wives. But Carrasco's main point is that the killing increased because of *underlying anxieties* which the Aztecs had about their position. This fits in with the views of Conrad and Demarest, discussed immediately below, but an alternative account will be given in the Conclusion. We have already glimpsed the argument in our outline of the Mixtec trajectory (chapters twenty and twenty-one).

In his discussion of flaying, Carrasco describes how, after sacrifice, and before victims were divided into parts which would be ceremonially eaten, their bodies were flayed 'and their skins worn by individuals who moved through the neighborhoods of the city and fought intense mock battles'.[29] This was of course an essentially shamanistic practice, or derived from shaman-type transformations, the skins in this case having a 'career', being passed from one warrior to another, as they gradually deteriorated, when they were buried. Men who didn't have a skin attempted to snatch pieces off those who did. Human skins were believed to have magical properties and when the skins were finally taken off, that symbolised renewal. Here too, though, we see the anxiety and dread which underlies the process. Xipe Totec was a god who flayed himself to provide food for humans. The 'Totec' part of his name translates as 'Awesome and Terrible Lord Who Fills One Up With Dread'.

Women victims – who were sacrificed in about a third of Aztec ceremonies – were sometimes flayed, and this is another area which Carrasco feels scholars have avoided. In considering this issue, he writes: 'Is the Aztec grossness really grossness, or just a more complex, sobering, and terrible story about some dimensions of the history of religions that is too hard to tell and very much too hard to sell? Or is there a level of sacred cruelty in Tenochtitlán that shocks our most effective categories of understanding?' He confesses that it had taken him a long time to be able to face reading the details of women and children sacrifice. Children were sacrificed in the first quarter of the Aztec year. There was apparently much weeping when this happened, and singing, but the killings went on to ensure that the rains came in abundance. Young girls were associated with young maize seeds, the feminine part of the cosmos, and their flayed skins were used to prepare men for war.[30]

Whether these practices are gross, they are certainly grisly. We described David Carrasco's argument, that human sacrifice increased greatly among the Aztecs in the period 1440–1521, on account – he said – of the 'anxiety' built into their society due to their tribute-gathering and the pressures this put on their rulers. But this explanation, while true enough as far as it goes, is at one remove from the fundamental reason. Why, after all, did this tribute-gathering need to increase? Why did the Aztec anxiety grow?

The explanation falls into three parts. The first part shows why anxiety levels were high in Mesoamerican society and is provided by the work of Arthur Joyce and Marcus Winter with the Mixtec. We saw in chapter twenty that, among the Mixtec, there was a marked link between religion and war. Why should this be? Joyce and Winter say that Mixtec society became much more stratified over time, that the elite deliberately and increasingly took on the role of ritual specialists and that part of their ideology was to *promote* war. This may not be unique in the history of the world but it was certainly unusual – what can account for it? Joyce and Winter's answer is that the main elements in Oaxacan religion were the *threats* that came from natural and supernatural forces. What this means of course is that supernatural forces were the names given to threatening natural forces that no one, in reality, understood. The truth is that the (religious) elite had no control over the (super)natural forces, such as volcanic eruptions, El Niño events, tsunamis, jaguars and the rest. Therefore, say Joyce and Winter, they sponsored an ideology that was, in effect, 'deceptive' (their word) on the non-elites, by promoting conflict, war, what was in effect a form of threat that they had *some* measure of control over. Not complete control, perhaps, but enough to reinforce their power and privileged position. Wars, calendrically determined and fought for religious reasons, could be won, while there could be no final victory over natural threats like volcanoes and hurricanes. Moreover, the sight of captives in a city was good evidence of the threat, and their sacrifice equally good evidence of the elite's ability to protect its population. The elite preserved their position by their exclusive access to sacred knowledge – calendrics, astronomy, writing and so on.[31]

It was a high-risk strategy, of course. Other elites, in other polities, were trying to do the same to them and this may be why so many Mesoamerican city-states met a sudden end: Monte Albán, Cerro de los Minas, Yucuñu, Chichén Itzá.

A second factor was that, as Geoffrey Conrad and Arthur Demarest point out, the Aztecs were worried that, in fact, the sun was *under*-nourished – this is why the demand for blood was greater than ever.[32] What might this mean? If the volcanoes of the area were erupting, especially perhaps if they were volcanoes out of sight but the ashes of which nevertheless obscured the sun, this would explain why it was felt to be undernourished – it was too weak to appear, to rise in the morning. In turn this would explain the Aztec worries and would have fuelled what Carrasco calls their cosmic paranoia.

The volcanic evidence supports this possibility. The Aztecs were not discovered by the Spanish until 1519. But in that year, as we saw in chapter five, Popocatépetl (which *could* be seen from Tenochtitlán) erupted and it did so again the following year, and the year after that, and again in 1522 and 1523. The Smithsonian survey confirms that other eruptions were taking place at much the same time in El Salvador, Guatemala and Nicaragua. We cannot be certain, because the records that far back are incomplete, but if Popocatépetl, and certain other volcanoes, were particularly active in, say, the last half of the fifteenth century, and the early years of the sixteenth, as the Smithsonian chronology strongly suggests, then this in itself might account for the increase in paranoia and violence in Aztec society. Either the sun was obscured a lot – undernourished – or people may have been killed by the eruptions, or crops may have failed because the sun was obscured or lava destroyed the fields. In such a scenario, earlier levels of worship would have been seen to have failed, as inadequate – the gods continued to be angry, they were unappeased by whatever level of sacrifice obtained at any particular point, and so efforts had to be redoubled, more victims had to be offered. Such a pattern of volcanic activity would certainly explain the profound level of anxiety which Carrasco believes underpinned Aztec society.

As all this showed, the Aztec ruler was the most powerful (and the most feared) man in Mesoamerica and his civilisation was at the peak of its power. But there were shadows not so far over the horizon. Under their system of beliefs, the Aztecs were forced to expand their territory simply to obtain more tribute and more victims. 'Increasingly the empire became vulnerable to sedition and rebellion. There are signs in Aztec proverbs and verse that their society was one of growing philosophical ambivalence, between ferocious militarism and human sacrifice on the one hand, and notions of benevolence, humility and mercy on the other.'[33] But, as with the Toltecs before them, the forces of militarism won the day.

Which in turn underlines the instability built into the system. With, at one point, perhaps 15,000 persons being sacrificed every year in Mexico, even if that figure is an exaggeration it meant that the Aztecs had to look further and further afield for victims. But they had no horses or chariots, only their fighting efficiency and their fanaticism to sustain them. Even so, they could not easily hold far-flung territories, which were constantly rebelling, or threatening to rebel. In an age before refrigeration, the outer provinces were too far off to contribute perishable foodstuffs. Instead, they provided luxury goods which boosted the self-image of the elite but did little to assuage the food shortages that existed nearer home – shortages which existed because the people who could labour in the fields had been already sacrificed.

Once the Aztec expansion faltered, and fewer victims for sacrifice were to be found, it followed that the gods would be even more dissatisfied. The Aztecs would then be required to redouble their efforts. But distant peoples had been encouraged by their victories, or successful resistance. If the Aztecs responded by requiring increased tribute, the outlying people could not easily be forced to produce more.

The fatal irony or paradox was that Aztec warfare destroyed the very system of subsistence on which it existed, because it killed thousands of food producers. The sacrificial cult, successful to begin with, turned increasingly maladaptive. It continued because the ritual calendar provided many dates suitable for sacrifice, by means of which the priests and warriors could move up the social scale. Ritual and subsistence were in conflict.

With the pace of expansion slowed, or even reversed, other ways to secure victims were needed. A class of slave-merchants emerged but part of the Aztec ideology required that *war captives only* be used in sacrifice, so slaves did not entirely fit the bill. So-called 'flowery wars' arose, in which states met periodically in a battle solely designed for each side to 'capture' warriors for sacrifice – a bizarre arrangement if ever there was one. But these were just a symptom of the inherent instability that lay at the heart of the system.

What would have happened to the Aztecs if the Conquest hadn't come along when it did? Moctezuma II recognised the predicament his nation faced but, as with Tolpitzin Quetzalcoatl and the Toltecs, the forces of fanaticism won out and the reforms he proposed were rejected. It is clear that the Triple Alliance was coming to the end of its prominence whether or not the Europeans arrived when they did.

The legend that Quetzalcoatl would one day return from overseas only made the Aztec collapse before Cortés – who for a short time was regarded as that god – easier than it might otherwise have been.

Immortality and Incest

The Incas called their empire *Tawantinsuyu*, the 'Land of the Four Quarters'. Accordingly, four great highways spread out from the central plaza of Cuzco, the capital, separating the kingdom into four *suyus* (quarters), which were in themselves modelled on the four quarters of the Inca heavens.[34] At its greatest extent, the empire extended from the borders of modern Colombia and the coastal regions of Ecuador and Peru, down through highland Bolivia as far as north-western Argentina, and on into central and southern Chile, some 4,300 kilometres or 2,700 miles from end to end. It was the largest empire of antiquity ever to arise south of the equator, with a population of some 10 million.

It was a long, thin, curving strip of land, of high mountains, coastal deserts, and tropical rainforests.[35] Its two most productive zones were at either end of this strip. The central Andes, around the capital city of Cuzco, was the heartland, comprised jointly of a flat basin north-west of Lake Titicaca, known as the altiplano, whose climate of very hot days and frosty nights made for productive potato fields where (as was discussed in chapter eleven) the foodstuffs were naturally freeze-dried, and with it extensive pastureland where large herds of domesticated llamas provided wool and carried loads. (In light of the overall thesis of this book, which is partly about the influence of large mammals on human history, we may note that it was reported in May 2011, as the final touches were being made to the typescript, that new research has shown it was the adoption of llama droppings as fertiliser, about 2,700 years ago, that allowed the Incas to switch from growing quinoa to the mass cultivation of maize, a more nourishing and easily stored crop. It was the ability to farm more efficiently that fuelled the territorial expansion of the Incas.)

Far to the north, the coastal desert would have been uninhabitable save for the fact that it was intersected by a number of relatively short, fast-flowing rivers, driving forceful irrigation systems that enabled the whole area to be richly populated. All this meant that the most populous areas of the empire were at opposite ends of Tawantinsuyu. The Inca achievement was to join these disparate regions together in a single civilisation, which was also the largest of its kind ever to exist using a technology based on bronze, copper and stone.

The Incas were later to give themselves a proud and long history but this is not supported by archaeology. (They never invented writing so there are no indigenous chronicles to consult.) They were in fact just one of several small groups – the Colla, Lupaca and Quechua were others – who existed in the wake of the fall of Tiwanaku (see chapter twenty) who competed for control of the productive altiplano. Like the Aztecs with the Toltecs, the Incas sought legitimacy by claiming descent from Tiwanaku when, in fact, they were just one among many local farming cultures.

The early Spanish chroniclers specified up to eight rulers between AD 1200 and 1438 but the archaeology does not support this either. The earliest Inca ruler worth the name emerged when the village of Cuzco began to grow. This man, the eighth ruler allegedly, and known as Viracocha Inca, appears to have been the first leader who actually held on to the new territories he conquered.

But it wasn't always as simple as that sounds. About 1438, Viracocha Inca's lands were besieged by the Chanca, yet another rival people to the north, who attacked Cuzco. The now-aged monarch escaped to the mountains, leaving the defence of the city to his son, Yupanqui. Against the odds, the young and inexperienced Yupanqui trounced the Chanca and, in recognition of this exploit, became the supreme ruler of the city, adopting the name by which he was always known afterwards, Pachakuti, 'he who remakes the world' (a phrase that could have applied to Tolpitzin Quetzalcoatl). A charismatic leader, among Pachakuti's achievements was the rebuilding of Cuzco, apparently in the form of a puma (a cat related to the cougar and the jaguar), with a great fortress as its head and a narrow triangle of streets in the east as its tail.[36] It was said that no fewer than twenty thousand workmen were drafted in for this project from outlying areas, to quarry the stones and haul them over huge distances, using ropes of hemp and harnesses of leather.

Cuzco was made up of single-storey houses with steeply pitched thatched roofs. Its paved streets had stone-lined channels with fast-flowing water draining down the middle, providing effective sanitation.

A river flowed through one of the central plazas, dividing the city into two halves or moieties, Cusipata to the west, Aucaypata to the east, which was larger and where the Incas' palaces and ceremonial buildings were located. The palace featured impressive gates, built of multi-coloured stone, and a large hall, which held 4,000 people.

South of the central plaza was the Temple of the Sun, the Coricancha, boasting walls lined in gold and encircling a courtyard. In this courtyard (it had fifteen-feet-high walls) the Conquistadores came face-to-face with 'many golden llamas, women, pitchers, jars and other objects'. There was in addition a garden of golden plants (replicas of maize with silver stems and ears of gold), and at the centre of the temple was a room with 'an image of the sun of great size, made of gold, beautifully wrought and set with many precious stones'.

Almost as impressive were the ashlars, which the palaces and great houses were built of, stones fitted so neatly together 'a knife could not be put between them'. Each had concave depressions carved out of their surfaces, to ensure the fit would be snug. Such elaborate stonework needed thousands of hours of work, 'but time and labour were of no concern to rulers with abundant manpower and no western notions of time'.[37]

In addition to everything else, Cuzco was a massive warehouse. In a society without money, the Incas kept their supplies (and their tribute) in a row of identical storehouses 'full of cloaks, wools, weapons, metal, cloth and all other goods that are grown in this country. There were weapons and thousands of tiny, coloured hummingbird feathers used to adorn the clothing of the nobility.' Most exotic of all, certainly for the Conquistadores, was the building which contained clothes, in particular cloaks, covered as they were in 'dense layers of gold and silver counters'.[38]

Above the level of the nuclear family, the basic unit of Inca society was the *ayllu*, essentially a kin group though it could extend beyond the village. Within it men were organised patrilineally and the women matrilineally, and though marriage was forbidden between couples who were closely related, the system was endogamous. The *ayllu* owned land communally and its leadership tended to be hereditary.[39]

Owing to the sheer extent of Inca territory, at so many different altitudes, a massive road system was developed that extended over some 19,000 miles. According to Spanish chroniclers, the roads were built into precipices, cut through rocks in the mountains, mounted as walls along rivers and were kept clean of refuse. They had lodgings and temples and storehouses spaced along them to aid travellers and were better than anything then available in Europe. Official runners used these roads to help administer the empire, the way-stations every mile and a half or so enabling messages to be sent over 125–150 miles a day – or Cuzco to Quito and back in 10–12 days. Llamas and armies moved along these roads in large numbers.

The Incas built large rafts of balsa wood, with rectangular sails, which enabled them to trade up and down the Pacific coast. Inca gold, silver and other valuable cargoes reached coastal communities far to the north. In fact, the first European contact with the Incas occurred when Francisco Pizarro encountered an Inca craft with cotton sails. 'They were carrying many pieces of silver and gold as personal ornaments ... including crowns and diadems, belts and bracelets ... They were taking all this to trade for fish shells from which they make counters, coloured scarlet and white.'[40]

Apart from Cuzco, with its temples of golden walls, brilliant sanitation, and the empire-wide network of efficient roads, Pachakuti's other main achievement was to begin to rebuild the state ideology, a project which was equally transformative. In the Inca context, Conrad and Demarest remind us once again that New World deities existed more as 'complexes' than as the more defined entities that Western gods were, who could vary their attributes and powers according to time and place. In the case of the Incas, their sky god had three important qualities (among others). He was Viracocha, the universal creator, Inti the sun god, and Illapa, 'the deity associated with thunderstorms and weather generally'.[41] They also point out that the Inca godhead seems to have been derived from the 'Gateway God' of Tiwanaku as a generalised creator/sky/weather deity (see above, chapter twenty). As Huitzilopochtli had been deliberately elevated among the Aztecs, so Pachakuti elevated Inti, the Sun God, as patron of the Incas. The people believed they were under the protection of Inti, and that their rulers were directly descended from him.

There was, perhaps, nothing especially noteworthy in all this, but there were two aspects of Inca religion that *were* remarkable. The first was the practice of *capac hucha*, or *capacocha*. This translates, roughly and literally, as something like 'solemn sacrifice' but in fact what was meant more often than not – and, again, this is a very new development in scholarship and research – is child sacrifice, and child sacrifice very often at extremely high altitudes, in the Andes mountains. There were several ceremonies which come under this heading. In one, boys and girls – especially good-looking boys and girls of five or six, from the families of chieftains as often as not, and up to a thousand at a time – were rounded up and brought to Cuzco 'to serve the Inca'.[42] Ceremonies in their honour were held in the capital but they were then sent back out, either to where they had come from, or to the *huacas* all over the empire, where they were sacrificed. In another ceremony, four children were sent to Cuzco every year, one each from each *suyu*. There they were, again, fêted, before being sent back to where they came from and sacrificed to the sun (by strangling). The families who provided the children were elevated in status afterwards and it was a major offence if the family showed anything other than complete satisfaction and delight when their offspring were killed.

Associated with this, the most recent research has established that many sacrifices, including those of children, took place at very high altitudes, at 17,000 feet or more. It is also now known that the Incas and other Andean civilisations who came before them worshipped mountains and maintained a hierarchy of summits, and mountain gods, according to altitude, with the highest being the most important. Mountains were regarded as the source of water, which they were felt to attract from the sea, of other spirits, and many of them were volcanoes. Many sacrifices at high altitude were clearly related to rain and the weather (victims were sometimes deliberately frozen to death).[43] Textiles, camelids, *chichi* beer and coca leaves all accompanied the sacrificial victims. Steve Bourget argues that worship in the mountains marked the start of the humid seasons – water again.[44]

Some colonial accounts say that the Inca sacrificed hundreds of people a year. Archaeology suggests they were exaggerating but the Incas did sacrifice children in cases of drought, earthquakes, famine, war, hailstorms, lightning storms and avalanches, among other episodes. The face of a *capac hucha* victim who froze to death on top of Cerro El Plomo, in Chile (altitude 17,716 feet), excavated in 1954, was painted with red ochre and four jagged yellow lines, which may represent lightning, a motif also found on the miniature clothing encasing the small figures that accompanied other *capac hucha* burials.[45]*

Sacrifice in Peru seems to have been weather-related but why children should be chosen more often than not isn't clear, unless they were felt to be either more easily replaceable (because they hadn't lived as long as adults), or more precious, so that the gods would take more notice.

The second remarkable belief that developed among the Inca was that the divine ruler himself should not die, moreover that his spirit should be kept alive by adoration of his royal mummy. This was an idea that had been originally conceived among the Chimu (in the Moche Valley of Trujillo, which Yupanqui conquered early in the fifteenth century), but was taken over and elaborated by the Inca.[46]

One can see how this idea of ancestor worship might have begun, and may have happened in one of two ways. We saw in chapter fifteen that some people in the dry deserts of western South America did not die in the accepted way but became, as we would say, mummified, apparently inhabiting a strange (and therefore sacred) half-way world between living and dying, when their body did not decay as 'normal'. Alternatively, in a catastrophe, an earthquake, say, some people would be 'taken' by supernatural forces, people who, almost by definition, would become ancestors chosen to join the gods. Whatever the reason, there grew up among the Inca the strange cult of royal ancestor worship. What was so idiosyncratic about this was that the central element was the concept of 'split inheritance' (again a Chimu idea originally). Under this system, when a ruler died, one of his sons inherited his position as ruler, his right to govern, to wage war, and to levy taxes. *But*, and it is an all-important 'but', he *received no material goods* from his father. All the recently deceased ruler's land – his buildings, chattels and servants – were held to be his property still and were entrusted to his *panaqa*, his *other* descendants in the male line. '*Panaqa* members served the

* Patrick Tierney, in his book, *The Highest Altar: The Story of Human Sacrifice* (1989), claimed that human sacrifice was *still* being carried out in remote parts of the Peruvian and Chilean Andes, and quoted evidence that seemed convincing in at least two cases. He also included in his book the following song after they had 'quartered' a boy-victim at Cerro Mesa, near the sea: 'Take this boy now,/We are helping you,/We are paying you with this boy./We are all orphans./Why do you punish us, God?/We sacrifice this boy to you,/We give him to you as a gift,/So that the tidal waves are calmed,/So that there are no more disasters.' Another informant told Tierney, a reporter for *Discovery* magazine, that 'People here are accustomed to sacrificing someone when the weather gets bad'.[47]

departed king, acted as his courtiers, maintained his mummy, and were fed by his perceived generosity.'[48] On this system of beliefs, an Inca ruler never died but continued to maintain a lavish court through the hands of his *panaqa*, his descendants. 'It was customary for the dead rulers to visit one another, and they held great dances and revelries. Sometimes the dead went to the houses of the living, and sometimes the living visited them,' wrote Pedro Pizarro in 1571. Fires were lit before the royal mummies, and food burned for them to consume; maize beer was offered and in one case the figure had a hollow stomach into which was placed a form of paste made of gold dust and the ashes of Inca kings' hearts. This particular mummy was held to govern meteorological phenomena, such as thunder, rain, hail and frost.[49]

More than that, during important ceremonies, the royal mummies – several of them, in order of seniority – were placed in special wall niches in the Coricancha temple in Cuzco, alongside images of Inti. Here too food was laid out for them to eat and they were venerated by everyone.

Because of this system, the new leader, who had inherited no material goods, had to acquire new wealth, and in a society with no money that meant land and compulsory labour. Inca law was based on the *mit'a* tax, an obligation whereby individuals provided to the state a certain amount of labour every year, in return for which the state provided produce. If a new leader were to be a success, therefore, he needed above all new land in order to fulfil his obligations. The institution of split inheritance for the rulers meant that, as time went on, more and more new land was almost by definition marginal and more and more fertile land was owned by the dead, and unavailable to the living rulers. Likewise, with more and more people working *panaqa* land, there was less and less labour available to work any new land that an Inca ruler might conquer.

Closely allied to ancestor worship was *huaca*, 'the great integrating concept of Inca religion'.[50] A *huaca* was a person, a place or a thing with sacred or supernatural associations and in practice applied to anything odd or unusual. According to some accounts, there were 328 *huacas* in the Cuzco area alone, each of which had to be 'fed' once a year.[51]

What advantages did such an apparently maladaptive system bring? It seems it was designed to keep the powers of the nobles in check. A noble who was faithful to a successfully expansionist Inca was rewarded with land, servants and other privileges such as being able to wear the headbands and ear plugs of nobility. (Commoners were forbidden to own luxury goods such as gold objects, anything excess to their immediate needs.) As with the Aztecs and in other societies, prowess on the battlefield was the key. The nobility, many of royal blood, held all important government posts in the military, priesthood and bureaucracy. As with the Aztecs, the Incas felt they had divine patronage, which helped strengthen their national identity, set them apart from their neighbours and assigned them a special place among the inhabitants of the Earth.[52]

Once fresh lands were conquered, the Incas would send out from Cuzco a range of trained officials, who would inventory everything, using their knotted tallies, or *quipus*, capable of storing very precise information. If the population was 100 or thereabouts, the Incas would install a hereditary *curaca*, a governor, often of local birth. For larger foreign populations, Inca nobles were sent in.

By the time of the Conquest, the Inca empire consisted of millions of people living all over the long strip of land, scattered in villages and larger political centres. This makes them sound successful and in many ways they were, judging by their population levels. But – another important 'but' – as with the Aztecs, there was an instability built into the Inca system. The practice of keeping a dead Inca's land in his *panaqa*, and requiring the new ruler to conquer fresh territory, worked only as long as there was land to conquer. As time went by, however, more and more land was in the hands of dead rulers, leaving only ever-more marginal and/or distant land available. To the east it was worse. On the far slopes of the Andes, the Amazonian rainforest began, for which the Incas were constitutionally unsuited. They could not cope with the humidity, the insects, the vegetation – the forest itself – or the style of guerrilla war waged by the Amazonian Indians who did not fight set pieces, as the Incas preferred and were used to, but attacked in ambushes and then melted quickly away. Campaigns there were a disaster.

This sparked a crisis and the idea gained ground – among some – to change the *panaqa* system. But this of course went directly against the interests and wishes of the warrior nobles.

In 1525, Pachakuti's grandson Huayna Capac died while making the last of the Inca expansions, in Ecuador. A bitter power struggle erupted between two of his sons, the half-brothers Huascar and Atahualpa.

This too highlighted the at times idiosyncratic nature of Inca society, because royal succession was not determined by primogeniture: instead, the emperor was supposed to bequeath his position to his most competent son by his principal wife. However, in an effort – in a rigidly hierarchical society – to keep as much power as possible in the royal line, the practice had developed whereby each emperor took one of his sisters as his principal wife, an incestuous cult that may have been an extension of the endogamous practices that existed in the *ayllu* (see above) and paralleled the system among the Mixtec (see chapter twenty). However it happened, it clearly carried dangers though in the case of Huascar and Atahualpa faulty genes do not appear to have played a role in what transpired.

Huascar was the legal heir, being the fruit of Huayna Capac's marriage to his sister, whereas Atahualpa was the progeny of a second marriage. Despite his genes, Huascar well understood that the state was overextended and he proposed ending the mummy cult. Incensed, the warrior nobility took Atahualpa's side and in the civil war that lasted for three years (1529–1532), he gradually beat down his rival.

As with the Toltecs, and the Aztecs, then, the warrior nobility won the day among the Inca, which shows how forcefully the nobles would strive to maintain their privileges in what was obviously an unstable – and ultimately unsustainable – system. As with the Aztecs, however, the Inca system was never played out towards its logical end because the victorious Atahualpa, on his way to Cuzco for his coronation, encountered none other than Francisco Pizarro and his 168 Spaniards, some of them mounted on horse-back.

Conclusion

The Shaman and the Shepherd: The Great Divide

The Iberian Moment

Towards the end of the fifteenth century, various historical forces came together to create a situation in which Europeans in general and the peoples of the Iberian Peninsula in particular felt impelled to venture overseas as both explorers and conquerors. They did so as a result of a complex of motives, of which two stood out: acquisitiveness and religious zeal. The search for a new route to the spices of the East was one factor but, as Bernal Díaz wrote, he went to the Indies, as he thought, 'to serve God and His Majesty, to give light to those who were in darkness, and to grow rich, as all men desire to do'.[1]

Spanish nobles were especially familiar with this ideology, if such it can be called, because they were accustomed to a long and successful war against the Muslim states in Spain that had offered both 'occasion and excuse'. The rest of Europe (i.e., Christendom) had by that time enjoyed a long respite from Muslim pressure on its eastern and southern edges due to the conquests of Genghis Khan whose swift victories, with a highly efficient cavalry operating over a vast area, and a remarkable religious tolerance, had made travel to the East safe and stimulated trade. The great monarchies of northern Europe had by now lost interest in crusading and had abandoned the fight against Islam to those who had Muslim neighbours in the Balkans and Byzantium, and the Iberian Peninsula.

The advent of the Ottoman Turks, who took Constantinople in 1453, was however a dangerous development. Newly Islamised, and a proud people who had also converted from being horsemen to sailors, the emergence of the Turks as the most powerful state of the Middle East (they invaded Italy in 1480), made them a distant but still menacing threat that required the Spanish to fear the only Muslim state surviving in Europe at the time, the ancient and highly civilised kingdom of Granada.

Within the Iberian Peninsula, Christian and Muslim states had co-existed side-by-side for centuries and had often formed alliances when it suited them. Moreover, the peninsula had become the principal point of contact between the two cultures, notably in Toledo where Jewish, Arab and Christian scholars had collaborated on a seminal series of works which ensured that the best of Greek thought survived and was translated and glossed into the languages of the new universities that sprang up from the twelfth century on. Philosophical, astronomical and medical works featured strongly in this tradition.

But Granada was not as strong as she looked – by now she paid tribute to Castile and the rulers of the Spanish/Christian state knew that it was only a matter of time before the former entity was incorporated into the latter. That moment came with the accession of Isabella, in 1474. Intensely religious, ascetic, fearful of the lurking danger in the East, Isabella set about subduing her Moorish neighbour, village by village and town by town, beginning in 1482, a campaign that took a decade to complete but finally succeeded when the capital fell in 1492.

There was also, as J.H. Parry has pointed out, a curious parallel between the Ottoman Turks, in the East, and the Castilians of Spain. 'The Castilians had never been as parasitic upon the horse in the same degree as the Turks, but they too, in Andalusia and elsewhere, employed mobile and largely mounted forces against sedentary communities. Among them, in the arid uplands of Castile, pastoral pursuits, the grazing of semi-nomadic flocks and herds, had long been preferred to arable farming ... The man on horseback, the master of flocks and herds, was best adapted to such conditions ... As the work of the conquest proceeded, the Castilians, or the upper classes, the fighting classes among them, retained their pastoral interests and possessions, their mobility and military effectiveness, and their respect for the man on horseback.'[2]

These were the people who settled the New World and it helps explain why small bands of mounted Spaniards could achieve such remarkable victories and could then settle as quasi-feudal overlords, retaining their pastoral interests and relying on conquered peasants to grow grain for them (because of course they had no domestic mammals of their own).

This account continues the theme of *The Great Divide*, that the domestic mammals of the Old World have had an important effect on the course of history, but it is not by any means the whole picture. There were many other factors which came into play: the plague, which shifted the centre of gravity of Europe to the west and north; the development of the north, due to the wool industry in the Netherlands and Britain, the routes between the Mediterranean and the English Channel being one of the elements in opening up the Atlantic; navigational and shipping innovations, introduced partly a result of the Crusades; the Crusades themselves, which sent men abroad to convert the unbelievers; the rediscovery of Ptolemy and his *underestimation* of the size of the earth; inaccurate and outdated travel literature which reinforced that view; the discovery that sailing in the tropics was easier than had been anticipated; deep-sea fishing exploits that had also helped familiarise and accustom sailors to the Atlantic; the discovery of islands in the ocean, which promised yet more land available to the brave.

All of these psychological and technical matters came together to produce what we might call the Iberian Moment, why it was that the Spaniards and Portuguese were the first to cross the Atlantic going from east to west, to discover the Americas, rather than the other way round, with Moctezuma's admirals travelling to Africa or Europe (or Japan, for that matter).

Patterns in *La Longue Durée*

The narrative of this book is not a straight line, not by any means; also, we must be careful not make it read like a 'Just So' story. We began, in the Introduction, by saying that we were to conduct a natural experiment, but one in which it would be impossible to verify all the details. Even so, as should now be clear, we *are* able to offer, in broad terms, a hypothesis as to why the two hemispheres diverged from each other. No less important, our theme also suggests a view on what, ultimately, it means to be human – it is, in a sense, a new version of what influences affect the broad sweep of human history.

Human societies developed in very similar ways, according to certain criteria. As several surveys confirm, they include the fact that most societies are egalitarian – and villages undefended – until they reach somewhere between 150 and 300 inhabitants; nowhere have foragers developed ceremonial architecture; above the level of the village, hereditary leadership, ceremonialism and warfare emerge, with approximately 25 per cent of males being killed violently; with warfare male 'superiority' emerges; elites everywhere cultivate a distinctive lifestyle. And so on. The fact that such similarities exist across the world is remarkable testimony to the underlying unity of human behaviour.[3]

It has not been my aim in this book to *deny* either the existence or the importance of the many similarities that exist across the range of human societies scattered over the Earth. Not at all. But it has been my aim, unlike earlier authors, to focus on the *differences*, and to show how fruitful contrasts can be, alongside the parallels. The work of Joyce Marcus is instructive here. It will be recalled from chapter twenty-one that she looked at hieroglyphic scripts in four Mesoamerican cultures and concluded that they were not literate in the accepted use of that term. Literacy, she decided, was not the aim of the Mixtecs, Zapotecs, Mayans and Aztecs, and that may have been true of all cultures with hieroglyphic script: its use was chiefly as propaganda, to boost a society's self-esteem, confirming the genealogy of the ruling regime and reinforcing social stratification. So all societies with writing are not necessarily literate. This is a useful gloss, and an advance.

We can now see that some profound differences have grown up between the peoples of the two hemispheres and we are at last in a position to put those differences into context.

We may say that, at its most basic, people 'become human' – become the rounded, integrated, reflective observers that they are – by means of a three-stage process.

The first element in this three-stage process is that people are placed, unavoidably, in a landscape, an environment. They live – or settle – on mountainsides, or in valleys, in the jungle, by rivers or next to the sea. They inhabit arid deserts, cold, tundra-like forests, or extensive grasslands. Some move between different landscapes.

They are surrounded by animals – by birds or fish, by predators perhaps. They share their landscape with plants – grasses, shrubs, trees and flowers, some more nutritious, more medically useful and more psychoactive than others. And people live among weather: they live surrounded by different and systematic admixtures of sunshine, rain, wind, hail, lightning, they suffer natural catastrophes such as earthquakes, volcanoes, hurricanes and tsunamis. They live under the heavens – the sun, the moon, and the stars including the Milky Way. And finally they live on land, on continents, that are scattered randomly across the spherical globe and are in different relationships with the great oceans. That land is primarily north-south in orientation, or east-west, configurations that are basic to weather and climate and to the *history* of weather and climate. All of these factors, we can now see, come together to create, broadly speaking, two great entities across the world, two configurations whose similarities and differences help explain the separate development of mankind on the two great hemispheres.

What is also clear from the story this book has told is that these combined factors – let's call them *environmental* factors – operate on human beings to produce in them, and this is the second stage of the three-stage process, an *ideology*, a way of looking at the world, a way of understanding and interpreting that world, a way of making sense of the Earthly phenomena that manifest themselves and surround human beings everywhere. It is straight away evident from this book that ideologies *vary* much more in the Old World than they do in the New, in ways that are discussed later.

In the third stage of the process, the ideologies that people adopt as a result of their surrounding environment, and the technologies they develop, continue to interact with that environment, which of course itself continues to change, partly as a result of the evolution of the Earth, of cosmological, astronomical and geological events, and partly as a result of the changes that overtake humankind itself as a result of the first two stages.[4]

And this perhaps is the second most interesting generalisation to emerge from our story, after the initial formulation of ideology based on the environment/climate/human conjunction: *this* is the determinant of what Fernand Braudel and the French historians would call *la longue durée*. History is in effect the narrative of humankind's changing ideology and its continuing interaction with the environment – economic, ecological, technological.[5]

If this analysis is correct, then it helps us understand the very different trajectories followed in the two hemispheres. As was outlined in chapter five, the Americas are a much smaller landmass than Eurasia, even without Africa added on. Moreover the New World is, as Hegel, Jared Diamond and others have pointed out, oriented in a north-south direction rather than east-west, as Eurasia is. This orientation in itself impeded development, relatively speaking, by slowing down the rate at which plants – and therefore animals and civilisation – can spread. This was not wholly bad, of course. It meant that in particular localities many species evolved. (The tropical rainforest, for example, occupies 7 per cent of the land surface of Earth but nurtures well over 50 per cent of the animal and plant species. Because there are so many insects and small mammals in the rainforest, energy is lost along the food chain, with the result that large mammals are relatively rare – and large mammals have played a vital role in our story.)[6] But the north-south orientation of the New World, in conjunction with other factors to be considered shortly, did undoubtedly slow down the development of humanity in the Americas. It was more a technical limit before anything else, but it had a knock-on effect too, as we shall see.

Alongside the general geographical alignment of the continents went an associated climatic variation, of which the most important elements were the Monsoon, the El Niño Southern Oscillation (ENSO), and the violent activity caused by volcanoes, earthquakes, winds and storms. The importance of the monsoon lies in the fact that, for the last 8,000 years, since the time of the last great flood, outlined in chapter two, the monsoon has been *decreasing* in strength. The varying strength of the monsoon and its temporal relation to the emergence (and subsequent collapse) of Old World civilisations was described in chapter five. All we need add here is that, given the domestication of cereal grasses in the Old World at about 10,000 years ago, the major environmental/ideological issue in Eurasia over that time has been *fertility*. The landmass, bit by bit, has been drying.

In the New World, on the other hand, the major factor affecting weather has been the *increasing* frequency of ENSO, from a few times a century about six thousand years ago, to every few years now. Besides the occurrences of ENSO itself, its relationship with volcanic activity, given the make-up of the Pacific Ocean (an enormous body of water over a relatively thin crust), also appears to have been important. We saw in chapter five that Meso- and South America are the most volcanically active mainland areas of the world where major civilisations have formed. Put all this together and the most important environmental issue in the Americas over the past few thousand years, which has had fundamental ideological consequences, has been the increasing frequency of *destructive weather*.

We cannot say with certainty that these differences were definitive, or that they ultimately account for the systematic ideological variations we shall be discussing shortly. We have already noted that in our natural experiment there are too many variables to satisfy purists. What we can say is that these systematic differences in climate across the hemispheres dovetail plausibly with the historical patterns that are observable between the New World and the Old World and to that extent may help us understand the different trajectories.

How and Why the Gods Smiled on the Old World

After the geographical and climatological factors that determined basic and long-term differences between the two hemispheres, the next most important factors lay in the realm of biology – plants and animals. In the plants realm, we may say that, again, there were two main differences between the hemispheres. The first concerned cereals: grain. In Eurasia in particuar there was a naturally occurring range of grasses – wheat, barley, rye, millet, sorghum, rice – susceptible of domestication and, because of the east-west configuration of the landmass, they were able to spread relatively rapidly once domestication was achieved. Surpluses were therefore built up relatively quickly and it was on this basis that civilisations were able to form. In the New World, on the other hand, what turned out to be the most useful grain there was evolved from teosinte which, in the wild, was, morphologically speaking, far more distant from the domesticated form than was the case with the Old World grasses. Furthermore, as we now know, because of its high sugar content (as a tropical rather than a temperate plant), maize was first used for its psychoactive properties rather than as a foodstuff. On top of everything else, maize – even when it did become a foodstuff – found it harder to spread in the New World because of the north-south configuration of the landmass, which meant that mean temperatures, rainfall and sunlight varied far more than they did in Eurasia. For this reason, the development of maize surpluses was much harder – and slower – to build up. As noted, possibly only Cahokia followed an Old World trajectory at all closely. Thus the domestication trajectory of the most important New World grain was very different from its more numerous counterparts in the Old World.

The second crucial area where plants differed between the Old World and the New was in the realm of hallucinogens. The influence of these plants on history has perhaps not been appreciated before to this extent, but it is now clear that the distribution of psychoactive plants across the world is curiously anomalous. The figures, as we saw in chapter twelve, are that between 80 and 100 hallucinogenic species occur naturally in the New World, compared with not more than eight or ten in the Old World.

It is also now clear that hallucinogens played a large and vital role in the religious thought of the New World but especially so in Central and South America, where the most advanced civilisations evolved.

The role and effect of hallucinogens was essentially two-fold. First, they made the religious experience in the Americas much more *vivid* than in the Old World. Second, because of their psychoactive properties, as shown by the experiments of Claudio Naranjo, discussed in chapter twelve, hallucinogens fostered ideas of *transformation*, between humans and other forms of life, and of travel, or soul flight, between the middle world and the upper and lower realms of the cosmos. Combined with a society in which, because of the lack of wheeled transport, or riding, and the north-south configuration, people found it relatively difficult to travel far, the journeys to the upper and lower realms were all the more important. The sheer vividness, and the fearsome nature of some of the transformations experienced in trance, the overwhelming psychological *intensity* of altered states of consciousness induced by hallucinogens, would among other things have made New World religious experiences far more *convincing* and therefore more resistant to change than those in the Old World where, as we have seen and shall see again shortly, horse-riding and wheeled transport – carts and chariots – meant that different groups, with different beliefs, came into contact with each other far more.

This is not to say that there were no hallucinogens in the Old World, or that they were not important. As was discussed in chapter ten, opium, cannabis (hemp) and *soma* were all widely used ritual substances in various regions of Eurasia. For a variety of reasons, however, the more powerful psychoactive substances gave way relatively early on to milder alcoholic beverages, whether this was because domesticated mammals needed to be controlled (riding, driving, ploughing and milking in particular needed concentration), or because the pastoral lifestyle was less communal, meaning people came together not so much for intense shamanistic ceremonies, but for more social bonding reasons, where milder euphoriants were more suitable, or because they came together to face threats from outside, when again strong psychoactive substances would have been inappropriate, while alcohol was acceptable in warrior-bonding. Beer and wine thus characterised the Old World whereas hallucinogens were more common in the Americas. This provoked a change in ideology in Eurasia, helping the relative demise of shamanism.

In the Old World what was worshipped instead were two aspects of fertility – the Great Goddess and the Bull. Though the Bull was worshipped as an aspect of fertility, we do well to remember that this animal was more often represented by his distinctive bucrania – its head and horns – than its sexual organs. Probably this was due – as many scholars have said – to the similarity between the bull's horns and the shape of the New Moon, added to which was the link between the phases of the moon and the menstrual cycle, especially the cessation of menstruation, which would have been noted. At this point in the history of the Old World, the Great Goddess is shown giving birth to a bull, with the bucrania emerging from her womb. Here we have a curious combination: no one can ever have *seen* a woman giving birth to a bull; there was clearly some confusion at this point over the mechanism of human reproduction. If the bull represented the powerful forces of nature, as well as fertility, as other scholars have maintained, these motifs may indicate that people were unaware of the real mechanics of reproduction and at that time believed instead that women were fertilised by one or other of the forces of nature, symbolically represented by bucrania. Whatever the exact beliefs, the essential thing is that throughout the Neolithic period in the Old World – whether the object of worship was the Great Goddess, the Bull, the cow, rivers or streams – the central issue was fertility, in particular human fertility.

What we now know, but previous scholars didn't, or didn't pay due attention to, is that there were *two* threats to fertility in the Neolithic Old World. There was the weakening monsoon, which affected the fertility of all living things, but there was also the fact that the *human* pelvic channel had grown narrower under a sedentary, grain-based diet, as compared with a hunter-gatherer one.

On top of all that was a developing interaction between humans and domesticated mammals that had immense ideological and economic consequences. Put succinctly and chronologically, those developments were as follows:

1. The domestication of cattle, sheep and goats enabled the exploitation of less good land. This brought about the development of pastoralism, as a result of which these kinds of farmers spread beyond village life and became more dispersed. This dispersal in turn had an effect on religious ideology, a move beyond shamanism. Among pastoralists the calendar was less important, because domesticated mammals give birth at different times of the year, unlike plants which, particularly in temperate zones, are more directly linked to the cycle of the seasons. (Cattle can give birth at any time of the year, goats in winter or spring; with sheep it depends on how near the equator they are – in temperate zones, sheep lamb in spring but in warmer climates the lambing season can extend throughout the year. For horses the natural breeding season is May to August.)

   A further aspect of domesticated mammals is that their whole life takes place above ground, as it were. Unlike plants, which need to be sown in the soil, and spend some time out of sight, before re-merging in a different form, animals are less mysterious. In a pastoral society the underworld is less important, less necessary, less ever-present. Together with the relative absence of hallucinogens, this development made the netherworld far less of an issue in the Old World than in the New.

   This may have had other consequences. Although people in the New World never developed the wheel, for good reasons, they did have the concept of roundness – they had rubber balls for their ball games, they sometimes formed balls out of human heads or captives' bodies, which they rolled down pyramid steps, and the combatants in boxing games fought with purely spherical carved stones in their fists. Furthermore, New World peoples undoubtedly witnessed the sun and moon in the day and night sky, and eclipses of both, without ever appearing to consider that the Earth itself was spherical. This is surely because, in a predominantly vegetal world, with the experience of the netherworld so vivid (and other 'realms' so accessible via hallucinogens), 'flatness' and layers were much more obvious than roundness. Not travelling great distances, particularly across the sea, aided by useful winds, they had less chance – and were therefore less prepared – to experience the world as a spherical object.

2. The domestication of the horse had a number of different consequences. It accompanied the development of the wheel and the chariot and led to riding. These were enormous advances, adding to the mobility of men and women in Eurasia, aiding in particular the creation of palace states, far larger than most of the New World states because the horse and chariot allowed larger territories to be conquered *and then held*. In the same way, the wheel and cart meant that more goods could be carried further, boosting trade and the prosperity and exchange of ideas that went with it. These factors all came together from time to time in great wars, which also displaced peoples, languages and ideas across vast areas in large numbers. The Old World was mobile in a way that the New World was not.
3. Horses and cattle in particular are large mammals, valued for their power. That power, however, meant that, as well as being useful, they were potentially dangerous. In such a context, the regular and frequent use of mind-altering substances was hazardous. A shaman in trance could not have handled a horse or a cow, let alone a bull. On top of that, as dispersed populations developed the habit of coming together for spouse selection, marriage, and to resist threats from outside (now greater, because wealth in the form of domestic mammals could be stolen as land couldn't), people were driven away from hallucinogens, which offered powerful, vivid (and at times threatening) but *private* experiences, and were led instead towards alcohol, which offered milder, euphoriant *communal* social-bonding experiences. This was a major move beyond shamanism.
4. In this way, pastoral nomadism emerges as one of the 'motors' of Old World history. This is because of its *inherent instability* as a way of life, because the weakening monsoon caused the drying of the

steppes – the natural home of pastoral nomads – meaning that they could no longer subsist as easily in their traditional fashion, and must disperse still further and invade the settled societies at the edges of the great grasslands. The predominantly east-west nature of the Central Asian steppes ensured that peoples and ideas travelled right across Eurasia. Since weather was more important to the nomads than vegetal fertility, and because they lived on milk, blood and meat, their gods were sky gods – storms and winds – and horses. Their religious ideology was very different from those of the more settled societies and the endemic conflict between nomads and settled peoples was both destructive and, in the long run, creative.

5. The virtually continuous conflict introduced into Eurasian history over 2,700 years, from 1200 BC to AD 1500, by the fact that highly mobile pastoral nomads were at all times more or less threatened by climatic factors (the weakening monsoon and the drying steppe) was one factor in bringing about the end of the Bronze Age, the destruction of the great palace states created on the back of the horse-drawn chariot, and eventually provoked the great spiritual change known as the Axial Age, the great turning away from (man-made) violence and the epochal turning in, which produced a new ideology, or morality, and culminated – this time among the pastoral nomadic Hebrews – in the idea of monotheism. It was, according to Daniel Hillel, the fact that the nomadic Hebrews wandered between so many different ecological habitats, that gave them the idea of one overarching God that governed *all* environments.
6. Greek rationalism, and Greek science, in particular the Greek concept of nature, partly brought about by a close examination of domesticated mammals and how their nature compared with human nature (whether they had souls, whether they had morals, whether they had language, whether they could suffer, chapter nineteen), when adapted to the Hebrew idea of a single, abstract god, eventually gave rise to the Christian idea of a rational God, who favoured order in the natural world, whose own nature could be gradually uncovered at some point in the future *because He favoured order*. And this idea, *of the possibility of progress*, of God revealing himself gradually, by means of linear history, as was outlined in chapter twenty-two, helped to create many of the innovations that would enable humankind to explore the Earth via its great oceans.

7. The many and varied tribes of pastoral nomads emerging on and then escaping from the central Asia steppes continued in the millennium-and-a-half after Jesus Christ. It assisted and hindered the east-west movement of goods and ideas, but above all maintained Eurasia as a landmass across which there was much rapid movement. The horse also proved to be a vector for the transmission of disease (the plague) across the same landmass, which had the dual effect – again in the long run – of promoting the wool industry in the north of Europe, the sheep providing the substance of the first great industry in the world, but also forced the inhabitants of the western Mediterranean to look for alternative routes to the East, where so many spices, silk and other luxuries came from. Together, these factors helped open up the Atlantic.

Again, we must emphasise that these developments were separated in time, location and ultimate effect; there was no inevitability about them – each of them, although they all involved domesticated mammals, was quite discrete. In a sense this is a meta-narrative of history but it is by no means a straight line, or even a line at all, more a series of punctuated events linked only by the involvement of domesticated mammals.

A further point is that most of this activity in the Old World took place in temperate zones (between seven degrees north and fifty degrees north), that is to say where the seasons were pronounced, where the planting and growing periods were carefully delineated. The seasonally characteristic nature of essentially fertility worship contributed to the organised nature of what was early religious life but it also had a far more important ideological corollary: *it worked.* The simple biology underlying a religion where fertility was the central issue, in temperate zones, was that – sooner or later – vegetation started to grow again. The cycles of planting and growth didn't invariably work, of course, when drought or rain bringing floods or other factors interfered with the rhythm (the bible's 'fat' and 'lean' years) but, essentially, far more times than it failed, fertility worship worked. In pre-literate societies, against a weakening monsoon, rituals would have grown more elaborate, and priests may have lost some credibility now and then but, by and large, until the development of monotheism and more abstract deities, largely removed from the seasons, fertility religion in temperate zones was a fairly predictable affair. Furthermore, the domestication of plants and animals took some of the *fear* out of life – fear of famine, for instance, though dependency on fewer varieties of plants carried its own risks.[7] It is in the nature of fertility worship that you want plants to grow, you want animals to give birth, you want something *to happen*. And, ultimately, the gods smiled on humankind.

'Tectonic Religion' in the New World

Ideological life in the Americas was very different. For a start, there were no domesticated mammals, save for the llama, vicuña and guanaco in South America. One effect of this absence of domesticated mammals was to make vegetal life far more salient in the Americas, and this brought with it certain ideas.

The simple, the most obvious, and the most powerful, is that plants need to be planted *underground*, where they undergo a transformation from seed into shoot. This, combined with hallucinogenic experiences, helps explain why, for the ancient Americans, the cosmos was divided vividly into three zones – the Upper, Middle and Underworld. It was convincingly reinforced by the experiences of the shaman who, in trance, underwent soul flight, in order to consult the gods and/or the ancestors, and who used hallucinogenic plants to achieve these feats. Fertility was an issue in the New World but, in the tropical rainforests, teeming with life and with plants growing in profusion all through the year (as manioc did, for example), and where the seasons hardly varied, it was never the *overwhelming* issue it was in the temperate Old World.

Much more important in the New World mindset were the feared and admired jaguar, and the weather gods – gods of lightning, rain and hail or violent winds, of thunderstorms, erupting volcanoes, earthquakes and tsunamis, 'dangerous weather', as Peregrine Horden and Nicholas Purcell put it.[8] Moreover, volcanic activity, 'tectonic religion', as they also say, comprise evidence of 'humanity's closeness to the underworld in general' and its sheer power in particular.[9] With the ENSO increasing in frequency, at least during the last 5,800 years (chapter five) the gods, far from smiling on humankind in the New World, have been getting angrier.

This outline is further supported by a third factor: copious evidence of violence inherent in ancient Meso- and South American religions generally. In chapter twelve we saw that the Cashinahua, under the influence of hallucinogens, saw snakes, falling trees, terrifying jaguars, anacondas and alligators. We saw that the Mayans dreaded the mushroom of the underworld and at other times worshipped storm gods. We saw that cacao was linked to volcanoes, sacred vessels being made of volcanic ash. In chapter fourteen, we saw that the jaguar was linked to lightning and thunder, how it was invariably depicted with its fangs showing, and snarling, its claws exposed as it raped or attacked or ate human hearts. We saw how, in some Mesoamerican cities, obsidian blades, used for ripping out the hearts of sacrificial victims, were metaphors for jaguar teeth. In chapter seventeen we noticed that the Olmec faced the problem of too much water, that they had 'inundation' cults, Lords of the Storms, Masters of Lightning, and that their shamans were known as 'men of hail'. We noted that the Chavín, though not a rainforest people, nonetheless depicted snarling jaguars in their art, that their architecture, as Richard Burger says, acted as a focus for 'dangerous supernatural forces'. In chapter twenty we recorded how volcanoes were treated as gods, the impact of earthquakes on the Chavín, the devastation that El Niños caused the Moche. We saw that the mountains were gods to the Mayans, that the Zapotec and Mixtec worshipped natural forces, rain and lightning respectively. And in chapter twenty-one we explored the concept of 'dark shamanism', the manipulation of threat, that negative events, events that need to be averted, were of immense importance in these New World societies.

In chapter twenty-one, Steve Bourget's work was discussed, in which he showed a direct association between sacrifice and torrential rain on the north coast of Peru, which he says may have been El Niño events (page 430). Children sacrificed nearby on mountain tops had zigzag decorations, as if they were dedicated to lightning (page 431). The very existence of weather shamans, as also discussed in chapter twenty-one, shows how important weather was, and the equation of weather with illness, as was also noted in that chapter, implies that it was the negative aspects of weather that were paramount. Among the Toltecs, Tezcatlipoca was a malevolent god – 'he caused plagues, droughts, frosts, food poisoning, starvation, the appearance of monsters and collective massacres'.[10] Enrique Florescano informed us in chapter twenty-three that, judging by the iconographic evidence, 'In the most ancient times, the important gods of Mesoamerica were those of the netherworld. These powers managed the forces of destruction, decadence and death . . .' (pages 481–2). We also noted in chapter twenty-one the description by Arthur Demarest and Geoffrey Conrad of the Aztec gods as 'ever-threatening'.[11] (Moctezuma Ilhuicamina was 'the angry lord, the archer of the skies' – aggression was built in.) As we saw in chapter twenty-three, each age of the Aztec cosmological system takes its name and character from its destructive elements, not from its creations. 'Each beginning is sure to result in a catastrophe and there appears to be no end in sight to this divine antagonism, these rains of fire, vigilante jaguars, deluges and hurricanes . . . One cannot help but be impressed by the persistence of the motif of change, sacrifice, death and destruction . . . '[12] And we noted in the same chapter that the second part of the name of the god Xipe Totec means 'dread'. Finally, the very fact that the Aztecs and other cultures featured jaguar warriors, that boxers wore jaguar masks, and that in the Aztec mythology the jaguar cult beat the eagle cult, all underline what was said earlier, that *fear* of the jaguar was the dominant emotion.

Put all this together and you have a crucial difference between New World and Old World gods.

If you worship angry gods, whether they be tsunamis or earthquakes, volcanoes or jaguars, your worship essentially takes the form of propitiation, of asking – petitioning – those gods *not* to do something, not to erupt if the god is a volcano, not to fall in torrents if the god is rain, not to produce destructive tsunamis and winds if the god is an ENSO event, not to attack humans if the god is a jaguar. In the New World – in Central and South America certainly – the predominant form of worship was directed towards making unpleasant things *not happen*.

And here is the crucial point: that form of worship doesn't work. That is to say, it didn't/doesn't work all the time, or to anything like the extent that fertility worship works. It no doubt works for *some* of the time: no one in the village is carried off by a jaguar for a certain number of weeks; there is no tsunami for a few years, or even decades; a volcano dies down, as the Icelandic ones did in 2010 and 2011. But, and it is again an important but, the angry gods are never totally appeased. Sooner or later, their wrath recurs. (There is some evidence that a raft of earthquakes, *circa* AD 1300, had an effect on what remained of the Mayan civilisation.)

We also know that, in the case of ENSO episodes, they have been getting *more* common, quite a lot more common, in fact. Looked at from the point of view of an Olmec or Mayan or Toltec or Aztec shaman, with their extremely accurate calendar, it would have seemed to them that worship wasn't working, that whatever traditional level of ritual had been practised in the past, it wasn't enough. This, after all, is why the Mixtec ritual specialists sponsored war, to manufacture threats that they *could* control.

In such circumstances, religious specialists would have decided that, if the current level of worship wasn't working, they must *either* manipulate threats they could control, war, *or*, they must redouble their efforts. And this is why the most profound and revealing difference between the Old World and the New occurs in the realm of human sacrifice. In the Old World, thanks to the proximity of domestic mammals, human sacrifice was gradually replaced by animal sacrifice and then, after AD 70, thanks again to the close similarity of domestic mammals to humans, blood sacrifice was abolished altogether. In the New World, however, far from being abolished, human sacrifice became more and more widespread, until in the fifteenth century tens of thousands of Aztec victims were being sacrificed each year. Inca sacrifice was not quite so numerous, but there were still hundreds of mountain *huacas* where people were sacrificed and, according to some accounts, hundreds of children were killed at a time (pages 490–2). We are now in a position to explain this striking anomaly and to discuss its central relevance to our story.

Ever-Angrier Gods

In South America there was a further factor, the idea that death – at least for some people – was not the end, that there was a form of continued existence midway between life and death, founded on the naturally occurring mummified remains that comprised part of the ritual life of the earliest inhabitants, and which found its fullest expression in the Inca system of split inheritance and *panaqa*, whereby dead kings were treated as, to all intents and purposes, still living (pages 492–3).

In such an environment, where death was apparently not always so 'final' as it is for us today, sacrifice would not have been seen as so terrible. This is not to say that it wouldn't have been without pain or suffering, but it *is* to say that it would not have been quite so terrible as it now sounds. We are reminded that, as discussed in chapter twenty-one, attitudes to death *were* different in the New World, where parents would give or sell their children as sacrificial victims, where gamblers at the ball game would wager their own lives on the result, where the *victors* in ball games were sometimes sacrificed (who in the modern world would 'win' under such circumstances?), or where the Inca parents who donated their children as sacrificial victims were not allowed to show any negative feelings. The inscriptions showing people with tears in their eyes (page 434) do seem to suggest that, despite the possibility that the psychoactive plants of the New World could help stupefy potential victims, pain was nevertheless very real in the sacrificial ceremonies. But it is probably wrong to see the pain of the victims in sacrificial rituals as separate from the pain of the captors and rulers who led the rituals, whose own auto-sacrifice was crucial. Pain had religious meaning.

One explanation we can give for this, though it perhaps betrays a modern, Western, post-Christian bias, is to say that asceticism, stoicism and fortitude were admired and valued in the New World civilisations. A better explanation, more functional, at least in this author's view, is that it reflected a subtle but marked change in ideology. Blood was important in the New World rituals and blood*letting*, as discussed in chapter twenty-one, was an evolution of the shamanistic system. Traditional shamans, entering trance via hallucinogens, had dominated small-scale societies, consisting of tens, or at most hundreds of people in villages. In the later great urban centres, with populations in the thousands, or tens of thousands, more theatre was needed, and the leaders needed to adopt a system that didn't break with tradition, not totally, but extended and improved it, which awed the larger populations and at the same time under-girded the shaman-kings' unique link to the gods. The deliberate shedding of their own blood in copious quantities, amid much self-inflicted pain, which produced trance – the traditional device of the shaman – was such a system. Pain, and the fear associated with it, became a form of authority – the worse the pain, the more blood that was shed, the more authority someone had. Sacrifice, self-sacrifice, even death itself, was in this scheme of things the ultimate form of power, in the Inca world as much as in the Aztec. Shamanism, and the vivid other worlds encountered in trance, convinced people of these other worlds much more than did the rituals of the Old World. Amid such a set of beliefs, one can see how attitudes to sacrifice would have been different: the more you are convinced that other worlds exist, the easier it is to forgo this one.

We don't know how this system came about and probably never shall. However, given that at least some New World wars were fought for captives, rather than for territory, and given that some rulers or nobles were tortured for considerable periods of time before being sacrificed – months or even years – it is at least possible, even likely one might think, that a small number of noble warriors was captured, tortured, during which time they lost so much blood that they entered trance, *and were then rescued.* Such individuals would have been able to recall their experiences once back in their own villages or towns and would have incorporated them as a feature of their own rituals.

We need to address the question as to why sacrifice was so strong in both Mesoamerica and the Andes. After all, we saw earlier that many ideas and practices did *not* travel easily (or at all) between the two regions (writing and the llama are two examples). This would seem to support the argument that sacrifice has its origins in catastrophe. Both Mesoamerica and the Andes, as is all too clear by now, were and are volcanically active, situated along the same tectonic rim, and both at the eastern end of the El Niño configuration. The practice of sacrifice grew up independently in the two regions, as it did in the Old World.

The fact that (animal) sacrifice effectively ended in the Old World in AD 70, while human sacrifice (and many other forms of painful violence) continued to grow in frequency in the Americas, is a salutary reminder of how environment and ideology can interact to produce marked differences in human behaviour, in the very meaning of humanity.

We shouldn't overlook the role of accident in history. This book has been about the systematic differences across the globe that account for the separate trajectories of the Old World and the New. But accident surely played a part too. A good example of this is provided by a comparison between the Aztecs and the Incas, on the one hand, and pastoral nomads on the other. As described in chapter twenty-three, both the Aztec and Inca societies were inherently unstable, the practice of securing ever-increasing numbers of captives for sacrifice, and worshipping dead kings who kept hold of their land, being ultimately maladaptive. We don't know where these maladaptive strategies would have led had not the Conquest intervened in either case, but the omens were not good.

On the other side of the world, the way of life of pastoral nomads was equally maladaptive – in the long run they could not continue to survive in their traditional lifestyle and this is why they were continually breaking out of the steppes. *But* they had somewhere to go, more settled societies to attack or trade with and make the most of. *Their* shortcomings, as it turned out, were in the long run productive. But there was nothing inevitable about it.

This book has been primarily about civilisations (not entirely, but mainly). Many groups of people, in both hemispheres, never developed into civilisations but this does not necessarily imply that those societies were *failures* in any way. To the contrary, such lifestyles as those of the Plains Indians in North America, whose coexistence with the bison endured for millennia, or the Native Americans of the north-west Pacific coast, who lived alongside rivers teeming with salmon for just as long, must be regarded as successful communities, thanks to the sheer abundance of food which surrounded them. The same may be said about the inhabitants of Australia, Melanesia, Micronesia and Africa, who also never developed 'high' civilisations. The people living in Australia in the seventeenth century, for example, when Europeans first arrived, had a Stone Age culture (with their own form of shamanism).[13] In his book, *Man's Conquest of the Pacific* (1979), the Australian archaeologist Peter Bellwood concludes that, though the South East Asia mainland did not develop an urbanised civilisation until the period of intensive Indian and Chinese influence, at about the time of Christ, that New Guinea had no sizeable animals – mammals or otherwise – and that Polynesia had only a 'half-civilisation', nevertheless 'the quality of life for the prehistoric South East Asian villager was probably no worse and perhaps much better than that of his Chinese, Sumerian or Egyptian urbanized counterpart'.[14] They had *adapted*: that is what people do. Civilisation is but one form of adaptation, as should be clear from this book.

*

We can now see that the main difference between the Old and the New World *civilisations* (leaving the smaller polities to one side) is in their patterns of adaptation to different environmental circumstances, and that the Old World ideologies changed more often and more radically than did the ideologies of the Americas. And that while this was due to some extent to differences in climate and geography – the weakening monsoon in the Old World and the increasing frequency of El Niño in the New World – it also had a great deal to do with the role in the Old World of domesticated mammals and in the New World of hallucinogenic plants. We may therefore say – exaggerating only slightly – that the core of Old World history was defined largely by the role of the shepherd, whereas in the New World an equivalent role was fulfilled by the shaman. As late as 1972, in Trujillo, Peru, there was an outdoor shamans' market, where folk medicines were traded.[15] The shaman and the shepherd epitomise the great divide.

In the Old World, the existence of domesticated mammals released humans from place and that mobility, in conjunction with the pattern of fertility, associated with the weakening monsoon, favoured the development of several ideologies, culminating in the Christian/Greek idea of an abstract but rational god, with ideas of linear time and 'progress'. In the New World, in Latin America at least, where civilisations appeared, the great violence and destructive capacity of the weather, indeed its *increasing* frequency of destruction, combined with the essentially vivid characteristics of trance-inducted shamanism, was much more difficult to cope with in a rational way. The gods of the New World were not as manageable, or anywhere near as friendly, as cooperative, as *understandable*, as those in the Old World. All these factors made the New World a harder place to adapt to than the Old World.

The natural experiment that has been the subject of this book enables us to say that the evidence presented here shows that religions, and worship, are *entirely natural* responses to the predicaments in which early peoples found themselves. Beliefs and practices such as shamanism, animal sacrifice, human sacrifice, bloodletting and the consumption of mind-altering drugs are clearly linked – and linked intimately – to the immediate landscape in which early peoples were located. Judaism, Christianity and Islam may be more 'developed' religions than most but, nevertheless, they are no exception to this general rule. Religion (or 'ideology', a better term) is therefore most fruitfully understood in an *anthropological* sense, as part of humankind's attempts to interpret his/her world and the enormous, mysterious forces that shape history, and account for the great divide.

Appendices

• Appendix 1 •

The (Never-Ending) Dispute of the New World

In the introduction, it was noted that, to an extent, people still wrangle over the relative merits of the Old and New Worlds in pre-Columbian times – which hemispheres had the biggest cities, and when, which civilisations were the most 'evil', whose shoes were the most comfortable. It has been the aim of this book to put these anecdotal and ad hoc observations and comparisons into a meaningful context. In this appendix, we describe the history of the 'dispute' about the New World. While it is not directly relevant to our main story, the chronology of the dispute does show how attitudes to the New World have themselves evolved over the centuries, often revealing more about the people making the comparisons than the civilisations that were the subject of their often very strongly held views.

Right from the word go, the Old World had a problem assimilating the New World – its history, its psychology, its very *meaning*. At first the rewards of empire were disappointing, just as that first momentous encounter on Guanahaní had been disappointing. Columbus's voyages sparked the spread of the Christian religion, the language and the culture of Spain, and initiated a transformative exchange of plants, animals and microbes, whose implications are still being unravelled.[1] As is now well known, Eurasian diseases such as smallpox and influenza, carried by colonists, ravaged American populations who lacked the immune systems that had been built up in the Old World over millennia. Syphilis appears to have gone the other way, though the latest evidence contradicts this.

Some measure of the initial impact of Columbus's discoveries can be had from the fact that his first letter was printed nine times in 1493, and reached twenty editions by the end of the century. The Frenchman Louis Le Roy wrote, 'Do not believe that there exists anything more honourable ... than the invention of the printing press and the discovery of the new world; two things which I always thought could be compared, not only to antiquity but to immortality.'[2] Yet John Elliott rightly warns us that there was another side, that many sixteenth-century writers had a problem seeing Columbus's achievement in its proper historical perspective. For example, when Columbus died in Valladolid, the city chronicle failed to mention his passing. Only slowly did Columbus begin to attract the status of a hero. A number of Italian poems were written about him but not until a hundred years after his death, and it was not until 1614 that he featured as the hero of a Spanish drama – this was Lope de Vega's *El Nuevo Mundo descubierto por Cristóbal Colón*.[3]

To begin with, interest in the New World was confined to the gold that might be found there and the availability of vast numbers of new souls for conversion to the Christian faith. The actual phrase, 'the Black Legend' wasn't used until 1912 by the Spanish journalist, Julián Juderiás, in *La leyenda negra*, a publication in which he protested the 'easy characterisation' of Spain by other European nations 'as a backward country of ignorance, superstition and religious fanaticism'.[4] The Spanish case is not so much that the Spanish colonists were *not* brutal – the evidence is just too overwhelming that they were – but that they were not *uniquely* so and that other European countries were envious of Spain's early successes in the New World and sought to spread the 'black legend' for their own purposes.

It is certainly true that the first condemnation of Spanish practices in the New World came from its own citizens, mainly missionaries, and of whom the most famous was the Dominican priest and bishop of Chiapas (in Mexico), Bartolomé de las Casas who, after the failure of his legal arguments against the *encomienda* system of slave labour, decided to publish his *A Very Short Account of the Destruction of the Indies* (Seville, 1552). 'This volume would quickly become a cornerstone of the Black Legend, to be translated and republished over the centuries with each new conflict involving Spain and its European rivals or American colonies.'[5]

The evidence is too well known to need any detailed rehearsing and reheating here – accounts of people being slowly roasted above a gibbet, on Hispaniola, women on Higuey being given only grass to eat, so that their milk dried up and they were unable to nourish their children, while in Florida, to coerce the 'natives' into obedience, their noses, lips and chins were 'sliced from their faces'.

Many contemporary historians now prefer to situate the Black Legend as part of the general rise of racism in the Renaissance, having to do with the discovery of many new parts of the world, and their inhabitants, as a result of the age of exploration, the re-conquest of Spain from the Arabs, and a militant form of Christianity that viewed itself as the one true faith, with all other religions being seen as inferior.

The Moral History of the Indies

The discovery of America was important intellectually for Europeans because the new lands and peoples challenged traditional ideas about geography, history, theology, even about the nature of man. Insofar as America proved to be a source of supply for goods for which there was a demand in Europe, it had an economic and therefore a political significance. 'It is a striking fact,' wrote the Parisian lawyer, Etienne Pasquier, in the early 1560s, 'that our classical authors had no knowledge of all this America, which we call New Lands.'[6] 'This America' was not only outside the range of Europe's experience but was beyond *expectation*. Africa and Asia, though distant and unfamiliar for most people, had always been known about. America was entirely unexpected and this helps explain why Europe was so slow in adjusting to the news. Anthony Pagden has said that it wasn't until the beginning of the eighteenth century that Europeans really began to get to grips with what the existence of the New World implied, or to consider the American Indian as a less developed form of life, like women or children.[7] There were constant battles over whether there were three or four or more stages of 'barbarianism' and where the various Indian polities existed on this progression.

There were people like Hernán de Santillán who thought the Incas had forms of government so good that they deserved to be imitated,[8] and Francisco de Vitoria who maintained that, since God would create nothing useless, efforts must be made to understand the Indian way of life.[9]

As the sixteenth century passed, Las Casas and José de Acosta insisted that *empirical* knowledge was essential if the Indians were ever to be understood, the former stressing the unique status of his voice, the primacy of his eye, the latter author dividing his *Historia* into two parts, the first dealing with 'works of nature' and the second with 'things of the free will' – normative behaviour, patterns of belief and the past history of the American man.[10] It was Acosta who surmised that the Indian could never have sailed to America, taking with him such fierce animals as the jaguar and the puma, and concluding therefore that he must have arrived via the Bering Strait. His astuteness was confirmed when he also observed that the Indians 'lived in fear of their gods', an important difference, he said, between them and Christians.[11]

There was a dispute within the dispute as to whether New World languages – of which there were many – were lexically as strong as Old World tongues, and capable of abstractions sufficient to describe the philosophical and theological concepts of Christianity. John Locke thought the Indian had no need of such words as 'treason', 'law' or 'faith' and could not count above a thousand because Amerindian languages were 'accommodated only to the few necessities of a simple and needy life unacquainted with Trade or Mathematics'.[12] Anthony Pagden attributes a watershed to the work of Joseph-François Lafiteau (d. 1740), the first man to pay full attention to Indian kinship terms and burial customs. This was a replacement of a psychological approach to the Indian (as we would say now) with a sociological one, a change in epistemology which represented a modest but decisive advance.[13]

If the problem of fitting the New World into the scheme of history as outlined in the scriptures was the most intractable of matters, explorers and missionaries alike found that, if evangelisation were to proceed, some understanding of the customs and traditions of the native peoples was required. Thus began their often-elaborate inquiries into Indian history, land tenure and inheritance laws, in a sense the beginning of applied anthropology.[14]

The fact that there was this need was brought about, of course, by the extensive destruction by the Spaniards of the Native American written sources wherever they were found. Any study of this period of history has to contend with the fact that, according to David Carrasco, the Princeton historian of religion introduced in the main text, only sixteen indigenous written (or painted) works have survived the destruction, perpetrated as often as not by friars intent on the extirpation of 'paganism' and the very people who, later on and realising the enormity of what they had done, did their best to rescue what knowledge they could. These sixteen documents are not the only material we can use to recreate pre-Columbian life in the Americas – Carrasco recognises another six types of entity more or less useful. These are: storybooks (i.e., with paintings) generated through Spanish patronage or written independently by Indians, with Spanish glosses; early prose works in Nahuatl (the language of the Aztecs) and Spanish, largely anonymous; prose writings of descendants of Indian elites; letters and histories by Spanish witnesses to the Conquest and its aftermath; priestly writings such as Sahagún's; the archaeological evidence. The *Handbook of Middle American Indians* contains four entire volumes (volumes 12–15, 1972–1975) providing a guide to ethnohistorical sources, which includes censuses of the prose, pictorial, and priestly writings and paintings, and text on a number of maps which were also compiled.[15] Gordon Brotherston, in his *Book of the Fourth World*, an attempt to recreate pre-Columbian history and understanding, lists some 163 sources of this kind.[16]

And even these, as Carrasco also points out, have to be treated with suspicion because of translation errors, the prejudices or beliefs of their European authors or editors, the hidden agendas of people who produced glosses to suit their own ends. Some were hidden in Indian communities or elsewhere and did not emerge into the full light of public scrutiny until much later. This applies, for example, to the *Codex Borgia*, the existence of which was not known about until 1792–97, or the *Codex Fejérváry-Mayer*, not rediscovered until 1829. All of which has made the whole enterprise of recreation of the pre-Columbian Indian world fraught with difficulties.

The early missionaries, fortified by a naïve belief in the natural goodness of man, assumed that native minds were 'simple, meek, vulnerable and virtuous' or, in the words of Las Casas himself, *tablas rasas*, blank slates, 'on which the true faith could easily be inscribed'.

The missionaries were to be disappointed. In his *History of the Indies of the New Spain* (1581), the Dominican Fray Diego Durán argued that the Indian mind could not be changed or corrected 'unless we are informed about all the kinds of religion which they practised ... And therefore a great mistake was made by those who, with much zeal but little prudence, burnt and destroyed at the beginning all their ancient pictures. This left us so much in the dark that they can practise idolatry before our very eyes.' Such a view became the justification for the detailed surveys of pre-Conquest history, religion and society undertaken by clerics in the later sixteenth century.[17]

The Spanish Crown was intimately involved and in the process introduced the questionnaire, bombarding their officials in the Indies with this new tool of government. The most famous were those drafted in the 1570s at the behest of the president of the Council of the Indies, Juan de Ovando. This was a time when the urge to classify was beginning to grow in every field of knowledge and knowledge about America was part of the trend. In 1565, Nicolás Monardes, a doctor from Seville, wrote his famous study of the medicinal plants of America, which appeared in John Frampton's English translation of 1577 under the title, *Joyfull Newes out of the New Founde Worlde*.[18] In 1571, Philip II sent an expedition to America under the leadership of the Spanish naturalist and physician Dr Francisco Hernández, to collect botanical specimens in a systematic way (but also to assess the capacity of the Indians to be converted). In the same year, the Spanish Crown created a new post, that of 'Cosmographer and Official Chronicler of the Indies', though there was a political as well as a scientific reason for this initiative: the political motive was to provide a detailed account of Spanish achievements in the New World, to counteract foreign criticisms, and at the same time it was felt that the science was necessary to reduce the widespread ignorance among the councillors of the Indies about the territory they had responsibility for.[19]

But it was not until 1590, a full century after Columbus's discovery, and the publication in Spanish of José de Acosta's great *Natural and Moral History of the Indies*, that the integration of the New World into the framework of Old World thought was finally cemented. This synthesis was itself the crowning achievement of a century of intellectual transformation, in which three very different aspects of the New

World were incorporated into the European mindset. There was first the American landmass, as a totally unexpected addition to the natural world. There was the American Indian, who had to be incorporated into the European/Christian understanding of humanity. And there was America as an entity in time, whose very existence transformed Europe's understanding of the historical process.[20] All this was, first and foremost, a challenge to classical learning. According to the bible, and to experience, there were three landmasses in the world – Europe, Asia and Africa – and to change this idea was as fundamental a break with tradition as the idea that there wasn't a torrid zone in the southern hemisphere. Moreover, until the Bering Strait was discovered in 1728, it was not clear whether America formed part of Asia or not. When, in 1535, Jacques Cartier encountered rapids in the St Lawrence River above the site of what would become Montreal, he named them '*Sault La Chine*', the Chinese Rapids. A century later, in 1634, Jean Nicolet, a French adventurer, was sent west to investigate rumours of a great inland sea, which led to Asia. When he reached Lake Michigan and saw ahead of him the cliffs of Green Bay he thought he had reached China and put on a robe of Chinese silk in their honour.[21]

One of the most powerful – if implicit – ideas at the time of the discovery of America was the dual classification of mankind, whereby peoples were judged in accordance with their religious affiliation (Judaeo-Christian, or pagan) or their degree of civility or barbarity.[22] Just how rational the Indians were was, however, open to doubt. Fernández de Oviedo was convinced the Indians were an inferior form of being, 'naturally idle and inclined to vice'. He discovered signs of their inferiority, he thought, in the size and thickness of their skulls, which he felt implied a deformation in a part of the body associated with a man's rational powers.[23] Fray Tomás de Mercado, in the 1560s, classified Negroes and Indians likewise as 'barbarians' because 'they are never moved by reason, but only by passion'. It was not far from there to the notorious theory of 'natural slavery'. This too was a major issue of the time. Pagans in the sixteenth century were divided into two, the 'vincibly ignorant' (Jews and Muslims, who had heard the true word, and turned away from it), and the 'invincibly ignorant', those like the Indians who had never had the opportunity to hear the word of God, and therefore couldn't be blamed. This soon became corrupted, however, as people like the Scottish theologian, John Mair, argued that some people were by nature slaves, and some by nature free.[24] In 1512 Ferdinand summoned a *junta* to discuss the legitimacy of employing native labour. Such documentation as has survived shows that many at the time argued that the Indians were barbarians and therefore 'natural slaves'. This was refined around 1530, by what came to be known as the 'School of Salamanca', a group of theologians that included Francisco Vitoria and Luis de Molina. They developed the view that if the Indians were not natural slaves then they were 'nature's children', a less developed form of humanity. In his treatise, *De Indis*, Vitoria argued that American Indians were a third species of animal between man and monkey, 'created by God for the better service of man'.[25]

Not everyone shared these views, however, and others, more sympathetic to the Indian, sought signs of his talent. The most accurate account of this clash of civilisations, on *either* side, says Ronald Wright, was written by some Aztecs for Friar Bernardino de Sahagún in the 1550s, and is now known as Book 12 of the *Florentine Codex*. The authors were anonymous, possibly to shield them from the Inquisition. However, the very search for these signs of Indian virtue and talent, says John Elliott, helped to shape the sixteenth-century idea of what constituted a civilised man. Bartolomé de las Casas, for instance, pointed out that God works through nature, and on these grounds alone Indians were God's creatures, 'men like us', and therefore available to receive the faith. He drew attention to Mexican architecture – 'the very ancient vaulted and primitive-like buildings' – as 'no small index of their prudence and good polity'. This was roundly rejected by Sepúlveda who pointed out that bees and spiders produced artefacts that no man could emulate.[26] But there were many other aspects of Indian social and political life which impressed European observers. 'There is,' wrote Francisco de Vitoria in the 1530s, 'a certain method in their affairs, for they have polities which are orderly arranged and they have definite marriage and magistrates and overlords, laws, and workshops, and a system of exchange, all of which call for the use of reason; and they also have a kind of religion.'[27]

This was more important than it might seem. Rationality, especially the ability to live in society, was held to be the criterion of civility. But if this could happen outside Christianity what happened to the age-old distinction between Christian and barbarian? 'Inevitably it began to be blurred, and its significance as a divisive force to decline.'[28]

Even when it didn't produce startlingly new ideas, the discovery of America forced Europeans back on themselves, causing them to confront concepts, traditions and problems which existed inside their own cultural traditions. For example, the veneration for classical antiquity meant that they were aware of other civilisations which had different values and attitudes to their own and in many ways had been superior. Some, like Joseph-François Lafiteau, even hoped that the 'modern' American 'savages' might provide an insight into what the classical world had been like. In fact, it was the existence and success of pagan antiquity which underpinned the two most notable treatises of the sixteenth century which attempted to incorporate America within a unified vision of history.

The first of these, Bartolomé de las Casas's massive *Apologética Historia*, was written during the 1550s, never published in his lifetime and indeed not rediscovered until the twentieth century. It was written in anger and in response to Sepúlveda's savage polemic against the Indians, *Democrates Secundus*, in which he compared Indians to monkeys. Las Casas argued that the Indian was an entirely rational individual, fully equipped to govern himself and therefore fit to receive the gospel. He paid proper due to the quality of Aztec, Inca and Mayan art and observed their ability to assimilate European ideas and practices that they found useful.[29]

José de Acosta's *De Procuranda Indorum Salute* was written a little later than Las Casas's treatise, in 1576. His most original contribution, which advanced the understanding of anthropology, was, first, to divide barbarians into three categories, and then to distinguish three kinds of 'native'. At the top, he said, were those who, like the Chinese and Japanese, had stable republics, with laws and law courts, cities and books. Next came those who, like the Mexicans and Peruvians, lacked the art of writing and 'civil and philosophical knowledge', but possessed forms of government. Lowest were those who lived 'without kings, without compacts, without magistrates or republic, and who changed their dwelling-place, or – if they were fixed – had those that resembled the cave of a wild beast'.[30] Acosta based his work heavily on research, as we would say, which enabled him to distinguish between the Mexica and the Inca, who formed empires and lived in settlements and did not 'wander about like beasts', and the Chuncos, the Chiriguanes, the Yscayingos and all the peoples of Brazil, who were nomadic and lacked all known forms of civil organisation. The fact that Indians had some laws and customs, but that they were deficient or conflicted with Christian practices, showed he said that Satan had beaten Columbus to it in the discovery of the New World.

Again, these arguments are more important than they look at first sight. The old theories, that geography and climate were primarily responsible for cultural diversity, were being replaced. A new issue was migration. 'If the inhabitants of America were indeed descendants of Noah, as orthodox thought insisted that they must be, it was clear that they must have forgotten the social virtues in the course of their wanderings. Acosta, who held that they came to the New World overland from Asia, believed that they had turned into hunters at some stage during their migration. Then, by degrees, some of them collected together in certain regions of America, recovered the habit of social life, and began to constitute polities.'

Age-Old Grudges

On the face of things, the discovery of millions of people living without the benefits of Christianity offered the church an unparalleled opportunity to extend its influence. But in practice the consequences were more complex. The Vatican had always claimed world-wide dominion, yet its scriptures showed no awareness of the New World and made no mention of it.[31] Certain sceptics expressed the view that America was so bad that she was nowhere near ready to be brought into the mainstream of history, not yet ready to be Christianised or civilised and that syphilis was a divine punishment for the 'premature' discovery and the great cruelty meted out by the Spanish during the Conquest.[32] The buffalo was an unsuccessful and pointless cross between a rhinoceros, a cow and goat. 'Through the whole extent of America, from Cape Horn to Hudson's Bay,' wrote the Abbé Corneille de Pauw in the

Encyclopédie, 'there has never appeared a philosopher, an artist, a man of learning.'[33]

Did the fact that America was not mentioned in the scriptures mean, perhaps, that she was a special creation, emerging late from the deluge, or had she perhaps suffered her own quite different deluge, later than the one that had afflicted the Old World and from which she was now recovering? Why was the New World's climate so different from Europe's? The Great Lakes, for example, were on the same latitude as Europe but their waters froze for half the year. Why were the New World's animals so different? Why were the people so primitive, and so thin on the ground? Why, in particular, were the people copper-coloured and not white or black? Most important of all, perhaps, where did these 'savages' come from? Were they descended perhaps from the lost tribes of Israel? Rabbi Manasseh Israel of Amsterdam believed that they were, finding 'conclusive evidence' in the similarity of Peruvian temples to Jewish synagogues. For some, the widespread practice of circumcision reinforced this explanation. Were they the lost Chinese perhaps who had drifted across the Pacific? Were they the descendants of Noah, that greatest of navigators? The American historian Henry Commager says that the most widely held theory, and the one that fitted best with common sense, was that they were Tartars, who had voyaged from Kamchatka in Russia to Alaska and had sailed down the western coast of the new continent, before spreading out.[34]

For some, America was a mistake, whose main characteristic was her backwardness. 'Marvel not at the thin population of America,' wrote Francis Bacon, 'nor at the rudeness and ignorance of the people. For you must accept your inhabitants of America as a young people; younger a thousand years, at the least, than the rest of the world.'[35] In France, the celebrated natural historian, the Comte de Buffon, no less, argued that America had emerged from the deluge later than the other continents, which explained the swampiness of the soil, the rank vegetation, and the density of the forests. Nothing could flourish there, he said, and the animals were 'stunted', mentally as well as physically, 'For Nature has treated America less as a mother than as a step-mother, withholding from [the Native American] the sentiment of love or the desire to multiply. The savage is feeble and small in his organs of generation ... He is much less strong in body than the European. He is also much less sensitive and yet more fearful and more cowardly.' Even Immanuel Kant thought that Native Americans were incapable of civilisation.[36]

After Buffon, however, the slanders on America reached 'a definitive climax' with *Recherches philosophiques sur les Américains ou Mémoires intéressants pour server à l'histoire de l'espèce humaine*, by the 'crotchety Prussian curate', Cornelius de Pauw, published in Berlin in 1768 despite its French title. De Pauw was a typical encyclopaedist, sarcastic and immodest to a degree in the display of his detailed knowledge, in which he claimed to find America 'degenerate', with less sensibility, humanity, taste and instinct than the Old World. In the American climate, he said, many animals lose their tails, dogs lose their bark, the genitals of certain beasts cease to function. Even iron, what little there is of it, loses its strength, all due to the earthquakes, floods and conflagrations suffered on the continent.[37]

Another abbé, Joseph Pernety, dismissed de Pauw's comments, arguing that even then these so-called 'accursed and unhappy lands' were providing Europeans with sugar, cocoa, coffee, cochineal and precious woods, that American men were 'better proportioned for the American women than Europeans', and that the 'savage creatures' of the Brazilian forest and Paraguay were more fearsome even than those of Africa.[38]

In fact, the number of well-known figures who felt impelled to intervene in this dispute was remarkable. Voltaire thought the 'marshy air' of the Americas was very unhealthy, producing 'a prodigious number of poisons', resulting in an extraordinary shortage of food.[39] In his poem, *The Deserted Village* (1769), Oliver Goldsmith, the author of *The Vicar of Wakefield*, described Georgia as a 'parched and gloomy' land, infested with scorpions, voiceless bats and rattlesnakes, 'ferocious tigers waiting to pounce and even more ferocious Indians'. The dominant vegetation, he said, was 'matted woods where birds forget to sing'. He repeated these calumnies in his *History of the Earth and Animate Nature* (eight volumes, 1774) which, despite its many mistakes, or because of them, enjoyed 'an unlikely success'.[40]

The Americans fought back. One man who did have some answers to this massive condescension was Thomas Jefferson. His answer to the charge that nature was sterile and emaciated in the New World was to point to Pennsylvania, 'a veritable garden of Eden, with its streams swarming with fish, its meadows with hundreds of song birds'. How could the soil of the New World be so thin when 'all Europe comes to us for corn and tobacco and rice – every American dines better than most of the nobles of Europe'. How could the American climate be so enervating when, by then, statistical tables showed a higher rainfall in London and Paris than in Boston and Philadelphia?[41] In 1780 a young French diplomat, the Marquis de Barbé-Marbois, had the idea to canvas opinion from several governors of American states and sent them a series of questions about the organisation and resources of their respective commonwealths. Jefferson's response was the most detailed, the most eloquent and by far the most famous – *Notes on Virginia*. There is something surreal about this book now but the issues it attacked were keenly felt at the time. Jefferson met Buffon and the European condescenders head on. He compared the work-rates of Europeans and Americans – as defined by actuarial statistics – to the advantage of the Americans. Buffon had claimed that the New World had nothing to compare with the 'lordly elephant' or the 'mighty hippopotamus', or the lion and the tiger. Nonsense, replied Jefferson, and pointed to the Great Claw or Megalonyx. 'What are we to think of a creature whose claws were eight inches long, when those of the lion are not 1½ inches?' Even by 1776, enough fossil bones of the mammoth had been found to show that it was indigenous to the New World and that it was a beast easily 'five or six times' larger than an elephant.[42]

Buffon and some of the other French *philosophes* had (from 3,500 miles away) called the Indians degenerate. Try fighting him, Jefferson responded. 'You will sing a different tune.'[43] He referred to the rhetoric and eloquence of Logan, chief of the Mingoes: this underlined that their minds, no less than their bodies, were as well adapted to their circumstances as were the Europeans.

Thomas Paine felt that the very 'wilderness' Europeans understood America to be would stimulate 'the development of a totally human and brotherly society without historical quarrels' to get in the way.[44] The Old World, in comparison, was the imperfect one, the backward one. From Spanish America, the Chilean economist Manuel de Salas defended his country as 'a privileged land . . . where wild beasts are not known, nor insects, nor poisonous reptiles'. He objected to the idea that Americans could not raise themselves to the level of the exact sciences, pointing to advances already being made in astronomy (Pedro Peralta), electricity (Benjamin Franklin) and history (Giovanni Ignazio Molina).

Similarly erudite was the Mexican Jesuit, Francisco Javier Clavigero who, in his *Historia antigua de México*, mounted a powerful counter-offensive against the sceptics of Aztec splendour. The history of the Toltecs, the Texcocans and the Aztecs offered as many examples of valour, patriotism, wisdom and virtue, he said, as did the histories of Greece and Rome. While he agreed that Aztec religion was 'puerile, cruel and superstitious', their architecture, though inferior to that of Europe, was 'superior to that of most Asiatic and African peoples'.[45]

Herder was kinder than many towards America. His basic doctrine of the equality of cultures, plus the generalised sympathy of the romantic movement for the vanquished and the exotic, all helped improve the popularity of the Aztecs. In his famous *Ideas for a Philosophy of the History of Mankind* (1784–91), Herder devoted chapter six of the sixth volume to the Americas, in which he noted the essential unity of humankind, argued that North America was better developed than Spanish America but said that the average American displayed an 'almost childish goodness and innocence'. Herder was sympathetic without being any better informed than those who were more critical. His main worry, shared by Diderot, was that the brutal policy of the Spanish would, eventually, reduce all cultures to one, levelling out the diversity of the world and, in doing so, obliterate much of what history had achieved and meant.[46]

Goethe thought the country a blessed land, without feudal remains or 'age-old grudges'. He was impressed that the country lived in the present and was 'not disturbed within thy inner self'.[47] His fellow Germans, Alexander von Humboldt, Georg Hegel, Friedrich von Schlegel and Arthur Schopenhauer, were equally forthright. Von Humboldt at least paid America the compliment of visiting it, both South and North, where he was much taken with the very great variety of nature, where he found the 'ancient noisy conflicts' felt closer, the plants differently distributed, the animals bigger, the rivers broader and deeper. He climbed mountains, explored rivers, collected unknown animals and plants, the latter growing – as he observed – with vivid hues unknown anywhere else.

In his history of the dispute of the New World, Antonello Gerbi says that the quarrel 'reaches its peak [until then] in the antithesis between Humboldt and Hegel, and at the same time the point of widest divergence between the two extremes'.[48] Hegel, known for perhaps the greatest overarching theory of history and philosophy ever produced, was always uncertain as to how to include America in the structure he thought he had identified. He could not ignore such a vast continent and he therefore found that the main division of the earth was into the New World and the Old, representing a philosophical division. The two worlds differed in everything, he said: the Old World is curled around the Mediterranean Sea like a horseshoe, the New World is elongated and in a north-south direction; the Old World is perfectly separated into three 'properly articulated' and integrated parts (Europe, Asia, Africa – Hegel often thought in threes), while the New World is split up, with a 'miserable hinge' joining the two parts.[49] The great mountain chains and bigger rivers run in different directions (mountains east-west in the Old World, north-south in the New; rivers north-south in the Old World, east-west in the New). Many of these observations were less than accurate, of course, but on top of that Hegel argued that the New World had a youthful look that was not entirely praiseworthy: everything there was new, and by new he meant 'immature and feeble', its fauna weaker, its flora more monstrous. Its civilisations lacked the two great 'instruments of progress', iron and the horse, and while no continent in the Old World ever allowed itself to be totally subjugated, 'the whole of America fell prey to Europe'.[50]

Friedrich Schlegel, a contemporary, also took the view that America was, biologically speaking, radically different from the Old World. He thought there were two kinds of humanity, those who had migrated from Asia, and the cannibals, 'the only autochthonous Americans'.[51] He also thought that the greatest division on earth was between the northern hemisphere and the southern, and that in both the Old World and the New the north was far more developed than the south. After the depredations of Napoleon, however, Schlegel did consider the possibilities of a European renewal in America, particularly for elites.

Finally, among the Germans, Schopenhauer argued that, nature-wise, 'America always shows us the inferior analogue in regard to the mammals, and in compensation the superior analogue in regard to the birds and reptiles.' He thought there were only three primitive (i.e., original) races – Caucasian, Ethiopian and Mongolian, all three belonging to the Old World. The Americans, therefore, he concluded, were 'climatically modified Mongolians ... In short, the will to live [*his* central idea], on its objectivisation in the Western hemisphere, felt itself very serpentine and very avian, not very mammiferous and not at all human.'[52]

Schopenhauer had added all this in a note appended to his major work, *The World as Will and Representation*, in 1859, the very year that Charles Darwin published *On the Origin of Species*. Darwin had familiarised himself with Alexander von Humboldt's work while he was a young man and was very conscious that there had once been a great range of megafauna in America that had gone extinct. This was a mystery – it could not have been a great geological catastrophe because it didn't show up in the record of the rocks, and if it had indeed occurred surely the small animals would have died out before the larger ones. Like Humboldt he used his powers of observation – to see that animals were abundant in some areas of the Americas, and not in others, as happened in Europe, meaning among other things that the great generalisations of people like Hegel had to be inaccurate oversimplifications. And some domesticated animals from Europe had gone wild in America, again due to environment. He did retain a few of Buffon's prejudices but, as the world now knows, his main ideas of evolution were matured during his voyage on the *Beagle*, in particular while he was in the Galapagos Islands, off Ecuador, and in Patagonia. This is where he conceived his concepts of variation, adaptation and the struggle for survival, leading to natural selection.[53]

The place of the native American in the great scheme of things was always likely to be the overriding issue. In America itself, by the nineteenth century, many people had direct experience of Indians and this tended to make them more passionate in whatever views they held. In his *Ancient Society*, 1875, based on an earlier lecture, 'Ethnical Periods', Lewis H. Morgan, the early anthropologist, introduced another level of scepticism into some of the early Spanish chronicles, and concluded that the Aztecs were in the middle stage of barbarism (he envisaged three phases: savagery, barbarism, civilisation), concluding that this made it impossible for the Aztecs to have had an empire in the true sense of that word – they had been in a simple confederacy of tribes at the time of the Conquest. Aided by research from Adolph F. Bandelier, the Aztec achievement was rendered both more realistic and smaller, and the Spanish achievement more impressive, presented as they were, at the time of the Conquest, with perpetual warfare between the Aztecs and their neighbours.[54]

Fish with Legs, Lions with Fins

The place of animals and plants attracted attention, too, and they were interesting philosophically and theologically but also commercially, plants especially. As for animals, understanding was hampered for a while by the inaccuracy of the early reports and an abiding interest in 'monsters' (fish with legs, iguanas with wings, 'lions' – i.e., pumas – with fins, pigs with navels on their backs). Some travellers claimed to find elephants in Mexico, and unicorns in Argentina. Other early accounts fastened on the armadillo ('an armoured horse' that, according to some accounts, did not eat) and the large herds of llama, the guanaco and vicuña in Peru, the latter's wool being regarded as of better quality than that of merino sheep in Spain.[55]

Animal remedies were another early interest (snake-skin as an aphrodisiac; an infusion of opossum tail as a cough medicine, or a laxative, or as an aid in childbirth; pounded porcupine spines as a cure for kidney stones). Not unreasonably, attempts were made to fit the new animals into existing taxonomies (the manatee was called an 'ox-fish'), as several writers saw themselves as a 'Pliny of the New World'. People did note that none of the large animals in the New World were found in the outlying Caribbean islands, which suggested therefore that they had not been brought there by previous travellers, that perhaps they were the result of a separate Creation.[56]

Because many of the more educated people who visited the New World early on were clerics a major concern was the religious significance of the natural world that was found there. Juan Eusebio Nieremberg, a Spaniard of German descent, considered the possibility that there was a category of creation that was intermediate between animals and plants. Athanius Kircher, a Jesuit polymath, devoted no little time to a consideration of where the newly discovered animals would have been accommodated in Noah's Ark. His answer was that animals can metamorphose into other forms and that the creatures which passed from Europe to the New World had degenerated, and to such a degree that it was very difficult to work out which was descended from which 'parent'. Others thought that the different climate of the Americas had produced different animal forms.[57]

But in general, as time went by, and as more scientists looked at the improving range of illustrations and descriptions, and even began dissecting specimens, either in America, or those examples which had been brought back, it was realised that there were fewer differences between the animals on the two continents than at first appeared. Clerics and others tried to fit New World animals into the Great Chain of Being (the dominant taxonomy before Darwin), and here some American animals fitted nicely, filling gaps for example between the monkeys and other species. Gradually it was realised that the animals of the New World had no marvellous, mysterious powers, that in fact, as Miguel de Asúa says, 'there was hardly anything new about them.'[58] This was a sensible enough conclusion that ought to have had more impact than it did. Scientifically, following von Humboldt, the continent became more fully understood and, following Darwin, more easily assimilable into general thought patterns. By the time of World War One, when America came to Europe's rescue, the polemic was dead.

'The Indians Die Easily'

Or not quite. By the twentieth century, science was beginning to take over from the grand theorists, this approach being associated first and foremost with the name of the German, Eduard Seler, whose careful, cautious commentaries on the main Native codices 'set standards of meticulous research that have not been surpassed'.[59] Seler was so cautious that he never wrote a general work of synthesis on Mesoamerica but he did point out continent-wide patterns and many Old World/New World analogies. In particular, he debunked any idea of Old World influence on the New and pictured the Aztecs as capable of complex astronomy and mathematics, sculpture, poetic imagination and self-expression. Not surprisingly, he was very popular in Mexico.

Mexico in any case was going through a particularly nationalistic phase just then, owing to the Mexican Revolution of 1910–1917, but in addition to that a number of discoveries in the early decades of the twentieth century also drew attention to ancient American civilisations. These discoveries included Machu Picchu (Inca) in 1911, Chavín de Huántar (Chavín) in 1919, La Venta (Olmec) in 1925, and San Lorenzo (Olmec) in 1945.

And if the scientific approach to the character of the New World was on the rise, this only meant that there were fresh areas to disagree about. After World War Two, as more and more became known about pre-Columbian civilisations, scholars became divided into Mayanists and Olmequistas, according to which civilisation they regarded as pre-eminent. Another – at times ferocious – area of dispute erupted over the population of the New World at the time of the Conquest. Indeed, according to William M. Denevan, this has become 'one of the great debates in history'.[60] There are (at least) two reasons why this figure should matter. Population size is a perhaps crude but easily understandable criterion by which to assess how successful a civilisation was or is – a reflection of what surplus of food it managed or manages to produce, so allowing more and more members of its society to break away from subsistence and engage in the 'higher' activities which comprise civilisation. And second, since we know the population of the Americas with a fair degree of accuracy for the later sixteenth and seventeenth centuries, the figure in 1492 gives us some idea of how many Native Americans were wiped out by the colonialists, either through warfare or disease.

Not only have the population figures been a source of deep disagreement but a knock-on effect of that has been a no less fierce debate as to whether epidemics spread widely and rapidly, or much less so. This has contemporary relevance because an accurate assessment of figures needs to take account of how quickly populations revive after catastrophes, catastrophes of the kind that can easily recur.

So far as pre-Conquest America is concerned, while there has been deep disagreement between those who favour a low population figure and those who prefer a higher one, the recent trend seems to be towards acceptance of higher figures. Any reader interested in the arguments is referred to *The Native Population of the Americas in 1492*, edited by William M. Denevan, first published in 1976 but updated for the five-hundredth anniversary of 1492 and republished in 1992. These are not the only estimates of recent years, as Denevan makes clear, some reaching in excess of 100 million. But his 1992 figure of 53.9 million is, perhaps, the one most scholars would now accept. Equally important, and just a shade less contentious, is the figure for by how much the aboriginal population of the Americas was reduced in the years following European contact, be it by war, starvation, or disease. This decline was massive but not uniform. Denevan says he found 'hundreds' of reports of rapid decline, single epidemics reducing villages by half and more with many tribes 'completely wiped out' in a few decades. William McNeill goes so far as to say that disease, chiefly smallpox, was a factor in why the New World was so easily conquered by the Europeans. Because of this, the Spaniards thought that 'The Indians die easily'.[61] Besides smallpox, the major killers were measles, whooping cough, chicken pox, bubonic plague, typhus, malaria, diphtheria, amoebic dysentery, influenza and a variety of helminthic infections. Diseases already present in the New World before 1492 included infectious hepatitis, encephalitis, polio, syphilis, Chagas' disease and yellow fever.

Some resistance was acquired by the seventeenth century but by then other matters had taken a toll, beyond disease, such as military action, malnutrition, starvation and the reduction of some cultural groups to a level at which traditional marriage pools were inadequate to provide eligible mates.[62] The slave ships from Africa also appear to have imported malaria, another blow.[63] As Denevan concludes, the discovery of America 'was followed by the greatest demographic disaster in the history of the world' and, unlike earlier crises, the Indian population recovered only slowly. The full extent of New World devastation may be seen from demographer Henry Dobyns' 1966 study which found that, in 1650, more than 150 years after the Conquest, the population of the Americas south of the United States which, according to him, had been between 80 million and 108 million in 1492, was now only 4 million, a decline of between 95 and 97.3 per cent. For once with statistics, the numbers need no elaboration.

Russell Thornton, in his *American Indian Holocaust and Survival: A Population History Since 1492*, adds to this picture by comparing the population of Native Americans in 1492 (his figure being 72+ million) with those in various parts of the Old World in 1500: Italy, ~10 million;

Portugal, 1.25 million; Spain, 6.5–10 million; British Isles, 5 million; France, 15 million; Netherlands, <1 million. Overall, there were, in 1500, seven times as many non-Americans in the world as 'native Indians'.

But he makes the point that numbers are not the whole picture. Estimates of the life expectancy of certain New World populations, and certain non-Americans, show that life expectancy was not so very different in the two hemispheres, but there are grounds for thinking, says Thornton, that the peoples of the Western Hemisphere were remarkably free of the serious diseases that afflicted Europeans. Several early European travellers to the New World observed the Indians to be of 'lusty and healthfull bodies', free of 'Feavers, Pleurisies ... Agues ... Consumptions ... Apoplexies, Dropsies, Gouts ... Pox, Measels, or the like' and to 'spinne out the threed of their days' to three-score, four-score and 'some a hundred yeares'.[64] William McNeill, in *Peoples and Plagues*, agreed. Many human infections, he points out, are derived from domesticated mammals (cattle share fifty diseases with humans, sheep and goats forty-six, the horse thirty-five), while the only domesticated mammals of similar size in the Americas – llamas and alpacas – live high in the cold Andes, where infections do not thrive, and in small herds, which cannot sustain disease chains. Maize and potatoes – unique to the Americas before 1492 – contain more calories than Old World cereals except rice.[65]

Native Indians did suffer from bacillary and amoebic dysentery, viral influenza and pneumonia, trypanosomiasis, non-venereal syphilis, pellagra and salmonella and other forms of food poisoning. But there is now little doubt, according to Thornton, that they were in better shape in regard to disease than Europeans and the reason is interesting.

The possibility is strong that, in migrating across Beringia, the inhospitably cold, harsh climate was a 'germ filter' whereby infectious diseases present in any human migrants were screened out. In hookworm, for example, neither the eggs nor the larvae can survive in soil temperature below ~59°. Furthermore, since the migrants existed in only small groups, their populations were also not large or dense enough to enable disease chains to develop.[66]

Not all diseases appeared straight away. For instance, Dobyns says that smallpox epidemics occurred in 1520–24, measles in 1531–33, influenza in 1559, typhus in 1586 and diphtheria not until 1601–1602. But the effect was still devastating, reducing the Native population by, in some cases, as we have seen, 97 per cent, and dwarfing the numbers killed in the forty-plus 'Indian wars'.[67]

One final figure that Thornton gives shows how much Indian life-ways were changed by the advent of Europeans: the population of buffalo fell from 60 million in 1492 to 40 million in 1800, to 14 million in 1870 and to *one million* five years later, when the buffalo stood in the way of the opening up of the Great Plains.[68]

No one was more acerbic in his condemnation of the European invaders than Kirkpatrick Sale, in *The Conquest of Paradise*.[69] For Sale, Columbus was a deceitful (and self-deceiving) opportunist, and Europe 'a civilisation that had lost its bearings, with failing soil and famine, epidemics and disease the order of the day'. Humanism had put the Christian Church in crisis, nationalism was on the rise and intimacy, sacredness and reverence had 'all but vanished'. Columbus wasn't the discoverer of America anyway, because at least twenty other fifteenth-century voyages had sighted or put ashore in the New World. Nor did Columbus care about nature – 'it was treasure he wanted'. The Tainos were not nearly as backward as he assumed from their dress. Their houses were more spacious and cleaner than the 'crowded and slovenly hovels' of south European peasantry and their crops and diet superior in yields than anything known in Europe at the time. The idea of the fierce and cannibalistic Caribs was 'merely a bogey'.[70]

Warming to his theme, Sale claimed there was no real intellectual explosion in Europe after the discovery, that the Indians were perplexed by the 'emotional coldness' of the Europeans and that the savage brutality of the Spanish was matched by that of the English who imposed the *plantation* on the north, a miniature of the Old World society.[71]

Sticking to his title, he argued that native Indian agricultural techniques, though less 'advanced' than European ones, were actually better suited to the American environment and ecologically more sound, as was proved, he said, by figures which showed that, within a few decades of the Conquest, 140 animal and bird species became extinct, and 200 plants, including 17 varieties of grizzly bears, seven forms of bat, cougars, auks and wolves. This all contrasted badly with the careful hunting protocols of the indigenous peoples who limited slaughter and understood the dangers of overkilling. In other words, for Sale pre-

Columbian North America 'was a prelapsarian Eden of astonishing plenitude.'[72]

Other scholars also collected Native American accounts of the invasion and Conquest, which underlined the brutality of the Spaniards alongside the hospitality shown by many of the indigenous peoples. Their hunting had been so spoiled, they noted, that their custom of offering animal skins to others when renewing treaties, could no longer be maintained, an important point of honour extirpated.[73] Everywhere, they were in want of deer. And they were equally upset when the invaders put out their Council Fire, not knowing how much it meant to them. After the Spanish invasion, *quipus*, like Mesoamerican books, were proscribed and burned as '*soguillas*' that preserved the memory of pagan ritual and dogma.[74] In Peru, the Spanish also spent years trying to locate and destroy the ancestor mummies. Friar Diego de Landa admitted that he and others had found 'a great number' of Mayan books 'and because they contained nothing but superstition and the devil's falsehoods, we burned them all, which upset [the Mayas] most grievously and caused them great pain'.[75]

Second to the brutality came the deceit, which rankled badly with the indigenous races, as did the newcomers' practice of eating cattle and pigs. The Aztec thought whoever ate such animals would turn into them. There were mixed feelings about children of mixed races: some kept it secret, others were proud of it. The Indians had their own language for the invaders – the frontiersmen were called 'the long knives' – and they could be dismissive: the North American Indians saw themselves as made of red clay, the white people as from white sand.[76]

At the same time these authors did not overlook the resilience of some aspects of Native American culture – for example, there were in 1992 twelve million people in the Andes still speaking the language of the Incas, and six million speakers of Maya, more or less the number of French-speakers in Canada.[77]

The dispute is an unedifying narrative, out of which very few people come well and which seems to prove that humans love nothing so much as taking sides. The sacrificial and self-sacrificial violence identified in this book still has the power to shock. But now, seen against the history of the differences between the two hemispheres, we have a much clearer understanding of why these practices developed. Pre-Columbian

America was not quite the paradise some writers have described it as. Nonetheless, it remains an unparalleled demographic tragedy that such *wholly different* societies, with a wholly different psychology, were so ruthlessly annihilated.

• Appendix 2 •

From 100,000 kin groups to 190 Sovereign states: Some Patterns in Cultural Evolution

This appendix, which explores the similarities between the civilisations of the Old World and the New, is available online at: www.orionbooks.co.uk/thegreatdivide

Notes and References

When two dates are given for a publication, the first refers to the hardback edition, the second to the paperback edition. Unless otherwise stated, pagination refers to the paperback edition.

INTRODUCTION: 15000 BC–AD 1500: A UNIQUE PERIOD IN HUMAN HISTORY

1 B.W. Ife, editor and translator, *Christopher Columbus: Journal of the First Voyage: 1492*, Warminster: Arts & Phillips, 1990, p. 13.
2 Ife, *Op. cit.*, p. 15.
3 *Ibid*, p. 25.
4 *Ibid*, p. 27.
5 *Ibid*, p. 246 *n*.
6 *Ibid*, p. xxi.
7 *Ibid*, p. xxiv.

CHAPTER I: FROM AFRICA TO ALASKA: THE GREAT JOURNEY AS REVEALED IN THE GENES, LANGUAGE AND THE STONES

1 S.J. Armitage *et al.*, 'The Southern Route "Out of Africa": Evidence for an Early Expansion of Modern Humans in Arabia', *Science*, Vol. 331, pp. 453–456, 28 January 2011; Michael D. Petraglia, 'Archaeology: Trailblazers across Africa', *Nature*, Vol. 470, pp. 50–51, 3 February 2011; Brenna M. Henn *et al.*, 'Characterizing the Time Dependency of Human Mitochondrial DNA Mutation Rate Estimates', *Molecular Biology and Evolution*, Vol. 26, Issue 1, 2008, pp. 217–230; Geoff Bailey, 'World Prehistory from the Margins: The Role of Coastlines in Human Evolution', *Journal of Interdisciplinary Studies in History and Archaology*, Vol. 1, No. 1 (Summer 2004), pp. 39–50. P.M. Masters and N.C. Flemming (editors), *Quaternary Coastlines and Marine Archaeology: Towards the Prehistory of Landbridges and Continental Shelves*, London and New York: Academic Press, 1983, *passim.*

2 Brian M. Fagan, *The Journey from Eden: The Peopling of Our World*, London and New York: Thames & Hudson, 1990, pp. 234–235. Spencer Wells, *Deep Ancestry: Inside the Genographic Project*, Washington DC: National Geographic Society, 2007, p. 93. Ted Goebel, 'The "Microblade Adaptation" and Recolonisation of Siberia during the Late Upper Pleistocene', Archaeological Papers of the American Anthropological Association, Vol. 12, Issue 1, January 2002, pp. 117–131.

3 Wells, *Op. cit.*, p. 96.

4 *Ibid*, p. 100.

5 *Ibid*, p. 99.

6 Sijia Wang *et al.*, 'Genetic variation and the population structure of Native Americans', http://dx.doi.org/10.1371%2Fjournal.pgen.0030185. Also: personal communication. Ugo A. Perego *at al.*, 'The initial peopling of the Americas: A growing number of founding mitochondrial genomes from Beringia', *Genome Research*, Vol. 20, 2010, pp. 1174–1179. Brenna M. Henn *et al.*, *Op. cit.*

7 Douglas Wallace, James Neel *et al.*, 'Mitochondrial DNA "clock" for the Amerinds and its implications for timing their entry into North America', *Proceedings of the National Academy of Sciences*, 1994; 91 (3), pp. 1158–1162.

8 Brian Fagan, *The Journey from Eden*, London and New York: Thames & Hudson, 1990, p. 198.

9 Fagan, *Op. cit.*, p. 205.

10 John Hemming, *Tree of Rivers: the story of the Amazon*, London and New York: Thames & Hudson, 2008, p. 278.

11 Tim Flannery, *The Eternal Frontier: An ecological history of North America and its peoples*, London: William Heinemann, 2001, p. 231–232. James Kari and Ben A. Potter (editors), *The Dene-Yeniseian Connection*, Anthropological Papers of the University of Alaska, New Series, Vol. 5, Nos. 1–2, 2010.

12 Nicholas Wade, *Before the Dawn: Recovering the Lost History of Our Ancestors*, London, Duckworth, 2007, p. 99.

13 Wade, *Op. cit.*, pp. 151–152. On the circumstances of men without wives siring children: Peter Bellwood, personal communication.

14 Brian Fagan, *The Great Journey: The Peopling of Ancient America*, London and New York: Thames & Hudson, 1987, p. 122.

15 Fagan, *The Great Journey*, *Op. cit.*, p. 127.

16 *Ibid*, p. 125.

17 Merritt Ruhlen, *The Origins of Language: Tracing the Evolution of the Mother Tongue*, New York: John Wiley, 1994, p. 295.

18 Ruhlen, *Op. cit.*, map 7, p. 90, and map 8, p. 108. Nelson Fagundes et al, 'Genetic, geographic, and linguistic variation among South American Indians: possible sex influence', *American Journal of Physical Anthropology*, vol. 117, 2002, pp. 68–78.

19 *Ibid*, pp. 134–137.

20 Johanna Nichols, *Linguistic Diversity in Space and Time*, Chicago and London: University of Chicago Press, pp. 9–10.
21 Nichols, *Op. cit.*, p. 298.
22 *Ibid*, p. 330.

CHAPTER 2: FROM AFRICA TO ALASKA: THE DISASTERS OF DEEP TIME AS REVEALED BY MYTHS, RELIGION AND THE ROCKS

1 John Savino and Marie D. Jones, *Supervolcano: The Catastrophic Event that Changed the Course of Human History*, Franklin Lakes, NJ: New Page, 2007, p. 123.
2 Savino and Jones, *Op. cit.*, p. 123.
3 *Ibid*, p. 125.
4 Jelle Zeitlinga de Boer and Donald Theodor Sanders, *Volcanoes in Human History: The Far-reaching Effects of Major Eruptions*, Princeton, NJ and Oxford: Princeton University Press, 2003, pp. 155–56.
5 Savino and Jones, *Op. cit.*, p. 125.
6 *Ibid*, pp. 132 and 144. See also: Michael D. Petraglia *et al.*, 'Middle Paleolithic Assemblages from the Indian Subcontinent Before and After the Toba super-eruption', *Science*, Vol. 317, 6 July 2007, pp. 114–116.
7 Michael Petraglia *et al.*, *Op. cit.* See also: Kate Ravilious, 'Exodus on the Exploding Earth', *New Scientist*, 17 April 2010, pp. 28–33.
8 Savino and Jones, *Op. cit.*, p. 47.
9 Stephen Oppenheimer, *Eden in the East: The Drowned Continent of South East Asia*, London: Weidenfeld & Nicolson, 1998, p. 17.
10 Oppenheimer, *Op. cit.*, p. 18.
11 *Ibid*, p. 19.
12 *Ibid*, p. 20. Peter Bellwood, *First Farmers: the Origins of Agricultural Societies*, Oxford: Blackwell, 2005, pp. 130 and 133.
13 *Ibid*, p. 21.
14 *Ibid*, p. 24.
15 *Ibid*, p. 32.
16 *Ibid*, p. 33.
17 *Ibid*, p. 35.
18 *Ibid*, p. 39.
19 *Ibid*, p. 62.
20 *Ibid*, p. 63.
21 *Ibid*. p. 64.
22 *Ibid*, p. 76.
23 *Ibid*, p. 77. For criticisms of Oppenheimer, see: Peter Bellwood, 'Some Thoughts on Understanding the Human Colonisation of the Pacific', *People and Culture in Oceania*, Vol. 16, 2000, pp. 5–17, especially note 5.
24 Oppenheimer, *Op. cit.*, p. 77. Geoff Bailey, 'World Prehistory from the Margin', *Op. cit.*, p. 43.
25 *Ibid*, p. 83.

26 David Frawley and Navaratna Rajaram, *Hidden Horizons: Unearthing 10,000 Years of Indian Culture*, Shahibaug, Amdavad-4: Swaminarayan Aksharpith, 2006, p. 61.
27 Frawley and Rajaram, *Op. cit.*, p. 65.
28 Georg Feuerstein *et al.*, *In Search of the Cradle of Civilization*, Wheaton, Illinois and Chennai, India: Quest Books, 2001, p. 91.
29 Oppenheimer, *Op. cit.*, p. 317.
30 *Ibid.*
31 *Ibid.*
32 *Ibid*, plate 1, facing p. 208.
33 Paul Radin, *The Trickster: A Study in American Indian Mythology*, London and New York: Routledge & Kegan Paul, 1956, p. 167.
34 Oppenheimer, *Op. cit.*, p. 359.
35 *Ibid*, p. 373.
36 Stephen Belcher, *African Myths of Origin*, London: Penguin Books, 2005, especially part 1.

CHAPTER 3: SIBERIA AND THE SOURCES OF SHAMANISM

1 Ronald Hutton, *Shamans: Siberian Spirituality and the Western Imagination*, Hambledon, UK and New York: 2001.
2 Piers Vitebsky, *The Shaman: Voyages of the Soul: Trance, Ecstasy and Healing from Siberia to the Amazon*, London: Duncan Blair, 2001, p. 86.
3 Vitebsky, *Op. cit.*, p. 11.
4 Hutton, *Op. cit.*, p. 59.
5 *Ibid*, p. 61.
6 *Ibid*, p. 74.
7 Hutton, p. 51.
8 Vitebsky, *Op. cit.*, p. 11.
9 Hutton, *Op. cit.*, pp. 11–12.
10 *Ibid*, p. 13.
11 *Ibid*, p. 26.
12 Mircea Eliade, *Shamanism: Archaic Techniques of Ecstasy*, Princeton, NJ: Princeton University Press, 1970, pp. 24 and 29.
13 Vitebsky, *Op. cit.*, p. 30.
14 *Ibid*, p. 32.
15 Hutton, *Op. cit.*, p. 107. See also: Tim Ingold, *The Perception of the Environment: Essays in livelihood, dwelling and skill*, London: Routledge, 2000, pp. 61ff.
16 Vitebsky, *Op. cit.*, p. 42.
17 *Ibid*, p. 45.
18 Peter Furst, *Hallucinogens and Culture*, San Francisco: Chandler & Sharp, 1988, p. 90.
19 Furst, *Op. cit.*, p. 91–2

CHAPTER 4: INTO A LAND WITHOUT PEOPLE

1 Dan O'Neill, *The Last Giant of Beringia: The Mystery of the Bering Land Bridge*, New York: Westview, 2004, p. 6.
2 Henry Steele Commager, *Empire of Reason: How Europe Imagined and America Realized the Enlightenment*, London: Weidenfeld & Nicolson, 1978. p. 106.
3 O'Neill, *Op. cit.*, p. 8.
4 *Ibid.*
5 *Ibid.*
6 *Ibid.*
7 *Ibid*, p. 11.
8 *Ibid*, p. 12.
9 *Ibid.*
10 *Ibid*, p. 13.
11 *Ibid*, p. 14.
12 *Ibid*, p. 15.
13 *Ibid*, p. 17.
14 *Ibid*, p. 64.
15 *Ibid*, p. 65.
16 J. Louis Giddings, *Ancient Men of the Arctic*, London: Secker & Warburg, 1968.
17 O'Neill, *Op. cit.*, p. 112.
18 *Ibid*, p. 114.
19 *Ibid*, pp. 121–122.
20 Jeff Hecht, 'Out of Asia', *New Scientist*, 23 March 2002, p. 12.
21 *Ibid*, p. 122.
22 *Ibid*, p. 123.
23 *Ibid*, p. 139.
24 O'Neill, *Op. cit.*, p. 141.
25 *Ibid*, pp. 145–147.
26 *Ibid*, p. 161. Renée Hetherington *et al.*, 'Climate, African and Beringian subaerial continental shelves, and migration of early peoples', *Quaternary International* (2007), DOI:10.1016/j.quaint.2007.06.033.
27 Gary Haynes, *The Early Settlement of North America: the Clovis Era*, Cambridge, UK: Cambridge University Press, 2002, p. 253.
28 Steven Mithen, *After the Ice: A Global Human History, 20,000–5000 BC*, London: Weidenfeld & Nicolson, 2003, p. 242.
29 Valerius Geist, 'Did large predators keep humans out of North America?', in Julia Clutton-Brock (editor), *The Walking Larder: Patterns of Domestication, Pastoralism and Predation*, London: Unwin Hyman, 1989, pp. 282–294.
30 Calvin Luther Martin, *In the Spirit of the Earth: Rethinking History and Time*, Baltimore and London: John Hopkins University Press, 1993, p. 88.
31 Jake Page, *In the Hands of the Great Spirit: the 20,000 year history of American Indians*, New York: Free Press, 2004, p. 37.
32 Anthony Sutcliffe, *On the Track of Ice Age Mammals*, London: British Museum

Publications, 1986, p. 167.

33 Sutcliffe, *Op. cit.*, p. 176.

34 Thomas Dillehay, *The Settlement of the Americas: A New Prehistory*, New York: Basic Books, 2000, p. 112.

35 Timothy Flannery, *The Eternal Frontier: an ecological history of North America*, London: William Heinemann, 2001, p. 117. Tom D. Dillehay, 'Probing Deeper into First American Studies', *Proc. Nat. Acad. Sciences*, 27 January 2009, pp. 971–978.

36 Flannery, *Op. cit.*, pp. 192–205.

37 Dillehay, *The Settlement of the Americas*, *Op. cit.*, pp. 164–165.

38 Haynes, *Op. cit.*, p. 91.

39 Dillehay, *The Settlement of the Americas*, *Op. cit.*, p. 110.

40 Haynes, *Op. cit.*, p. 32.

41 *Ibid.*, p. 91. Michael R. Waters *et al.*, 'Redefining the age of Clovis: Implications of the peopling of America', *Science*, Vol. 315, No. 5815, 23 February 2007, pp. 1122–1126; Michael R. Waters *et al.*, 'The Buttermilk Creek Complex and the Origins of Clovis at the Debra L. Friedkin Site, Texas', *Science*, Vol. 331, No. 6024, 25 March 2011, pp. 1599–1603; Jon M. Erlandson *et al.*, 'Paleoindian Seafaring, Maritime Technologies and Coastal Foraging on California's Channel Islands', *Science*, Vol. 331, No. 6021, 4 March 2011, p. 1122.

42 Flannery, *Op. cit.*, p. 92.

43 Dillehay, *The Settlement of the Americas*, *Op. cit.*, p. 267.

44 Haynes, *Op. cit.*, pp. 249–250.

45 *Ibid*, pp. 113 and 115.

46 *Ibid*, p. 112.

47 *Ibid*, p. 269.

48 Haynes, *Op. cit.*, p. 183.

49 *Ibid*, p. 198.

50 *Ibid*, p. 161.

51 *Ibid*, p. 163. Waters *et al.*, 'Redefining the Age of Clovis', *Op. cit.*, p. 1125.

52 *Ibid*, p. 166.

53 Paul S. Martin, *Twilight of the Mammoths: Ice Age Extinctions and the Rewilding of America (organisms and environments)*, Los Angeles and Berkeley: University of California Press, 2005 (reissue).

54 Haynes, *Op. cit.*, p. 166. R. Dale Guthrie, 'New carbon dates link climatic change with human colonization and Pleistoene extinctions', *Nature*, Vol. 441, 11 May 2006, pp. 207–209, DOI:10.1038/nature04604.

55 *Ibid.*

56 Jared Diamond, *Guns, Germs and Steel: the Fates of Human Societies*, New York and London: W.W. Norton, 2005, p. 47.

57 *New Scientist*, 26 May 2007, pp. 8–9. Briggs Buchanan *et al.*, 'Peleoindian demography and the extraterrestrial impact hypothesis', *Proc. Natl. Acad. Sci.* Vol. 105, 2008, pp. 11651–11654; James Kennett and Allen West, 'Biostratigraphic evidence supports Paeloindian disruption at ~12.9 ka',

http://www. pnas.org?content/105/50/E110.full.
58 Haynes, *The Early Settlement*, *Op. cit.*, p. 161.
59 *Ibid.*, p. 163.
60 *Ibid.*, p. 166.
61 Kate Ravilious, 'Messages from the Stone Age', *New Scientist*, 20 February 2010, pp. 30–34.
62 Diamond, *Op. cit.*, p. 67.

CHAPTER 5: RINGS OF FIRE AND THERMAL TRUMPETS

1 David Landes, *The Wealth and Poverty of Nations*, London: Abacus, 1998, p. 7.
2 Landes, *Op. cit.*, p. 17.
3 *Ibid*, p. 19.
4 *Ibid.*
5 Peter D. Clift and R. Alan Plumb, *The Asian Monsoon: causes, history, and effects*, Cambridge, UK: Cambridge University Press, 2008, p. 136.
6 Brian Fagan, *The Long Summer: How Climate Changed Civilization*, London: Granta, 2004, p. 170.
7 Fagan, *The Long Summer, Op. cit.*, p. 171.
8 Clift and Plumb, *Op. cit.*, p. 203.
9 *Ibid*, p. 204. T. J. Wilkinson, *Archaeological Landscapes of the Near East*, Tucson: University of Arizona Press, 2003, p. 210.
10 Brian Fagan, *From Blackland to Fifth Sun*, Reading, MA: Perseus Books, 1998, p. 278.
11 Clift and Plumb, *Op. cit.*, p. 207.
12 *Ibid.*
13 *Ibid*, p. 204.
14 *Ibid*, p. 212.
15 *Ibid*, p. 214.
16 *Ibid*, pp. 214–215.
17 *Ibid*, p. 215.
18 Tom Simkin *et al.*, *Volcanoes of the World*, Stroudsberg, PA: Hutchinson Ross Publishing for the Smithsonian Institution, Washington DC, 1981, *passim.*
19 David K. Keefer *et al.*, 'Early Maritime Economy and El Niño events at Quebrada Tacahuay, Peru', *Science*, 18 September 1998, Vol. 281, No. 5384, pp. 1833–1835.
20 Jelle Zeitlinga de Boer and Donald Theodor Sanders, *Earthquakes and Human History*, Princeton and Oxford: Princeton University Press, 2004, p. 16.
21 Paul Wheatley, *The Pivot of the Four Corners*, Edinburgh: Edinburgh University Press, 1971, pp. 478 and 481.
22 Wheatley, *Op. cit.*, p. 228.
23 Kerry Sieh and Simon LeVay, *The Earth in Turmoil: Earthquakes, Volcanoes and their Impact on Humankind*, New York: W.H. Freeman, 1998, pp. 146–151.
24 Jelle Zeitlinga de Boer and Donald Theodor Sanders, *Volcanoes in Human*

History, Princeton and Oxford: Princeton University Press, 2003, p. 4. For details about the Planchón-Peteroa landslide, see: *Global and Planetary Change*, DOI: 10.1016/j.gloplacha.2010.08.003, quoted in: Kate Ravilious, 'How climate change could flatten cities', *New Scientist*, 16 October 2010, p. 14.

25 De Boer and Sanders, *Volcanoes in Human History*, *Op. cit.*, pp. 6–7.

26 Kerry Emanuel, *Divine Wind: The history and science of hurricanes*, Oxford and New York: Oxford University Press, 2005, pp. 187–189.

27 Emanuel, *Op. cit.*, p. 32.

28 Art Wolf and Ghillean Prance, *Rainforests of the World: Water, Fire, Earth and Air*, London: Harvill, 1998, p. 245.

29 Timothy Flannery, *The Eternal Frontier*, London: William Heinemann, 2001, pp. 83–118.

30 Clift and Plumb, *Op. cit.*, p. 223.

31 *Ibid.*

32 *Ibid*, p. 225. T. J. Wilkinson, personal communication, May 2011.

33 *Ibid*, p. 226.

34 *Ibid*, p. 227.

35 Jared Diamond, *Guns, Germs and Steel: The Fates of Human Societies*, London: Jonathan Cape, 1997, p. 177.

36 Diamond, *Op. cit.*, p. 367.

37 Jared Diamond, *The Third Chimpanzee: The Evolution and Future of the Human Animal*, New York: Harper Perennial, 1992, pp. 222–223.

38 Richard Keatinge, *Peruvian Prehistory: An Overview of Pre-Inca and Inca Society*, Cambridge, UK: Cambridge University Press, 1988, p. 38.

39 Diamond, *Guns, Germs and Steel*, *Op. cit.*, p. 581.

40 *Ibid*, pp. 190 and 370.

41 Peregrine Horden and Nicholas Purcell, *The Corrupting Sea: A Study of Mediterranean History*, Oxford: Blackwell, 2000, p. 346.

42 Horden and Purcell, *Op. cit.*, p. 381.

43 *Ibid*, p. 141.

44 *Ibid*, pp. 185–215, *passim*.

45 Oppenheimer, *Eden in the East*, *Op. cit.*, p. 32.

CHAPTER 6: ROOTS *v.* SEEDS AND THE ANOMALOUS DISTRIBUTION OF DOMESTICABLE MAMMALS

1 Diamond, *Guns, Germs and Steel*, *Op. cit.*, p. 128.

2 *Ibid*, p. 149.

3 *Ibid*, p. 101.

4 Carl O. Sauer, *Agricultural Origins and Dispersals*, Cambridge, MA: MIT Press, 1952/1969, p. 73.

5 Diamond, *Guns, Germs and Steel*, *Op. cit.*, p. 125.

6 *Ibid*, p. 142.

7 *Ibid*, pp. 150–151.

8 *Ibid*, p. 418.
9 Jeff Hecht, 'Out of Asia', *New Scientist*, 23 March 2002, p. 12.
10 Diamond, *Guns, Germs and Steel*, *Op. cit.*, p. 173. The earliest record of domesticated dogs may be at Oberkassel in Germany, dated to 14,000 years ago. See: Juliet Clutton-Brock, *A Natural History of Domesticated Mammals*, Cambridge UK: Cambridge University Press, 1999, p. 58.
11 Graeme Barker, *The Agricultural Revolution in Prehistory: Why did foragers become farmers?*, Oxford: Oxford University Press, 2006, 145.
12 Diamond, *Guns, Germs and Steel*, *Op. cit.*, p. 400.

CHAPTER 7: FATHERHOOD, FERTILITY, FARMING: 'THE FALL'

1 David Lewis-Williams, *The Mind in the Cave*, London and New York: Thames & Hudson, 2002, pp. 199–200 and 216–217.
2 Lewis-Williams, *Op. cit.*, pp. 224–225.
3 *Ibid.*, pp. 285–286.
4 *Ibid.*
5 Mircea Eliade, *A History of Religious Ideas*, Vol. 1, London: Collins, 1979, p. 20.
6 Anne Baring and Jules Cashford, *The Myth of the Goddess: Evolution of an Image*, Arkana/Penguin Books, 1991/1993, pp. 9–14.
7 Enrique Florescano, *The Myth of Quetzalcoatl*, trs. Lysa Hochroth, Baltimore and London: Johns Hopkins University Press, 1999, p. 199.
8 Elizabeth Wayland Barber and Paul T. Barber, *When They Severed Earth from Sky: How the Human Mind Shapes Myth*, Princeton, NJ and Oxford: Princeton University Press, 2004.
9 *Nature*, DOI:10.1038/nature07995. See also: Haim Ofek, *Second Nature: Economic Origins of Human Evolution*, Cambridge, UK: Cambridge University Press, 2011, especially the maps on pp. 185 and 188.
10 *Nature*, DOI:10.1038/nature08837.
11 Malcolm Potts and Roger Short, *Ever Since Adam and Eve: The Evolution of Human Sexuality*, Cambridge, UK: Cambridge University Press, 1999, p. 85.
12 Baring and Cashford, *Op. cit.*, p. 6.
13 *Ibid*, p. 30.
14 David R. Harris (editor), *The Origin and Spread of Agriculture and Pastoralism in Eurasia*, London: University College London Press, 1996, p. 135.
15 Harris (editor), *Op. cit.*, p. 166.
16 Chris Scarre, 'Climate change and faunal extinction at the end of the Pleistocene,' chapter 5 of *The Human Past*, edited by Chris Scarre, London: Thames & Hudson, 2006, p. 13. See also: Peter Bellwood, *First Farmers: the Origins of Agricultural Societies*, Oxford: Blackwell, 2005, p. 65.
17 Peter Watson, *Ideas; A History from Fire to Freud*, London: Phoenix/Weidenfeld & Nicolson, 2006, p. 77.
18 See also: Jared Diamond, *Guns, Germs and Steel*, *Op. cit.*, p. 105.
19 Mark Nathan Cohen, *The Food Crisis in Prehistory*, New Haven, CT: Yale

University Press, 1977.
20 Les Groube, 'The impact of diseases upon the emergence of agriculture', in Harris (editor), *Op. cit.*, pp. 101–129.
21 Jacques Cauvin, *The Birth of the Gods and the Origins of Agriculture*, Cambridge, UK: Cambridge University Press, 2000 (French publication, 1994, translation: Trevor Watkins), p. 15.
22 Cauvin, *Op. cit.*, pp. 16 and 22.
23 *Ibid*, pp. 39–48.
24 *Ibid*, p. 128.
25 Fagan, *The Long Summer*, *Op. cit.*, p. 103.
26 Michael Balter, *The Goddess and the Bull: Çatalhöyük: An archaeological journey to the dawn of civilization*, New York: Free Press, 2005, pp. 176ff.
27 http://wholehealthsource.blogspot.com/2008/08/life-expectancy-and-growth-of.html. Posted 5 August, 2008.
28 Elaine Pagels, *Adam and Eve and the Serpent*, London: Weidenfeld & Nicolson, 1988, p. 29.
29 Pagels, *Op. cit.*, p. 27.
30 Potts and Short, *Op. cit.*, p. 46.
31 Pagels, *Op. cit.*, p. xiv.
32 Jean Delumeau, *The History of Paradise: The Garden of Eden in Myth and Tradition*, trs. Matthew O'Connell, New York: Continuum, 1995, p. 196.
33 Delumeau, *Op. cit.*, p. 197.
34 Delumeau, *Op. cit.*, p. 7; Potts and Short, *Op. cit.*, p. 152.
35 Timothy Taylor, *The Prehistory of Sex*, London: Fourth Estate, 1997, p. 144.
36 Taylor, *Op. cit.*, p. 132.
37 Cauvin, *Op. cit.*, p. 69.

CHAPTER 8: PLOUGHING, DRIVING, MILKING, RIDING – FOUR THINGS THAT NEVER HAPPENED IN THE NEW WORLD

1 Andrew Sherratt, *Economy and Society in Prehistoric Europe: Changing Perspectives*, Edinburgh: Edinburgh University Press, 1998, p. 158.
2 Diamond, *Guns, Germs and Steel*, *Op. cit.*, pp. 132 and 162.
3 Sherratt, *Op. cit.*, p. 161.
4 *Ibid*, p. 165.
5 *Ibid*, p. 170.
6 *Ibid*, p. 171.
7 Robert Drews, *The End of the Bronze Age: changes in warfare and the catastrophe ca. 1200 BC*, Princeton, NJ: Princeton University Press, 1993.
8 Sherratt, *Op. cit.*, p. 173.
9 *Ibid.*
10 *Ibid*, p. 178.
11 *Ibid*, p. 180.
12 *Ibid*, p. 181.
13 *Ibid*, p. 184.

14 *PLos Computational Biology*, DOI:10.1371/journal.pcbi.1000491; http://www.livescience.com/2751-love-milk-dated-6000.html).
15 Sherratt, *Op. cit.*, p. 188.
16 *Ibid*, p. 191. S.K. McIntosh (editor), *Beyond Chiefdoms*, Cambridge: Cambridge University Press, 1999, pp. 73–75; Susan Keech McIntosh, 'Floodplains and the Development of Complex Society: Comparative Perspectives from the West African Semi-arid Tropics', in Elisabeth Benson & Lisa Lucero (editors), *Complex Polities in the Ancient Tropical World*, Archaeological Papers of the American Anthropological Association, Number 9, 1999, pp. 151–165.
17 *Ibid*, p. 192.
18 *Ibid.*
19 *Ibid*, p. 194.
20 *Ibid*, p. 195.
21 *Ibid.*
22 *Ibid*, p. 198.

CHAPTER 9: CATASTROPHE AND THE (ALL-IMPORTANT) ORIGINS OF SACRIFICE

1 Sherratt, *Op. cit.*, p. 336.
2 *Ibid*, p. 337.
3 *Ibid*, p. 334.
4 *Ibid*, p. 351.
5 *Ibid*, p. 353.
6 Chris Scarre, 'Shrines of the Land: religion and the transition to farming in Western Europe'; paper delivered at the conference, 'Faith in the past: Theorising an archaeology of religion', in Kelley Hays Gilpin and David S. Whitley (editors), *Belief in the Past: theoretical approaches to the archaeology of religion*, Walnut Creek, CA: Left Coast Press, 2008, p. 6.
7 Sherratt, *Op. cit.* p. 355.
8 *Ibid*, p. 356.
9 Colin Renfrew, *Before Civilization*, London: Cape, 1973, pp. 162–163.
10 Eliade, *Op. cit.*, p. 117.
11 Immanuel Velikovsky, *Ages in Chaos*, London: Sidgwick & Jackson, 1953; *Earth in Upheaval*, Garden City, NY: Doubleday, 1955; *Worlds in Collision: terror and the future of global order*, Basingstoke (UK): Palgrave, 1950 and 2002, edited by Kim Booth and Tim Dunne.
12 Benny J. Paiser, *et al.* (editors), *Natural Catastrophes During Bronze Age Civilisations: Archaeological, Geological, Astronomical and Cultural Prospectives*, Oxford: British Archaeological Reports, International Series, 728, 1998., p. 28.
13 Paiser, *Op. cit.*, p. 23.
14 *Ibid.*
15 *Ibid*, p. 55.
16 *Ibid*, pp. 60–61.

17 *Ibid*, p. 64.
18 *Ibid*, p. 42.
19 *Ibid*, p. 46.
20 *Ibid*, p. 174.
21 Robert G. Hamerton-Kelly (editor), *Violent Origins: Walter Burkett, Rene Girard and Jonathan Z. Smith on Ritual Killing and Cultural Formations*, Stanford: Stanford University Press, 1987, *passim*, but especially p. 204.
22 Hamerton-Kelly, *Op. cit.*, p. 179.
23 Fagan, *From Black Lands to Fifth Sun*, *Op. cit.*, pp. 93–94.
24 *Ibid*, p. 245.
25 Jan N. Bremner (editor), *The Strange World of Human Sacrifice*, Leuven, Paris, Dudley, MA: Peeters, 2007, p. 230.
26 Marija Gimbutas, *The Gods and Goddesses of Old Europe: 6500 to 3500 BC*, London: Thames & Hudson, 1982, p. 236.
27 *Ibid.*
28 *Ibid.*

CHAPTER 10: FROM NARCOTICS TO ALCOHOL

1 Sherratt, *Op. cit.*, p. 406.
2 Mark David Merlin, *On the Trail of the Ancient Poppy*, London and Toronto: Fairleigh University Press and Associated Universities Press, 1984, *passim.*
3 Sherratt, *Op. cit.*, p. 408.
4 *Ibid*, p. 409.
5 *Ibid*, p. 410.
6 *Ibid*, p. 411.
7 *Ibid.*
8 *Ibid*, pp. 414–416.
9 *Ibid*, p. 417.
10 *Ibid*, pp. 419–421.
11 *Ibid*, p. 423.
12 *Ibid*, p. 422.
13 *Ibid.* p. 424.
14 *Ibid.*
15 *Ibid.* p. 380.
16 *Ibid*, pp. 386–387.
17 *Ibid*, p. 391.
18 *Ibid*, p. 392.
19 *Ibid*, p. 393.
20 *Ibid*, p. 396.
21 Andrew Sherratt, 'Alcohol and Its Alternatives', in Jordan Goodman *et al.* (editors), *Consuming Habits: Drugs in History and Anthropology*, London and New York: Routledge, 1995, pp. 16–17.
22 Sherratt, 'Alcohol and Its Alternatives,' *Op. cit.*, pp. 17–18.
23 *Ibid*, pp. 18–19.

24 *Ibid*, p. 20.
25 Merlin, *Op. cit.*, p. 269.
26 *Ibid*, pp. 212 and 220–221.
27 Sherratt, 'Alcohol and Its Alternatives,' *Op. cit.*, p. 30.
28 *Ibid.*
29 *Ibid.* p. 31.
30 Mott T. Greene, *Natural Knowledge in Pre-Classical Antiquity*, Baltimore and London: Johns Hopkins University Press, 1992, chapter 6.

CHAPTER II: MAIZE: WHAT PEOPLE ARE MADE OF

1 John Reader, *Propitious Esculent: The Potato in World History*, London: William Heinemann, 2008, p. 32.
2 Redcliffe Salaman, *The History and Social Influence of the Potato*, Cambridge, UK: Cambridge University Press, 1949, p. 2.
3 Reader, *Op. cit.*, p. 26.
4 *Ibid*, p. 11.
5 Salaman, *Op. cit.*, p. 38.
6 Reader, *Op. cit.*, p. 16.
7 *Ibid*, pp. 27–28.
8 *Ibid*, p. 32.
9 Diamond, *Guns, Germs and Steel*, *Op. cit.*, p. 137.
10 John Staller *et al.* (editors), *Histories of Maize: Multidisciplinary Approaches to the Prehistory, Linguistics, Biogeography and Evolution of Maize*, Amsterdam: Elsevier/Academic Press, 2006, p. 55.
11 Bruce F. Benz *et al.*, 'El Riego and Early Maize Agricultural Evolution', in Staller *et al.* (editors), *Op. cit.*, pp. 74–75.
12 Michael Blake, 'Dating the Initial Spread of *Zea Mays*', in Staller *et al.* (editors), *Op. cit.*, p. 60.
13 Bruce Benz *et al.*, 'The Antiquity, Biogeography and Culture History of Maize in the Americas', in Staller *et al*, (editors), *Op. cit.*, p. 667.
14 Benz *et al.*, 'El Riego and Early Maize . . . ', *Op. cit.*, pp. 68–69.
15 *Ibid.*
16 Benz *et al.*, 'The Antiquity, Biogeography and Culture History of Maize . . .', *Op. cit.*, p. 671.
17 Sergio J. Chávez *et al.*, 'Early Maize on the Copocabana Peninsula: Implications for the archaeology of the Lake Titicaca Basin', in Staller *et al.* (editors), *Op. cit.*, p. 426.
18 Christine H. Hastorf *et al.*, 'The Movements of Maize into Middle Horizon Tiwanaku, Bolivia', in Staller *et al.* (editors), *Op. cit.*, p. 431.
19 Henry P. Schwarcz, 'Stable Carbon Isotope Analysis and Human Diet: A Synthesis', in Staller *et al.* (editors), *Op. cit.*, p. 319.
20 John E. Staller, 'The Social, Symbolic and Economic Significance of *Zea Mays* L. in the Late Horizon period', in Staller *et al.* (editors), *Op. cit.*, p. 449.
21 *Ibid*, p. 452; and see p. 454 for elaborate *chichi* rituals.

22 Nicholas A. Hopkins, 'The Place of Maize in Indigenous Mesoamerican Folk Taxonomies', chapter 44 of Staller *et al.* (editors), *Op. cit.*; and Jane H. Hill, 'The Historical Linguistics of Maize Cultivation in Mesoamerica and North America', chapter 46 of Staller *et al.* (editors), *Op. cit.*
23 Robert L. Rankin, 'Siouan Tribal Contacts and Dispersions Evidenced in the Terminology for Maize and Other Cultigens', chapter 41 of Staller *et al.* (editors), *Op. cit.*
24 William E. Doolittle *et al.*, 'Environmental Mosaics, Agricultural Diversity, and the Evolutionary Adoption of Maize in the American Southwest', in Staller *et al.* (editors), *Op. cit*, pp. 109ff.
25 Thomas P. Myers, 'Hominy Technology and the Emergence of Mississippian Societies', in Staller *et al.* (editors), *Op. cit.*, p. 515.
26 Brian Stross, 'Maize in Word and Image in Southeastern Mesoamerica', in Staller *et al.* (editors), *Op. cit.*, p. 584.
27 Stross, *Op. cit.*, p. 585.
28 *Ibid*, p. 587.
29 Gordon Brotherston, *Book of the Fourth World*, *Op. cit.*, p. 139.

CHAPTER 12: THE PSYCHOACTIVE RAINFOREST AND THE ANOMALOUS DISTRIBUTION OF HALLUCINOGENS

1 Peter T. Furst, *Hallucinogens and Culture*, Novato, CA: Chandler & Sharp, 1976/1988, p. 2.
2 Furst, *Op. cit.*, p. 3.
3 *Ibid*, p. 6.
4 *Ibid.*
5 *Ibid*, p. 8.
6 *Ibid.*
7 *Ibid*, p. 9. See also the paintings and drawings in: Thomas Donaldson, *The George Catlin Indian Gallery in the U.S. Museum*, Annual Report of the Smithsonian Museum for 1885. Washington DC: US Government Printing Office, 1886.
8 Furst, *Op. cit.*, pp. 10–11.
9 *Ibid*, p. 11.
10 *Ibid*, p. 44.
11 *Ibid*, pp. 45–46.
12 Gerardo Reichel-Dolmatoff, 'The cultural contexts of an Aboriginal Hallucinogen: *Banisteriopsis Caapi*', in Peter T. Furst (editor), *Flesh of the Gods: The Ritual Use of Hallucinogens*, New York: Prager, 1972, pp. 84–113.
13 Furst, *Hallucinogens and Culture*, *Op. cit.*, p. 55.
14 *Ibid*, p. 62.
15 *Ibid*, p. 65.
16 Mott T. Greene, *Natural Knowledge in Pre-Classical Antiquity*, *Op. cit.*, chapter 6.
17 Furst, *Hallucinogens and Culture*, *Op. cit.*, p. 67.

18 *Ibid*, p. 81.
19 *Ibid*, pp. 77–78.
20 *Ibid*, pp. 79–80.
21 See also: Gordon R. Wasson, 'Ololiuhqui and other Hallucinogens of Mexico', in *Summa Anthropológica en homenaje a Roberto J. Weitlaner*, Mexico, DF: Instituto Nacional de Antropoligia e Historia, 1967, pp. 328–348.
22 Furst, *Hallucinogens and Culture*, *Op. cit.*, p. 87.
23 *Ibid*, p. 109.
24 *Ibid*, p. 110.
25 *Ibid*, p. 111.
26 *Ibid*, p. 113.
27 For first-hand accounts, see: Barbara G. Myerhoff, *The Peyote Hunt: The sacred journey of the Huichol Indians*, Victor Turner (editor), Ithaca, NY: Cornell University Press, 1974; and: Fernando Benítez, *In the Magic Land of Peyote*, trs. John Upton, Austin, TX: The University of Texas Press, 1975.
28 Furst, *Hallucinogens and Culture*, *Op. cit.*, pp. 131–132.
29 *Ibid*, p. 134.
30 *Ibid*, p. 138.
31 *Ibid*, p. 139.
32 Lowell J. Bean and Katherine Siva Saubel, *Temalpakh: Cahuilla Indian Knowledge and Usage of Plants*, Banning, CA: Malki Museum Press, 1972.
33 Richard Evans Schultes, 'Ilex Guyana from 500 A.D. to the Present', Gothenburg Ethnographic Museum, *Etnologiska Studier*, No. 32, 1972, pp. 115–138.
34 Furst, *Hallucinogens and Culture*, *Op. cit.*, p. 152.
35 *Ibid*, p. 156.
36 *Ibid*, p. 158.
37 *Ibid*, p. 160.
38 Michael D. Coe, 'The shadow of the Olmecs', *Horizon*, Vol. 13, No. 4, 1971, pp. 970–973.
39 Julian H. Steward (editor), *Handbook of South American Indians*, 6 vols, Washington DC: Smithsonian Institution, Bureau of American Ethnology, Bulletin 143, 1963; Reprint: New York, Cooper Square. See especially Vol. 1, pp. 265, 275, 424 and Vol. 3, pp. 102, 414.
40 Furst, *Hallucinogens and Culture*, *Op. cit.*, pp. 166–169.
41 Michael J. Harner, *Hallucinogens and Shamanism*, Oxford and New York: Oxford University Press, 1973, p. xv.
42 Harner, *Op. cit.*, p. 12.
43 *Ibid*, pp. 16–17.
44 *Ibid*, pp. 23–25.
45 *Ibid*, pp. 30–31.
46 *Ibid*, p. 38.
47 *Ibid*, p. 46.
48 Claudio Naranjo, *The Healing Journey: New Approaches to Consciousness*, New York: Pantheon, 1973, p. 122.
49 Harner, *Op. cit.*, p. 129.

CHAPTER 13: HOUSES OF SMOKE, COCA AND CHOCOLATE

1 W. Golden Mortimer, *History of Coca: 'The Divine Plant' of the Incas*, San Francsico: And/Or Press, 1974, p. 22.
2 Dominic Steatfeild, *Cocaine: An Unauthorized Biography*, London: Virgin, 2001, p. 3.
3 Steatfeild, *Op. cit.*, p. 6.
4 *Ibid*, p. 8.
5 *Ibid*, p. 10.
6 Mortimer, *Op. cit.*, p. 155.
7 Steatfeild, *Op. cit.*, p. 27.
8 *Ibid*, pp. 28–29.
9 *Ibid*, p. 31.
10 Francis Robicsek, *The Smoking Gods: Tobacco in Mayan Art, History and Religion*, Norman, OK: Oklahoma University Press, 1978, pp. 1–4.
11 Robicsek, *Op. cit.*, p. 23.
12 *Ibid*, pp. 27–29.
13 *Ibid*, pp. 31–35.
14 *Ibid*, pp. 37–38.
15 *Ibid*, p. 43.
16 Johannes Wilbert, *Tobacco and Shamanism in South America*, New Haven, CT: Yale University Press, 1993, pp. 16–17.
17 Diego Durán, *Book of the Gods and Rights*, Oxford and New York: Oxford University Press, 1975 (originally published 1574–76).
18 Robicsek, *Op. cit.*, pp. 104–106.
19 *Ibid*, pp. 120–121.
20 *Ibid*, p. 157.
21 Cameron L. McNeil, *Chocolate in Mesoamerica: A Cultural History of Cacao*, Gainesville, FL: University Press of Florida, 2006, p. 1.
22 McNeil, *Op. cit.*, p. 8.
23 *Ibid*, p. 12.
24 *Ibid*, p. 14.
25 *Ibid*, p. 17.
26 Robicsek, *Op. cit.*, p. 118.
27 *Ibid*, p. 141.
28 *Ibid*, p. 154.
29 *Ibid*, p. 163.
30 Sophie Coe and Michael D. Coe, *The True History of Chocolate*, London and New York: Thames & Hudson, 1996, pp. 98–99.
31 Furst, *Hallucinogens and Culture*, *Op. cit.*, p. 156.
32 *Ibid*, p. 158.
33 *Ibid*, p. 160.
34 Michael D. Coe, 'The shadow of the Olmecs', *Horizon*, Vol. 13. No. 4, 1971, pp. 970–973.

CHAPTER 14: WILD: THE JAGUAR, THE BISON, THE SALMON

1 Wolf and Prance, *Rainforests of the World*, *Op. cit.*, p. 214.
2 Nicholas J. Saunders, *People of the Jaguar: The Living Spirit of Ancient America*, New York and London: Souvenir Press, 1989, p. 94.
3 Elizabeth P. Benson (editor), *The Cult of the Feline*, Washington DC Dumbarton Oaks Research Library, 1972, p. 2.
4 Saunders, *Op. cit.*, p. 31.
5 Benson (editor), *The Cult of the Feline*, *Op. cit.*, p. 51.
6 *Ibid*, p. 52.
7 *Ibid*, pp. 54–56.
8 *Ibid*, p. 57.
9 Gerardo Reichel-Dolmatoff, *Desana: Simbolism de los Indios Tukano del Vaupés*, Bogotá: 1968, p. 99.
10 Gerardo Reichel-Dolmatoff, 'La cultura material de los Indios Guahibo', *Revista de Instituto Etnológico Nacional* (Bogotá), Vol. 1, No. 2., 1944, pp. 437–506.
11 Benson (editor), *The Cult of the Feline*, *Op. cit.*, p. 69.
12 *Ibid*, p. 158.
13 Saunders, *Op. cit.*, pp. 80–82.
14 Benson (editor), *The Cult of the Feline*, *Op. cit.*, p. 139.
15 Saunders, *Op. cit.*, p. 135.
16 Benson, *The Cult of the Feline*, *Op. cit.*, p. 137.
17 *Ibid*, p. 138.
18 *Ibid*, p. 140.
19 Saunders, *Op. cit.*, p. 144.
20 *Ibid.*
21 *Ibid*, p. 147.
22 *Ibid*, p. 148.
23 *Ibid*, p. 150.
24 Brotherston, *Book of the Fourth World*, *Op. cit.*, p. 242.
25 Saunders, *Op. cit.*, p. 151.
26 *Ibid*, p. 152.
27 *Ibid*, p. 154.
28 Robert Wrangham, *Catching Fire: How Cooking Made Us Human*, London: Profile Books, 2009, p. 101.
29 Brian Fagan, *Ancient North America*, *Op. cit.*, p. 91.
30 *Ibid*, p. 93.
31 *Ibid*, pp. 116–120.
32 Dennis Stanford, 'The Jones Miller site: An example of Hell Gap Bison Procurement Strategy', in L. Davis and M. Wilson (editors), 'Bison Procurement and Utilization: A Symposium', *Plains Anthropological Memoir*, Vol. 16, 1978, pp. 90–97.
33 Fagan, *Ancient North America*, *Op. cit.*, p. 130.
34 *Ibid.*

35 G.C. Frison, *Op. cit.*, pp. 77–91.
36 Fagan, *Ancient North America, Op. cit.*, p. 298.
37 *Ibid.*
38 *Ibid*, p. 300.
39 Jake Page, *In the Hands of the Great Spirit, Op. cit.*, p. 51.
40 Fagan, *Ancient North America, Op. cit.*, p. 368.
41 *Ibid*, pp. 369–370.
42 *Ibid*, p. 372.
43 *Ibid*, p. 373.
44 S. Struever and F. Holton, *Koster: Americans in Search of the Prehistoric Past*, New York: Anchor Press, 1979.
45 Fagan, *Ancient North America, Op. cit.*, p. 375.
46 Melvin Fowler, 'Cahokia and the American Bottom: Settlement Archaeology', in Bruce D. Smith (editor), *Mississippian Settlement Patterns*, New York: Academic Press, 1978, pp. 455–478.

CHAPTER 15: ERIDU AND ASPERO: THE FIRST CITIES SEVEN AND A HALF THOUSAND MILES APART

1 Bernardo T. Arriaza, *Beyond Death: The Chinchorro Mummies of Ancient Chile*, Washington DC: Smithsonian Institution Press, 1995, pp. 12ff.
2 Bernardo T. Arriaza, 'Arsenias as an environmental hypothetical explanation for the origin of the oldest mummification practice in the world', *Chungara Revista de Antropologia Chilene*, Vol. 37, No. 2, December 2005, pp. 255–260.
3 *Ibid.*
4 Arriaza, *Beyond Death, Op. cit.*, pp. 61–62.
5 *Ibid*, p. 144.
6 Juan P. Ogalde *et al.*, 'Prehistoric psychotropic consumption in Andean Chilean mummies', *Nature Proceedings*: hdl:10101/npre.2007,1368.1: Posted 29 November 2007.
7 Michael Moseley, *The Maritime Foundations of Andean Civilization*, Menlo Park, CA: Cummings, 1975.
8 Ruth Shady Solis *et al.*, 'Dating Caral: a pre-ceramic site in the Supe Valley on the central coast of Peru', *Science*, Vol. 292, No. 5517, 27 April 2001, pp. 723–726.
9 *Ibid.*
10 Roger Atwood, 'A monumental feud', *Archaeology*, Vol. 58, No. 4, July/August 2005.
11 Discovermagazine.com/2005/sep/showdown-at-caral. By Kenneth Miller, p. 5 of 19.
12 Ruth Shady Solis *et al.*, *Op. cit.*
13 Hans J. Nissen, *The Early History of the Ancient Near East*, Chicago: University of Chicago Press, 1988, pp. 5 and 71; Petr Charvát, *Mesopotamia Before History*, London: Routledge, 2002, p. 134. Douglas H. Kennett *et al.*, 'Early State Formation in Southern Mesopotamia: Sea Levels, Shorelines, and Climate

Change', *Journal of Island & Coastal Archaelogy*, Vol. 1, Issue 1, 2005, pp. 67–99; DOI 10:1080/15564890600586283. T.J. Wilkinson, *Archaeological Landscapes of the Near East*, *Op. cit.*, especially pp. 17–31 and 152–210.

14 Nissen, *Op. cit.*, p. 69.
15 *Ibid.*
16 Gwendolyn Leick, *Mesopotamia*, London: Penguin, 2002, p. 2.
17 Charvát, *Op. cit.*, p. 93.
18 *Ibid.* See also: 'Oldest image of god in Americas found', *New Scientist*, 19 April 2003, p. 13.
19 Nissen, *Op. cit.*, p. 72.
20 Charvát, *Op. cit.*, p. 134.
21 Ruth Shady Solis *et al.*, *Op. cit.*
22 Mason Hammond, *The City in the Ancient World*, Cambridge, MA: Harvard University Press, 1972, p. 39.
23 Kenneth Miller (*Discover* magazine), *Op. cit.*, 4 of 19.
24 *Ibid.*
25 Miller, *Op. cit.*, 5 of 19. See also: Jeffrey Quilter *et al.*, *El Niño, Catastrophism and Culture Change in Ancient America*, Dumbarton Oaks Precolumbian Studies, Cambridge, MA: Harvard University Press, 2009.
26 Brian Fagan, *From Black Lands to Fifth Sun*, *Op. cit.*, p. 63.
27 Michael E. Moseley, 'Punctuated Equilibrium: Searching the ancient record for El Niño', *Quarterly Review of Archaeology*, Vol. 8, No. 3, 1987, pp. 7–10. See also: David K. Keefer *et al.*, 'Early maritime economy and El Niño events at Quebrada Tacahuay, Peru', *Science*, Vol. 281, No. 5384, 18 September 1998, pp 1833–35.
28 Moseley, *Punctuated Equilibrium*, *Op. cit.*, and Keefer *et al.*, *Op. cit.*

CHAPTER 16: THE STEPPES, WAR AND A 'NEW ANTHROPOLOGICAL TYPE'

1 Hans J. Nissen, *The Early History of the Ancient Near East*, *Op. cit.*, pp. 132–133.
2 H.W.F. Saggs, *Before Greece and Rome*, London: B.T. Batsford, 1989, p. 62.
3 D. Schmandt-Besserat, *Before Writing*, Vol 1: *From Counting to Cuneiform*, Austin, TX: University of Texas Press, 1992.
4 Richard Rudgley, *Lost Civilizations of the Stone Age*, London: Orion, 1998, p. 50.
5 *Ibid.*
6 *Ibid*, p. 54. The French scholar who has cast doubt on this reconstruction is: Jean-Jacques Glassner, in *The Invention of the Cuneiform: Writing in Sumer*, Baltimore and London: Johns Hopkins University Press, 2003.
7 Leick, *Op. cit.*, p. 75.
8 Nissen, *Op. cit.*, p. 136.
9 Saggs, *Op. cit.*, p. 105.

10 *Ibid*, p. 111.
11 Lionel Casson, *Libraries in the Ancient World*, New Haven, CT and London: Yale University Press, 2001, p. 4.
12 *Ibid*, p. 13.
13 Saggs, *Op. cit.*, pp. 156–158.
14 Fredrick R. Matson (editor), *Ceramics and Man*, London: Methuen, 1966, pp. 141–143.
15 Leslie Aitchison, *A History of Metals*, London: Macdonald, 1960, p. 37.
16 *Ibid*, p. 40.
17 *Ibid*, p. 41.
18 Theodore Wertime *et al.* (editors), *The Coming of the Age of Iron*, New Haven, CT: Yale University Press, 1980, p. 36.
19 Aitchison (editor), *Op. cit.*, p. 78.
20 *Ibid*, p. 82.
21 *Ibid.*
22 *Ibid.*
23 *Ibid*, p. 98.
24 Stuart Piggott, *Wagon, Chariot and Carriage*, London and New York: Thames & Hudson, 1992, p. 16.
25 *Ibid*, p. 21.
26 Robert Drews, *The End of the Bronze Age: Changes in Warfare and the Catastrophe ca. 1200 BC*, Princeton, NJ: Princeton University Press, 1994, p. 104.
27 Drews, *Op. cit.*, p. 106.
28 *Ibid*, p. 112.
29 *Ibid*, p. 119.
30 *Ibid*, p. 125.
31 Anne Baring and Jules Cashford, *The Myth of the Goddess*, *Op. cit.*, pp. 115–116.
32 *Ibid*, p. 140. Deborah Valenze, *Milk: a Local and Global History*, Yale, CT and London: Yale University Press, 2011, p. 17.
33 *Ibid.*
34 *Ibid*, p. 190.
35 *Ibid*, p. 209.
36 *Ibid*, p. 234.
37 *Ibid*, p. 277.
38 *Ibid*, p. 278.
39 Elena Efimovna Kuzmina, *The Prehistory of the Silk Road*, editor Victor H. Mair, Philadelphia, PA: University of Pennsylvania Press, 2008, p. 10.
40 Gérard Chaliand, *Nomadic Empires: From Mongolia to the Danube*, trs. A.M. Berrett, Rutgers, NJ: Transaction, 2005, pp. 8–10.
41 Kuzmina, *Op. cit.*, pp. 88 and 100.
42 *Ibid*, p. 4.
43 Braudel, *Op. cit.*, pp. 110–111.
44 A.M. Khazanov, *Nomads and the Outside World*, trs. Julia Crookenden, Cambridge, UK: Cambridge University Press, 1984, p. 92.

45 John Larner, *Marco Polo and the Discovery of the World*, New Haven, CT: Yale University Press, 1999, p. 25.
46 Chaliand, *Op. cit.*, p. 7.
47 Khazanov, *Op. cit.*, p. 96.
48 *Ibid*, p. 43.
49 *Ibid*, p. 32.
50 *Ibid*, p. 51.
51 *Ibid*, p. 69.
52 M.L. Ryder, *Sheep and Man*, London: Duckworth, 1983, p. 10.
53 Ryder, *Op. cit.*, p. 80.
54 *Ibid*, pp. 652–655.
55 Hannah Velten, *Cow*, London: Reaktion Books, 2007, p. 13.
56 Velten, *Op. cit.*, p. 34.
57 *Ibid*, p. 77.
58 *Ibid*, p. 106.
59 Nicola di Cosmo, *Ancient China and Its Enemies: the rise of nomadic power in East Asian history*, Cambridge, UK: Cambridge University Press, 2004, p. 32.
60 Khazanov, *Op. cit.*, p. 71.
61 Kuzmina, *Op. cit.*, p. 62.
62 Khazanov, *Op. cit.*, p. 82.
63 Chaliand, *Op. cit.*, p. xii.
64 Ernest Gellner, *Plough, Sword and Book: The Structure of Human History*, London: Collins Harvill, 1988, p. 154.
65 Chaliand, *Op. cit.*, p. 11.
66 Kuzmina, *Op. cit.*, p. 161.
67 Di Cosmo, *Op. cit.*, p. 31.
68 *Ibid*, p. 32.
69 Kuzmina, *Op. cit.*, p. 65.
70 Baring and Cashford, *Op. cit.*, p. 156.
71 *Ibid.*
72 Chaliand, *Op. cit.*, p. 12.
73 Baring and Cashford, *Op. cit.*, pp. 156–158.
74 *Ibid.*
75 Joseph Campbell, *The Masks of God: Occidental Mythology*, London: Secker & Warburg, 4 vols, 1960–68, Vol. 1, pp. 21–22.

CHAPTER 17: THE DAY OF THE JAGUAR

1 Brian Fagan, *Kingdoms of Gold, Kingdoms of Jade: The Americas Before Columbus*, London and New York: Thames & Hudson, 1991, p. 96.
2 *Ibid*, p. 97.
3 *Ibid*, p. 98.
4 *Ibid*, p. 99.
5 John E. Clark and Pary E. Pye (editors), *Olmec Art and Archaeology in*

Mesoamerica, Washington, DC: National Gallery of Art/Yale University Press, 2000, p. 219.
6 Fagan, *Op. cit.*, p. 103.
7 David Grove, *Chalcatzingo: Excavations on the Olmec Frontier*, London and New York: Thames & Hudson, 1984, pp. 104–105.
8 Clark and Pye (editors), *Op. cit.*, p. 23.
9 *Ibid*, p. 164.
10 *Ibid*, p. 88.
11 *Ibid*, p. 89.
12 Grove, *Op. cit.*, p. 126.
13 *Ibid*, p. 116.
14 *Ibid*, p. 208.
15 *Ibid*, p. 209.
16 *Ibid*, p. 186.
17 *Ibid*, p. 165.
18 *Ibid*, p. 164.
19 William J. Conklin and Jeffrey Quilter (editors), *Chavin Art, Architecture and Culture*, Los Angeles and Berkeley: University of California Press/Cotsen Institute of Archaeology, 2008, p. 119.
20 *Ibid*, pp. 158–159.
21 *Ibid*, p. 154.
22 *Ibid*, pp. 275–277.
23 *Ibid.*
24 Clark and Pye (editors), *Op. cit.*, p. 167.
25 Richard L. Burger, *Chavin and the Origin of Andean Civilization*, London and New York: Thames & Hudson, 1995, p. 128.
26 Conklin and Quilter (editors), *Op. cit.*, p. 152.
27 Burger, *Op. cit.*, p. 167.
28 Conklin and Quilter (editors), *Op. cit.*, p. 210.
29 Burger, *Op. cit.*, p. 170.
30 Conklin and Quilter (editors), *Op. cit.*, p. 80.
31 *Ibid*, p. 135.
32 *Ibid*, p. 170.
33 Burger, *Op. cit.*, p. 171.
34 *Ibid*, p. 216.
35 *Ibid*, p. 157.
36 Conklin and Quilter (editors), p. 259.
37 Burger, *Op. cit.*, p. 157.
38 *Ibid.*
39 Conklin and Quilter (editors), *Op. cit.*, pp. 259–260.
40 Burger, *Op. cit.*, p. 159.
41 *Ibid.*
42 *Ibid*, p. 189.
43 Conklin and Quilter (editors), *Op. cit.*, p. 112.
44 *Ibid*, p. 26.

45 *Ibid*, p. 30.
46 *Ibid*, p. 195.
47 *Ibid*, p. 196.
48 *Ibid*, p. 198.
49 Burger, *Op. cit.*, p. 202.
50 *Ibid*, p. 203.

CHAPTER 18: THE INVENTION OF MONOTHEISM AND THE END OF SACRIFICE IN THE OLD WORLD

1 Karen Armstrong, *The Great Transformation: The World in the Time of the Buddha, Socrates, Confucius and Jeremiah*, London: Atlantic/Knopf, 2006, p. xii.
2 V. Gordon Chile, *Prehistoric Migrations in Europe*, Cambridge, MA: Harvard University Press, 1950, p. 180.
3 Drews, *Op. cit.*, p. 97.
4 Armstrong, *Op. cit.*, pp. xiii-xiv.
5 *Ibid.*
6 *Ibid*, pp. 3–4.
7 *Ibid*, pp. 5–7.
8 *Ibid*, pp. 8–10.
9 *Ibid*, p. 11.
10 Edward Bryant, *The Quest for the Origins of Vedic Culture*, Oxford: Oxford University Press, 2001.
11 Armstrong, *Op. cit.*, pp. 24–25.
12 *Ibid*, p. 79.
13 *Ibid*, p. 84.
14 Paul Dundas, *The Jains*, London and New York: Routledge, 2002, p. 17.
15 Patrick Olivelle, *Upanihads*, Oxford and New York: Oxford University Press, 1996, pp. xxxiv-xxxv.
16 Armstrong, *Op. cit.*, p. 133.
17 *Ibid*, pp. 196–199.
18 *Ibid*, p. 234.
19 *Ibid*, p. 239.
20 *Ibid*, p. 274.
21 Joseph Campbell, *The Masks of God, Oriental Mythology*, London: Penguin Books, 1991, p. 236.
22 Armstrong, *Op. cit.*, p. 284.
23 Edward Conze, *Buddhism: Its Essence and Development*, Oxford: Oxford University Press, 1951, p. 125.
24 Armstrong, *Op. cit.*, p. 361.
25 *Ibid*, p. 366.
26 Jacques Gernet, *Ancient China: From the beginning to the Empire*, trs. Raymond Rudorff, London: Faber, 1968, pp. 37–65.
27 Armstrong, *Op. cit.*, p. 35.

28 *Ibid*, p. 73.
29 *Ibid*, pp. 77 and 114.
30 *Ibid*, p. 119.
31 *Ibid*, p. 154.
32 *Ibid.*
33 Gernet, *Op. cit.*, pp. 83–84.
34 A.C. Graham, *Disputers of the Tao: Philosophical Arguments in Ancient China*, La Salle, Illinois: Illinois University Press, 1989, pp. 9ff.
35 Armstrong, *Op. cit.*, p. 205.
36 *Ibid*, pp. 207–211.
37 Sima Qian, *Records of the Grand Historian* 124, in Fung Yu-Lan, *A Short History of Chinese Philosophy*, ed. and trs. Derk Bodde, New York, 1976, p. 50.
38 Armstrong, *Op. cit.*, pp. 272–274.
39 *Ibid*, p. 292.
40 Graham, *Op. cit.*, pp. 111–130.
41 Mencius 7A 1, taken from D.C. Lau, trs., *Mencius*, Hong Kong: Chinese University Press, 1970.
42 Armstrong, *Op. cit.*, pp. 340–347.
43 *Ibid*, p. 372.
44 *Ibid*, p. 43.
45 *Ibid*, p. 63.
46 S. David Sperling, 'Israel's religion in the Near East', in Arthur Green (editor), *Jewish Spirituality*, 2 vols, London and New York, The Crossroad Publishing Company, 1986, 1988, Vol. 1, pp. 27–28.
47 Armstrong, *Op. cit.*, p. 80.
48 *Ibid*, p. 93.
49 *Ibid*, p. 94.
50 *Ibid*, p. 99.
51 R.E. Clements, *God and Temple*, Oxford: Oxford University Press, 1965, pp. 90–95.
52 Ezekiel 2:12–15.
53 Armstrong, *Op. cit.*, p. 182.
54 *Ibid*, p. 382.
55 Guy G. Stroumsa, trs. Susan Emanuel, *The End of Sacrifice: Religious Transformation in Late Antiquity*, Chicago: University of Chicago Press, 2009, p. 5.
56 Polyminia Athanassiadi and Michael Frede, *Pagan Monotheism in Late Antiquity*, Oxford: Oxford University Press/Clarendon Press, 1999.
57 Athanassiadi and Frede, *Op. cit.*, pp. 8–9.
58 *Ibid*, pp. 17–20.
59 *Ibid*, pp. 24–25.
60 *Ibid*, pp. 31–38.
61 *Ibid*, pp. 41–43.
62 *Ibid*, p. 55.
63 *Ibid*, pp. 69–70.
64 *Ibid*, p. 110.

65 Daniel Hillel, *The Natural History of the Bible: An Environmental Exploration of the Hebrew Scriptures*, New York: Columbia University Press, 2006, pp. 16–18.
66 Hillel, *Op. cit.*, pp. 56–62.
67 *Ibid*, p. 67.
68 *Ibid*, pp. 244–245.
69 *Ibid*, pp. 103–104.
70 *Ibid*, p. 133.
71 *Ibid*, pp. 173–179.
72 *Ibid*, pp. 181 and 208.
73 Stroumsa, *Op. cit.*, p. 71.
74 Bremner, *Op. cit.*, p. 252, note 63.
75 Walter Burkert, *The Orientalizing Revolution: Near Eastern Influences on Greek Culture in the Early Archaic Age*, trs. Margaret E. Pindar, Cambridge, MA: Harvard University Press, 1992, pp. 73–75.
76 Stroumsa, *Op. cit.*, p. 33. Miranada Aldhouse Green, *Dying for the Gods: Human Sacrifice in Iron Age and Roman Europe*, Stroud: Tempus, 2001, p. 31.
77 *Ibid*, pp. 57–60.
78 Ingvild Saelid Gilhus, *Animals, Gods and Humans: changing attitudes to animals in Greek, Roman and early Christian ideas*, London: Routledge, 2006, p. 2.
79 Gilhus, *Op. cit.*, p. 152.
80 *Ibid*, pp. 38–40.
81 *Ibid*, p. 61.
82 *Ibid*, pp. 97–98.
83 *Ibid*, p. 126.
84 *Ibid*, pp. 144–148. Caroline Grigson and Juliet Clutton-Brock, *Animals and Archaeology, volume 4, Husbandry in Europe*, Oxford: BAR International Series, 202, 1984, p. 186.
85 *Ibid*, p. 171.
86 *Ibid*, pp. 263–267.
87 Stroumsa, *Op. cit.*, pp. 67–69.
88 *Ibid*, p. 30.
89 *Ibid*, p. 39.
90 *Ibid*, pp. 53–54.
91 René Girard, *Violence and the Sacred*, trs. Patrick Gregory, Baltimore and London: Johns Hopkins University Press, 1977. Quoted in Stroumsa, *Op. cit.*, p. 81.
92 Stroumsa, *Op. cit.*, p. 91.
93 *Ibid*, pp. 101–102.
94 *Ibid*, p. 124.

CHAPTER 19: THE INVENTION OF DEMOCRACY, THE ALPHABET, MONEY AND THE CONCEPT OF NATURE

1 Karen Armstrong, *Op. cit.*, p. 168.

2 *Ibid*, p. 169.
3 *Ibid*, p. 144.
4 *Ibid*, p. 139.
5 *Ibid*, p. 145.
6 Walter Burkert, *Greek Religion*, trs. John Raffar, Cambridge, MA: Harvard University Press, 1983, pp. 44–49.
7 Oswyn Murray, *Early Greece*, Brighton: Harvester Press 1990 (reprint 1999), pp. 173–185.
8 Armstrong, *Op. cit.*, p. 184.
9 *Ibid*, p. 183.
10 *Ibid*, p. 223.
11 *Ibid*, p. 224.
12 Murray, *Op. cit.*, pp. 236–246.
13 John Keane, *The Life and Death of Democracy*, London: Simon & Schuster, 2009, p. 10.
14 Keane, *Op. cit.*, pp. 15–18.
15 *Ibid*, pp. 45–50.
16 *Ibid*, p. 52.
17 *Ibid*, p. 60.
18 Leonard Shlain, *The Alphabet and the Goddess*, London: Penguin, 1998, p. 65.
19 Diamond, *Guns, Germs and Steel*, *Op. cit.*, p. 226.
20 Shlain, *Op. cit.*, p. 66.
21 *Ibid*, p. 68.
22 Ernest Gellner, *Plough, Sword and Book: The Structure of Human History*, London: Collins Harvill, 1988, p. 72.
23 Gellner, *Op. cit.*, p. 77.
24 Diamond, *Guns, Germs and Steel*, *Op. cit.*, p. 231.
25 Shlain, *Op. cit.*, p. 70.
26 Robert K. Logan, *The Alphabet Effect*, Boston: William Morrow, 1986, pp. 34–35.
27 Logan, *Op. cit.*, p. 40.
28 *Ibid*, p. 97.
29 *Ibid*, pp. 104 and 114–115.
30 Tim Ingold, *The Perception of the Environment: Essays in livelihood, dwelling and skill*, London: Routledge, 2000, chapter 4, 'From trust to domination: an alternative history of human-animal relations', pp. 61–76.
31 Erwin Schrödinger, *Nature and the Greeks and Science and Humanism*, Cambridge, UK: Cambridge University Press, 1954/1996, pp. 55–58.
32 Geoffrey Lloyd and Nathan Sivin, *The Way and the Word: Science and Medicine in Early China and Greece*, New Haven, CT and London: Yale University Press, 2002, pp. 242–248.
33 Greene, *Op. cit.*, pp. 78ff.
34 Gerard Naddaf, *The Greek Concept of Nature*, Albany, NY: State University of New York Press, 2005, p. 15.
35 H.D.F. Kitto, *The Greeks*, London: Penguin, 1961, p. 177.

36 A.R. Burn, *The Penguin History of Greece*, London: Penguin, 1966, p. 131.
37 *Ibid*, p. 138.
38 David C. Lindberg, *The Beginnings of Western Science*, Chicago: University of Chicago Press, 1992, p. 34.
39 Burn, *Op. cit.*, p. 248.
40 Lindberg, *Op. cit.*, p. 31.
41 E.R. Dodds, *The Greeks and the Irrational*, Los Angeles and Berkeley: University of California Press, 1951.
42 Lloyd and Sivin, *Op. cit.*, p. 241.
43 Michael Grant, *The Classical Greeks*, London: Weidenfeld & Nicolson, 1989, p. 70.
44 *Ibid*, p. 72.
45 Armstrong, *Op. cit.*, p. 108.
46 Jack Weatherford, *The History of Money*, New York: Three Rivers Press (Crown), 1997, p. 27.
47 *Ibid*, p. 30.
48 *Ibid*, pp. 34–35.

CHAPTER 20: SHAMAN-KINGS, WORLD TREES AND VISION SERPENTS

1 Lido Valdez, 'Walled settlements, Buffer Zones and Human decapitation in the Acari Valley, Peru', *Journal of Anthropological Research*, Vol. 65, No. 3, 1969, pp. 386–416.
2 Fagan, *Kingdoms of Gold*, *Op. cit.*, p. 188.
3 *Ibid.*
4 *Ibid*, p. 189.
5 Helaine Silverman and Donald A. Proulx, *The Nasca (Peoples of America)*, London and New York: Blackwell-Wiley, 2002.
6 Andy Coghlan, 'Chop-happy Nazca learned hard lesson', *New Scientist*, 17 November 2009, p. 16.
7 Fagan, *Kingdoms of Gold*, *Op. cit.*, p. 189.
8 *Ibid*, p. 192.
9 *Ibid.*
10 *Ibid*, p. 194.
11 *Ibid*, p. 172.
12 *Ibid*, p. 173.
13 Steve Bourget and Kimerly L. Jones, *The Art and Archaeology of the Moche: An ancient Andean Society of the Peruvian North Coast*, Austin, TX: University of Texas Press, 2008, pp. 202–203.
14 Bourget and Jones, *Op. cit.*, p. 56.
15 Fagan, *Kingdoms of Gold*, *Op. cit.*, p. 180.
16 Bourget and Jones, *Op. cit.*, p. 35; Fagan, *Op. cit.*, p. 196.
17 Bourget and Jones, *Op. cit.*, p. 260.
18 Moseley *et al.*, *Op. cit.*, p. 89.
19 Bourget and Jones, *Op. cit.*, p. 210.

20 Linda Schele and David Freidel, *A Forest of Kings: The Untold Story of the Ancient Maya*, New York: Quil/William Morrow, 1990, p. 112.
21 Schele and Freidel, *Op. cit.*, pp. 46 and 61. Roderick J. McIntosh *et al.* (editors), *The Way the Wind Blows: Climate, History and Human Action*, New York: Columbia University Press, 2000, p. 244.
22 David Freidel, Linda Schele and Joy Parker, *Maya Cosmos: Three Thousand Years on the Shaman's Path*, New York: Quil/William Morrow, 1993, pp. 81–95. See also: Anthony Aveni, *People and the Sky*, London and New York: Thames & Hudson, 2008, p. 49.
23 Schele and Freidel, *Op. cit.*, p. 117.
24 *Ibid*, p. 207.
25 Aveni, *Op. cit.*, p. 208.
26 Peter S. Rudman, *How Mathematics Happened: The First 5,000 Years*, Amherst, MA: Prometheus, 2007, pp. 129–130.
27 Schele and Freidel, *Op. cit.*, p. 87; Roderick J. McIntosh *et al.* (editors), *Op. cit.*, pp. 275–277.
28 *Ibid*, p. 121.
29 *Ibid*, pp. 85 and 380.
30 *Ibid*, pp. 123–126.
31 Fagan, *Kingdoms of Gold*, *Op. cit.*, p. 126.
32 DOI: 10.1016/j/jas.2009.01.020.
33 Freidel, Schele and Parker, *Op. cit.*, pp. 123 and 131.
34 *Ibid*, p. 145.
35 Kent Flannery and Joyce Marcus (editors), *The Cloud People: Divergent Evolution of the Zapotec and Mixtec Civilizations*, New York and London: Academic Press, 1983, pp. 218, 340, 357–359.
36 Flannery and Marcus (editors), *Op. cit.*, pp. 38–39.
37 *Ibid*, pp. 347–350.
38 Arthur Joyce and Marcus Winter, 'Agency, Ideology and Power in Oaxaca', *Current Anthropology*, Vol. 37, No. 1, February 1996, pp. 33–47.
39 Flannery and Marcus (editors), *Op. cit.*, p. 152.
40 Aveni, *Op. cit.*, p. 134.
41 *Ibid*, p. 136.
42 *Ibid*, p. 153.
43 Fagan, *Kingdoms of Gold*, *Op. cit.*, p. 195; Roderick J. McIntosh *et al.* (editors), *Op. cit.*, p. 273.
44 Jake Page, *In the Hands of the Great Spirit: the 20,000-year history of the American Indians*, New York: Free Press, 2003, p. 2.
45 Fagan, *Kingdoms of Gold*, *Op. cit.*, p. 203.
46 *Ibid*, pp. 204–205.
47 Fagan, *From Black Lands to Fifth Sun*, *Op. cit.*, pp. 169 and 215. William E. Doolittle, *Cultivated Landscapes of Native North America*, Oxford: Oxford University Press, 2001, pp. 39, 194 and 254.
48 Fagan, *Kingdoms of Gold*, *Op. cit.*, pp. 204–211.
49 *Ibid*, p. 209.

50 *Ibid.*
51 *Ibid*, p. 212.
52 *Ibid*, p. 213.
53 *Ibid*, pp. 213–214.
54 *Ibid*, p. 216.
55 *Ibid.*
56 *Ibid*, p. 217.
57 *Ibid*, p. 220.

CHAPTER 21: BLOODLETTING, HUMAN SACRIFICE, PAIN AND POTLATCH

1 Linda Schele and Mary Ellen Miller, *The Blood of Kings: dynasty and ritual in Mayan art*, Fort Worth,TX: Kimbell Art Museum, 1986, p. 42.
2 Maria Longhena, *Maya Script*, trs. Rosanna M. Giammanco Frongia, New York: Abbeville, 2000, p. 65.
3 Schele and Miller, *Op. cit.*, p. 45.
4 *Ibid*, p. 175.
5 *Ibid*, p. 177.
6 *Ibid*, p. 178.
7 *Ibid*, p. 179.
8 *Ibid*, p. 180.
9 *Ibid*, p. 193.
10 *Ibid*, p. 210.
11 *Ibid*, p. 214.
12 *Ibid*, p. 216.
13 *Ibid*, pp. 215–218.
14 *Ibid*, p. 241.
15 E. Michael Whittington, *The Sport of Life and Death: The Mesoamerican Ballgame*, London and New York: Thames & Hudson, 2001, pp. 71–75.
16 Whittington, *Op. cit.*, p. 39.
17 *Ibid*, p. 81. See also: Schele and Miller, *Op. cit.*, p. 243.
18 Whittington, *Op. cit.*, p. 21.
19 *Ibid.*
20 *Ibid*, p. 120.
21 *Ibid*, p. 29.
22 *Ibid*, p. 30.
23 Schele and Miller, *Op. cit.*, p. 243.
24 Whittington, *Op. cit.*, p. 76.
25 Schele and Miller, *Op. cit.*, p. 245.
26 *Ibid*, p. 248.
27 *Ibid*, p. 249.
28 Whittington, *Op. cit.*, pp. 42–48.
29 *Ibid*, pp. 55–63.
30 *Ibid*, p. 110.

31 Elizabeth Benson and Anita G. Cook (editors), *Ritual Sacrifice in Ancient Peru*, Austin, TX: University of Texas Press, 2001, p. 183.
32 Heather Orr and Rex Koontz (editors), *Blood and Beauty: Organized Violence in the Art and Archaeology of Mesoamerica and Central America*, Los Angeles: The Cotsen Institute of Archaeology at UCLA, 2009, p. 128.
33 Benson and Cook (editors), *Op. cit.*, pp. 12–13.
34 Orr and Koontz (editors), *Op. cit.*, p. 115.
35 *Ibid*, p. 297.
36 *Ibid*, p. 305.
37 *Ibid.*
38 Orr and Koontz, *Op. cit.*, pp. 47 and 53.
39 Benson and Cook (editors), *Op. cit.*, pp. 8 and 41.
40 Orr and Koontz (editors), *Op. cit.*, p. 287.
41 *Ibid*, p. 258.
42 *Ibid*, p. 243.
43 Aveni, *Op. cit.*, p. 169.
44 Aveni, *Op. cit.*, p. 136.
45 Longhena, *Op. cit.*, pp. 23–24.
46 Joyce Marcus, *Mesoamerican Writing Systems: Propaganda, Myth and History in Four Ancient Civilizations*, Princeton, NJ and Oxford: Princeton University Press, 1992.
47 Marcus, *Op. cit.*, p. 435.
48 *Ibid*, p. 7.
49 *Ibid*, p. 441.
50 Fagan, *From Black Lands to Fifth Sun*, *Op. cit.*, p. 293.
51 William C. Sturtevant (general editor), Wayne Suttle, volume editor, *Handbook of North American Indians: Volume 7, Northwest Coast*, Washington DC: Smithsonian Institution Press, 1990, p. 84.
52 Aveni, *Op. cit.*, p. 223.
53 *Handbook of North American Indians*, *Op. cit.*, p. 85.
54 Marcel Mauss, *The Gift: forms and functions of exchange in archaic societies*, trs. Ian Cunninson, London: Coehn & West, 1954.
55 *Handbook of North American Indians*, *Op. cit.*, pp. 85–86.

CHAPTER 22: MONASTERIES AND MANDARINS, MUSLIMS AND MONGOLS

1 Rodney Stark, *The Victory of Reason: How Christianity Led to Freedom, Capitalism and Western Success*, New York: Random House, 2005, p. 5.
2 Gellner, *Plough, Sword and Book*, *Op. cit.*, p. 89.
3 Stark, *Op. cit.*, pp. 6–7.
4 *Ibid*, p. 9.
5 Gellner, *Op. cit.*, p. 84.
6 Stark, *Op. cit.*, p. 11.
7 *Ibid*, pp. 15–17. But see: Charles Freeman's untitled and undated review of

Stark's book on Amazon.com. And also: Mott T. Greene, *Natural Knowledge in Pre-classical antiquity*, *Op. cit.*, p. 143.

8 Stark, *Op. cit.*, p. 17.
9 *Ibid*, p. 22.
10 *Ibid*, pp. 28–29.
11 *Ibid*, p. 59.
12 *Ibid*, p. 64.
13 Freeman, *Op. cit.*
14 Stark, *Op. cit.*, p. 81.
15 *Ibid*, pp. 83–84.
16 Anthony Pagden (editor), *The Idea of Europe*, Cambridge, UK and Washington DC: Cambridge University Press/Woodrow Wilson Center Press, 2002, p. 81.
17 R.W. Southern, *Scholastic Humanism and the Unification of Europe*, Vol. 1, *Foundations*, Oxford: Basil Blackwell, 1995, p. 1.
18 *Ibid*, p. 5.
19 Herbert Musurillo SJ, *Symbolism and the Christian Imagination*, Dublin: Helicon, 1962, p. 152.
20 Southern, *Op. cit.*, p. 22.
21 *Ibid*, p. 64.
22 Stark, *Op. cit.*, p. 82.
23 *Ibid*, p. 113.
24 *Ibid*, pp. 35–39.
25 *Ibid*, p. 41.
26 Douglas North and Robert Thomas, *The Rise of the Western World*, Cambridge, UK: Cambridge University Press, 1953, p. 33.
27 North and Thomas, *Op. cit.*, p. 43.
28 Carlo M. Cipolla, *Before the Industrial Revolution: European Society and Economy, 1000–1700*, London and New York: Routledge, 2003, p. 141.
29 D.A. Callus (editor), *Robert Grosseteste*, Oxford: Oxford University Press, 1955, p. 98.
30 Robert Pasnau, *Aquinas on Human Nature*, Cambridge, UK: Cambridge University Press, 2003.
31 Robert Benson and Giles Constable, *Renaissance and Renewal in the Twelfth Century*, Oxford: Oxford University Press, 1982, p. 45.
32 Stark, *Op. cit.*, pp. 106ff.
33 Janet Abu-Lughod, *Before European Hegemony: The World System AD 1250–1350*, Oxford: Oxford University Press, 1989, pp. 3–4.
34 Abu-Lughod, *Op. cit.*, pp. 16–17. On the Arabic/Islamic spread, see: Peter Bellwood, *First Farmers*, *Op. cit.*, p. 192.
35 *Ibid*, p. 155.
36 *Ibid*, p. 158.
37 *Ibid*, p. 170.
38 William McNeill, *Plagues and People*, Oxford: Blackwell, 1977, p. 19.
39 Norman Cantor, *In the Wake of the Plague: The Black Death and the World It*

Made, London: Simon & Schuster, 2001, p. 191.
40 Abu-Lughod, *Op. cit.*, p. 174.
41 *Ibid*, p. 237. Terence Ranger and Paul Slack (eds.), *Epidemics and Ideas: Essays in the Historical Perception of Pestilence*, Cambridge UK: Cambridge University Press, 1992, p. 83.
42 Cantor, *Op. cit.*, pp. 15–16.
43 Stark, *Op. cit.*, pp. 148ff.
44 Cantor, *Op. cit.*, p. 210.

CHAPTER 23: THE FEATHERED SERPENT, THE FIFTH SUN AND THE FOUR *SUYUS*

1 Fagan, *Kingdoms of Gold*, *Op. cit.*, p. 18.
2 *Ibid*, p. 154.
3 Geoffrey W. Conrad and Arthur A. Demarest, *Religion and Empire: The Dynamics of Aztec and Inca Expansionism*, Cambridge UK: Cambridge University Press, 1984, pp. 26 and 29. Richard A. Diehl, *Tula: the Toltec Capital of Ancient Mexico*, London and New York: Thames & Hudson, 1983, p. 141.
4 David Carrasco, *Quetzalcoatl and the Irony of Empire; Myths and Prophecies in the Aztec Tradition*, Revised Edition, Boulder, CO: University Press of Colorado, 2000, pp. 104ff.
5 Carrasco, *Quetzalcoatl*, *Op. cit.*, p. 132.
6 *Ibid*, pp. 63ff.
7 *Ibid*, p. 156.
8 *Ibid*, p. 18. See also: Fagan, *From Black Land to Fifth Sun*, *Op. cit.*, p. 364.
9 Carrasco, *Quetzalcoatl*, *Op. cit.*, p. 20.
10 Conrad and Demarest, *Op. cit.*, p. 22.
11 *Ibid*, p. 23.
12 *Ibid.*
13 *Ibid*, p. 38.
14 *Ibid*, p. 23.
15 Aveni, *People and the Sky*, *Op. cit.*, p. 141.
16 Conrad and Demarest, *Op. cit.*, pp. 17 and 29.
17 Carrasco, *Quetzalcoatl*, *Op. cit.*, p. 199.
18 *Ibid*, pp. 44–45.
19 *Ibid*, p. 36.
20 *Ibid*, pp. 160–165.
21 Carrasco, *City of Sacrifice*, *Op. cit.*, p. 78.
22 Carrasco, *Quetzalcoatl*, *Op. cit.*, pp. 93–94.
23 Carrasco, *City of Sacrifice*, *Op. cit.*, p. 79.
24 *Ibid*, p. 32.
25 Florescano, *Quetzalcoatl*, *Op. cit.*, pp. 73–74.
26 Fagan, *Kingdoms of Gold*, *Op. cit.*, p. 33.
27 Carrasco, *City of Sacrifice Op. cit.*, p. 56.
28 *Ibid*, p. 74.

29 *Ibid*, p. 141.
30 *Ibid*, pp. 193 and 198.
31 Arthur Joyce and Marcus Winter, 'Ideology, Power and Urban Society in Pre-hispanic Oaxaca', *Current Anthropology*, Vol. 37, No. 1 (February 1996), p. 37.
32 Conrad and Demarest, *Op. cit.*, pp. 185–186.
33 Fagan, *Kingdoms of Gold*, *Op. cit.*, p. 25.
34 *Ibid*, p. 41.
35 *Ibid.*
36 *Ibid*, p. 44. *The* (London) *Times*, 23 May 2011, p. 12.
37 *Ibid*, p. 46.
38 *Ibid*, p. 48.
39 Conrad and Demarest, *Op. cit.*, p. 97.
40 Fagan, *Kingdoms of Gold*, *Op. cit.*, p. 53.
41 Conrad and Demarest, *Op. cit.*, p. 100.
42 Benson and Cook (editors), *Ritual Sacrifice in Ancient Peru*, *Op. cit.*, p. 17.
43 Tierney, *The Highest Altar*, *Op. cit.*, p. 28.
44 Tierney, *Op. cit.*, p. 117.
45 Benson and Cook (editors), *Op. cit.*, p. 17.
46 Conrad and Demarest, *Op. cit.*, p. 91.
47 Fagan, *Op. cit.*, p. 48.
48 Tierney, *Op. cit.*, pp. 178 and 203.
49 Conrad and Demarest, *Op. cit.*, p. 115.
50 *Ibid*, p. 102.
51 Tierney, *Op. cit.*, p. 30.
52 Conrad and Demarest, *Op. cit.*, p. 110.

CONCLUSION: THE SHAMAN AND THE SHEPHERD: THE GREAT DIVIDE

1 J.H. Parry, *The Age of Reconnaissance: Discovery, Exploration and Settlement, 1450–1650*, London: Cardinal/Sphere, 1973, p. 35.
2 Parry, *Op. cit.*, p. 46.
3 Appendix 2, available online, discusses the literature on the similar development of complex societies.
4 Conrad and Demarest, *Religion and Empire*, *Op. cit.*, p. 196.
5 *Ibid*, p. 206.
6 Art Wolf and Ghillean Prance, *Rainforests of the World: Water, Fire, Earth and Air*, *Op. cit.*, p. 281.
7 Calvin Luther Martin, *In the Spirit of the Earth*, *Op. cit.*, p. 58.
8 Peregrine Horden and Nicholas Purcell, *The Corrupting Sea*, *Op. cit.*, p. 417.
9 *Ibid*, p. 419.
10 Florescano, *Quetzalcoatl*, *Op. cit.*, p. 42.
11 Conrad and Demarest, *Op. cit.*, pp. 72–74.
12 Florescano, *Op. cit.*, pp. 93–94 and 98.

13 Aveni, *Op. cit.*, p. 191.
14 Peter Bellwood, *Man's Conquest of the Pacific*, New York and London: Oxford University Press, 1979, p. 198.
15 Bourget and Jones, *The Art and Archaeology of the Moche*, *Op. cit.*, pp. 43–44.

APPENDIX I: THE (NEVER-ENDING) DISPUTE OF THE NEW WORLD

1 Geoffrey Simcox and Blair Sullivan, *Christopher Columbus and the Enterprise of the Indies: A Brief History with Documents*, Boston and New York: Bedford/St Martin's Press, 2005, p. 31. The 'Columbian exchange', and its consequences, is now the subject of a new study: Charles C. Mann, *1493: How the Ecological Collision of Europe and the Americas Gave Rise to the Modern World*, London, Random House, 2011.
2 J.H. Elliott, *The Old World and the New*, Cambridge, UK: Cambridge University Press/Canto, 1970/1992, pp. 9–10.
3 Elliott, *Op. cit.*, p. 11.
4 Margaret R. Greer, *et al.*, *Rereading the Black Legend: The Discourse of Religion and Racial Difference in the Renaissance Empires*, Chicago and London: Chicago University Press, 2007, p. 1.
5 Greer *et al.*, *Op. cit.*, p. 5.
6 Anthony Pagden, *European Encounters with the New World: from the Renaissance to Romanticism*, New Haven, CT and London: Yale University Press, 1993, p. 6.
7 Anthony Pagden, *The Fall of Natural Man: The American Indian and the origins of comparative ethnology*, Cambridge, UK: Cambridge University Press, 1982, pp. 99 and 104.
8 *Ibid*, p. 84.
9 *Ibid*, p. 151 and Pagden, *European Encounters*, *Op. cit.*, p. 167.
10 Pagden, *The Fall of Natural Man*, *Op. cit.*, pp. 174 and 195.
11 Pagden, *European Encounters*, *Op. cit.*, p. 127.
12 *Ibid*, p. 5.
13 Elliott, *Op. cit.*, p. 25.
14 Robert Wauchope (general editor), *Handbook of Middle American Indians*, 16 vols, Austin, TX: University of Texas Press, 1964–76.
15 Gordon Brotherston, *Book of the Fourth World: reading the native Americans through their literature*, Cambridge, UK: Cambridge University Press, 1992.
16 Elliott, *Op. cit.*, p. 34.
17 Leithäuser, *Op. cit.*, pages 165–166 for Indian drawings of these activities.
18 Elliott, *Op. cit.*, p. 38.
19 Acosta had a theory that minerals 'grew' in the New World, like plants.
20 Evgenii G. Kushnarev (edited and translated by E.A.P. Crownhart-Vaughan), *Bering's Search for the Strait*, Portland: Oregon Historical Society Press, 1990 (first published in Leningrad [now St Petersburg], 1968).
21 Bodmer, *Op. cit.*, p. 67.
22 Elliott, *Op. cit.*, p. 43.

23 Pagden, *The Fall of Natural Man*, *Op. cit.*, p. 39.
24 This view envisaged the Indian as one day becoming a free man but until that time arrived he must remain 'in just tutelage under the king of Spain'. Pagden, *The Fall of Natural Man*, *Op. cit.*, p. 104.
25 Wright, *Op. cit.*, p. 23. Also: Bodmer, *Op. cit.*, pp. 143–144.
26 Pagden, *The Fall of Natural Man*, *Op. cit.*, p. 45.
27 *Ibid*, p. 46.
28 *Ibid*, p. 119.
29 Elliott, *Op. cit.*, p. 49.
30 *Ibid*, pp. 81 and 86.
31 *Ibid*, p. 95.
32 Benjamin Keen, *The Aztec Image in Western Thought*, New Brunswick, NJ 1971/1990, p. 261.
33 Henry Steele Commager, *The Empire of Reason: how Europe imagined and America realized the enlightenment*, London: Weidenfeld & Nicolson, 1978, p. 83.
34 Jack P. Greene, *The Intellectual Construction of America: exceptionalism and identity from 1492 to 1800*, Chapel Hill, NC: University of North Carolina Press, 1993, p. 128.
35 Antonello Gerbi, *The Dispute of the New World: The History of a Polemic, 1750–1900*, trs. by Jeremy Moyle, Pittsburgh, PA: University of Pittsburgh Press, 1973, pp. 52ff.
36 Keen, *Op. cit.*, pp. 58–60.
37 *Ibid*, p. 88.
38 Gerbi, *Op. cit.*, p. 42.
39 *Ibid*, p. 163.
40 Merrill D. Peterson, *Thomas Jefferson and the New Nation*, Oxford: Oxford University Press, 1970, pp. 159–160.
41 Commager, *Op. cit.*, p. 98.
42 *Ibid*, p. 99.
43 Commager, *Op. cit.*, p. 246.
44 Keen, *Op. cit.*, p. 297.
45 Pagden, *European Encounters*, *Op. cit.*, p. 167.
46 Keen, *Op. cit.*, p. 359.
47 *Ibid*, p. 417.
48 *Ibid*, p. 425.
49 *Ibid.*
50 *Ibid*, p. 445.
51 *Ibid*, p. 458.
52 *Ibid*, p. 456.
53 Commager, *Op. cit.*, p. 394.
54 Miguel Asúa and Roger French, *A New World of Animals: Early Modern Europeans on the Creatures of Iberian America*, Aldershot: Ashgate, 2005, pp. 36–37.
55 *Ibid*, p. 82.

56 *Ibid*, p. 188.
57 *Ibid*, p. 229.
58 Keen, *Op. cit.*, p. 448.
59 William M. Denevan (editor), *The Native Population of the Americas in 1492*, Madison, WI: University of Wisconsin Press, 1976/1992.
60 William H. McNeill, *Plagues and Peoples*, Oxford: Blackwell, 1977, p. 211.
61 Denevan, *Op. cit.*, p. 7.
62 McNeill, *Op cit.*, pp. 211–212.
63 Russell Thornton, *American Indian Holocaust and Survival: A Population History Since 1492*, Norman, OK and London: Oklahoma University Press, 1987.
64 Thornton, *Op. cit.*, p. 39.
65 McNeill, *Op. cit.*, pp. 50 and 201–202.
66 Thornton, *Op. cit.*, pp. 40–41.
67 *Ibid*, p. 48.
68 *Ibid*, p. 52.
69 Kirkpatrick Sale, *The Conquest of Paradise*, New York: Knopf, 1991.
70 Sale, *Op. cit.*, pp. 97–99.
71 *Ibid*, p. 248.
72 *Ibid*, p. 316.
73 Ronald Wright, *Stolen Continents: The 'New World' Through Indian Eyes*, Boston: Houghton Mifflin, 1992, p. 128.
74 Brotherston, *Op. cit.*, p. 77.
75 Wright, *Op. cit.*, p. 168.
76 *Ibid*, p. 210.
77 Brotherston, *Op. cit.*, p. 4.

Sources for Figures

1. Anne Baring and Jules Cashford, *The Myth of the Goddess: Evolution of an Image*, Viking Arkana, 1991, p. 33.
2. Juliet Clutton-Brock (editor), *The Walking Larder: Patterns of Domestication, Pastoralism, and Predation*, Unwin Hyman, 1989, p.285.
3. Anne Baring and Jules Cashford, *The Myth of the Goddess: Evolution of an Image*, Viking Arkana, 1991, p. 34.
4. Benny J. Peiser *et al.* (editors), 'Natural Catastrophes during Bronze Age Civilizations: Archaeological, Geological, Astronomical and Cultural Perspectives', *British Archaeological Reports, International Series*, 1998, p. 51.
5. Benny J. Peiser *et al.* (editors), *Natural Catastrophes During Bronze Age Civilizations: Archaeological, Geological, Astronomical and Cultural Perspectives*, British Archaeological Reports, 728, 1998, p. 61.
6. Benny J. Peiser *et al.* (editors), *Natural Catastrophes During Bronze Age Civilizations: Archaeological, Geological, Astronomical and Cultural Perspectives*, British Archaeological Reports, 728, 1998, p. 61.
7. Andrew Sherratt, 'Alcohol and Its Alternatives', in Jordan Goodman *et al.* (editors), *Consuming Habits: Drugs in History and Anthropology*, Routledge, 1995, p. 414.
8. Mark David Merlin, *On the Trail of the Ancient Opium Poppy*, Associated Universities Press, 1984, p. 233.
9. Peter T. Furst, *Hallucinogens and Culture*, Chandler & Sharp, 1976, p. 71.
10. Nicholas J. Saunders, *The People of the Jaguar, The Living Spirit of Ancient America*, Souvenir Press, 1989, p. 74.

11 Nicholas J. Saunders, *The People of the Jaguar, The Living Spirit of Ancient America*, Souvenir Press, 1989, p. 72.
12 Richard L. Burger, *Chavin and the Origins of Andean Civilization*, 1995, p. 157.
13 Brian Fagan, *Kingdoms of Gold, Kingdoms of Jade*, 1991, p. 119; and/or Linda Schele and David Freidel, *A Forest of Kings: The Untold Story of the Ancient Maya*, 1990, p. 267.
14 Heather Orr and Rex Koontz (editors), *Blood and Beauty: Organized Violence in the Art and Archaeology of Mesoamerica and Central America*, The Cotsen Institute of Archaeology at the University of California at Los Angeles, 2009, p. 108.
15 Heather Orr and Rex Koontz (editors), *Blood and Beauty: Organized Violence in the Art and Archaeology of Mesoamerica and Central America*, The Cotsen Institute of Archaeology at the University of California at Los Angeles, 2009, p. 129.
16 Heather Orr and Rex Koontz (editors), *Blood and Beauty: Organized Violence in the Art and Archaeology of Mesoamerica and Central America*, The Cotsen Institute of Archaeology at the University of California at Los Angeles, 2009, p. 271.
17 Heather Orr and Rex Koontz (editors), *Blood and Beauty: Organized Violence in the Art and Archaeology of Mesoamerica and Central America*, The Cotsen Institute of Archaeology at the University of California at Los Angeles, 2009, p. 199.
18 Heather Orr and Rex Koontz (editors), *Blood and Beauty: Organized Violence in the Art and Archaeology of Mesoamerica and Central America*,The Cotsen Institute of Archaeology at the University of California at Los Angeles, 2009, p. 274.
19 Enrique Florescano, *The Myth of Quetzalcoatl*, 1999, pp. 166, 169 and 170 respectively.

Index